NEUSIEDLERSEE: THE LIMNOLOGY OF A SHALLOW LAKE IN CENTRAL EUROPE

MONOGRAPHIAE BIOLOGICAE

Editor

J. ILLIES

Schlitz

VOLUME 37

Dr. W. Junk bv Publishers The Hague – Boston – London 1979

NEUSIEDLERSEE: THE LIMNOLOGY OF A SHALLOW LAKE IN CENTRAL EUROPE

Edited by

H. LÖFFLER

Dr. W. Junk bv Publishers The Hague – Boston – London 1979

The distribution of this book is handled by the following team:

for the United States and Canada

Kluwer Boston, Inc.
160 Old Derby Street
Hingham, MA 02043
USA

for all other countries

Kluwer Academic Publishers Group
Distribution Center
P.O. Box 322
3300 AH Dordrecht
The Netherlands

Library of Congress Cataloging in Publication Data CIP
Main entry under title:

Neusiedlersee, the limnology of a shallow lake in
central europe.
(Monographiae biologicae)
Includes index.
1. Limnology—Neusiedler See. 2. Neusiedler
See. i. Löffler, Heinz. ii. Series.
QP1. P37 (QH145) 574'.08s (551.4'82'0943615)
ISBN 90-6193-089-8 79-26230

Contents

Biology, ecology and production

Foreword

H. Löffler

Interest aroused in the past by Neusiedlersee, located in what was earlier the western part of Hungary, was mainly due to the enormous fluctuations in water level and the consequent threat to the livelihood of the local population, who lived chiefly from stock-farming. Practically no tourism touched the area until 1920, about a century later than Schubert's visit to the Salzkammergut lake district, the traditional resort area in Upper Austria. Probably because of its small popularity there is almost no documentation in the form of engravings or paintings. One of the exceptions is an oil painting by Schnorr-Carolsfeld, from about 1820. It is obvious that there was no *Phragmites* at the time he painted the view near Neusiedl.

When the shallow Neusiedlersee was discovered by tourists it began to attract scientific interest, although some earlier work by mineralogists, geographers and hydrographers does exist. But it was only after the Second World War that this interest could be intensified, the lake now being an alkaline, turbid body of water confined by the vast areas of *Phragmites* which had come into existence after the lake last dried out in 1868. Scientific work gained considerable impetus from international activities such as the International Biological Programme, International Hydrological Decade, and Man and Biosphere, the results of which will be included in this volume. But in spite of all such efforts Neusiedlersee will undoubtedly remain, in many respects, a mystery.

Not only to satisfy the curiosity of limnologists but in order to ensure proper management of the lake area, far more information will have to be gathered in the future concerning, for example, the tectonic origin of the lake, the extent of the catchment area, water and salt budgets, the seiches and current systems, movements of the sediment in the open lake and the function of the large *Phragmites* belt within the nutrient cycle.

The menaces to the lake by tourism, continuous pressure for more housing on the shore, as well as other activities connected with agriculture and fishery are manifold and obvious. There is also the problem of preventing a repetition of the drying out of the lake which might result from a series of very dry years. Some steps have recently been taken to prevent further damage to the lake area: it has been accorded the status of a *Landschaftsschutzgebiet* and in 1977 the lake was made a Biosphere Reserve by UNESCO, which should ensure control of building activities. There are also good hopes that in the near future Hungary will cooperate in such efforts, thus putting an end to the dangers with which this unique lake and bird sanctuary has been faced in the past. There are obvious shortcomings in this book. Hydrophysics (currents and seiche properties), water and nutrient budget, bacteriology and protozoology can be listed as examples of the most prominent. It is therefore hoped that this report will have a challenging effect on Neusiedlersee research. The authors express their thanks for continuous support by the Federal Government and to the authorities of the communities round the lake, especially to the community of Neusiedl, which provided a small station at the lake. They are obliged to their Hungarian colleagues, to the national committees of IBP and MaB, to the Burgenländische Forschungsinstitut Illmitz for support of some research projects, the Austrian Mineral Oil Company for some two thousand samples from Seewinkel (Löffler, Chapter 3), to Dipl. Ing. I. Krönke for experiments in the wind tunnel (Jungwirth, Chapter 8), to Miss Susan Powell for data on Kjeldahl nitrogen and aminoacids (Neuhuber, Chapter 10), to Dr. Annemarie Schmid and Dr. M. Dokulil for data on oxygen (Neuhuber & Hammer, Chapter 11), to Prof. Preisinger for information on the sediment (Jungwirth, Chapter 13), to the Center of Electron Microscopy, Graz, Hofrat F. Grasenik, Dipl. Ing. Johanna Blaha, Dr. P. Golob and Ing. H. Waltinger (Schroll, Chapter 14), to Prof. Agnes Ruttner-Kolisko for data (Dokulil, Chapter 18; Herzig, Chapter 22), to Dr. Hans Winkler for establishing computer programmes (Herzig, Chapter 22), to Burgenländische Fischereigenossenschaft (Hacker, Chapter 28), to Mrs. Joy Wieser for her patience in translating German as well as correcting poor English manuscripts, to Wolfgang Klejch for his excellent drawings, to Miss Ingrid Gradl for typing the manuscripts, to Molden Publishers for providing clichés, and to Dr. W. Junk bv, publishers for continuous cooperation.

To
Professor W. Kühnelt

'Time and space, however, would fail me to tell of all the marvels of the world beneath the waters. They would sound like the wild fancies of a child's fairy tale, and yet they are all literally true' (HUDSON & GOSSE Vol. I, p. 4).

Physiography

1 The catchment area, a geographical review

H. Nagl

The largest lake of Austria, Neusiedlersee, is situated in the lowest plain of the country (113 m above sea level). Geomorphology, climate and man have formed a region, only 40 km southeast of Vienna, that is rare in Central Europe, unique in Austria and without its like in Western Europe. Together with countless saline pools the lake borders the Small Hungarian Plain, with its typical form of landscape known as the Puszta. Early settlement and exploitation of the region by man, and unfavourable soil conditions, the hot, dry summer, have led to the almost complete disappearance of forests except for small scattered woods. The following pages will give a short survey of the landscape itself, its economic exploitation and its inhabitants (Fig. 1.1).

1.1 Geomorphological development

Entering the Neusiedlersee region via the escarpment separating it from the Vienna Basin one descends by 40 m (Gate of Bruck) or even 100 m (Gate of Wiener Neustadt). The Leitha mountains (450 m), Rosalia mountains (748 m) and the Ödenburger mountains near Sopron (554 m) constitute the western border of this region, which can be split up as follows (Fig. 1.2): central region: Neusiedlersee (300 km^2, reed belt included) and the Seewinkel (400 km^2); marginal regions: Parndorf plateau (200 km^2), Wulka basin or Eisenstadt basin (150 m^2), the hill range of Rust and Balf (75 km^2), the Ikva region (approximately 225 km^2). In the southeast large portions of the Seewinkel and the Ikva region are taken up by the Waasen (Hanság), a swamp now largely cultivated. All these, however, are of related origin. In contrast to the lake region and its surroundings (Seewinkel, Parndorf plateau, Wulka valley), which is characterized by young fluviatile forms of accumulation and a Quaternary tectonic, the marginal areas are marked by remnants of old terrestrial and marine forms. Even before the dismemberment of the Carpathian system in the Carpathian it had been levelled to a system of peneplains. Their derivates cover the ridges of the mountains rising in the west. In the Young Tertiary (Badenian and Pannonian), marine deposits occurred similar to those at the east end of the Alps: Leitha limestones, sands and clays and after desalination of the sea, freshwater limestones. Therefore very different conditions for the subsequent soil formation were given from the very beginning. Thus beside crystalline material and sediments with a disposition to karst phenomena, water-retaining clays and water-permeable sands can be found. Apart from the hills of Rust which are completely covered by Leitha limestone (a popular building stone in historical Vienna), the marginal areas were only of real importance to their eastern neighbourhood up to the Upper Miocene/Pliocene. At that time pediments reached

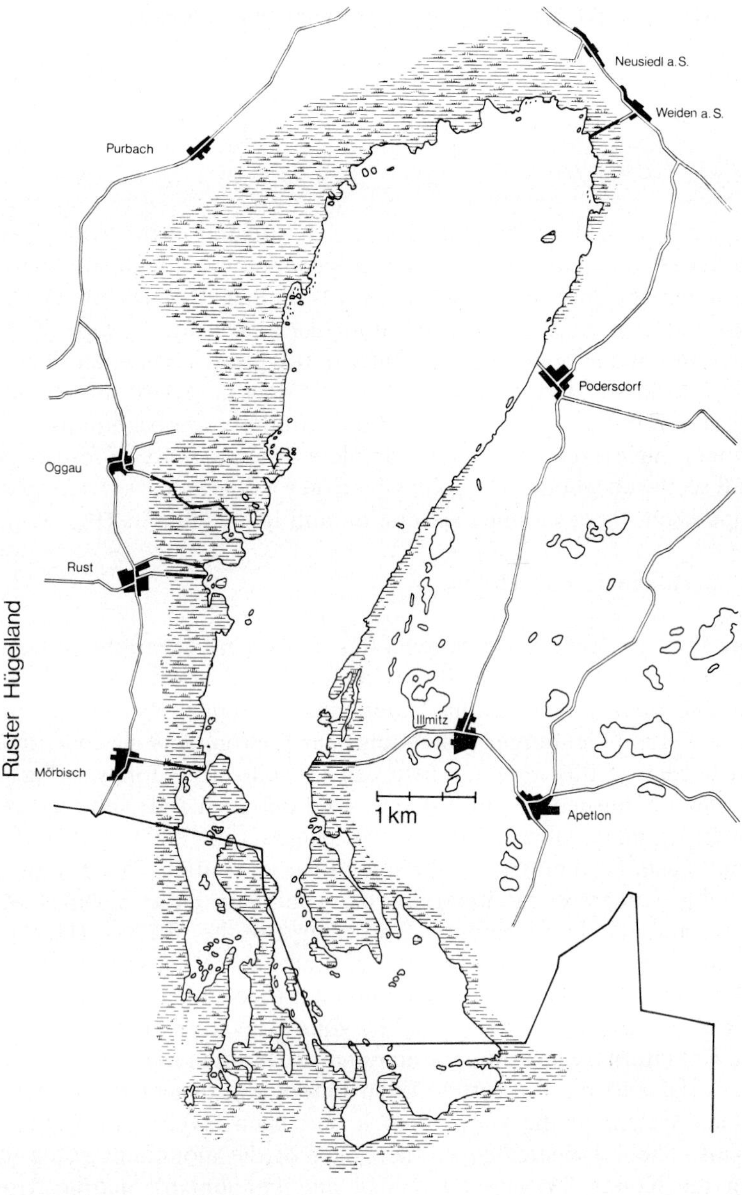

Fig. 1.1. The area of Neusiedlersee.

from elevated areas eastward to the descending Pannonian region. The entire area was continental, the Pannonian lake had receded to the lowest region in southern Hungary. At the beginning of the Quaternary, at 1.8 to 2.0 mio. b.p. according to palaeomagnetic studies, the development of the present surface forms of the lake region proper set in, the Danube playing an important part in this process.

In the Pliocene the Danube crossed the area of the present mountains of Hainburg and at the beginning of the Quaternary it turned towards the depression area in the south, flowing through the Gate of Bruck.

The deeper-lying part of the pediments formed under semi-arid conditions were covered by gravel. Much the same is true for the Pitten and Wulka, which turned east, passing through the Gate of Wiener Neustadt. In the

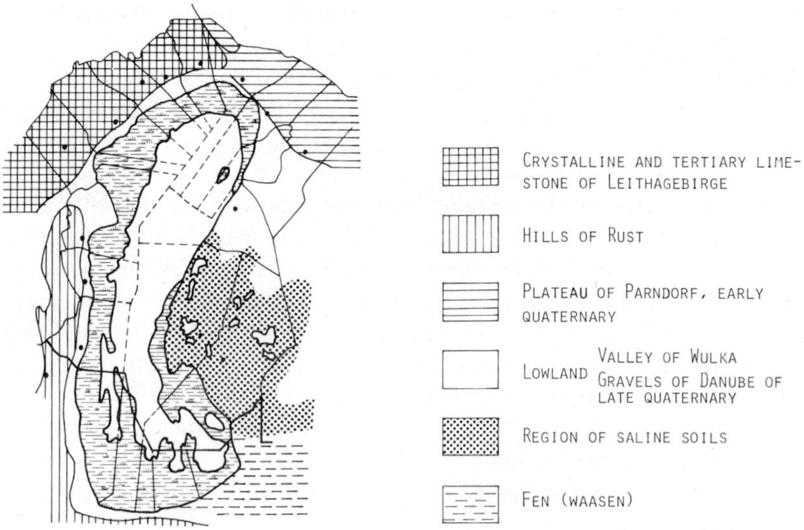

Fig. 1.2. Landscape of the Neusiedlersee area.

southern part of the present lake there was an elevated area, however, as evidenced by the detour of the early Pleistocene Ikva.

Thus the present lake region contains two completely different elements: the elevated area and the extensive plains. The first-mentioned dates back to two very different epochs and different forces, the latter to very recent, chiefly fluviatile processes. Despite this, the genesis of the surface formation can only be explained in its overall context; the differentiation on the basis of age allows of a systematic interpretation, which is, however, because of divergent opinion, not generally accepted. Higher and older forms existing in the Leitha, Rosalia and Ödenburger mountains are not directly connected with the lake and the Seewinkel and will therefore not be discussed: instead the few articles on the subject will be reviewed briefly.

In his fundamental study of the morphology of the Vienna Basin Hassinger (1918) interpreted the numerous flat forms in the Leitha mountains as marking the Pannonian lake shore, and compared them to those found on the western border of the Vienna Basin. He believed the lowest, at 240 to 250 m altitude, to be connected with the Levantine lake, a theory now rejected; moreover he considered the quartz-gravel terraces in the zoological garden east of Eisenstadt to be fluviatile formations, whereas in fact they are part of the Wulka system.

Roth-Fuchs (1926) studied the Leitha mountains and put the border between the Leitha mountains and the foothills too far eastward. Down to 240 m she assumes abrasion terraces left by retreating phases of the Pannonian sea, but correctly recognizes the fluviatile processes below 230 m. Special mention must be made of Winkler-Hermaden (1957), who, among others, gave a modern description of the genesis and temporal classification of the levelling processes in the Alps and their foothills.

Kapounek (1938) sets the upper limit of fluviatile accumulations in the Wulka region at a level of 200 to 210 m; according to Riedl (1963) this is not the absolute limit of fluviatile forms, but corresponds to the upper limit of the Pleistocene surface systems.

The lower terraces descending from the edge of the mountains to the Wulka basin and the lake region constitute a more serious problem. Riedl (1960, 1963, 1963b, 1964) especially stressed the need for an integrated study, the more so since the Wulka is the only surface river connected with the lake. The Danube, however, moulded the surface of the land through which it flowed to a greater extent than the Wulka and its morphological influence continued up to the present geological age (this still exists from the hydrological point of view). The genesis and the age of the lake (see Chapters 2 and 3) itself are closely linked to the events on the mountain margins and the neighbouring river systems, so that the problems can be approached in a variety of ways.

Geomorphology and palaeoclimate will be dealt with in the following (see also Chapter 2): as mentioned above, the landscape was formed chiefly by exogenic and endogenic Pleistocene processes. The first decisive change in the opinion that Neusiedlersee is a remnant of the Pannonian lake was brought about by Hassinger (1918). He believed that the lake was a depressed former riverbed of the Danube, winding around the Parndorf plateau. This opinion was only tenable until the lake sediments were investigated.

Szadecky-Kardoss (1938) proved that the gravels of the Seewinkel are recent Quaternary Danube gravels (in the south possibly also of the Raab-Rabnitz system) and that the Seewinkel and Waasen were interlaced by fen systems and have partly been influenced by tectonic movements until very recent times.

Wiche (1951) disproved the deflation theory widespread among Hungarian authors, as well as the theory of the shore terraces of the Leitha mountains whose existence is traced back to systems of subaeolian forces. Büdel (1933), Winkler-Hermaden (1957) and Tollmann (1955) proved a late Pannonian lift of the Leitha mountains and related the altitude levels to the various prehistoric climates. Since the mapping by Franz & Husz (1961) growing importance has been given to soil science, especially to the palaeogeographical aspects of Seewinkel (see below).

As to the lake itself Küpper (1957) represented an opinion completely different from that of Hassinger and said that the lake's primary tectonic formation was caused by more recent subsidences, a statement now generally recognized (see Chapter 3).

The evaluation of borings led Tauber (1959) to the belief that the terrace border of the Parndorf plateau is erosive, a theory already put forward by Hassinger. However it seems likely that it is tectonic in origin. Moreover, in his opinion when the gravels of Seewinkel were deposited, the area of the present lake must have been higher because it mostly lacks gravel.

Sauerzopf (1956) was the first to review the history of the Quaternary development. Frasl (1961) succeeded in furnishing definite proof that Riss gravels are deposited – heavy minerals similar to the Gänserndorf terrace, or rather the Gänserndorfer plain, according to Fink[1] – below the Würm gravels (spectrum of heavy minerals similar to the Prater terrace).

Among the geomorphological studies those of Riedl warrant special mention. In 1963 he studied the marginal areas of Neusiedlersee which, although deserving priority, had so far been neglected (Riedl 1963b). He used besides geomorphological and pedological methods the morphology of profiles, which enabled him to achieve a more exact classification and his results will be considered in more detail.

Riedl differentiates the area descending from 200 m to the level of the reed belt (1 to 3 per cent incline) into Quaternary systems, which must correspond to the development of the Danube. As mentioned above the higher pediments, abrasion terraces and elements of oldland played no role in the formation of the lake area. The Quaternary terraces are connected with solifluction pediments on the eastern edge of the Leitha mountains, which, as their indentation shows, are to be traced back to a stepwise uplift of these mountains during the Pleistocene. The highest uplift (200 m) is formed over Leitha limestone and Pannonian marl, all of the lower ones are Tertiary. The contact layers of the rubble and gravel in the higher systems show an intensive reddish weathering. This appears very distinctly at the level of 140/145 m. The middle Pleistocene plane at 120 m exhibits subsidences which descend to the Würm age plane occupied finally by the lake (117 m). All planes end at the northern edge of the Leitha range because of a lack of the necessary mountain area for the formation of pediments: the Parndorf plateau could thus remain, although its borders are exogenic forms they were built by denudation. The hills near Rust are characterized by the Pitten's oldest accumulations of gravel, indented with solifluction rubble. The depression limiting the south and the east of the hill range was probably first formed by a bifurcation of the Pitten and in its second phase by separate development of the Pitten, the Wulka and the Lateinerbach.

Although some of these hypotheses cannot be absolutely proved, the classification of the terraces by means of weathering horizons and palaeosoils was a definite advance.

In 1965 Riedl studied the region east of the lake, following the Würm pediment across the lake area. The solifluction rubble covers large parts of the Riss-Würm interglacial salt horizon postulated by Franz. After detailed

1. See the end of this chapter for modern concepts of the age and date of the accumulations.

studies Riedl traced the lakes of Seewinkel to ice laccoliths. The sediments of Seewinkel were dealt with by Husz (1962a, 1962b). On the basis of studies by Szadecky-Kardoss (1938) and the findings of Sauerzopf (1959) and Tauber (1959) he is convinced that the present surface and the pools of Seewinkel are to be traced back to minute differentiations in accumulation (shallow grooved and delta rubble cones) of the most recent Quaternary; moreover he believes that under Würm gravels Riss gravels of the Danube can also be found, which, due to subsidences, are no longer terraces. Beneath the Riss gravels, sandstone concretions separate out a margin with organic remnants (wood splinters). This sandstone zone constitutes the limit of a higher and a lower groundwater table. The author uses stratigraphical, pedological, mineralogical and palynological methods of dating. Frasl and Klaus deserve special mention in connection with the last two methods. All sediments of the lake dam (most recent accumulations formed in modern times) and even the Riss gravels and Riss-Würm soils could be dated. The stratigraphical situation gives important hints on the formation of saline soils.

In recent times the landscape has also been influenced by aeolian and limnic forces of the last ice age and the Holocene. In the higher regions quicksands (for example, in Parndorf Plateau) or loesses (Wulka region, hills of Rust) can be found. The lake dam may be a special limnic form bordering large parts of the eastern shore. Formerly it was thought to be a beach ridge. Löffler's (1974) opinion that it is a consequence of the repeated ice shoves combined with a raised water level and a strong west wind seems to be correct. Finally mention must be made of the saline soils and swamps which are a characteristic feature of the Seewinkel. Before discussing the soils something should be said concerning the age of the gravels.

Dating of gravels along larger rivers has become more precise in recent years thanks to the possibility of absolute dating (particularly of wood finds and dendrochronology) in the case of the Würm gravels. The results thus obtained led to the dropping of the stratigraphical term 'low terrace' for the Danube and some of its tributaries. Fink (1973) calls these accumulations, which started in the Würm ice age (cryoturbations) and continued through the Holocene up to the present time, the Prater terrace. The gravels of Seewinkel must be regarded as part of this Prater terrace even though so far no C^{14} datings have been made.

The gravels of the Riss terrace (stratigraphical: high terrace) are found on the northern border of the Parndorf plateau as well as in the deep strata of Seewinkel below its gravels. They were comparable with the gravels of the Gänserndorf plain by Frasl.

It is highly uncertain whether the gravels of the Parndorfer plateau correspond to those of the 'Arsenal terrace' because the Quaternary tectonic does not allow of a direct connection.

1.2 Soils

As is to be expected, chernozems are dominant in dry areas except where humid soils with a high percentage of humus, or swamps, occur. On the

Parndorf plateau, which is characterized by a cover of drifting sand, lime-free parachernozems are prevalent. A peculiarity of the Seewinkel is the frequent occurrence of saline soils, as described by Franz & Husz (particularly 1961a).

According to Franz the existing saline soils are not the result of the present climate but have their origin in a deeper saliferous horizon. The latter is a zone of a salt-containing chalky substrate similar to loess, rich in reddish quartz gravels and forming a sticky layer impermeable to water. It fills hollows in particular and is responsible for the salt pools (often occurring periodically), which disappear in periods of drought when salt efflorescences dominate the scenery. The stratigraphy of the area and pollen analyses

Table 1.1. Systematics of saline soils

Hilgard	Gedroiz	De Sigmond	U.S. system	Szabolcz, Földvari	Austrian proposal[a]
white alkali	solonchak	alkali-salt soils= saline soils	saline soil >400 $\mu S\ cm^{-1}$ <15% pH 8.5	solonchak >300 mg% <15%	solonchak >300 mg% <15%
		salt-alkali soils = saline alkali soils	saline-alkali soil >400 $\mu S\ cm^{-1}$ >15%	solonchak-solonetz 300–500 mg% 15–20%	solonchak-solonetz >300 mg% >15%
black alkali	solonetz	desalinated ("worn out") alkali soil	nonsaline-alkali soil <4,000 $\mu S\ cm^{-1}$ >15% pH 8.5–10	meadow solonetz <300 mg% >20%; solonetz-like meadow soil <200 mg%, <15%	solonetz <300 mg% >15%
	solod	degraded alkali soil			

Conductivity in $\mu S\ cm^{-1}$ (microsiemens), percentage sodium in the sorption complex, quantity of soluble salts in mg%. According to J. Fink: Blätter zur Physischen Geographie 57 (1974).
[a]Cp. nr. 13 Mitt. Ö.B.G., p.81.

indicate that the saliferous horizon came into being in the Riss-Würm interglacial when a shallow lake dried out. Depending upon the thickness of the superimposed layer and further development of the soil either solonetz or solonchak developed, or chernozems in the case of thick superimpositions.

Fink (1969) classified the salt soils according to their profile morphology as follows: solonchak, with free salts; solonchak-solonetz, with free and bound salt, according to Husz (1966), following Hungarian nomenclature: and solonetz, with bound salts. All belong to the chemical group of saline-alkaline soils (United States, internal) and to the salt-alkali soils of De Sigmond (Table 1.1).

The saline soils with their white efflorescences in autumn and their halophytic flora are a characteristic feature of the landscape. In addition to their marked influence on the scenery the saliferous horizon and the salt lakelets are an essential factor in the development of the typical flora and fauna of the region.

The Pannonian environment, which extends much further to the east, influences not only the nature of the landscape surrounding Neusiedlersee and its climate, but also its economy. The wide stretches of uninterrupted horizon, the hot, dry summers, extensive stock-farming on natural pastures, intensive wine-growing, the typical farmhouses with their drying frames for maize have all played their part in determining the character of the cultivated countryside and its inhabitants. Further, Burgenland, the province in which this region is situated, belonged to the Hungarian portion of the Austro-Hungarian Empire following the agreement of 1867, a fact that has influenced its development.

1.3 Climate

Next to the morphological elements of the landscape, climate is one of the most decisive factors influencing the environment. As a consequence of the unhampered eastern influences in this area and the lack of major differences in relief the Pannonian climate is very pronounced (as compared with the usual Central European climate). The Pannonian climate is a more continental type of the European transitional climate, which is centred in Alföld (the Great Hungarian Plain). The continental character is manifest in a strong fluctuation of the monthly temperature averages (usually 20 to 23°C, in some years much higher, for example 26.7°C in 1963 at Neustift/Rosalia; 27.3°C in 1964 at Donnerskirchen), in generally high summer and annual temperatures (daily average in summer frequently above 25°C, Eisenstadt up to 34.6°C, annual average about 10°C, at Donnerskirchen frequently above 11°C), and on the other hand, in a relatively dry climate with little rainfall. The dryness and the high evaporation values in summer (see Chapter 6) often cause a short but marked period of drought, especially if July shows a distinct second rain minimum. Thus rains in July at Rust vary between 248 mm (221 per cent) and 20 mm (30.3 per cent), the average being 66 mm (1961–1970). Sandy soils completely dry out, even their deeper layers (Parndorf plateau), and local substrate steppes develop on Leitha limestones. The extremely intermittent nature of the humidity is also unfavourable for soils dependent on groundwater (gley and saline soils). This situation, which is critical for most of the cultivated plants, is mitigated by the lake. The latter reduces the daily temperature fluctuations (very important for wine growing), and the early frosts are usually delayed by some three weeks. The shallowness of the lake is, however, a disadvantage in that temperatures are lowered in spring by the high energy consumption involved in melting the winter ice. In addition to its direct effects on agriculture, climate also exerts a considerable indirect influence on cultivated land. The negligible drainage factor (10 to 25 per cent) means a scarcity of running water; groundwater is often deep-lying or influenced by the saliferous layer. Thus one of the typical features of the Seewinkel is the draw wells. In recent decades the swamps have been drained and the natural agricultural land developed.

It should not be forgotten that what we are trying to preserve as the characteristic elements of northern Burgenland are partly the products of man's own activity. The following pages will deal with the occupations pursued by the inhabitants of this region.

1.4 The present cultivated landscape

Calculated on the basis of the space available for settlement, the population density of the lakeside communities is 200 to 300 km^{-2} in the area of Mörbisch-Rust, 80 to 120 in the northern Seewinkel and less than 40 around Illmitz. The population is mainly rural and consists of 89 per cent whose mother tongue is German (including peasants in steppe areas), 9 per cent Croatians and 1.5 per cent Magyars. The name Burgenland itself (previously German West Hungary) was derived from the four citadels Ödenburg, Wieselburg, Pressburg and Eisenburg. None of these belong to the present federal province of Burgenland.

In addition to Neusiedl am See and the free town of Rust the following 17 communities border the lake: Mörbisch, Oggau, Donnerskirchen, Purbach, Breitenbrunn, Winden am See, Jois, Weiden am See, Gols, Podersdorf, Illmitz, Apetlon, and five Hungarian communities for which no up-to-date statistics are available.

Apart from the extensive exploitation of swamps, steppes and zick soils, especially in earlier times, the possibilities offered by intensive cultivation methods have given rise to a very high population density in the adjoining areas. These factors have combined to render the region one of high agrarian density.

Although these communities include practically no foreign ethnic groups nowadays, some of the neighbouring villages have a high percentage of minorities (Oslip/Uslop, Neudorf near Podersdorf). The Croatian groups can be traced back mainly to recolonization following the havoc caused by the Turkish wars, in which whole villages such as Moschado or Zitzmannsdorf were wiped out. In the eighteenth century in particular new colonists arrived from the Vojvodina (the Croatian name for Neusiedlersee is *Niuzaljsko jezero*) and the Austro-Hungarian agreement of 1867 led to greater Hungarian influence (Hungarian *fertö*, swamp lake). It is therefore not surprising that even today cultural elements such as architecture and operating systems of all three population groups are interwoven. Before the Second World War gypsies accounted for 2.2 per cent of the population.

Fluctuations in population tend to be slight; the only places exhibiting higher rates of population increase are Podersdorf and Breitenbrunn with 11 and 9 per cent respectively (1961 to 1971). Neusiedl am See, Winden am See and Gols showed an average increase of about 5 per cent, whereas the population of Donnerskirchen underwent the largest decrease (3 per cent). The increased numbers of births, between 30 and 50 per thousand, is therefore partially or even completely balanced by emigration.

This situation is largely the result of the important role played by the primary economic sector. On the eastern side of the lake over 60 per cent of the population is dependent upon employment in agriculture or forestry; west of the lake (Rust included) the number lies between 40 and 60 per cent and even in Neusiedl, the main town of the district, it amounts to more than 20 per cent, a very high figure for such a central community. A large percentage of the working population (secondary sector) commutes to other places of employment, and only in Neusiedl are the numbers of incoming commuters higher than of the outgoing: 44.3 per cent of all employees come from other communities. Table 1.2 gives a survey of population development and the

Table 1.2. Population movements (1961–1971) and number of commuters (1971)

Community	Residents		Increase/ decrease (%)	Employ-ment	Commuters	
	1961	1971			out	in
Mörbisch	2,333	2,302	−1	1,245	322	43
Rust	1,690	1,704	+1	796	252	158
Oggau	1,790	1,847	+3	753	371	27
Donnerskirchen	1,646	1,593	−3	709	226	32
Purbach	2,184	2,159	−1	979	394	114
Breitenbrunn	1,211	1,321	+9	587	250	91
Winden/See	1,004	1,071	+7	446	190	45
Jois	1,295	1,270	−2	583	224	62
Neusiedl/See	3,826	3,999	+5	1,664	478	737
Weiden/See	1,685	1,701	+1	935	198	75
Gols	3,120	3,287	+5	1,329	416	83
Podersdorf	1,627	1,814	+11	829	133	48
Illmitz	2,316	2,376	+3	798	310	42
Apetlon	1,925	1,893	−2	787	221	26

numbers of commuters (see also Fig. 1.3). While the decrease in outward-bound commuters partially reflects the rapid growth of population, other changes clearly indicate that upheavals are in progress. More than 30 per cent of the population is agrarian, which is three times the average value for Austria. The great decline in employment figures in most communities, as well as the large number of outward-bound commuters (almost 50 per cent in Oggau) is remarkable.

Although rural settlements dominate they often have the appearance and character of small towns. This is due to the large, often several-storeyed houses of the wine growers (also found in the wine-growing outskirts of Vienna, and in Wachau), and to the compact arrangement of the angular or elongated houses along a main street or surrounding a village square. On the west of the lake the large part of the typical Burgenland houses, their arches often decorated with agricultural produce, and the thatched houses of Seewinkel, which were one of the main tourist attractions, have been destroyed in the past two decades due to overzealous and misjudged modernization. Even worse are the monstrous empty apartment houses that disrupt the beauty of the horizon.

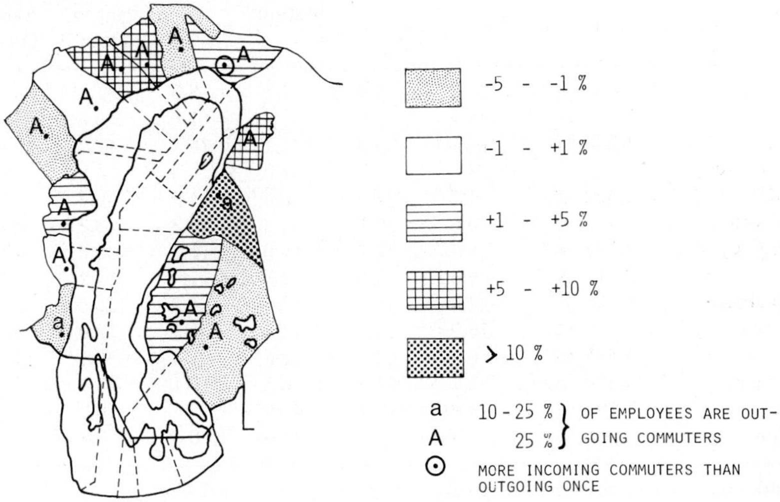

Fig. 1.3. Population movements (1961–1971) and number of commuters (1971), increase and decrease in %.

1.5 Soil exploitation

Full-time employment is largely provided in agrarian communities with little or no industry, in the form of enterprises covering less than 10 hectares.

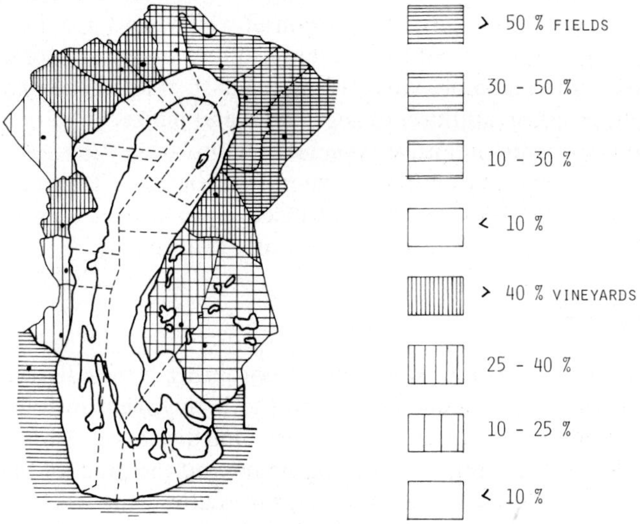

Fig. 1.4. Land use as a percentage of community area (excluding the lake): fields (1973) and vineyards (1974).

Table 1.3. Land use in 1969, and area devoted to vineyards in 1974[a]

Community	Total area (ha)	Lake (%)	Fields[b]	Meadows, pastures[b]	Forest[b]	Vine-yards[b] (1969)	Vine-yards[b] (1974)[c]
Mörbisch	2,867.96	30.9	7.9	47.5	8.8	31.6	37.3
Rust	1,977.19	35.9	8.9	35.7	1.0	36.6	40.0
Oggau	1,243.45	n.d.	23.4	21.1	16.7	41.2	43.1
Donnerskirchen	9,103.47	3.0	8.3	59.2	4.8	4.6	5.2
Purbach	1,418.41	n.d.	27.9	6.6	14.9	40.7	54.2
Breitenbrunn	1,125.90	0.6	30.1	4.3	29.3	28.3	46.8
Winden/See	1,347.74	19.3	32.7	4.6	11.2	20.0	41.3
Jois	2,710.19	9.6	35.6	2.8	15.4	19.2	33.5
Neusiedl/See	4,756.06	24.4	53.3	7.0	4.2	16.7	21.2
Weiden/See	4,320.34	26.0	55.0	12.5	1.1	23.0	25.3
Gols	4,544.53	0.3	62.9	1.6	1.3	30.3	32.4
Podersdorf	4,590.84	42.7	54.3	6.5	0.1	28.5	41.7
Illmitz	9,177.55	29.9	14.7	49.9[d]	0.1	16.7	39.0
Apetlon	8,266.66	24.3	42.1	42.9[d]	0.6	9.8	12.3

[a] Apart from the vineyards the distribution was much the same in 1974.
[b] As a percentage of the total area, excluding that of the lake.
[c] The area covered by vineyards shows enormous increase (in Illmitz 131.2 per cent, Winden/See 106.5 per cent) despite a temporary expansion ban. Unfortunately almost all vineyard expansion is at the cost of natural uncultivated land (pasture and meadows). Moreover the fields now penetrate into the (former) swampy areas of the Waasen (Hanság).
[d] Swamp included (in Illmitz 32.6 per cent of the total).

Tourist traffic is the only other source of income within the communities themselves. As Table 3 shows, a high percentage of the total area is often occupied by the lake or the reed belt, or by other bodies of water such as salt lakelets. The greater part of the remaining utilizable area consists of arable land (Figs. 1.4 and 1.5), a large proportion of which is occupied by vineyards and orchards in the west, and by meadows, fields, or semi-natural grassland and, recently, more and more vineyards in the east. Wheat, barley and rye are the dominant crops, in addition to sugar beet and maize. High quality soil in some regions permits intensive vegetable cultivation (lettuce, tomatoes, cucumbers and peppers) or berry growing (strawberries). Plantations of sour cherries and chestnut trees, and in some places even tobacco and spices, are to be found on the eastern slopes of the Rosalia (Table 1.3).

1.6 Tourist traffic

A structural analysis of the communities bordering Neusiedlersee reveals a lack of industrial employment since even the highly mechanized farms only require additional help in certain seasons. Furthermore, there is a trend towards well-paid jobs, regular working hours and the manifold possibilities of entertainment offered by cities. Few have realized that the disruption of the structure of rural communities and spoliation of natural or cultivated landscapes in attempts at modernization merely bring temporary profit, if any at all, but can in no case be of permanent value. Clearly, a well-planned

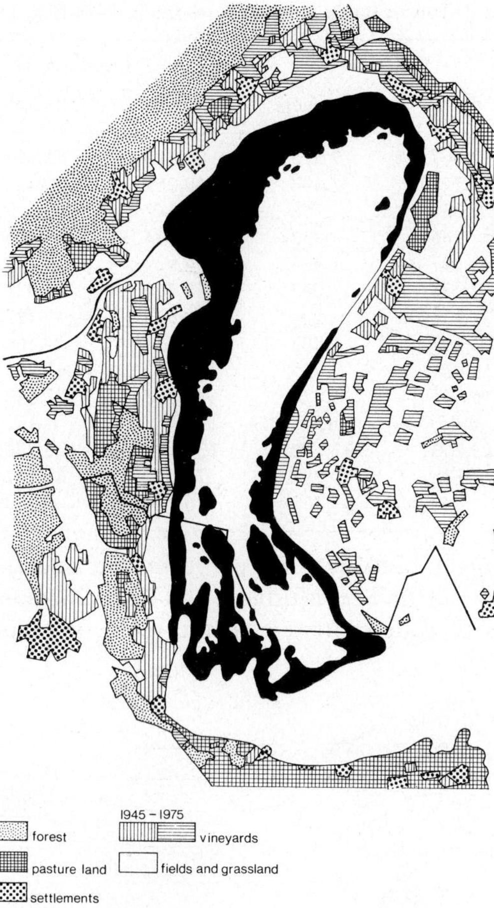

Fig. 1.5. Land use of the Neusiedlersee area.

tourist trade, which does not include the erection of apartment houses or bungalows, is of the utmost importance. On the one hand, it provides the individual communities with the financial means for extension and development, and on the other hand it ensures the preservation of the natural and cultural treasures which themselves represent tourist attractions. This is all the more desirable in the case of a landscape that is practically unique in Europe. The rapid development of tourism, as reflected in the large numbers of overnight bookings and in the high percentage of foreign visitors, gives some indication of the popularity enjoyed by the region. Table 1.4 and Fig. 1.6 summarize the figures for the year 1974. The figure of 70,000 tourists with more than 450,000 overnight bookings clearly proves the attraction of a lake region with bathing facilities. Even though figures are far below the Austrian average, two guests per inhabitant is still well above the European average. The high percentage of foreign tourists (70 to 90 per cent and even 92 per cent in Podersdorf, with 100,000 overnight bookings) is striking for the non-Alpine parts of Austria. Winden am See was least frequented (138

Table 1.4. Tourist traffic in the lake resorts in 1974 (Fig. 1.6)

Community	Inquiries		Overnight bookings		
	total	foreigners	total	foreigners	percentage
Mörbisch	12,667	7,575	68,719	55,993	81.5
Rust	10,575	7,494	48,085	40,411	84.0
Oggau	1,238	758	4,642	3,527	76.0
Donnerskirchen	385	332	2,164	1,905	88.0
Purbach	2,307	1,401	16,927	12,121	71.6
Breitenbrunn	1,738	1,003	8,256	5,934	71.9
Winden/See	138	104	949	722	76.1
Jois	1,607	1,083	5,733	4,566	79.6
Neusiedl/See	14,051	7,985	60,277	47,392	78.6
Weiden/See	6,441	3,024	29,875	22,542	75.5
Gols	n.d.	n.d.	n.d.	n.d.	n.d.
Podersdorf	13,233	11,644	136,174	126,270	92.7
Illmitz	7,885	7,002	72,812	68,642	94.3
Apetlon	890	763	6,316	5,880	93.1

visitors, 949 overnight bookings, 76 per cent foreigners). Whereas the average stay at Neusiedl am See, Mörbisch and Rust amounts to four or five days, it is 10 days at Podersdorf due to the reedless bathing beach. The

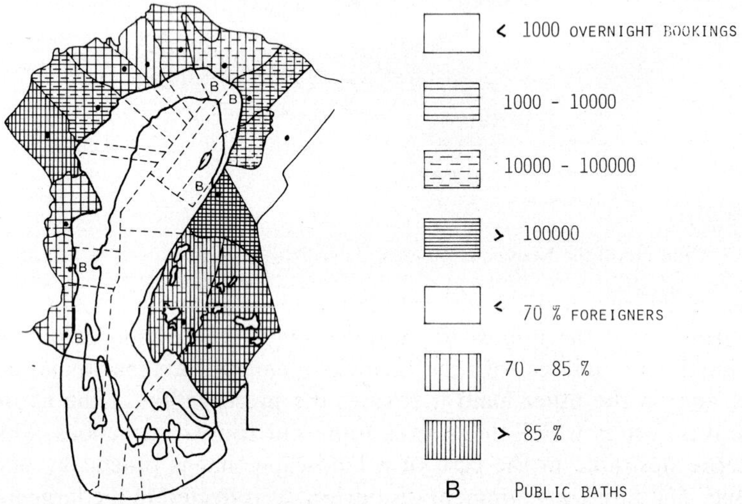

Fig. 1.6. Tourist traffic (1974): overnight bookings and percentage of foreigners.

remarkably high percentage of foreign visitors (chiefly from the German Federal Republic) in the Seewinkel should be borne in mind before developing, or in other words destroying, such a unique landscape.

1.7 Future prospects

The future development of the Neusiedlersee region depends upon the way in which the following three main conflicting spheres of interest can be reconciled:

(1) Promotion of agriculture, including conversion of pastureland to vineyards, soil improvement and so on, which would involve claims upon areas set aside as nature reserves;

(2) Increased attractions for the tourist trade, such as prefabricated houses, holiday settlements, bathing facilities, approach roads, especially down to the lake itself. Developments of this nature are incompatible with the preservation of scenic beauty and wildlife, and are in conflict with agricultural interests;

(3) The preservation of scenery and wildlife, including the creation of a Neusiedlersee national park. Protected areas extended; spoiled areas restored as far as possible, and the existing (semi-natural) areas retained.

Excessive zeal in the first two points would so much harm the scenic beauty that tourism itself would sooner or later sustain considerable losses. An extreme regard for the third point would entail substantial restrictions on economic development. Here, too, a compromise has to be made: This should aim at:

(1) Improvement of living conditions throughout the region (great decrease in employment figures during recent years): optimum but not maximum economic development;

(2) Where possible, complete preservation of the characteristic scenery, flora and fauna, that is, development of a national park, and the utmost preservation of landscape compatible with well-being in all spheres of human activity.

References

Allgemeine Landestopographie des Burgenlandes 1954. I. Bgld. Landesregierung.

Aumüller, S & Litschauer, G. S. 1964. Allgemeine Bibliographie des Burgenlandes II: Geographie. Hgg. Bgld. Landesarch. u. bgld. Landesbibl., Eisenstadt.

Bauer, H. 1965. Ein Entwicklungsprogramm für das 'Planungsgebiet Neusiedlersee' in Hinblick auf seine fremdenverkehrsmäßige Erschliessung. Diss. TH, Vienna.

Bernhauer, A. 1962. Zur Verlandungsgeschichte des Burgenländischen Seewinkels. Wiss. Arb. Bgld. 29: 143–171.

Bernt, D. 1970. Die Raumansprüche von Wirtschaft, Siedlung, Verkehr, Naturschutz, Bundesheer. Österr. Inst. f. Raumplanung, i.A. der Neusiedlersee Planungsgesellschaft.

Burgenland Entwicklungsprogramm 1968. Österr. Inst. f. Raumplanung u. Inst. f. empirische Sozialforschung, Vienna.

Burgenland, 50 Jahre: 1921–1971. Hsg. Amt der Bgld. Landesreg., Eisenstadt, 1971.

Büdel, J. 1933. Die morphologische Entwicklung des südlichen Wiener Beckens und seiner Umgebung. Berliner Geogr. Studien 4.

Fink, J. 1955. Exkursion zwischen Salzach und March, Abschnitt Wien-Marchfeld-March mit Beiträgen von E. Frasl und F. Brandtner. Beitr. Pleistozänforschung in Österr., Sonderheft 4.

Fink, J. 1957. Quartärgeologisch-bodenkundliche Karte Mattersburg-Oberpullendorf. Erl. zur geol. Karte: Mattersburg-Deutschkreutz 1:50,000. Geol. Bundesanst., Vienna.

Fink, J. 1960. Leitlinien einer österreichischen Quartärstratigraphie. Mitt. Geol. Ges. Wien 53: 249–266.

Fink, J. 1961. Zur Gliederung des Jungpleistozäns in Österreich. Mit. Geol. Ges. Wien 54: 1–25.
Fink, J. 1965. The Pleistocene in eastern Austria. Inter. Stud. of the Quaternary., Spec. Rep. Soc. America 84: 179–199.
Fink, J. 1966. Die Paläogeographie der Donau. In: Limnologie der Donau II, pp. 1–50. Schweizerbarth, Stuttgart.
Fink, J. 1969. Nomenklatur und Systematik der Bodentypen Österreichs. Mitt. Österr. Bodenkundl. Ges. 13.
Fink, J. 1970. Österreichs Böden im Spiegel der bodenbildenden Faktoren. In: Mem. N. C. Cernescu et M. Popovat, pp. 7–34. Geol. Inst. Ser. C.-Pedol. 8, Bucharest.
Fink, J. 1973. Zur Morphogenese des Wiener Raumes. Z. Geomorph. 17: 91–117 Berlin-Stuttgart.
Franz, H. & Husz, G. 1961a. Die Salzböden und das Alter der Salzsteppe im Seewinkel. Mitt. Österr. Bodenkundl. Ges. 6.
Franz, H. & Husz, G. 1961b. Neusiedl/See-Podersdorf-Illmitz, Exkursion C. Mitt. Österr. Bodenkundl. Ges. 6.
Frasl, G. 1961. Zur Petrographie der Sedimente des Seewinkels. Mitt. Österr. Bodenkundl. Ges. 6: 62–67.
Gerabeck, K. 1952. Die Gewässer des Burgenlandes. Burgenländische Forschung 20.
Hassinger, H. 1918. Beiträge zur Physiogeographie des inneralpinen Wiener Beckens und seiner Umrahmung. Festb. A. Penck, Bibl. Geogr. Hand. Stuttgart 160–197.
Husz, G. 1962a. Untersuchungen über die Entstehung von Salzböden im Seewinkel (Burgenland) als erste Grundlage ihre Melioration. Diss. Hochsch. f. Bodenkultur, Vienna.
Husz, G. 1962b. Zur Bodenkartierung im Salzbodenbereich des Seewinkels. Wiss. Arb. Bgld. 29: 172–180.
Kapounek, J. 1938. Geologische Verhältnisse der Umgebung von Eisenstadt. Jahrb. Geol. R.-A. Wien 88: 49–98.
Klaus, W. 1962. Zur pollenanalytischen Datierung von Quartärsedimenten im Stadtgebiet von Wien, südliches Wr. Becken und Burgenland. Verh. Geol. BA., 20–38.
Kopf, F. 1963. Wasserwirtschaftliche Probleme des Neusiedler Sees und des Seewinkels. Österr. Wasserwirtsch. 15 (9–10): 190–203.
Küpper, H. 1955a. Exkursionen zwischen Salzach und March, Abschnitt Wien-Neusiedlersee mit Beiträgen von H. Küpper und A. Papp. Beitr. Pleistozänforschung Österr. Sonderheft D. Verh. Geol. BA.
Küpper, H. 1955b. Exkursionen im Wiener Becken Südlich der Donau mit Ausblicken in den Pannonischen Raum. Beitr. Pleistozänforschung, Verh. Geol. BA., 127–157.
Küpper, H. 1957. Zur Kenntnis des Alpenabbruches zwischen südlichem Wiener Becken und dem Ostrand der Rechnitzer Schieferinsel. Erläuterungen zur geol. Karte Mattersburg-Deutschkreuz 1:50,000. Geol. BA. Wien 59–67.
Küpper, H. 1958. Zur Geschichte der Wiener Pforte. Mitt. Geogr. Ges. Wien 100: I/II.
Küpper, H. 1961. Erläuterungen zur Aussicht vom Hackelsberg. Mitt. Bodenk. Ges. 6: 53–55.
Kuras, P. 1971. Wirtschaftsstruktur und Wirtschaftsaufbau des Burgenlandes. Diss. HfW, Vienna.
Landschaft Neusiedlersee. 1959. Grundriß der Naturgeschichte des Großraumes Neusiedlersee. Wiss. Arb. Bgld. 1–168.
Löffler, H. 1974. Der Neusiedlersee, Naturgeschichte eines Steppensees. Molden, Vienna, Munich, Zurich.
Pecsi, M. 1964. Ten years of physicogeographic research in Hungary. Ung. Akad. Wiss.
Riedl, H. 1960. Zur Geomorphologie des Kogels und dessen Umgebung. In: R. Riedl et al., Die befahrbaren Klüfte im Steinbruch St. Margarethen (Bgld.). Wiss. Arb. Bgld. 25: 3–45.
Riedl, H. 1963a. Bemerkungen zur Altersfrage eiszeitlicher Terrassen im östlichen Hügelland.– Unsere Heimat, 33–37.
Riedl, H., 1963b: Beiträge zur Morphogenese der Randgebiete des Neusiedlersees und des Gebietes der Wr. Neustädter Pforte. Festschr. S. Morawetz, Mitt. Naturwiss. Ver. Stmk. 93: 73–88.
Riedl, H. 1964. Erläuterungen zur Morphologischen Karte der eiszeitlichen Flächensysteme der Wulka. Wiss. Arb. Bgld. 31: 175–195.

Roth-Fuchs, G. 1926. Erklärende Beschreibung der Formen des Leithagebirges. Geogr. Jahreber. Österr. 13: 29–106.
Sauerzopf, F. 1956. Das Werden des Neusiedlersees. Bgld. Heimatblätter 18 (1): 1–6.
Sauerzopf, F. 1959a. Zur Entwicklungsgeschichte des Neusiedlersees. Wiss. Arb. Bgld. 23: 107–111.
Sauerzopf, F. 1959b. Problem Neusiedlerseedamm. Bgld. Heimatblätter 21 (1).
Scherf, E. 1935. Geologische und morphologische Verhältnisse des Pleistozäns und Holozäns der großen ungarischen Tiefebene und ihre Beziehung zur Bodenbildung der Alkalibodenentstehung. Jahresber. kgl. ung. geol. Anstalt 1925–1928.
Schroll, E. 1959. Zur Geochemie und Genese der Wässer des Neusiedler Seegebietes. Wiss. Arb. Bgld. 23: 55–64.
Schroll, H. & Tauber, A. 1959. Geochemisch-stratigraphische Beziehungen in den Neogensedimenten des Seewinkels. Bgld. Wiss. Arb. Bgld. 23.
Statistisches Zentralamt, Wien. 1973. Agrar- und Fremdenverkehrs-statisik. Volkszählungsergebnisse, Vienna.
Staudacher, C. 1974. Vergleichenden Strukturuntersuchungen von Neusiedl, Podersdorf und Rust: Versuch einer planungsbezogenen Darstellung von Landesgemeinden. Phil. Diss. Vienna.
Swarowsky, A. 1920. Die hydrographischen Verhältnisse des Burgenlandes. Burgenland-Festschrift 49–61.
Szadecky-Kardoss, E. 1938. Geologie der rumpfungarischen kleinen Tiefebene. Mitt. berg. u. hüttenm. Abt. Sopron 10:1.
Tauber, A. F. 1952. Grundzüge der Geologie vom Burgenland. Bgld. Landeskunde.
Tauber, A. F. 1959a. Zur Oberflächengeologie des Seewinkels. Wiss. Arb. Bgld. 23: 24–26.
Tauber, A. F. 1959b. Geologische Stratigraphie des Neusiedlerseegebietes. Wiss. Arb. Bgld. 23: 18–24.
Tauber, A. F. 1959c. Grundzüge der Tektonik des Neusiedlerseegebietes. Wiss. Arb. Bgld. 23: 26–31.
Tauber, A. F. & Wieden, P. 1959. Zur Sedimentschichtfolge im Neusiedlersee. Wiss. Arb. Bgld. 23: 68–73.
Tollmann, A. 1955. Das Neogen am Nordwestrand der Eisenstädter Bucht. Wiss. Arb. Bgld. 10: 5–75.
Wessely, G. 1961. Geologie der Hainburger Berge. Jahrb. Geol. BA., 104(2).
Wiche, K. 1951. Die Oberflächenformen des Burgenlandes. Bgld. Landeskunde 98–136.
Wiche, K. 1970. Die Flächentreppe des mittleren Burgenlandes. Wiss. Arb. Bgld. 44: 5–38.
Winkler-Hermaden, A. 1957. Geologisches Kräftespiel und Landesformung. Springer, Vienna.

2 The hydrogeology of Neusiedlersee and its catchment area

T. Gattinger

Only during the past two decades have the application of new methods and modern techniques in systematic studies brought us nearer to the solution of

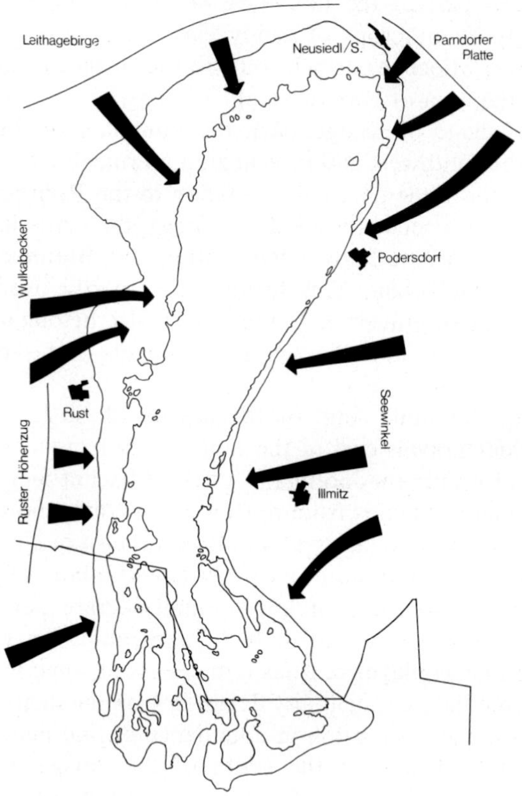

Fig. 2.1. Schematic presentation of groundwater and its flow direction.

the hydrogeological riddles presented by the lake. In this chapter an attempt is made to provide a survey of present knowledge concerning the hydrogeology of Neusiedlersee and its catchment area (Fig. 2.1).

2.1 The boundaries of the hydrogeologic catchment area

In the south, the boundary of the hydrogeologic catchment area of Neusiedlersee runs on Hungarian territory south of the small town of Balf at the southwestern corner of the lake, within sandy clays and sands of the Pannonian (uppermost Pont). Except for local deviations the strata in this area generally dip towards the lake. The continuation of this formation of low permeability is excluded from the recharge area as the escarpment of Pannonian rocks east of Fertöboz dips to the south, thus directing the groundwater in the overlying Quaternary gravels to the adjacent Ikva river.

From the vicinity of Balf the boundary of the recharge area takes a northwesterly direction, passing the northern part of the town of Ödenburg (Sopron) and running along a group of hills on the northern side of the Ikva river to the border between Hungary and Austria south of the village of Klingenbach. This part of the boundary of the catchment area is marked hydrogeologically by outcrops of sandy clays and sands of the Badenian (Tortonian, Lower Miocene), which come to the surface from below Tertiary gravels west of the area of Rakosi Patak.

South of the village of Klingenbach the boundary of the recharge area bends towards the southwest and runs in an impermeable formation of upper Tertiary clays of the Badenian and Sarmatian to the Rohrbach forest. From there, the boundary reaches the saddle of Sieggraben crossing the mountain of Brentenriegel in a southerly curve within the Brennberg boulder-bed formation of the Helvetian. This formation is partly overlain by marine sediments consisting of gravels and sands which also belong to the Helvetian. Both the boulder beds and the overlying gravels and sands are of high permeability.

After crossing the fault zone of the saddle of Sieggraben, where the Brennberg boulder gravels east of the fault are in tectonic contact with the crystalline rocks forming the mountains of the adjacent Sieggraben forest to the west, the boundary bends from northwest to north and then follows the ridge of the Rosalia mountains up to a point southeast of Neudörfl village. In this area the main rock types are mica schists and quartz phyllonites. These are laminated schistose rocks of low to intermediate permeability in the direction of their schistosity and of high permeability in tectonic fracture zones. Their weathering layer reaches a thickness of some metres and shows also low to intermediate permeability depending on the degree of decomposition and the consequent variation in abundance of fine material.

From the Neudörfl area to the south of the village of Hornstein the boundary runs through gravels of Quaternary terraces which overlie gravels and sand of the Pont. Its continuation is mainly marked by the culmination zone of crystalline rocks of the Leitha mountain range anticline. This anticline shows an axial depression at Schweinberg with overlying Leitha limestone (Miocene).

Northwest of the village of Breitenbrunn the crystalline rocks of the Leitha mountain range dip towards the northeast and disappear under Tertiary sedimentary rocks. Up to the village of Parndorf the first part of the boundary of the hydrogeologic recharge area runs through Leitha limestone and finally through silty and fine-grained sandy sediments of the Tertiary overburden. From the village of Parndorf to the east, and to the southeast as far as the village of Halbturn, it runs through gravels and sands of the Parndorf terrace, then turns southwestwards in an area of Quaternary gravels and sands of the Seewinkel and reaches the southern end of Neusiedlersee southwest of the village of Apetlon.

2.2 The geological formations of the catchment area and the permeability of their lithologies

The crystalline basement (mainly granitic gneisses, schistose gneisses and micaschists) which outcrops at the periphery (Rosalia and Leitha mountain range) and the inner part (hills of Rust) of the recharge area forms the deepest aquiclude of this region. Intermediate permeability is restricted to the joints between the rock layers and to the clefts of fault zones.

The Tertiary sequence of sedimentary rocks and sediments partly consists of permeable gravels, sands and sandstones and partly of impermeable or semipermeable clays, siltstones and silts forming the floor or roof layers of aquifers.

In the southwest (the area of Brennberg mountains) the lower Auwald gravels, which are coarse gravels in a matrix of sands and fine grained gravels (with freshwater sediments and coal seams at their base), are the deepest formation of the Tertiary. Occasional lenticular intercalations of clay of local extent can be found. The pebbles are well rounded with diameters of 10 to 15 centimetres and consist mainly of gneisses, micaschists and granites, originating from the adjacent area to the south. The overlying upper Auwald gravels show a similar composition except for an increase in detritus from the calcareous Alps. Freshwater sediments consisting of a basal coal seam and sands and argillaceous marls are intercalated between the upper Auwald gravels and the overlying Brennberg boulder beds.

The Brennberg boulder beds consist of barely rounded blocks of local origin (quartz, granite gneiss, schistous gneiss, micaschist) with diameters from half to one metre and a matrix of sands, clays and occasionally sandstones and loam. They are overlain by marine gravels and sands.

In the hills of Rust, the basement series of the Tertiary consists of the so-called Rust gravels (Fuchs 1965), the composition of which resembles that of the Auwald gravels, except for a higher sand content in the matrix.

Lower and upper Auwald gravels, Rust gravels and Brennberg boulder beds belong to the Helvetian formation of the Upper Miocene (Küpper 1957a, 1957b; Fuchs 1965). The Auwald gravels and the Brennberg boulder beds reach a thickness of some hundred metres, the Rust gravels more than one hundred metres. In general, the thickness of the clastic series decreases from west to east and from southwest to northeast.

The whole complex of the Helvetian formations is of high permeability, which also characterizes the 500-metre-thick gravels and sands of the Badenian formation which outcrop in the southwest of the recharge area around the town of Mattersburg. They are overlain by clays with a thickness up to 700 metres, showing intercalations of fine-grained sands which increase towards the top. The clays form a considerable aquiclude.

In the hills of Rust and in the Leitha mountain range, the Badenian is predominantly represented by Leitha limestone which partly occurs in a detrital facies. This means that it was eroded and broken up soon after its formation and redeposited with very little diagenetic compaction. Besides these limestones, sands and sandstones may be of local importance as aquifers, and marls as aquicludes.

The high permeability of the Leitha limestone complex, some hundred metres thick, is due, on the one hand, to high porosity caused by its detrital nature and, on the other, to tectonic fracturing and subsequent karstic solution.

Like the Badenian the Sarmatian formation in the west shows thick gravel and sand complexes, the most western part around Wiesen and Mattersburg consisting of very coarse gravels with components of crystalline rocks, besides material from the calcareous Alps. Further to the east, in the Wulka river basin and around Rohrbach-Drassburg, silts occur in addition to the gravels and sands, especially in the upper parts of the Sarmatian formation.

The Sarmatian of the Rust hills also consists of gravels with a predominance of components derived from southwest of the area and originally the calcareous Alps (Fuchs 1965). They overlie sandstones of the Upper Sarmatian, which partly rest on beach conglomerates at the western edge of the hills of Rust.

Conglomerates, gravels and sands of the Sarmatian also occur in the Leitha mountain range.

The thick clastic series of the Sarmatian are highly permeable. Fine-grained quartz sands and, at the eastern flank of the hills of Rust, debris of the Pannonian formation, occur at the periphery of the Wulka river basin. Apart from this, the Pannonian sediments consist of clays and silts, which form the impermeable capping to the permeable Upper Miocene rock series described above.

Of the quaternary sediments of the area (besides the sand and gravels covering the Parndorf plateau) the gravels of the Seewinkel are of major importance. Their origin was established by Fuchs (1974). This complex of fine-grained to intermediate quartz gravels, partly well and partly poorly rounded, show varying contents of crystalline and limestone components. It conducts an extensive groundwater stream from the east towards Neusiedlersee. The bottom of this shallow aquifer is formed by upper Pannonian clays.

2.3 Hydrogeologic conditions in the different parts of the recharge area

On account of the different geological history four following regions can be distinguished, each exhibiting special morphological and structural features: the hills of Rust, the Wulka river basin, the Leitha mountain range, and the Parndorf plateau.

2.3.1 *The Rust hills*

The basement of the hills of Rust consists of crystalline rocks forming an uplifted fault block (Fuchs 1965) which emerges through its Neogene overburden southwest of Mörbisch village (granitic gneisses), south of the Silberberg near Oslip village, and on Goldberg and Seeberg near Schützen village (albite chlorite gneisses).

These impermeable rocks form the aquiclude for the overlying Tertiary rock series, parts of which are of high permeability.

To begin with, mention must be made of the so-called Rust gravels, which, despite their name, consist of coarse- and fine-grained sands. They constitute a complex, more than 100 metres thick, overlying the crystalline basement and dipping smoothly east to southeast in the direction of Neusiedlersee. The interbedded gravels consist of crystalline components, especially coarse gneiss. Boulders of diameters up to one metre may occur. Coarse-grained gravel layers alternate with layers of sands and fine-grained gravels, the differences in grain size very distinctly revealing the bedding of the sediments.

The Rust gravels have been classified stratigraphically as fluvial sediments of the upper Helvetian (Fuchs 1965). Distribution of grain sizes and thickness as well as orientation of the axes of the pebbles and the composition of the material suggest that transport took place from southwest towards northeast.

The following series overlay the Rust gravels in the higher part of the Upper Tertiary: coral reefs with intercalations of fine-grained sands, marls, masses of detrital lithothamnic limestones of the mid-Badenian and, finally, conglomerates, sandstones and delta sediments of the upper Sarmatian.

Of all the rock series forming the hills of Rust, the Rust gravels are the most important for the groundwater supply to Neusiedlersee, due to their high permeability and inclination towards the lake basin. Additional aquifers are provided by the highly permeable masses of lithothamnic limestone, the conglomerates, the delta sediments and the masses of debris of the Pannonian on the eastern flank of the hills of Rust.

Groundwater infiltrated in the area of the hills of Rust reaches the subsoil of the lake through these aquifers and ascends to the surface water through clefts and cracks of tectonic fracture zones which form the eastern border of the fault block of the hills of Rust. This is elevated about 600 metres above the adjacent lake basin lying to the east. These fracture zones run parallel to the western shore of the lake and are connected with the extensive Neusiedl fault zone. The so-called 'boiling fountains' or subaquatic springs located on this structure, are evidence of the fact that the ascent of groundwater to the lake is predominantly connected with this fault zone. The mechanism of groundwater descent through aquifers dipping to the subsoil of the lake basin and re-ascent through fracture zones is typical for the western and northern parts of the recharge area of the lake.

2.3.2. *The Wulka river basin*

Overlying the crystalline basement of the Wulka river basin, just as in the hills of Rust, are masses of fluvial sediments which were deposited by the same river system flowing from south to north in the Helvetian. This river system, which originated in the area of Ödenburg (Sopron), brought crystalline gravels into what was then the continental Wulka basin and crossed the Leitha mountain range which had not yet been uplifted. Corresponding sediments are to be found today in the depression of crystalline rocks around Burgstall-Hoher Stein, 400 metres above sea-level. They provide evidence of the uplift of this complex, and thus of the formation of the northern boundary of the Wulka river basin during a subsequent orogeny. Prior to this uplift the basin was repeatedly subject to transgressions with a variety of consequences. The first of the transgressions during the late lower Badenian did not give rise to any notable deposition of marine sediments but primarily to the destruction of the fluviatile sedimentary structures of the gravel and sand masses and their resedimentation. This partly destroyed the original bedding of the sediments. The gravel components are better rounded than those of the hills of Rust, yet grain size and distribution are very similar in the two areas.

During the middle Badenian, marine sedimentation of reef limestones, fine-grained sands and marls also took place in the Wulka basin. It continued during the Sarmatian with the development of beach conglomerates on the periphery of the basin and with the sedimentation of Leitha limestone, sandstones and delta deposits. At the periphery of the basin sparse remnants of lower Pannonian sediments are also preserved. Unlike the hills of Rust, marine sedimentation continued in the Wulka basin area during the middle Pannonian. Whitish-grey quartz sands were deposited in the southern part, yellow sandy silts in the northern part.

As in the hills of Rust, fluvial gravel and sand masses at the base of the Wulka basin form the most important aquifer. On the southern periphery in particular, they occur in conjunction with the stratigraphically lower Auwald gravels and boulder beds which are of even higher permeability than the Rust gravels. Other aquifers on the southern periphery are the beach conglomerates and, in the central part of the basin, the Leitha limestones, besides the fine-grained sands of the middle Pannonian in the southern part of the basin. The middle Pannonian sediments of the northern part of the basin, consisting mainly of silts, are impermeable.

The groundwater movement in the Wulka basin is governed by the considerable inclination of the aquifers away from the peripheral recharge areas, where they outcrop (Brennberg, Rosalia mountains, Leitha mountain range), to the central part of the basin where they are found at a depth of some hundred metres below the surface.

Of special importance for the movement of these groundwaters towards Neusiedlersee is the fact that the area north of the hills of Rust, that is, the area of the Wulka river mouth, had already formed as a depression during the lower part of the upper Sarmatian and that this depression was on the whole preserved throughout the following periods of geological history. Furthermore, there are distinct indications of a fault zone beginning in the crystalline formations of the Rosalia mountains and striking through the basin of Mattersburg. It continues at some depth below the Wulka valley, running parallel with the southern marginal fault zone of the western part of the Leitha mountain range up to the lake, where it meets the southern continuation of the Neusiedl fault zone. This structure provides an additional zone of high permeability for the groundwaters recharging Neusiedlersee from the Wulka basin. The southern continuation of the Neusiedl fault zone and the fault system east of Mörbisch both enable the groundwater to ascend to the lake. Further permeable zones are provided by faults cutting the hills of Rust from west to east and constituting another exit for groundwater from the Wulka basin to the subsoil of the lake, where it again ascends through the southern continuation of the Neusiedl fault zone.

The Quaternary deposits are of minor importance for the recharge of groundwater from the west and north to Neusiedlersee. The lower and upper terrace gravels generally overlie Pannonian sediments with little or no permeability. The groundwater recharged in these gravels does not descend to greater depths as it is mostly forced into the surface water by shallow aquiclude layers after flowing a relatively short distance underground.

It must also be mentioned that there is no groundwater connection between the southern Vienna Basin and the Wulka river basin. An impermeable barrier consisting of Pannonian clay sediments effects the hydrogeologic separation of the two basins.

2.3.3. *The Leitha mountain range*

The Leitha mountain range as far as the area of the village of Donnerskirchen forms the northern part of the recharge area of the Wulka basin. To the northeast and east of this area the hydrogeologic situation is as follows. The southern flank of the crystalline anticline of the Leitha mountain range is overlain by Leitha limestones and conglomerates generally dipping southwards under marls and clays of the Upper Tertiary. From the section of Breitenbrunn eastward the limestones and conglomerates increase considerably in extent and cover the culmination of the crystalline anticline. In this part of the recharge area the limestones and conglomerates dip under clays and marls of the Upper Tertiary.

Step faults run from Donnerskirchen towards the northeast as well as from Purbach to Breitenbrunn and beyond Jois, and downthrow to the south. In addition, the Wulka fault already mentioned forms an important member of this step-fault system, and is an essential element of the northern boundary of the lake basin.

Like the fault zone in the area of Mörbisch and Rust the Wulka fault zone in the northern part of the lake forms a structure especially suitable for the ascent of groundwater recharging from the southern flank of the Leitha mountain range.

Because of its low permeability the crystalline basement series is of major importance as an aquiclude. Moreover, in the region where it is exposed to the surface it functions as a precipitation collector directing the surface runoff to the Leitha limestone formation.

The gradient of this formation, the high permeability of which is at least in part attributable to karst cavities, favours the transmission of groundwater to the fault system, allowing groundwater re-ascent as mentioned above.

2.3.4. *The Parndorf plateau and the Seewinkel*

The Parndorf plateau consists of a base of upper Pannonian silt which is covered by elder Pleistocene gravels. The riser of the terrace follows a tectonic pattern.

The gravels of the adjacent Seewinkel are fluvial deposits from the Würm and Riss ice age. The groundwater shed between the Neusiedlersee and the Leitha river with respect to the Danube river in the north was formed by upliftings in the pre-Würm stage. Evidence and arguments given prove that the gravels of the Seewinkel were deposited during the younger Pleistocene.

Besides the groundwater of the Parndorf plateau and the Seewinkel, groundwater is recharged to a limited extent in the clastic parts of upper Pannonian sediments at the western end of the Parndorf plateau. These aquifers dip towards the southeast. They pass the groundwater over a distance of four to twelve kilometres to the area of Neusiedl, Weiden and Gols where it is exploited from wells and boreholes of fifteen to a hundred metres depth. Locally, the confined aquifers meet the overlying Quaternary gravel sheet of the Seewinkel and cause the artesian water to seep into the unconfined Quaternary aquifer which conveys the groundwater to the lake.

Some former investigators of the region made the assumption that groundwater in the area of the Leitha river around the village of Zurndorf, north of the Parndorf plateau, would be recharged to Neusiedlersee. The aquifer in question consists of sand and gravel intercalations known from boreholes in the clay formation of the Pannonian. These boreholes were separated by distances of five to six kilometres.

Quite apart from the fact that the assumed aquifers have no access to the surface, and infiltration there is impossible, they only form unconnected lens-like beds. Thus, groundwater recharge from the Leitha area through the Pannonian base of the Parndorf plateau can be disregarded.

2.4. The magnitude of the groundwater recharge to the lake basin

Calculations of the groundwater recharge to the lake from the areas of Parndorf plateau and Seewinkel revealed a minimum annual average of 24 million and a maximum of 30 million m^3.

At an annual average precipitation of 650 mm, this means that a minimum of 20 per cent, and a maximum of 25 per cent of the precipitation infiltrates, giving an average minimum groundwater recharge of 4 to 5 l. sec^{-1} km^{-2} or a minimum total recharge of 760 l. sec^{-1} and a maximum of 950 l. sec^{-1} for the recharge area of 184 km^2. These values are in good agreement with those upon which the representation of this area on the hydrogeological map of Austria is based.

In view of this agreement and knowledge of the hydrogeologic properties of the entire recharge area it appears justifiable to use the same infiltration rates to estimate the groundwater replenished from other points of the recharge area. It must, however, be emphasized that the results of such estimates merely indicate orders of magnitude.

Parts of the total recharge area are unsuitable for infiltration. Only the areas covered by formations permitting infiltration and conveyance of groundwater towards the lake basin can be termed effective in a hydrogeological sense. The portion of the total area meeting these conditions covers 444 km^2, constituted as follows:

Parndorf plateau and Seewinkel	184 km^2;
hills of Rust	33 km^2;
Wulka river basin	163 km^2;
Leitha mountain range	64 km^2.

Assuming that 20 per cent of the average annual precipitation of 650 mm infiltrates, the total annual recharge of these four regions amounts to 58 million m^3. Data for the other parts of the recharge area are as follows:

hills of Rust: annual recharge 4.318 million m^3 (137 l. sec^{-1});
Wulka river basin: annual recharge 21.284 million m^3 (675 l. sec^{-1});
Leitha mountain range: annual recharge 8.398 million m^3 (266 l. sec^{-1}).

The total recharge (without the contribution of the Parndorf plateau and the Seewinkel) amounts to 34 million m^3 or 1078 l. sec^{-1}.

Comparing the various parts of the recharge area it appears that the Parndorf plateau plus the Seewinkel yield 41.4 per cent, and the Wulka river basin a comparable quantity of 36.7 per cent of the total recharge whereas the Leitha mountain range and the hills of Rust, with 14.5 and 7.4 per cent respectively, yield relatively less.

The question of whether all groundwater recharged in the Tertiary aquifers reaches the lake by ascent in the fault zones or whether part of it underflows the lake and continues its course towards deeper parts of the Pannonian basin (Hungarian territory), remains undecided. However, the general decrease of grain size towards the centre of this basin suggests that the aquifers of high permeability do not continue over long distances and that there is no substantial transport of groundwater to the subsurface of the Hungarian Plain. Moreover, the regional fault system of the lake basin produces a groundwater trapping effect by decompression which forces the groundwater to ascend from the deep aquifers to the surface, that is, to the lake bottom.

2.5 The groundwater discharge in the lake

As a consequence of the different hydrogeological conditions prevailing in the northeastern and eastern area, on the one hand (Parndorf plateau, Seewinkel), and the northern and western area on the other (Leitha mountain range, Wulka river basin, hills of Rust), distinct differences in the groundwater discharge in the two parts of the lake are observable.

For this reason, special methods had to be employed in recording the discharge areas and in locating the subaquatic groundwater springs in the lake. The aquifers of the Parndorf plateau and the Seewinkel were investigated by well-logging and the groundwater table depicted by isohypses. Groundwater discharge from the rest of the replenishment area, however, had to be located by infrared aerial photography since only few of the subaquatic springs, the conspicuous 'boiling springs', were known so far.

2.5.1. *Groundwater discharge in the northeastern and eastern part of the lake*

The groundwater from the aquifers of the Parndorf plateau and from the Seewinkel is discharged to the lake along a continuous line on the northeastern and eastern margins of the lake. Due to the different hydraulic gradients in various sections of the aquifer the average groundwater velocity varies over a relatively wide range.

In the section of Neusiedl (northeast) it is about 1 metre, in the section between Weiden and Podersdorf (east-northeast) about 0.5 metres, and in the section between Podersdorf and Illmitz (east) 0.2 down to 0.01 m per day at an average K_f-value of 10^{-3}.

The entire groundwater discharge along the edge of the lake from Neusiedl to south of Illmitz occurs as seeping inflow from the aquifer. Concentrated groundwater streams are absent.

2.5.2. *Groundwater discharge in the northern and western part of the lake*

The interpretation of the infrared aerial photographs has proved that the groundwater discharge in the northern and western parts of the lake is entirely different from that in the northeastern and eastern parts.

The photographs cover the western half of the lake, the northern and western margins of the lake and, for general orientation, parts of the eastern lake shore in the vicinity of Podersdorf. The survey was carried out by the Austrian Bundesamt für Eich- und Vermessungswesen at a time when the lake was frozen after a prolonged frost with temperatures of $-15°C$, and with a temperature difference between the lake surface and the discharged groundwater of at least 10°C.

The first step in the interpretation of the pictures was the identification of 'warm' water inflow from sewage channels and surface streams after which they could be disregarded. The remaining groundwater discharge points and areas were subsequently checked by more than eight hundred temperature measurements using a thermistor during three summer periods when the temperature of the surface water of the lake was about 20 to 23°C and that of the discharged groundwater about 10°C.

The most significant result of the infrared aerial photography was the finding that the extent of the Neusiedl fault zone is accurately marked by a great number of concentrated and diffuse groundwater springs. The pictures not only show the line of the fault zone but also its bifurcation northeast of Oggau. Moreover, the eastern part of the Wulka fault zone, the fault zone crossing the hills of Rust east of Oggau and Rust, as well as the fault zone east of Rust and Mörbisch running parallel with the lake shore, are distinctly marked by series of subaquatic groundwater springs.

The relation of a great number of larger and smaller subaquatic groundwater springs in the lake to the subsurface tectonic structures has not yet been entirely clarified. This would require a better knowledge of structural details, only to be gained by further drilling and by geophysical investigations, especially seismic surveying.

Finally, a few words concerning the drying out of the lake during the last century.

Taking into account the complexity of the recharge situation, a considerable number of factors, but two in particular, must be in congruence over a number of years or even decades for this event to occur. First of all, infiltration must be low due to low precipitation in various parts of the recharge area in different years, which in turn must correspond to the differences in flow times necessary to cover the different distances between recharge areas and discharge zones. Secondly, a period of high evaporation must exactly coincide with the period when the differentiated low infiltration makes itself felt on the discharge side.

Other factors, such as temporary changes in the permissivity of the aquifers or in the zones of groundwater ascent, the causes of which we do not know, are theoretically possible. In any case although the seemingly accidental drying out of the lake may be a rare episode, it should nevertheless as a hydrogeological phenomenon be the main topic of further hydrogeologic investigations.

References

Baranyi, S. & Urban, J. 1978. A Fertö tó hidrológiai jellemzöi (Zusammenfassende Darstellung der hydrologischen Parameter des Neusiedler Sees) Vituki (Budapest), Közlem. 7, 1–89.

Bistritschan, K. 1939. Ein Beitrag zur Geologie des Wechselgebietes. Verh. R.-A.f. Bodenf. Wien 4: 11.

Bobies, C. A. 1958. Über die Pedalion-Korallenfazies im Wiener und Eisenstädter Becken. Verh. Geol. B.-A. Wien 1(38).

Böhm, A. 1883. Über die Gesteine des Wechsels. Tscherm. min. petr. Mitt. 5: 197.

Bundesversuchs- und Forschungsanstakt Arsenal 1968: Bericht über die Grundwasserströmung am Ostufer des Neusiedler Sees.

Czjzek, J. 1852. Geologische Verhältnisse der Umgebung von Hainburg, des Leithagebirges und der Ruster Berge. Jahrb. Geol. R.-A. Wien 4: 35.

Fuchs, W. 1965. Geologie des Ruster Berglandes (Burgenland). Jahrb. Geol. B.-A. Wien 108.

Fuchs, W. 1974. Bericht über Exkursionen in die Oststeiermark, in das südliche Burgenland und nach Westungarn zur Klärung der Herkunft der Seewinkelschotter. Verh. Geol. B.-A. Wien 4: 118–121.

Häusler, H. 1939. Über das Vorkommen von Windkanten am Westrand des Neusiedler Sees. Verh. Geol. R.-A. Wien 5–6: 185.

Kapounek, J. 1939. Geologische Verhältnisse der Umgebung von Eisenstadt. Jahrb. Geol. R.-A. Wien 88: 49–98.

Kieslinger, A. 1955. Rezente Bewegungen am Ostende des Wiener Beckens. Geol. Rundschau Stuttgart 43: 178.

Kümel, F. 1952. Über Untersuchungen entland der burgenl. Nord-Südstraße (Bericht 1951). Verh. Geol. B.-A. Wien 1: 57.

Küpper, H. 1955. Beiträge zur Pleistozänforschung Österreichs, Abschnitt Wien-Neusiedler See. Verh. Geol. B.-A. Wien 127.

Küpper, H. 1957a. Geologische Karte von Mattersburg-Deutschkreuz, 1:75.000. Geol. B.-A., Vienna.

Küpper, H. 1957b. Erläuterungen zur geol. Karte Mattersburg-Deutschkreuz. Geol. B.-A., Vienna.

Küpper, H., Prodinger, W. & Weinhandl, R. 1955. Geologie und Hydrologie einiger Quellen am Ostabfall des Leithagebirges. Verh. Geol. B.-A. Wien 133.

Mohr, H. 1912. Versuch einer tektonischen Auflösung des NO-Spornes der Zentralalpen. Denkschr. Öst. Akad. Wiss., math.-nat. Kl. 88: 1.

Prey, S. 1949. Zur Geologie der NW-Abdachung des Leithagebirges. Verh. Geol. B.-A. Wien 72.

Richarz, P. S. 1908. Über die Geologie der Kleinen Karpathen, des Leithagebirges und des Wechsels. Mitt. Geol. Ges. Wien 1: 26.

Roth-Fuchs, G. 1929. Beiträge zum Problem 'Neusiedler See'. Mitt. Geogr. Ges. Wien 72: 47.

Roth-Telegd, L. 1879. Geologische Skizze des Kroisbach-Ruster Berges und des südl. Teiles des Leithagebirges. Földt. Közl. Budapest 20 (3–4): 139.

Roth-Telegd, L. 1905. Geologische Spezialkarte der Länder der ung. Krone, Umgebung Kistmarton (Eisenstadt), Sekt.-Blatt Zone 14, Col. XV, 1:75.000. Budapest.

Sauerzopf, F. 1956. Das Werden des Neusiedlersees. Bgld. Heimatblätter 18(1): 1–6.

Siel, A. 1957. Das Jungtertiär in der näheren Umgebung von Hornstein im Burgenland. Mitt Ges. Geologie- u. Bergbaustud. Wien 8: 60.

Szadecky-Kardoss, E. 1938. Geologie der rumpfungarischen kleinen Tiefebene. Mitt. berg. u. hüttenm. Abt. Sopron 10: 1.

Tauber, A. F., et al. 1959. Landschaft Neusiedler See. Wiss. Arb. Bgld. 23: 55.

Tollmann, A. 1955. Das Neogen am Nordwestrand der Eisenstädter Bucht.-Wiss. Arb. Bgld. 10: 5–75.

Vacek, M. 1892. Über die krystallinischen Inseln am Ostende der alpinen Centralzone. Verh. Geol. R.-A. Wien 15: 367.

Vendl, M. 1928. Geologische Karte der Umgebung von Sopron, 1:25.000. Sopron.

Vendl, M. 1929. Geologie der Umgebung von Sopron I. Mitt. ber. u. hüttenm. Abt. Sopron 1: 225.

Vendl, M. 1933. Daten zur Geologie von Brennberg und Sopron. Mitt. berg. u. hüttenm. Abt. Sopron 5(2): 386.

Winkler-Hermaden, A. 1957. Geologisches Kräftespiel und Landformung. Springer, Vienna.

3 Origin and geohistorical evolution

H. Löffler

The modern lake lies on top of 400 to 600 m of Pannonian sediments, produced by the final stage of the Paratethys which disappeared during the later Pliocene. Below the open lake area these sediments directly cover the crystalline material which culminates in the Leitha mountain range (see Chapter 1). It is of some interest that between the crystalline and the Pannonian cover, sediments of the earlier stages of the Paratethys (brackish Sarmatian and marine Badenian) are lacking, although they are present further west and east of the lake area. Their absence is thought to be due to the shallowness of the earlier stages within this area.

When Hassinger (1905) put forward his hypothesis of Neusiedlersee presenting an old oxbow of the Danube system it was already well established that at least the whole Pleistocene fell in between the last stages of the Pannonian lake and Neusiedlersee. More recently it could be demonstrated that not only the oxbow hypothesis has to be dropped but also the idea that the Neusiedlersee basin was formed by deflation. Küpper (1957) gave proof of young tectonic movements mainly in the southwest of the lake and along tectonic fracture zones which run south-north. According to him the recent basin took its final shape due to these tectonic events at the end of the last cold period and thus occupies a Young Pleistocene pediplain (Riedl 1964). Another proof against the oxbow hypothesis is the distribution of gravel bordering the present lake in the east but not extending at all into its area save west of Illmitz and southeast of Weiden. These gravels, clearly recognized since Szadecky-Kardoss (1938) as Young Pleistocene Danube gravels – influences of the Raab-Rabnitz System in the southern part of the Seewinkel have been described by Fuchs (1974) – give further proof of the tectonic origin of the lake. Their absence from most of the basin indicates not only the lack of any fluviatile activity but probably also that the lake basin was significantly higher. If this were not the case it is difficult to understand why the gravel fans produced by the Danube after the Würm, and almost as likely also after the Riss glaciation, should not have reached what is at present the deepest depression of the area (gravels of the older Pleistocene have been recorded on top of the Parndorfer plateau and, of problematic origin, also on the hills of Rust (Küpper 1961): at that time the Danube riverbed was approximately situated where the lower section of the Leitha river now runs). Although the horizontal distribution and thickness of the Seewinkel gravels were already described in early papers by Szadecky-Kardoss (1938) and Tauber (1959a, 1959b) and more recently by the Austrian Oil Company, ÖMV, in an unpublished report, opinions differ as to the extent of the portion originating from the penultimate and the last glaciations. Moreover, the origin and age of the gravel underlying the Hanság have not yet been elucidated (Sauerzopf 1959a: Husz 1965).

On the whole, however, the distribution of the gravels together with the analysis of the fracture zones in the southwest and west of the lake basin (Küpper 1957) suggest that the final subsidence of the present basin must have occurred at the end of the Würm glaciation. Some indications of a much more recent age were given by Schroll (1959) who attempted to calculate the geochemical budget of calcium and magnesium. Since, however, such calculations are likely to give arbitrary results when applied to astatic lakes, they can rarely be considered as valuable arguments in the estimation of their age.

Another indication of the time of the final formation of the present lake basin is given by the distribution of certain ostracods. Since, however, sedimentation processes within this highly astatic lake have been disturbed by droughts floods and wind it is unlikely that any core represents more than the last century which corresponds to the sedimentation period since the last drought. And as will be shown further on, even these core fragments may have undergone disturbances. From the analysis of ostracods it appears, however, that the earliest stages of the lake included areas which have become dry during the most recent subsidences in the southwestern area and are thus still able to give information about the lake's early fauna.

The most important area of this nature is the Hanság plain, which stretches along the artificial outlet of Neusiedlersee, extends over approximately 100 km^2 and was covered until very recently by a bog which was flooded when the water level of the lake was exceptionally high (Fig. 3.1). So far, cores have only been taken within the Austrian part, but there is no reason to suppose that profiles from the Hungarian part, where the major area of the Hanság is covered by the elder forest of Kapuvar, would be any different. The gravels mentioned above are covered by about 30 cm of almost entirely inorganic lacustrine sediments of the gyttja type commencing at somewhat more than 113 m above sea level. They are succeeded by organic gyttja, marsh soils and finally by peaty material covered by a thin layer (about 20 cm) of terrestrial soil. Pollen analysis shows that all of the core section below the peaty material belongs to the late Pleistocene but is pre-Alleröd (Bobek, Löffler & Schultze 1978). The total length of cores taken southeast of Tadten amounts to approximately 130 cm. With respect to sea level, lacustrine material in the Hanság begins slightly higher (113.7 m?) than in the Austrian Neusiedlersee basin (112.9 m?). The inorganic gyttja of the Hanság profile is characterized by *Cytherissa lacustris* and *Limnocythere sanctipatricii*. Both species avoid warm water. *Limnocythere sanctipatricii* even avoids any stagnant water of more than 15°C and in dimictic lakes it is a hypolimnetic species. *Cytherissa lacustris* is somewhat more tolerant with respect to thermal conditions but is, however, absent from periodic lakes on account of its exceptionally long generation time of approximately one to two years. Moveover, as is indicated by their distribution, both species are hardly fit for passive dispersal, probably due to the absence of resting eggs. *Limnocythere sanctipatricii* has been found so far only in the plains southeast of Neusiedlersee (Fig. 3.2), including the Hanság, whereas subfossil *Cytherissa lacustris* also occurs east and west (Herrmann 1970) of the northern lake and, as has been mentioned, very rarely in the sediments of the lake itself. Many samples have also been taken along the southwestern shore of the lake but none of the species mentioned has been found. Thus the distribution of the two species in the vicinity of the lake suggests that a late Würm stage of Neusiedlersee must have existed in the plains southeast of the present lake basin. This is

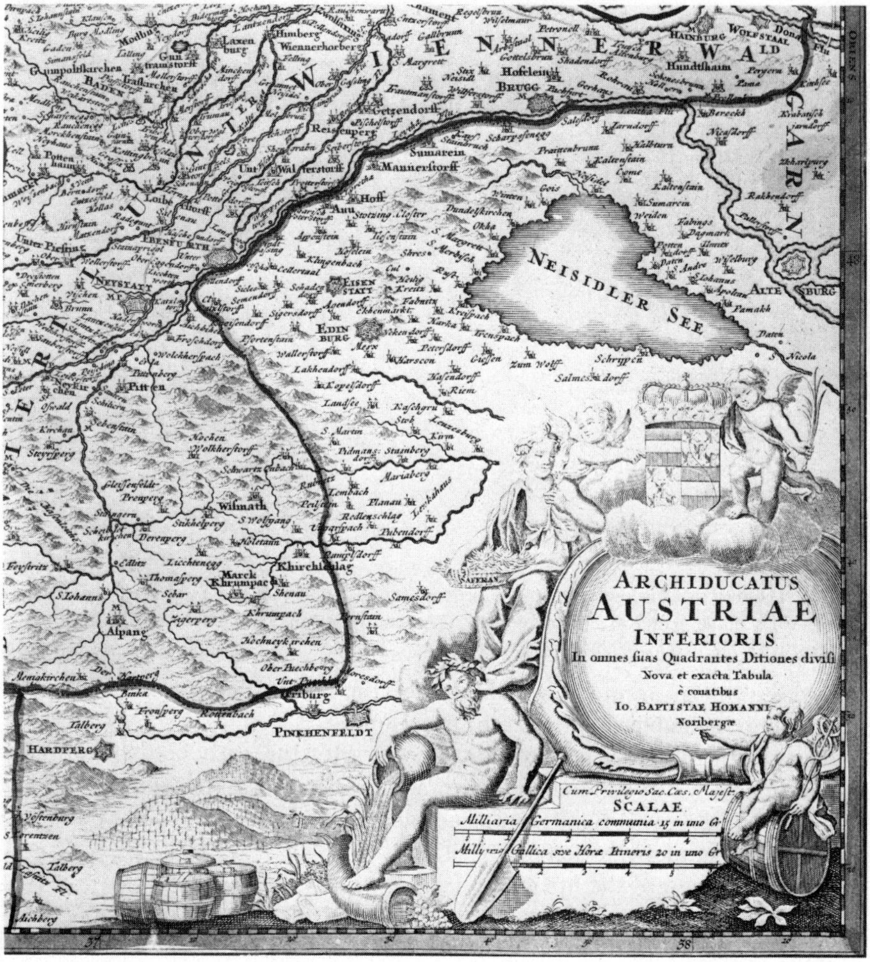

Fig. 3.1. Map of Neusiedlersee indicating a high level situation. The Hanság here is considered as a part of the lake.

supported by pollen analysis. It probably also extended into the area of Gols where most recently an ostracod fauna with *Cytherissa lacustris* and *Limnocythere sanctipatricii* (besides *Ilyocypris*, *Cypridopsis* and *Candona*) has been found at a depth of 170 cm (upper 90 cm clay covering 70 cm terrestrial black soil which lies upon a yellowish stratum rich in molluscs and ostracods). This indicates either a level almost 2 m above the present one during the last cold period or tectonic movements. The occurrence of *Cytherissa* without

Limnocythere sanctipatricii in the northern section of the lake indicated that this part may be younger, although older than the southwestern portion. This would agree with the geological statement that the most recent subsidences

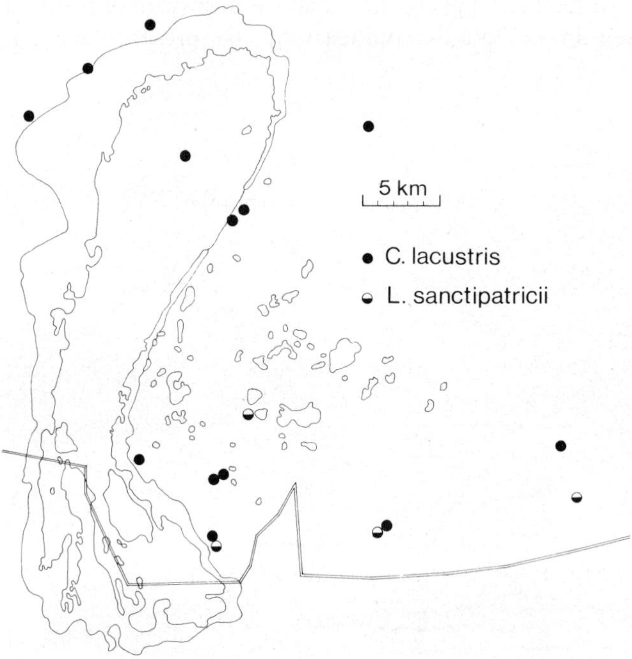

Fig. 3.2. Localities of subfossil *Cytherissa lacustris* (black circles) and *Limnocythere sanctipatricii* (black-white circles) in Neusiedlersee area. The configuration of Neusiedlersee is presented together with the extent of the *Phragmites* belt.

took place in this area. Coring along the Hungarian shores would definitely bring highly valuable additional information.

Analysis of the Hanság cores shows that after the early cold stage of the lake a typical warm water fauna with *Metacypris cordata*, *Ilyocypris sp.* and *Candona candida* was finally replaced by a swamp fauna not unlike the one found in the modern *Phragmites* belt.

Within the area of the open lake (Fig. 3.3) sediment layers of zero to sixty centimetres cover the Pannonian sands which contain almost no organisms. Most of the sediment is a mixture of allochthonous (washed-in terrestrial) and autochthonous lacustrine origin (see Chapter 13). The latter almost exclusively contains organisms which lived in the lake subsequent to its last drying out in 1868. In addition, remnants of plants and animals from undeterminable fragments of earlier lake periods can be recognized. One specimen of *Cytherissa lacustris* has been found west of Podersdorf. There is no doubt that this species belongs to a very early lake stage. The possibility cannot be excluded that in some parts undisturbed sediments of older lake stages might be found on top of the Pannonian material although, so far, proof of this is lacking[1]. Another argument in favour of considering the sediments on top of

1. Very recently such undisturbed sediments have been found at the western edge of the *Phragmites* belt near Breitenbrunn.

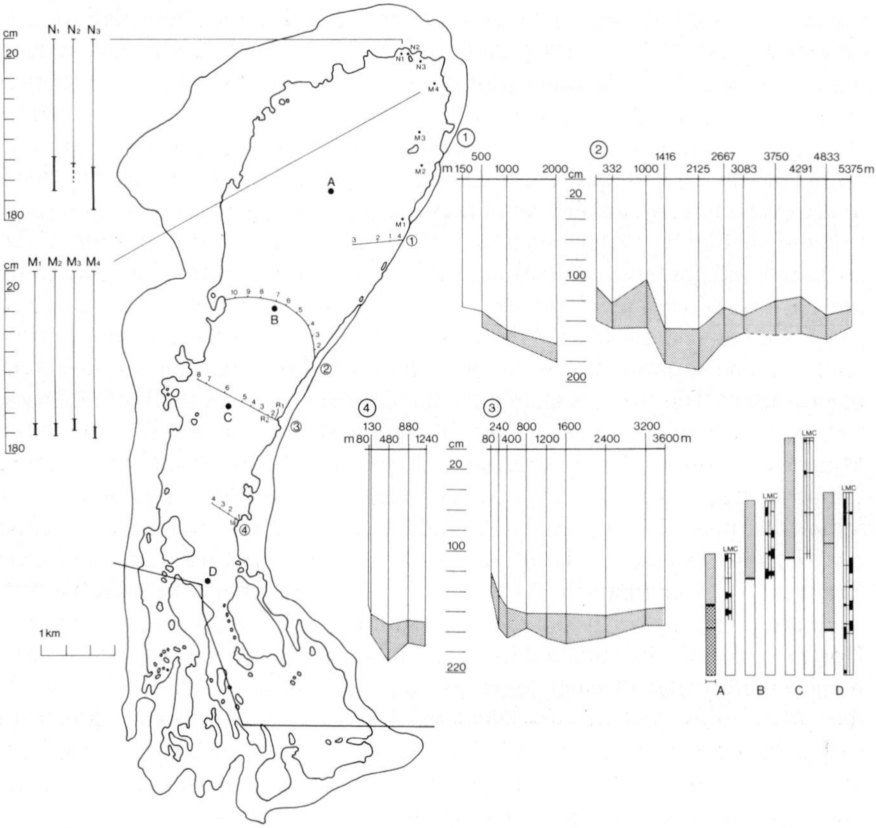

Fig. 3.3 Depth of the non-Pannonian sediment in Neusiedlersee shown in profiles 1 to 4 (shaded area: non-Pannonian sediment), in location M1–M4, N1–N3 and in locations A–D. In locations A–D the left-hand column shows the distribution of sediments (dotted: non-Pannonian sediments, large dots: non-Pannonian sediments rich in mica), the right-hand column represents the remnants of organisms counted at depths indicated (L: *Limnocythere inopinata*, M: ephippia of *Moina*, C: *Campylodiscus clypeus*). The scale below A means 2 cm^3 soil. *Moina rectirostris* has been absent from the lake since 1950 (see Chapter 22).

the Pannonian material as having been deposited after the last desiccation is Moser's (1866) description of the dry lake surface. He not only gave data of analyses which correspond to properties of the Pannonian sediment (see Chapter 12), but he also stated that the uppermost layers hardly differ from deeper ones. Moreover he reports that 'humic substances' have been encountered only in areas where *Phragmites* was growing before the desiccation of the lake.

Attempts to analyse the stratification of material sedimented over approximately the last hundred years have been a failure. Within the northern portion of the lake (north of the profile Podersdorf-Breitenbrunn) whitish sediments poor in organic material prevail. They are characterized by *Limnocythere inopinata* (see Chapter 24), shells of chydorids (*Chydorus* and *Alona*) and ephippia of *Moina*, absent from the lake for at least the past decade. The finding of males of *Limnocythere inopinata* within this area (mainly west of Podersdorf) is thought to be indicative of higher alkalinities as observed from 1885 to 1904 and from 1927 to 1935. Sediments rich in organic substances prevailing within the central area of the lake are very often situated on a layer of sand likewise of recent origin. The sandy horizon could be due to sorting by wind during the period when the lake was dry. Cores of these sediments reveal that ephippia of *Moina* are absent from their upper portions or are even entirely lacking. All of them contain *Campylodiscus*. With respect to sediments in the eastern section of the northern and central portion of the lake sand and gravel are usually directly exposed to the water. Currents along the eastern shore (see Chapter 8), wave action and ice drifting may be considered as responsible for the lack of any muddy sediment in this part of the lake. The situation in the southern portion of the lake is a very different one (south of 'Ratzenböck channel'). Its sediment is rich in shells of molluscs, lacking in *Limnocythere inopinata* or generally poor in subfossil material. Much more information on this part will be needed before an overall picture can be formed. This is also the case with respect to the *Phragmites* belt, whose sediments may provide us with additional valuable information. Some questionable statements about the age of the *Phragmites* have been made by Kral (1970). A comparative stratigraphy within the open lake area has also been attempted by analysing the distribution of different elements in cores. However, the results obtained by neutron activation analysis for Mn, Na and K show merely that although there are considerable differences within each of the five cores taken off Weiden, Podersdorf, Rust and Mörbisch (Mg 0.25–1.55 mg g^{-1}, Na 2.7–10.0 mg g^{-1} and K 8.2–26.8 mg g^{-1} of samples dried at 105°C), no identification of individual strata throughout the open lake has been possible (Hedrich 1975).

In summarizing, it can be said that the lake has been formed tectonically. The Hanság plain and areas southeast of the present lake were definitely part of the earliest stage at the end of Würm. A later stage may have included the northern portion of Neusiedlersee and areas east and west of it as far as Gols and Breitenbrunn. Its recent addition seems to be the southwestern part of the modern lake. How far man has contributed to the gradual desiccation of the Hanság (above all by regulation of the Ikva system in 1568) or to what extent the latest subsidences influenced the loss of the lake areas in the southeast and northeast remain a matter of discussion.

There is little doubt that the origin of the salinity (see Chapter 9) is closely related to the marine sediments of the Neusiedlersee area. Influences from the marine sediments are recognizable as connate waters and have contributed to the presence of mineral waters in most of the lake and Seewinkel area. Sometimes these mineral waters have concentrations of more than 17 per cent (Hock 1956, Weis 1956), NaCl and Na_2SO_4 dominating. Due to a southeasterly decline of the Pannonian layers and the structure of the sediments some of the mineral waters are of an artesian nature and used as such in villages bordering the slope of the Parndorf plateau and in the southwest of the Seewinkel (Tauber et al. 1958). There is good evidence that such saline waters enter the lake along the tectonic fracture zones mentioned above. Fig. 3.4 shows the chlorinity of two 30 m profiles taken near the lake

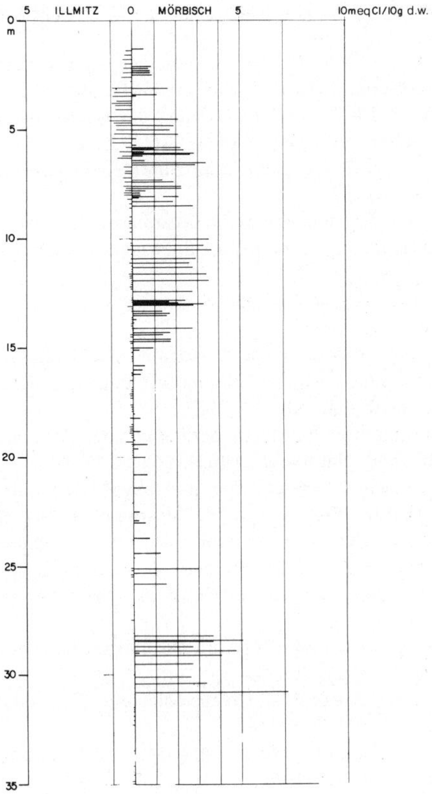

Fig. 3.4. The distribution of chloride in 35 m cores taken at the shore of Neusiedlersee near Illmitz and Mörbisch, taken in 1971.

shore at Illmitz and Mörbisch. Remarkably high concentrations have been found near the fracture system at Mörbisch whereas the Illmitz profile indicates only low or very moderate values. Increased values of sodium, magnesium and chloride are also observable in water samples taken in the area near Mörbisch and during periods of little water movement, especially under ice (see Chapter 9). Obviously the part played by this more saline upwelling groundwater becomes dominant during periods of low lake level. Estimations of the quantity, however, have not so far been made. Upwelling groundwater along the eastern shore seems to contain far less sodium and chloride whereas alkaline earth cations are dominant.

References

Anonymous 1865–1871. Die letzte Trockenperiode des Neusiedler Sees. An anonymous old chronicle.

Bobek, M., Löffler, H. & Schultze, E. 1978. Neue Daten zur Geschichte des Neusiedlersees. Biol. Forsch. Inst. Burgenld. Ber. 29, 5–10.

Fritsch, V. 1961. Die Geoelektrische Aufschließung eines Heilwasservorkommens im Bereiche des Neusiedlersees im österreichischen Burgenland. Boll. Geofis. Teor. Appl. 3(9): 1–15.

Fritsch, V. & Tauber, A. F. 1959. Die Mineralwässer des Neusiedlerseegebietes. Wiss. Arb. Bgld. 21: 34–39.
Fritsch, V. & Tauber, A. F. 1966. Beitrag zur Interpretation geoelektrischer Messungen in Mineralwassergebieten. Gerlands Beitr. Geophys. 75(4): 301–312.
Fritsch, V. & Tauber, A. F. 1967. Die Veränderung der geoelektrischen Struktur des Untergrundes durch Mineralwässer. Acta Hydrophys. 12(1): 5–26.
Fritsch, V. & Tauber, A. F. 1969. Geoelektrische Untersuchungen von Salzwasservorkommen. Zeitschr. Geophysik 15: 161–173.
Fuchs, W. 1974. Bericht über Exkursionen in die Oststeiermark, in das südliche Burgenland und nach Westungarn zur Klärung der Herkunft des Seewinkelschotter. Verh. Geol. B.-A. Wien 4: 118–121.
Gattinger, T. E. 1975. Das hydrogeologische Einzugsgebiet des Neusiedlersees. Verh. Geol. B.-A. Wien 4: 311–346.
Hedrich, E. 1975. Verteilung von Mn, Na und K in den Neusiedlersee-Sedimenten. Unpubl. rep.
Herrmann, P. 1970. Pleistozäne Ostracodenfauna aus dem nördlichen Neusiedlerseebecken. Mitt. Öst. Akad. Wiss., math.-nat. Kl. 221–223.
Husz, G. 1965. Zur Kenntnis der quartären Sedimente des Seewinkelgebietes (Burgenland, Österreich). Wiss. Arb. Bgld. Naturwiss. (1963–1964): 147–205.
Husz, G. 1967. Ein Vergleich österreichischer und ungarischer Salzböden hinsichtlich ihres Chemismus und ihrer Textur. Wiss. Arb. Bgld. Naturwiss. (1966–1967): 161–174.
Klaus, W. 1962. Zur Pollenanalytischen Datierung von Quartärsedimenten im Stadtgebiet von Wien, südlichen Wiener Becken und Burgenland. Verh. Geol. B.-A. Wien 1: 20–34.
Kopf, F. 1974. Der neue Wasserhaushalt des Neusiedler Sees. Österr. Naturwissenschaft 7–8 (170–180).
Kral, F. 1970. Ergebnisse pollenanalytischer Untersuchungen im nördlichen Burgenland. Mitt. Ostalpin-Dinarischen Sekt. 10(2): 20–30.
Löffler, H. 1971. Beitrag zur Kenntnis der Neusiedler See-Sedimente. Sitz. Ber. Öst. Akad. Wiss., math.-nat. Kl. 179(8–10): 313–318.
Löffler, H. 1972. The distribution of subfossil ostracods and diatoms in pre-alpine lakes. Verh. Internat. Verein. Limnol. 18(1019–1050).
Molnár, B. & Szonoky, M. 1974. On the origin and geohistorical evolution of the natron lakes of the Bugac region. Móra Ferenc Múz. Évkönyve 75(1): 257–270.
Moser, I. 1866. Der abgetrocknete Boden des Neusiedler See's. Jahrb. Geolog. Reichsanst, 16: 338–344.
Müller, G. 1969. Sedimentbildung im Plattensee, Ungarn. Naturwiss, 56: 606–615.
Papp, A. 1951. Das Pannon des Wiener Beckens. Mitt. Geol. Ges. Wien 39–41: 99–193.
Riedl, H. 1964. Erläuterungen zur morphologischen Karte der eiszeitlichen Flächensysteme der Wulka. Wiss. Arb. Bgld. 31: 175–195.
Riedl, H. 1965. Beiträge zur Morphogenese des Seewinkels. Wiss. Arb. Bgld. 32: 5–28.
Sauerzopf, F. 1956. Das Werden des Neusiedler-Sees. Bgld. Heimatblätter 18(1): 1–6.
Sauerzopf, F. 1959a. Zur Entwicklungsgeschichte des Neusiedlerseegebietes. Wiss. Arb. Bgld. 23: 107–111.
Sauerzopf, F. 1959b. Der Wasserhaushalt des Neusiedlersees. Wiss. Arb. Bgld. 23: 101–104.
Sauerzopf, F. 1959c. Die Wasserstandsschwankungen des Sees. Wiss. Arb. Bgld. 23: 92–101.
Schmid, H. 1968. Das Jungtertiär an der SE-Seite des Leithagebirges zwischen Eisenstadt und Breitenbrunn (Burgenland). Wiss. Arb. Bgld. 41: 1–74.
Schroll, E. 1959. Zur Geochemie und Genese der Wässer des Neusiedler Seegebietes. Wiss. Arb. Bgld. 23: 55–64.
Sebestyén, O. 1968. Remains of *Pediastrum kawraisky* Schmidle (Chlorophyta, Protococcales) in the sediments of Lake Balaton. Annal. Biol. Tihany 35: 203–226.
Szadecky-Kardoss, E. 1938. Geologie der rumpfungarischen Kleinen Tiefebene. Mitt. berg. u. hüttenm. Abt. Sopron 10: 1.
Tauber, A. F. 1959a. Grundzüge der Tektonik des Neusiedlerseegebietes. Wiss. Arb. Bgld. 21: 26–31.
Tauber, A. F. 1959b. Hydrogeologie und Hydrochemie der Parndorfer Heideplatte. Bgld. Heimatblätter 21(1): 7–22.

Tauber, A. F. 1963. Neusiedlersee – Mineralwässer und Mineralwasserlagerstätte. Allg. Landestopogr. Bgld. 2: 785–809.

Tauber, A. F. et al. 1958. Die artesischen Brunnen des Seewinkels im Burgenland. Wasser und Abwasser 1–54.

Tauber, A. F. & Wieden, P. 1959. Zur Sedimentschichtfolge im Neusiedlersee. Wiss. Arb. Bgld. 23: 68–73.

Wieden, P. 1959. Sediment-petrographische Untersuchung des Schlammes vom Neusiedler See (Bgld). Wiss. Arb. Bgld. 23: 73–80.

4 Morphology and morphometry

H. Löffler

At the present time Neusiedlersee, including the areas overgrown with *Phragmites*, is a kidney-shaped body of water extending approximately from 47° 38′ to 47° 57′ N and from 16° 41′ to 16° 52′ E. At its deepest, the modern Neusiedlersee basin is about 113.5 m above sea level

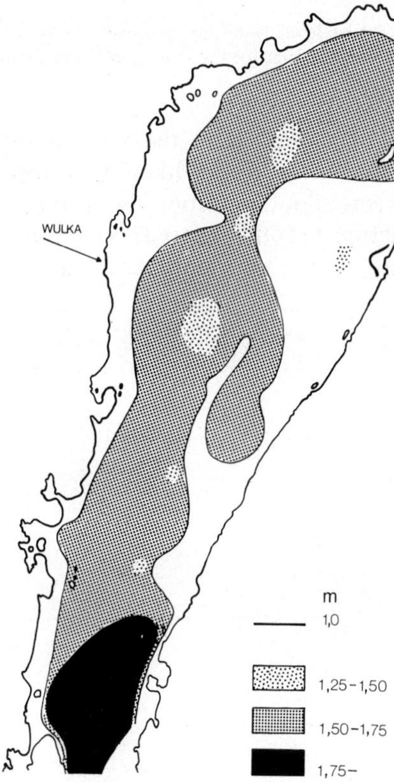

Fig. 4.1. Bathymetric map of Neusiedlersee. Data were obtained by an echograph (Atlas, AN 6014), 12–13 May 1969.

(Fig. 4.1). Being a highly astatic lake its area and volume fluctuate considerably from virtually zero to values far beyond those prevailing at present. Fig. 4.2 presents the fluctuations in water level during the last twenty years; Fig. 4.3 the approximate volume at different water levels. With the water surface at 115.5 m above sea level the total area approaches 300 km^2, of which about 200 km^2 are covered by *Phragmites* (see Chapter 20). This has been the situation for the last ten years, the mean water depth amounting to about

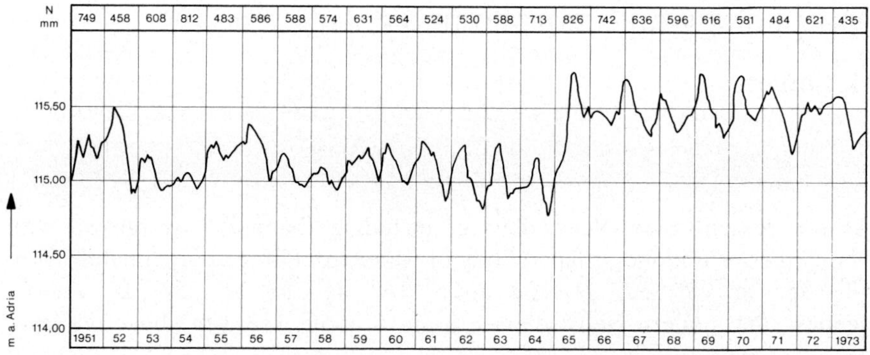

Fig. 4.2 Fluctuations of the water level (monthly averages, 1951–1973) according to Kopf (1974). From 1965 onward new regulations for the management of the sluice of the Einserkanal were set up.

one metre. There have been reports in the past, however, of lake areas exceeding 500 km² (1786) and it is very likely that even larger areas were covered at times of severe flooding from the Danube or Raab-Rabnitz system, before its regulation (1568). Apart from the tectonic events which contributed to the formation of the present lake basin (see Chapter 3), many

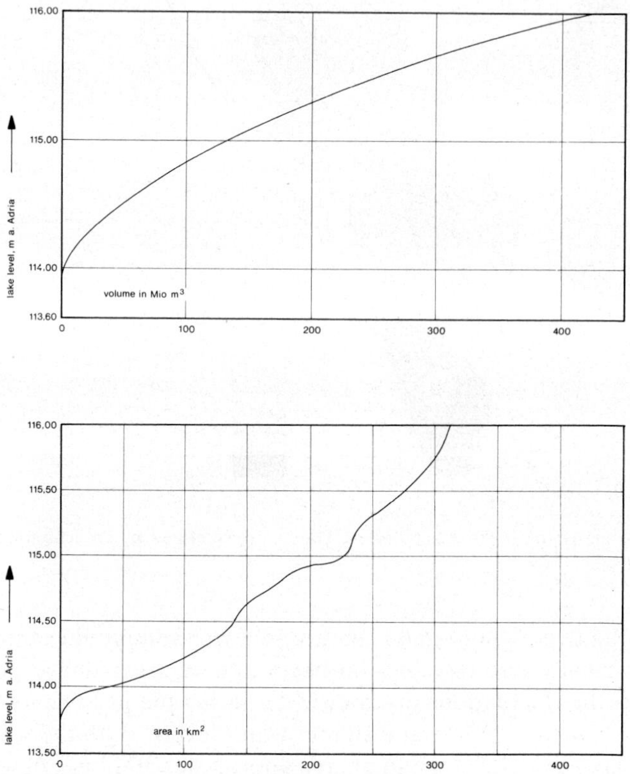

Fig. 4.3. Volume-water level ratio of Neusiedlersee, above, and area-water level ratio of Neusiedlersee, below (slightly modified after Kopf 1974).

other parameters have to be taken into consideration. For example, the fact that the lake basin is situated on top of Tertiary sand sediments from the Pliocene, with a thickness varying between 400 and 600 m (see Chapter 2) means that any eroding force may influence its shape. Currents, seiches, drifting ice, formation of vegetation belts and floods within the catchment area all play a part. If it is taken into account that within the last two centuries the lake has almost or completely dried up on several occasions (1740, 1773, 1811–1813 and 1864–1868) it seems justifiable to estimate the total of such events since the lake's formation about 20,000 years ago as being between 100 and 400. Any intermediate refilling of the basin must have resulted in considerable changes in its surface, especially if a very high water level was achieved rapidly. Such events must have provided an input of eroded terrestrial material from the environment. On the other hand, floods from the Raab-Rabnitz-Ikva system (far less from the Wulka) certainly contributed to the lake's southern and northwestern shorelines respectively. Since the prevailing wind is northwest, drifting ice mainly exerts its shearing forces on the slightly sloped eastern shore extending from north of Podersdorf to south of Illmitz. Drifting ice may accumulate up to a height of ten metres and can thus exert a considerable pressure. Little doubt remains that the formation of a sand wall of as much as three metres (with some gravel) is the result of such ice drifts. A hypothesis which regards this wall as a lake tongue built up by water currents at a stage when the lake covered a large area (Tauber 1959) is definitely unacceptable. Southeast winds which are also frequent in winter obviously exert a much smaller shearing force. This is partly because the western shore area is much flatter. One of the most important factors influencing the shape of the lake basin is the *Phragmites* belt, whose development since the last desiccation of the lake, during which it was nonexistent, is described in Chapter 20. Almost nothing is known about the possible extent of such belts before the lake dried out in the last century, although there is little doubt that such *Phragmites* formations were also characteristic of earlier stages of the lake.[1] An indication of this was given by Winkler (1923). In its rapid progress, chiefly from the south and west towards the centre of the lake, the *Phragmites* belt is a recipient of turbid material stirred up in the open lake by water movements. Such material drifts into the reed belt where a part of it eventually sediments. Since this is a very frequent process (see Chapters 10 and 15) it has resulted in the deposit of considerable amounts of sediment within the *Phragmites*-covered areas and hence in a quite obvious change of the bathymetric properties of the basin. Moreover, any alternation in configuration of the *Phragmites* belt also leads to deviations of the current system imposed upon the lake by wind (see Chapter 8). This in turn results in different patterns of sedimentation and erosion in the open basin. Changes in the *Phragmites*-covered reed belt often originate lakeward of the main belt in new, small and isolated stands which may also contribute

1. A painting by Schnorr-Carolsfeld in 1835, however, does not indicate any *Phragmites* stands. Maps by Walter (1754/55) and Mollo (1808) show that during the lake stage prior to 1868 the Phragmites belt was poorly developed, probably due to trampling by stock.

to the current pattern, although little is known about these secondary influences. Neither do we possess any information concerning the influence of the *Phragmites* belt on seiches, which together with the other water movements certainly deserve more attention (see Chapter 8). Two attempts have so far been made to describe the bathymetric situation of the Neusiedlersee. A geodetic survey was made in 1963 (Kopf 1964), and in 1969 an echographic investigation of the lake was carried out (Löffler 1971). The results are contradictory, perhaps on account of the rapidly changing relief of the lake (which would be worth further investigation) or due to shortcomings in the methods employed. In geodetic studies it is not unlikely that errors due to soft mud account for differences in the order of centimetres. Further, optical disturbances may have contributed slight errors. Since echograph measurements were carried out during a long period of calm (Beaufort scale less than 1) major errors are unlikely to have occurred. In addition, many of the profiles taken were rechecked and examined using an ordinary corer. One of the main differences between the results obtained obviously concerns the area of maximum depth. According to Kopf this is situated in the northern portion of the lake, whereas echograph readings indicate it as being situated in the southern portion (Fig. 4.1). It should, however, be kept in mind that the discrepancies do not exceed 10 to 20 cm. Because of possible changes in bathymetry, frequent echographical readings would be highly desirable.

References

Kopf, F. 1964. Die wahren Ausmaße des Neusiedlersees. Österr. Wasserwirtschaft 12: 255–262.

Kopf, F. 1974. Der neue Wasserhaushalt des Neusiedler Sees. Österr. Naturwiss. 7–8 (170–180).

Löffler, H. 1971. Beitrag zur Kenntnis der Neusiedler See-Sedimente. Sitz. Ber. Öst. Akad. Wiss., math.-nat. Kl. 179(8–10): 313–318.

Sauerzopf, F. 1959. Introduction. In: Landschaft Neusiedlersee. Wiss. Arb. Bgld. 23: 5–18.

5 Climatic conditions

H. Dobesch and F. Neuwirth

The following description of the meteorological conditions prevailing in the climatic region of Neusiedlersee is based upon long-term observations combined with an analysis of measurements published by the Central Bureau of Meteorology and Geodynamics, Vienna, within the framework of the International Hydrological Decade. In addition to the usual description of the climate, research carried out during the IHD (Dobesch & Neuwirth 1974; Mahringer 1966; Mahringer & Motschka 1968) has yielded detailed information concerning the specific properties of the main body of water itself and its interactions with the atmospheric layers immediately above it. The project has especially benefited from observations made over a number of years at a meteorological station in the middle of the lake, during the ice-free periods.

5.1 General climatic conditions of the Neusiedlersee region

The climate of the Neusiedlersee region is to a large extent determined by the lake's position in the Pannonian climate zone (see Chapter 1) which is manifest, on the one hand, in a more continental character and, on the other, in a distinct leeward effect in the spurs of the Eastern Alps in the zone of the prevailing west winds (Steinhauser 1965). As a consequence, in the warmer seasons the higher temperatures and relatively little rainfall exert a greater influence than the low temperatures in the cooler season. In the latter season the cooling caused by terrestial radiation is mitigated by cloud and high fog cover which frequently last for long periods at a time in autumn and winter, just as in the Danube Basin. The region is one of the warmest in Austria, with an annual mean temperature of almost 10°C, mainly due to the high summer values. An average of more than 240 days annually for the 5°C limit, which is important for the development of the vegetation, explains the favourable thermal position of the region. In Austria similar values are measured only in the southernmost part of the Mur in southern Styria.

The climate data from the meteorological station at Neusiedl am See presented in Table 5.1 characterize the climatic conditions of the lake region. According to this table the absolute maximum is 37.8°C, the absolute minimum −23°C. The average number of frost days (minimum air temperature below 0°C) per year is 91, of which 25 are to be expected in January and 21 in December. The sum of the ice days (daily maximum air temperature below 0°C) amounts to 26. In the course of the year ice days occur from November until March, with a maximum of 10.8 in January. The annual average of summer days is 60.6 (daily maximum at least 25°C). The annual mean relative humidity amounts to 76 per cent with a maximum of 85 per cent in December and minimum of between 69 and 70 per cent from April to

Table 5.1. Essential climatic elements measured at Neusiedl am See in the years 1951–1970 (monthly and annual averages).

	Jan.	Feb.	Mar.	Apr.	May	Jun.	Jul.	Aug.	Sep.	Oct.	Nov.	Dec.	Year
Air temperature (°C)	−1.5	0.4	4.4	10.5	14.6	18.5	20.0	19.4	15.9	10.6	5.3	0.7	9.9
Air temperature, absolute maximum (°C)	12.7	17.5	24.0	30.0	32.6	33.5	37.8	36.0	32.0	27.2	22.0	15.5	37.8
Air temperature, absolute minimum (°C)	−23.0	−22.4	−17.5	−3.7	−0.5	3.6	4.3	6.1	0.0	−3.2	−9.2	−17.3	−23.0
Air temperature, average maximum (°C)	1.3	3.8	8.9	15.9	20.4	23.9	25.6	25.1	21.6	15.8	8.5	3.2	14.5
Air temperature, average minimum (°C)	−4.5	−3.1	0.3	5.2	9.1	13.1	14.4	14.0	10.8	6.1	2.2	−2.0	5.5
Number of frost days	25.2	18.6	14.0	2.9	0.2	0.0	0.0	0.0	0.0	1.9	7.3	21.0	91.0
Number of ice days	10.8	6.4	1.2	0.0	0.0	0.0	0.0	0.0	0.0	0.0	0.5	7.2	26.1
Number of summer days	0.0	0.0	0.0	1.5	5.1	13.1	17.4	16.0	7.0	0.5	0.0	0.0	60.6
Relative humidity (%)	83	80	75	69	69	70	69	72	75	80	84	85	76
Cloud (tenths)	7.1	6.9	6.2	5.5	5.5	5.2	5.0	4.7	4.6	5.2	7.3	7.8	5.9
Number of foggy days	3.6	3.3	1.7	0.2	0.2	0.1	0.1	0.3	0.9	3.5	3.8	4.8	22.4
Clear days	2.5	2.7	4.4	5.0	4.7	4.8	7.1	6.9	7.9	7.0	2.7	1.6	57.3
Cloudy days	15.1	12.6	11.6	8.1	7.2	5.7	5.5	5.1	5.6	7.6	16.0	17.5	117.6
Relative humidity 0700 hrs	87	86	85	81	81	79	80	84	88	92	90	88	85
Relative humidity 1400 hrs	77	72	63	53	53	55	53	54	55	60	70	78	72
Relative humidity 2100 hrs	84	81	77	72	74	76	74	76	79	84	87	87	79

July. As for cloud, 5.9 tenths of the sky are covered on an average, the maximum being 7.8 tenths in December and the minimum 4.6 tenths in September. The average number of foggy days, that is, with al least temporary fog, limiting visibility at the observation site to less than 1 km, is 22.5, the highest values are reached in December with 4.8 days, the lowest values in June and July with 0.1 days. The average number of clear days is 57.3, with 7.9 in September and 1.6 in December. Clear days are days when between 0.0 and 1.9 tenths of the sky are covered by cloud. On the other hand, the number of cloudy days (8.1 to 10.0 tenths of the sky covered by cloud) is 117.6; 45.2 in winter, 29.2 in autumn, 26.9 in spring and 16.3 in summer. Most of the cloudy days are in November and December, the least in August.

5.2 The individual components of the climate and their specific pattern in the lake region

5.2.1 *Temperature and humidity of the air*

A modification of the distribution of temperature in the boundary layer of the atmosphere above the lake is the result of differences in the heat budget as

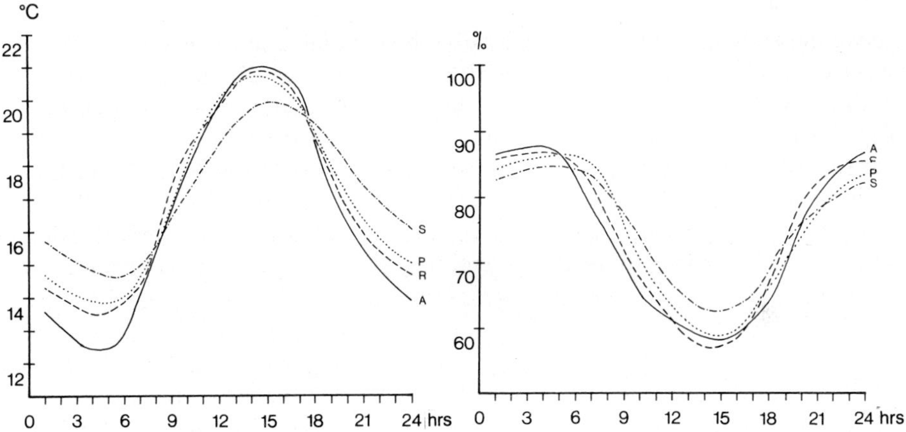

Fig. 5.1. Daily courses of temperature and relative humidity (means: May–October), R: Rust, P: Podersdorf, S: mid-lake station, A: Althof.

compared with a land surface. Since the temperature of the water surface during the day is lower than that of the land, but higher during the night, it has a mitigating effect on the temperature curve in the air mass above the water surface (Central Bureau of Meteorology and Geodynamics 1973). Figure 5.1 clearly shows this for the period between May and October. But this effect is much weaker in the riparian zones, as a comparison of the mid-lake station with Rust and Podersdorf shows. The figure demonstrates a typical feature of an inland station (Althof) that is not influenced by the lake; compared to the stations near the lake the latter is characterized by a larger temperature amplitude. The relative humidity behaves in an analogous manner: during the night the land values are higher than those measured above the open lake. During the day the situation is reversed (Central Bureau of Meteorology and Geodynamics 1975).

These differences are also manifest in the course taken by the stability of the atmospheric density stratification: according to Fig. 5.2 the stratification

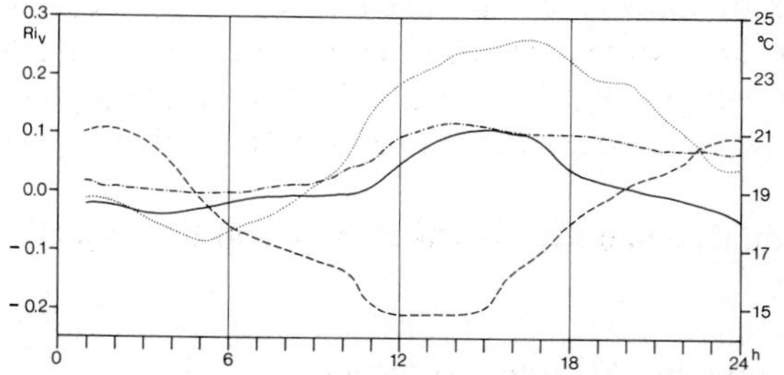

Fig. 5.2. Mean daily course of Ri_v (July 1969) above water (—mid-lake station) and above grain (– – – Althof); daily course of water temperature (·—·—· mid-lake station) and air temperature (· · · · · mid-lake station). After Dobesch (1973).

above the lake tends to be more stable in the daytime due to the colder water surface (in Fig. 5.2 Ri_v, the virtual Richardson number, is a measure for the stability of an air stratum; $Ri_v > 0$ means stable, $Ri_v = 0$ means neutral, $Ri_v < 0$ means unstable stratification); a maximum is measured around 4 p.m. because of the strong shift of the temperature wave over the water. Over the land the lower air strata are more unstable during the day owing to the strong warming up of the land surfaces, whereas by night the atmospheric density stratification becomes more stable, creating an inversion near the ground.

5.2.2 *Winds*

Throughout the year northwest and southeast winds prevail in the region around Neusiedlersee (Steinhauser 1970a, 1970b). Nevertheless owing to the large area of the lake certain differences are observable between the western and the eastern shores. In the winter months northwest winds on the western shore are most frequent in the second half of the night and least frequent in the afternoon (Fig. 5.3). Winds from southeasterly directions behave in exactly the reverse manner. At the stations along the eastern shore northwesterly and southeasterly winds are of the same frequency in the afternoon, whereas in the second half of the night a distinct northeasterly component becomes noticeable. It is remarkable that in the course of the day the difference between the highest and the lowest frequency of the two prevalent wind directions is much smaller than on the western shore. The probable explanation for this phenomenon is that in the hills of the western shore a downwind system develops during the night and an upwind during the day. This system has a tendency to be superimposed upon the land-lake breeze

a.

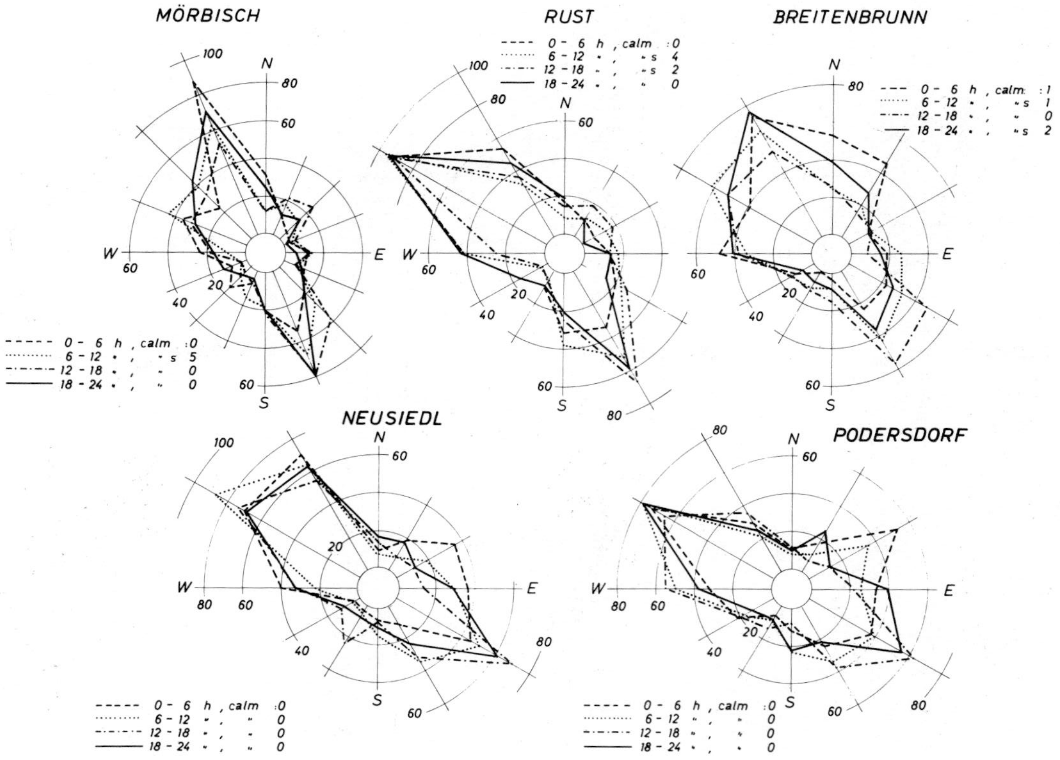

Fig. 5.3. Frequency distributions of the wind direction: (a) January and February 1967; (b) July and August 1967 (Steinhauser 1970a, 1970b).

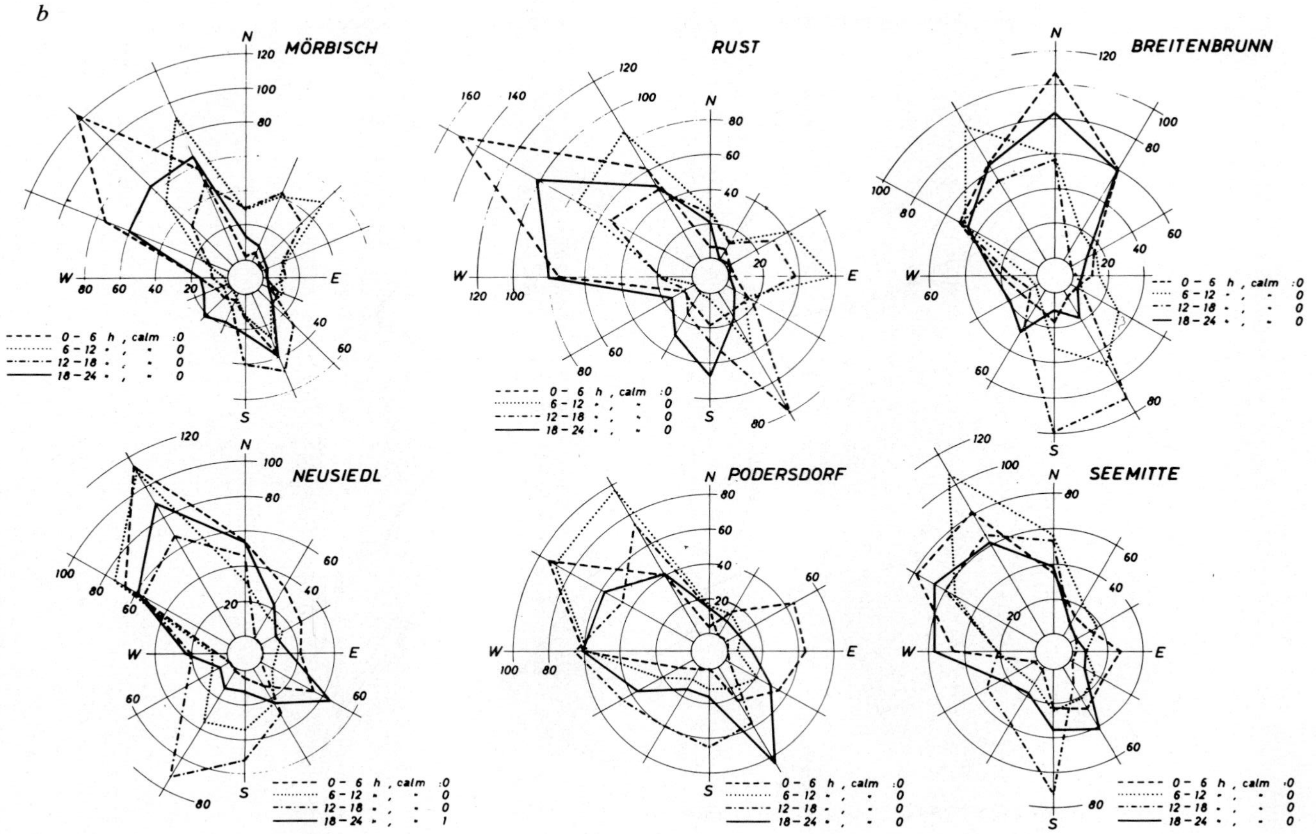
b
MÖRBISCH
RUST
BREITENBRUNN
NEUSIEDL
PODERSDORF
SEEMITTE
N
E
S
W
0 - 6 h , calm :0
6 - 12 " , " 0
12 - 18 " , " 0
18 - 24 " , " 0
18 - 24 " , " 1

Fig. 5.3. Continued

system, whereas the region east of Neusiedlersee is perfectly flat and is influenced exclusively by the land-lake wind system. These distinctions between the eastern and western lake shore are extremely well marked in the warm seasons. It is striking that on the eastern shore the frequencies of both northwesterly and southeasterly winds change only little during the night, whereas by day the southerly and southeasterly winds are less frequent in the morning than in the afternoon, the northwesterly winds, however, being more frequent before noon. Accordingly, the tendency for a landbreeze to develop enhances the frequency of the southeast winds during the night and the tendency of the lake breeze to develop enhances the northwest wind during the day. The wind rose of the western shore indicates a strong prevalence of the northwest winds by night with an increasing frequency from the first to the second half of the night, corresponding to the cooling down of the land and the mountain slopes. By day the easterly winds coming from the lake are equally predominant.

In general, wind direction is only little influenced by the lake, but the prevailing wind directions, northwest and southeast, result from the general circulation (Steinhauser 1970a, 1970b; Dobesch & Neuwirth 1975).

A distinct difference between wind speeds in the shore zones and the open lake is observable. Winds above the open lake have nearly twice the speed of those on the shores; here again the eastern lake shores are windier than the western zone, a ratio that is at its clearest when the winds are northwesterly, whereas in the case of southeast winds the eastern shore may be calmer than the western shore.

The daily fluctuations of the wind velocity above the open lake are smaller than those in the shore zones. The differences in speed above the open lake and land are therefore greater during the night than during the day.

Table 5.2 (Steinhauser 1970) represents the monthly averages of the wind velocity, clearly showing the trifling differences within the region of Neusiedlersee. The monthly averages of wind speed are about 3.4 km $hour^{-1}$ lower on the western shore than in the east.

The average differences are 3.5 km $hour^{-1}$ in winter, 3.7 in spring, 4.0 in summer and 2.6 in autumn. The average velocity is highest in the middle of the lake throughout the year. The monthly averages of the wind speed are 10.2 km $hour^{-1}$ on the western shore, 13.7 in the east and 16.6 in the middle of the lake. The higher speed in the middle of the lake is due partly to less friction with the water surface and partly to a stronger interaction of the upper winds with the air layers near the surface above the lake by night, when the water surface is warmer than the land surface. This is also reflected in the fact that in the middle of the lake wind speeds are usually not much lower during the night than during the day. On the western shore the wind speeds reach their daily maximum in the early afternoon. The daily fluctuation distinctly shifts in the course of the year: on the western shore it is 3.7 km $hour^{-1}$ in winter and 5.2 km $hour^{-1}$ in summer. In winter the daily course of the wind velocity on the eastern shore is as regular as on the western shore, whereas in spring and summer the maximum flattens out and shifts toward the later afternoon. On this shore daily fluctuations are slighter than on the west

Table 5.2. Monthly and annual average wind velocities in km $hour^{-1}$ (Steinhauser 1970).

	Jan.	Feb.	Mar.	Apr.	May	Jun.	Jul.	Aug.	Sep.	Oct.	Nov.	Dec.	Year
Mörbisch													
1967	11.9	12.3	15.1	14.9	14.5	11.3	9.7	7.7	8.5	9.2	10.3	11.1	11.4
1968	11.9	11.2	11.9	12.2	12.8	10.1	11.2	10.6	10.5	7.8	10.5	9.4	10.8
1969	11.6	11.0	10.5	11.6	11.5	12.1	10.0	8.7	7.3	8.5	10.1	13.4	10.5
Mean	11.8	11.5	12.5	12.9	12.9	11.2	10.3	9.0	8.8	8.5	10.3	11.3	10.9
Rust													
1967	13.7	13.5	17.1	16.6	14.5	12.6	10.2	7.8	8.5	9.0	10.2	12.4	12.2
1968	14.1	12.2	13.6	12.1	13.0	10.8	12.4	11.2	10.7	7.5	10.5	9.5	11.3
1969	11.5	11.9	10.7	12.8	12.3	14.7	11.9	9.3	6.9	9.0	9.8	14.7	11.3
Mean	13.1	12.5	13.8	13.8	13.3	12.7	11.5	9.4	8.7	8.5	10.2	12.2	11.6
Breitenbrunn													
1967	14.3	12.3	15.8	15.6	13.7	10.9	9.5	7.6	9.2	7.9	—	—	—
1968	16.2	11.0	14.1	12.2	12.3	10.1	11.2	11.2	11.7	6.6	10.8	7.2	11.2
1969	11.1	11.6	10.7	10.5	10.7	13.2	10.0	8.8	6.1	7.6	8.8	15.4	10.4
Mean	13.9	11.6	13.5	12.8	12.2	11.4	10.2	9.2	8.7	7 4	—	—	—
Neusiedl am See													
1967	14.3	14.1	17.1	16.9	15.8	13.1	11.1	8.5	10.0	7.7	11.6	12.8	12.7
1968	13.0	13.5	—	—	—	—	—	—	—	—	—	—	—

Neusiedl-Berg													
1967	19.4	18.4	22.1	21.2	19.6	16.0	14.7	11.6	13.8	13.4	16.0	19.8	17.2
1968	21.3	18.5	19.6	18.7	18.1	13.7	15.9	15.3	15.3	11.4	17.4	14.1	16.6
1969	17.9	18.7	18.6	16.1	15.5	17.7	15.9	12.0	11.6	12.7	14.8	23.0	16.2
Mean	19.5	18.5	20.1	18.7	17.7	15.8	15.5	12.9	13.6	12.5	16.1	19.0	16.7
Podersdorf													
1967	16.3	15.9	22.6	20.5	17.1	16.3	13.7	11.4	11.8	10.7	11.9	17.5	15.5
1968	18.1	17.0	18.7	14.5	16.0	14.3	16.4	14.2	14.0	9.5	13.5	9.7	14.7
1969	11.5	13.5	14.3	14.2	13.5	18.1	14.5	12.3	9.2	12.8	14.0	21.0	14.1
Mean	15.3	15.5	18.5	16.4	15.5	16.2	14.9	12.6	11.7	11.0	13.1	16.1	14.7
Mid-lake													
1967	—	—	—	—	21.0	18.9	16.3	13.2	14.5	14.4	—	—	—
1968	—	—	—	16.5	19.4	16.9	19.5	18.1	17.5	12.6	—	—	—
1969	—	—	—	—	16.8	21.5	17.4	15.1	12.1	13.9	—	—	—
Mean	—	—	—	—	19.1	19.1	17.7	15.5	14.7	13.6	—	—	—

shore throughout the year, especially in the summer period. They amount to 3.2 km hour^{-1} in winter, and to 3.0 in summer. That the daily fluctuation is more marked on the western shore than in the east is partly explained by the fact that during the day, when the hills in the west warm up, a system of winds due to the slopes is superimposed upon the lake breeze tendency; thus by day the winds from the east show a greater increase in speed than in the night. The daily fluctuation in the middle of the lake is very small, its peak value shifting towards late afternoon in summer. In this zone the daily fluctuation amounts to 2.0 to 2.3 km hour^{-1}, the daily courses being irregular.

The differences in the total number of slight breezes with speeds between 0.0 and 5.0 km hour^{-1} are alike in all months and therefore seem to be typical of this region. Another interesting phenomenon is the scarcity of slight winds on the eastern lake shore. The frequency of strong winds (over 30 km hour^{-1} is very different for the eastern and the western shore and is in fact twice as high in the east as in the west. Strong winds in the middle of the lake are as frequent as, or more frequent than on the eastern shore.

Storms generally last for one day in the west (Dobesch & Neuwirth 1975, Felkel 1972) whereas on the eastern shore they more frequently last for two days and come almost exclusively from the northwest, whereas on the western shore northwest and southeast winds are equally frequent. In the winter season such strong winds occur more often and last longer.

5.2.3. *Radiation*

In the summer period temperature conditions in the Neusiedlersee region are favourable compared with other Austrian provinces, chiefly as a result of a greater amount of sunshine. Fig. 5.4 shows the annual course of the average number of daily hours of sunshine for the stations in northern Burgenland and some other stations. The values prove that from May to September the region around Neusiedlersee has even more sunshine than the sunny province Carinthia. In summer and spring Neusiedl am See outdoes all other Austrian registration stations, with 795 and 571 hours of sunshine respectively. The IHD program also included the individual components of radiation (Koch 1975; Mahringer 1969): the daily maximum of global radiation can be calculated at 600 cal. cm^{-2} in June and July, the highest possible value at noon being 70 cal. cm^{-2}. Being extremely opaque, Neusiedlersee has a higher albedo than other Austrian lakes. At a sunshine level above 17° the average albedo of the open lake area is between 11 and 13 per cent. According to the degree of turbidity, wind and waves, this value may even exceed 16 per cent. On clear days a distinct daily curve can be observed, showing a minimum at noon and maxima in the morning and in the evening. During the winter months the albedo values depend on the extent to which the lake is covered by ice and snow. With an ice cover, values between 40 and 50 per cent were found, and in the case of fresh snow even as much as 70 per cent. In summer the net radiation, i.e. the total of short-wave and long-wave components, may even reach 70 per cent of the global radiation, which means that in July, for example, 10,000 cal. cm^{-2} and more are quite possible. This quantity is

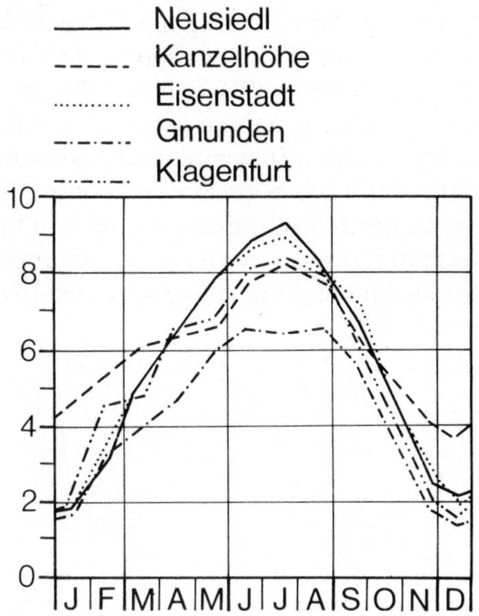

Fig. 5.4. Daily duration of sunshine (Steinhauser 1965).

available for the energy budget in the surface boundary layer. In the summer period the net radiation values above the water surface are higher than those above the land, but in the winter period the situation is reversed. The highest daily total of net radiation (up to 530 cal. cm^{-2}) is measured above the water in July. Here too, the absolute maxima of the hourly values are measured at 62 cal. cm^{-2}, whereas the peak values are 20 per cent lower above land.

5.2.4. *The light climate of Neusiedlersee*

The light climate of Neusiedlersee is influenced by the large degree of turbidity arising from the extremely high quantities of suspended inorganic and organic matter (Panosch 1973). These features are attributable to its shallowness: even light breezes whirl up mud and organic and inorganic substances suspended in the water and thus render the water cloudy. This is why the lake often changes colour, transmission and extinction within short periods of time. Thus the back-scattered light is about 5.5 per cent of the surface-incident light, which is a very high value. 87 per cent of the upper light is scattered and 12.5 per cent is absorbed (in 1 m of water depth). This high degree of scatter is due to the great turbidity of the water. Moreover the lake has a strong extinction in blue and violet. Of the 44 per cent of the total extinction of the incoming light 38 per cent is due to scattering and only 6 per cent to the absorption in all wavelengths of the spectrum. The average vertical extinction coefficient for a water stratum of 1 m is 0.678, the dispersion coefficient 1.06. Measurements of the spectral transmission proved that transmission was optimal in the orange range (625 nm) and in some series even in the red range (675 nm). This corresponds with the fact that with an increasing opacity the maximum shifts towards the red wavelengths. The minimum transmission is clearly in the ultraviolet wavelengths with the optical centre at 375 nm. Under a 15 cm ice cover the extinction coefficient is reduced to 0.4, i.e. the water has a greater transmission under the ice than without ice, since the suspended substances can deposit under the ice cover. Measurements in the reed showed a maximum transmission between 550 and 775 nm (Fig. 5.5). In the visible range of the spectrum the back scattering of

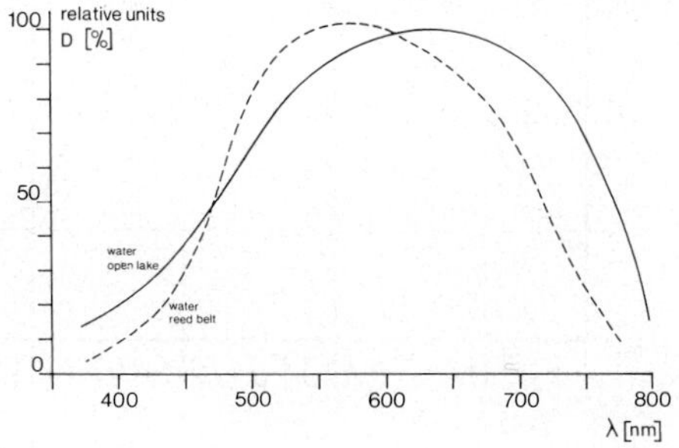

Fig. 5.5. The spectral transmission in open lake and reed belt water, in 1 m depth (Panosch 1973).

the incoming radiations lies between 7 per cent and 28 per cent and is highest at 520 nm and lowest at 480 nm. Summarizing, Neusiedlersee is one of the most turbid, opaque, inland waters in Europe, with a low degree of transmission and a very high dispersion.

5.2.5. *Precipitation*

Precipitation in the region of Neusiedlersee is characteristically small and low in frequency as compared with the rest of Austria (Steinhauser 1965). In the summer months rainfall is about half that of the Salzkammergut (Table 5.3); accordingly the number of rainy days is also much smaller than in the other regions listed.

The average amount of precipitation is about 600 mm, with a maximum in summer and a minimum in winter; the former chiefly caused by frequent thunderstorms. The distribution is as follows (Neuwirth, 1976.): least rainfall is measured directly above the middle of the lake and on the eastern shore, whereas the southern part and that part of the western shore east of the Leitha mountains give the highest values. Very rarely do more than 50 mm of rain fall on one day and then only in the southern part of the lake.

In the winter period snowfall is possible from November to April. The average value is 30–35 days with snowfall leading to a closed snow cover for 45–50 days.

Table 5.3. Precipitation, 1901–1950 (Steinhauser 1965).

	Jan.	Feb.	Mar.	Apr.	May	Jun.	Jul.	Aug.	Sep.	Oct.	Nov.	Dec.	Year
Mean precipitation (mm)													
Eisenstadt	33	29	41	54	67	57	79	64	70	66	49	42	651
Neusiedl a. S.	37	33	42	47	63	54	68	61	56	55	49	44	609
Andau	34	31	32	42	50	64	67	63	52	47	49	45	576
Gmunden	81	75	72	101	116	142	166	137	112	81	69	75	1227
Bregenz	83	72	84	108	138	182	199	178	153	106	86	85	1474
Klagenfurt	42	40	54	78	93	117	113	117	101	97	82	57	991
Number of days with precipitation $\leqslant 0.1$ mm													
Eisenstadt	8.4	8.5	10.5	10.8	11.3	12.1	10.8	11.8	9.6	10.6	10.3	11.1	125.8
Neusiedl a. S.	9.7	8.8	9.0	10.9	10.8	10.9	11.2	10.7	8.8	10.6	11.8	10.9	124.1
Andau	7.8	7.3	8.3	8.6	10.5	9.8	10.3	9.6	8.6	10.0	9.7	9.3	109.8
Gmunden	16.6	13.8	15.6	17.7	17.1	18.2	16.7	15.6	13.6	13.3	13.0	16.0	187.2
Bregenz	11.4	10.8	12.5	14.9	15.8	16.0	15.6	14.5	13.2	12.2	12.0	13.0	161.9
Klagenfurt	7.6	7.3	9.3	11.7	13.6	13.8	13.7	12.4	10.8	12.1	11.0	9.9	133.2
Number of days with precipitation $\geqslant 0.1$ mm													
Eisenstadt	6.7	5.7	6.4	8.7	8.7	9.2	9.6	8.7	7.0	7.6	7.3	8.1	93.7
Neusiedl a. S.	6.9	6.0	5.9	8.4	8.3	8.4	8.3	8.3	6.4	7.4	8.1	8.0	90.4
Andau	6.0	5.1	5.4	7.7	7.6	8.6	8.0	7.2	6.3	6.4	6.6	8.1	83.0
Gmunden	11.8	10.9	11.4	14.1	13.9	14.1	15.0	14.0	10.8	10.4	10.6	12.7	149.7
Bregenz	10.0	9.5	10.8	13.1	13.2	14.5	14.5	13.5	11.6	10.4	10.4	11.5	143.0
Klagenfurt	5.3	5.7	7.0	9.3	10.4	10.8	10.6	10.0	8.4	8.2	7.9	7.3	100.9

5.2.6. *Water temperatures*

Water temperatures in Neusiedlersee are influenced, on the one hand, by its shallowness, and on the other hand, by the high frequency of strong winds, which thoroughly mix the water. Only on calm days can a vertical stratification of temperature develop, the maximum differences between the bottom and the water surface being 10°C. Even average wind speeds cause a complete mixing of the entire water body. Due to the shallowness the mean horizontal differences tend to be slight (Darnhofer 1971).

Fig. 5.6 shows the average annual course of the water temperature at a depth between 20 and 30 cm. The maximum is reached in the first half of

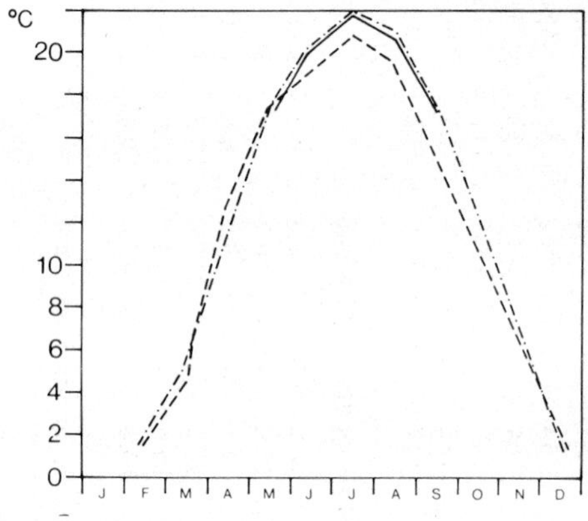

Fig. 5.6. Mean annual course of water temperature, 1967–1970, in 20–30 cm depth (– – – reed belt station, ·–·–· Podersdorf, —— mid-lake station). After Hammer (1975).

July, a very specific peculiarity of Neusiedlersee. On account of its small water volume the lake cannot store much heat and therefore the maximum water temperature scarcely lags behind the air temperature, whereas in deeper lakes, the maximum temperature is attained in late summer. Yet another difference is due to the fact that Neusiedlersee warms up faster in spring than is the case in deeper lakes but also cools down faster in fall so that it tends to freeze relatively early. Table 5.4 presents some data regarding the

Table 5.4. Freezing data (Steinhauser 1965).

	Period of freezing			Complete ice cover		
	beginning	end	duration	beginning	end	duration
Neusiedlersee						
average	13 Dec.	7 Mar.	78 days	22 Dec.	20 Feb.	54 days
earliest	26 Nov.	22 Jan.	31 days	30 Nov.	22 Dec.	10 days
latest	24 Dec.	4 Apr.	106 days	12 Jan.	26 Mar.	97 days
Mondsee (average)	15 Jan.	15 Mar.	59 days	28 Jan.	1 Mar.	40 days
Wolfgangsee (average)	9 Jan.	17 Mar.	75 days	31 Jan.	17 Feb.	31 days
Wörthersee (average)	16 Jan.	18 Mar.	59 days	25 Jan.	7 Mar.	42 days

onset, the end and the duration of the ice cover as compared to a number of other lakes.

5.2.7. *Evaporation*

Evaporation is an essential element of a lake's water balance. On the open water surface of Neusiedlersee the annual evaporation amounts to approximately 900 mm of water per unit of area (Neuwirth 1974). The average annual course of evaporation given in moving averages (order five) in Fig. 5.7

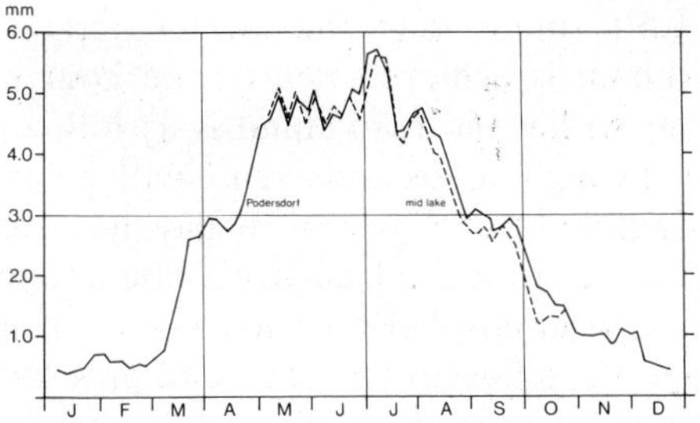

Fig. 5.7. Mean course of evaporation, 1968–1969 (— Podersdorf, - - - mid-lake station). After Neuwirth (1974).

shows the large increase in evaporation in the second half of March when the ice cover begins to melt. Evaporation reaches its maximum as early as the first ten days of July; this peak, which is reached very early as compared with deeper waters, is due to the shallowness of the lake and its consequently low capacity to store heat. In a few instances the daily total evaporation may exceed 10 mm, as is the case, for example, when after a period of fair weather with high water temperatures, cold air masses, with strong northwest winds and cool, relatively dry air, are advected (Mahringer 1970). Chapter 6 refers in detail to the problems of analysing evaporation in the special case of Neusiedlersee.

5.3. The local climate in the reed belt

One of the most specific features of the lake is its extensive reed belt, which is characterized by a special microclimate (Chapter 20). For example, at the bottom of the crop irradiation already reaches peak values in April, when most reeds are no higher than 30–80 cm, so that incoming radiation is only blocked by old stalks from the previous year. However, in summer only 25–30 per cent of the radiation at the top of the reed penetrate the crop, whereas in winter the corresponding values are around 60–70 per cent (Darnhofer 1971).

In drawing up the short-wave radiation budget the following standard values can be taken for the individual components of radiation: about 40 per cent of the entire radiation reaches the bottom of the reed stand, that is, the water surface, 16 per cent is reflected by the reed surface, so that 44 per cent of the radiation is available for the energy balance in the reed.

A characteristic of the temperature conditions in the reed stand is that in spring the monthly averages of the air temperature within the stand lie below those above it, whereas in summer they exceed them and remain higher till autumn. This is due to the fact that in the clear summer and autumn nights the fully developed reed prevents a good deal of black body radiation and thus keeps the air in the reed crop warmer than the air above. In the course of the day, too, the temperature in the reeds differs much from that above it. The reed warms up more and faster before noon and the highest temperature recorded can be 1.5°C more than the air temperature above the reed; moreover this maximum is achieved nearly one hour earlier. The vertical temperature gradient within the reed stand is as follows. During the warm seasons temperature by night tends to decrease with height above the ground. The difference may amount to 1°C per m. By day, however, temperatures are highest in the upper zone of the reed, clearly a case of temperature inversion. This means that by day atmospheric exchanges are confined to the upper zones of the reed and the adjacent air mass, and only by night are the lower zones also able to participate in this exchange (Dobesch 1975). The vapour pressure in the reed stand is distinctly higher than the vapour pressure of the air above, and decreases only slightly from the bottom to the top. A minimum is reached in the uppermost zone, where interaction with the drier air above the reeds commences. The vertical and horizontal distribution of the water temperature in the reed stand depends on the density and height of the reed, on the water level and the exact location in the reed belt, since areas closer to the open lake are subject to better horizontal mixing. Vertical temperature gradients of up to 5°C/75 cm lasting for longer periods have been observed. In this case the nocturnal cooling of the water causes mixing of only an uppermost layer of 20–30 cm. Below this level no important fluctuations are measured in the course of the day. Only a strong cooling of the surface due to a long period of bad weather can mix the entire water in the reed belt, this, however, with an appropriate time lag showing that the reed masses have an insulating effect on the water in the reed belt.

References

Central Bureau of Meteorology and Geodynamics 1973. Klimadaten des Neusiedlerseegebietes. Teil I: Tabellen der Lufttemperatur, 1966–1970. Arbeiten der Zentralanstalt für Meteorologie und Geodynamik 13.

Central Bureau of Meteorology and Geodynamics 1975. Klimadaten des Neusiedlerseegebietes. Teil II: Tabellen der Stundenwerte der relativen Luftfeuchte, 1966–1970. Arbeiten der Zentralanstalt für Meteorologie und Geodynamik 15.

David, A. & Koszma, F. 1973. Strahlungshaushalt des Neusiedler Sees. Idöjárás 77: 325–337.

Darnhofer, T. 1971. Verdunstungsstudien im Schilfgürtel des Neusiedler Sees. Dissertation, University of Vienna.

Dobesch, H. 1973. Das Wind-, Temperatur- und Feuchteprofil über einer freien Wasserfläche. Arch. Met. Geoph. Biokl., Ser. A. 22: 47–70.

Dobesch, H. 1974. Die numerische Bestimmung der Transporte fühlbarer und latenter Wärme mittels verschiedener Methoden über einer freien Wasserfläche. Arch. Met. Geoph. Biokl., Ser. A, 23: 263–284.

Dobesch, H., 1975. Das Bestandsklima im Schilfgürtel des Neusiedlersees. Verhandl. Ges. Ökologie 173–176.

Dobesch, H., 1976. Rauhigkeitsparameter und Verdrängungshöhe über verschiedenen natürlichen Unterlagen. Arch. Met. Geoph. Biokl., Ser. A, 25: 125–130.

Dobesch, H. & Neuwirth, F. 1974. Übersicht über die Ergebnisse aus den hydrometeorologischen Untersuchungen im Gebiet des Neusiedler Sees im Rahmen der Internationalen Dekade 1966–1974. Wetter und Leben 26: 151–156.

Dobesch, H. & Neuwirth, F. 1975. Kleinräumige Unterschiede des Windfeldes im Südteil des Neusiedler Sees. Wetter und Leben 27: 38–46.

Felkel, H. 1972. Stürme in Podersdorf am Neusiedler See (1969–1971). Wetter und Leben 24: 198–207.

Hahn, F. 1972. Mikroklimatologische Studien im Gebiet des Neusiedlersees. Dissertation, University of Vienna.

Hammer, N. 1975. Ergebnisse von Registrierungen der Wassertemperatur im Neusiedler See unter Berücksichtigung der Beeinflussung durch meteorologische Faktoren. Dissertation, University of Vienna.

Hann, W. 1975. Seespiegelschwankungen des Neusiedlersees. Dissertation, University of Vienna.

Harflinger, O. 1969. Hydrometeorologische Studien im Gebiet des Seewinkels. Dissertation, University of Vienna.

Koch, E. 1975. Der Strahlungshaushalt des Neusiedler Sees. Dissertation, University of Vienna.

Kozma, F. & Tóth, E, 1975. Methoden zur Berechnung der Verdunstung des Neusiedler Sees. Arch. Met. Geoph. Biokl., Ser. B, 23: 99–109.

Mahringer, W. 1966. Über die Einrichtung meteorologischer Stationen zur Bestimmung der Verdunstung des Neusiedlersees. Wetter und Leben 18: 223–229.

Mahringer, W. 1969. Der Strahlungshaushalt des Neusiedler Sees im Jahre 1967. Arch. Met. Geoph. Biokl., Ser. B, 17: 51–72.

Mahringer, W. 1970. Verdunstungsstudien am Neusiedler See. Arch. Met. Geoph. Biokl., Ser. B, 18: 1–20.

Mahringer, W. & Motschka, O. 1968. Meteorologische Untersuchungen am Neusiedlersee im Jahre 1967 im Rahmen der Internationalen Hydrologischen Dekade. Wetter und Leben 20: 159–163.

Mahringer, W. & Motschka, O. 1971. Verbesserungen an Sternpyranometern und Strahlungsbilanzmessern. Arch. Met. Geoph. Biokl., Ser. B, 19: 149–156.

Motschka, O. & Zimmerman, K. 1972. Der Aufbau einer Registrieranlage zur Bestimmung des vertikalen Windprofiles. Arch. Met. Geoph. Biokl., Ser. B, 20: 345–352.

Neuwirth, F. 1971. Ergebnisse von vergleichenden Messungen mit Verdunstungswannen im Gebiet des Neusiedler Sees. Arch. Met. Geoph. Biokl., Ser. A, 20: 361–382.

Neuwirth, F. 1973a. Die Bestimmung der Verdunstung aus einer Class-A-Wanne durch empirische Verdunstungsformeln. Arch. Met. Geoph. Biokl., Ser. A, 22: 97–118.

Neuwirth, F. 1973b. Experiences with evaporation pans at a shallow steppe-lake in Austria. Proceedings of the Int. Symp. on the Hydrology of Lakes, Helsinki. IAHS Publication 109: 290–297.

Neuwirth, F. 1974. Über die Brauchbarkeit empirischer Verdunstungsformeln dargestellt am Beispiel des Neusiedler Sees nach Beobachtungen in Seemitte und in Ufernähe. Arch. Met. Geoph. Biokl., Ser. B, 22: 233–246.

Neuwirth, F. 1975. Die Abhängigkeit der Verdunstung einer freien Wasserfläche (Neusiedler See) von meteorologischen Einzelelementen. Arch. Met. Geoph. Biokl., Ser. A, 24: 53–67.

Neuwirth, F., 1976. Kleinräumige Niederschlagsverhältnisse im Gebiet des Neusiedler Sees. Wetter und Leben, 28: 166–177.

Panosch, K. 1973. Das Lichtklima des Neusiedlersees. Dissertation, University of Vienna.

Steinhauser, F. 1965. Klimatologische Gesichtspunkte für die Kurorteplanung im Burgenland. Wiss. Arb. Bgld. 30: 125–137.

Steinhauser, F. 1970a. Kleinklimatische Untersuchung der Windverhältnisse am Neusiedler See. Part one: Die Windrichtungen. Idöjárás 74: 76–88.
Steinhauser, F. 1970b. Kleinklimatische Untersuchung der Windverhältnisse am Neusiedlersee. Part two: Die Windstärken. Idöjárás 74: 324–345.
Wihl, G. 1975. Ein Beitrag zur Erfassung der Windverhältnisse im Gebiet des Neusiedler Sees. Part one: Perzentile der Windgeschwindigkeit. Wetter und Leben 27: 47–62.

6 Evaporation

H. Dobesch and F. Neuwirth

One of the main goals of the International Hydrological Decade programme was the determination of the water and heat balance of the lake with

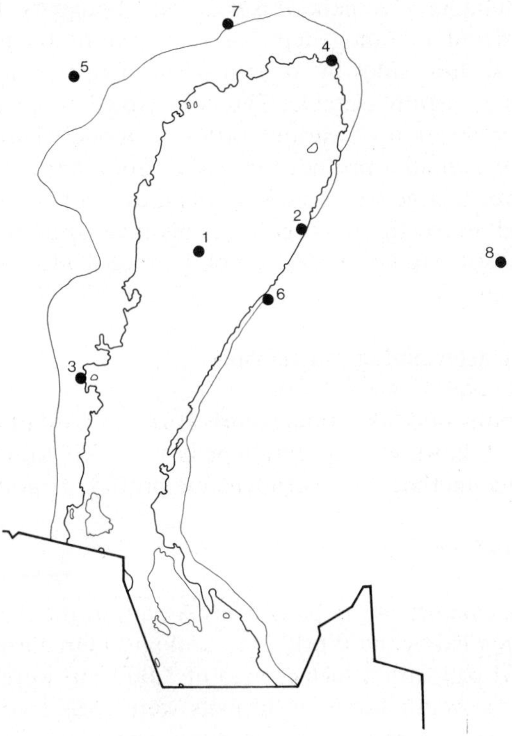

Fig. 6.1. Positions of the stations on Neusiedlersee: 1. Mid-lake station, 2. Heat budget station Podersdorf, 3. Reed station Rust, 4. Wind recording station Neusiedl, 5. Wind recording station Purbach, 6. Wind recording station 'Hölle', 7. Station Breitenbrunn, 8. Land station Althof. Lake limits according to Kopf (1964).

particular emphasis on the exact estimation of evaporation. For this purpose a number of meteorological stations were set up in the lake region (Fig. 6.1).[1]

In addition to such stations for measuring the fundamental meteorological parameters influencing climate, others were established to analyse the total energy fluxes in the boundary layer of the air. The sites of these four stations were chosen so as to include the three types of landscape characteristic of Neusiedlersee, the open lake (mid-lake and Podersdorf), the reed belt (Rust)

1 Most of the data recorded in this chapter were obtained during the course of the IHD of the Zentralanstalt für Meteorologie und Geodynamik, Wien (Central Bureau of Meteorology and Geodynamics, Vienna) between 1966 and 1974.

and the cultivated areas (Althof). Some forty elements were registered at these, including air temperature and humidity, wind speed at different levels, radiation components, water and ground temperatures at different depths, evaporation in evaporation pans, precipitation, and in addition, at the mid-lake station, turbidity of water, height of the waves, and currents. All these elements were continuously registered over a period of several years, and the data obtained were automatically processed.

Even though in action only from April to November, when the lake was free of ice, the mid-lake station played a particularly important role. The theory for calculating turbulent fluxes in the boundary layer of the air demands a horizontal homogeneity of the area to be measured. Being surrounded on all four sides by at least 4 km of open water, the mid-lake station was thus eminently suitable. The data gained at this station therefore permitted the application of various diffusion models based on the wind, temperature and humidity profiles measured. Long-term registration of the microclimate over a large water surface was thus rendered feasible.

As mentioned above the main goal of this investigation was to estimate evaporation. In order to cover this complex process adequately a variety of methods was used.

6.1 Methods for determining evaporation

At the four stations mid-lake, Podersdorf, Rust and Althof evaporation was measured by the following: evaporation pans, empirical formulae of evaporation, heat balance method, and aerodynamic profile methods.

6.1.1 *Evaporation pans*

To ascertain the evaporation in pans, class A pans of the American Weather Bureau recommended by the World Meteorological Organization (1966) and the Russian GGI pan with a surface area of 3,000 cm^2 were used. The daily fluctuations of the water levels in all pans were registered by evaporation recorders, and control measurements were carried out according to the recommendations of the WMO. The temperature of the water surface both in the pans and in the lake was also continuously recorded.

At the mid-lake station a floating GGI-3000 pan was installed which, unlike the class A pan, permitted a fair approximation between the temperature in the pan and that in the lake (Table 6.1). On account of waves only those evaporation values from the floating GGI pan that were measured at a wind speed below 25 km $hour^{-1}$ could be used (Table 6.2). To avoid these difficulties a rinsed GGI-3000 pan was installed on the shore near Podersdorf. The evaporation pan was mounted in a large outer tank which was filled with lake water thermostated to the temperature of the lake (Neuwirth 1973).

Fig. 6.2 depicts the mean decade values of evaporation in mm/day measured from the class A pans at the mid-lake, Rust and Podersdorf stations between 1967 and 1970 (Neuwirth 1971). According to the figure the highest evaporation values were measured in the middle of the lake, whereas they

Table 6.1. Diurnal variation, monthly averages and extremes of the surface temperature in the GGI-3000 pan, at Neusiedlersee and in the class A pan, July and October 1968 at the mid-lake station, in °C (Neuwirth 1973).

		Time (hours)										
		3	6	9	12	15	18	21	24	Average	Max.	Min.
Jul.	GGI-3000	18.9	18.7	19.7	21.4	22.0	21.0	20.0	19.4	20.1	31.0	13.8
	Lake	19.4	19.1	19.7	21.0	21.5	21.0	20.3	19.8	20.2	30.3	14.4
	Class A	15.9	15.7	18.4	23.1	24.4	22.5	19.1	17.0	19.5	38.7	11.3
Oct.	GGI-3000	10.8	10.5	10.9	11.8	12.1	11.6	11.2	11.0	11.2	19.4	4.1
	Lake	11.1	10.9	11.1	11.7	11.9	11.6	11.3	11.2	11.3	17.9	6.3
	Class A	8.9	8.4	8.7	11.8	13.8	12.8	10.9	9.7	10.6	24.0	1.9

Table 6.2. Evaporation totals (mm) from the GGI-3000 pan and from the class A pan during calm days and their ratio; mid-lake station 1969 (Neuwirth 1973).

Month	No. of suitable days	GGI-3000	Class A	Ratio
May	5	25.1	33.3	0.75
June	13	50.2	93.6	0.53
July	17	79.3	119.7	0.66
August	13	61.5	88.3	0.69
September	18	46.3	77.3	0.59
October	12	16.3	22.0	0.73
Totals	78	278.7	434.2	0.64

were about 15 per cent lower at Podersdorf and Rust. Table 6.3 comprises the monthly pan coefficients, the ratio of pan evaporation to lake evaporation, of the class A pan at the mid-lake station from May to October 1967 to 1969.

Table 6.3. Monthly pan coefficients for the class A pan at the mid-lake station, May to October 1967 to 1969 (Neuwirth 1973).

Year	May	Jun.	Jul.	Aug.	Sep.	Oct.	May–Oct.
1967	0.70	0.70	0.70	0.71	0.75	0.68	0.71
1968	0.75	0.77	0.73	0.77	0.81	0.82	0.78
1969	0.80	0.77	0.68	0.61	0.59	0.61	0.67
Average	0.75	0.75	0.70	0.70	0.72	0.70	0.72

The coefficients are derived from a comparison with the evaporation values from the heat balance method (see Section 6.1.3).

Fig. 6.3 illustrates, by means of four typical weather situations, the extent to which the daily course of evaporation in the lake and in the class A pan may vary according to the weather conditions, *a* showing the difference between the class A pan and the lake before the passage of a front approaching from the west during bright weather with strong south winds.

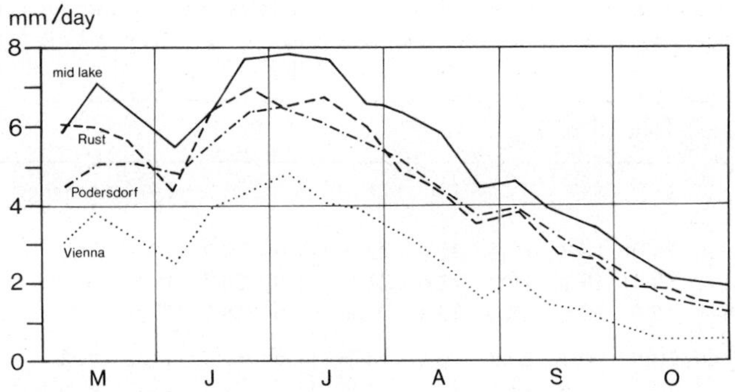

Fig. 6.2. Mean course of evaporation from class-A pans at mid-lake, in Podersdorf, Rust and Vienna, from May to October 1967–1970.

Due to strong overheating in the pan, evaporation is considerably higher than from the lake. During northwest weather following the passage of a cold front, *b*, the evaporation from the lake remains higher all day, since, due to the greater heat capacity, the water temperature is higher than in the pan. During high pressure conditions with light breezes, *c*, the values measured in the class A pan are higher than those of the lake, but on the whole are rather low because of the low wind velocity. The evaporation curve during a cold front which passed between two and three p.m. is shown by *d*. During its passage the evaporation in the class A pan strongly increased due to the high

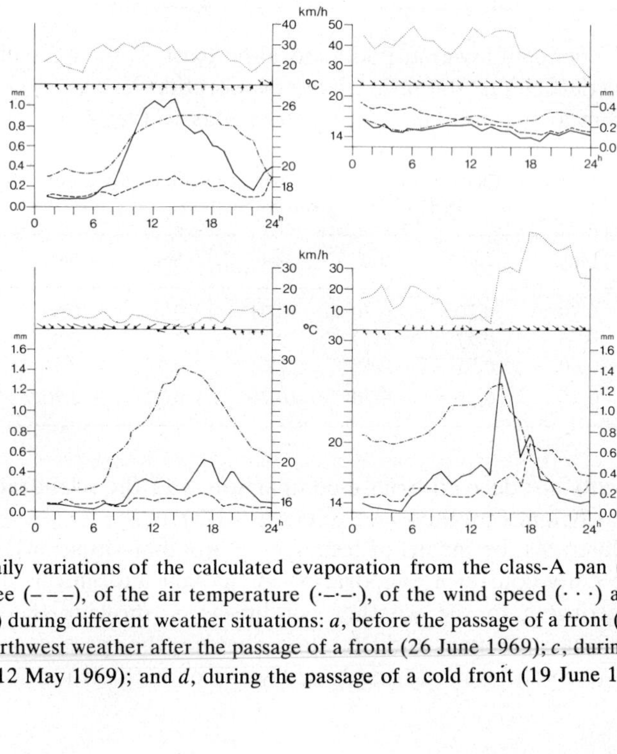

Fig. 6.3. Daily variations of the calculated evaporation from the class-A pan (——) and from Neusiedlersee (– – –), of the air temperature (·–·–·), of the wind speed (· · ·) and of the wind direction (↑) during different weather situations: *a*, before the passage of a front (24 June 1969); *b*, during northwest weather after the passage of a front (26 June 1969); *c*, during high pressure conditions (12 May 1969); and *d*, during the passage of a cold front (19 June 1969).

deficit of saturation and the onset of heavy winds. During the following hours evaporation is lower in the pan than from the lake, the situation being the same as that shown by *b*.

6.1.2 *Empirical formulae of evaporation*

Empirical formulae for the evaporation of a free water surface tend to be written as follows:

$$\frac{E_e}{L} = V = \mathrm{f}(u)\ (e_0 - e_z),$$

V being evaporation, $\mathrm{f}(u)$ a function of the wind speed u, e_0 the saturation vapour pressure at the temperature of the water surface undergoing evaporation, and e_z the atmospheric vapour pressure at a height of z above the water level. The difference in vapour pressure $e_0 - e_z$ is later termed the deficit of saturation. E_e indicates latent heat flux and L the heat of evaporation.

In the case of Neusiedlersee the wind function has been represented as follows: $\mathrm{f}(u) = a + bu^c$; a and b are constants that must be empirically determined. The successful application of the empirical formula for the evaporation largely depends on the determination of these constants. Consequently it appears unjustifiable simply to take over the constants valid for one lake and apply them to another, since they depend not only on the capacity of the measuring devices used, but also on local environmental influences in the widest sense. The exponent c was either assumed to be 1 or determined from Sutton's Turbulence Number (Sutton 1953; Richter 1969) based on profile measurement of the wind speed above the lake. According to this method c was 0.85. The constants a and b were obtained by comparisons with the evaporation results estimated by the heat balance method (Neuwirth 1974): $a = 0.13$ and $b = 0.028$ if $c = 1$; $a = 0.07$ and $b = 0.047$ if $c = 0.85$; wind speed was measured in km hour^{-1}, 3 metres above the water surface; vapour pressure was measured in mb, 2 metres above the lake. Table 6.4 sums up the results gained from (1) with the aid of the above constants as well as the monthly totals of evaporation ascertained using the heat balance method. Table 6.5 presents the evaporation figures per hour calculated from the hourly values of the wind velocity and the saturation deficit. Here the slight daily fluctuation in evaporation of Neusiedlersee becomes evident. This empirical method also made possible an estimation of the lake's annual evaporation at the shore station Podersdorf (Table 6.6), and the figures obtained (an annual total of 900) were comparable with the results of earlier calculations (Szestay 1958; Kopf 1964). A comparison of the evaporation values for the open lake area and the littoral zone shows that the total evaporation near the shore is only slightly higher than in the open lake near Podersdorf (May and October): 698 mm in 1968, 697 mm in 1969; in the middle of the lake: 655 mm in 1968 and 670 mm in 1969. Marked differences may be seen on certain days depending on the weather. In Fig. 6.4, *a* and *b* represent the daily curves of evaporation under particular weather conditions typical of Neusiedlersee. During prefrontal southerly weather, *a*, on a clear

Table 6.4. Monthly evaporation values in mm estimated using the empirical evaporation formulae (V_1, $V_2 = (a+bu)(e_0-e_z)$, V_3, $V_4 = (a+bu^{0.85})(e_0-e_z)$; V_1, V_3 from daily values, V_2, V_4 from hourly values), and the heat budget method (V) at mid-lake, May to October, 1967 to 1970 (Neuwirth 1974).

Month	Year	V_1	V_2	V_3	V_4	V
May	1967	132.7	132.3	129.5	128.4	130.6
	1968	140.3	140.0	137.7	136.0	139.5
	1969	157.3	156.5	151.8	151.9	171.8
	1970	122.0	119.4	119.2	115.1	—
Jun.	1967	145.6	144.0	142.9	143.0	142.7
	1968	145.8	146.0	143.5	141.5	148.5
	1969	133.6	132.5	130.9	128.8	143.6
	1970	150.6	149.4	148.6	145.5	—
Jul.	1967	175.0	171.9	172.4	167.2	166.0
	1968	150.6	148.7	147.4	144.1	161.3
	1969	144.2	143.8	141.2	138.9	152.5
	1970	149.6	134.0	131.9	129.7	—
Aug.	1967	132.5	130.5	130.8	127.3	118.2
	1968	104.8	106.0	103.2	103.1	121.4
	1969	109.5	109.6	108.4	107.5	107.7
	1970	134.3	124.1	124.6	121.1	—
Sep.	1967	74.0	75.3	72.7	72.9	77.2
	1968	80.4	81.3	79.0	78.9	87.1
	1969	80.6	81.0	80.1	79.3	77.6
	1970	106.7	106.9	105.6	104.3	—
Oct.	1967	53.1	53.6	52.1	51.9	48.4
	1968	33.5	34.4	33.0	33.1	39.8
	1969	44.6	45.3	43.7	43.7	50.5
	1970	41.2	42.5	40.5	41.2	—
Mean for 1967–1970 (May–Oct.)		685.7	677.3	666.9	658.6	

day with a fairly strong southeast to south wind blowing all day the wind velocity at Podersdorf is lower than in the middle of the lake, and thus the evaporation there remains lower than at the mid-lake station throughout the day. A cold front following on a period of undisturbed weather does not elicit any essential differences in evaporation between Podersdorf and the midlake station. The front between two and three p.m. led to a strong increase of the saturation deficit over the water surface, which together with a heavy northwest wind caused increased evaporation at both stations (*b*). The conditions after the passage of a front during continuous northwest weather are reflected in *c*. The heavy northwest wind throughout the day and a considerable saturation deficit resulted in high evaporation values at both stations. The fact that the noctural evaporation at Podersdorf is lower than at the mid-lake station, whereas by day the situation is the reverse, is due to the similar course taken by the temperature in the water layer near the surface. Conditions during high pressure where a changeable light wind keeps

Table 6.5. Daily course, monthly averages and hourly extremes of evaporation of Neusiedlersee, estimated from $V = (0.13+0.28u)(e_0-e_z)/24$ at mid-lake (May–October, 1967–1970) in mm.

Month	Year	Hour 03.00	06.00	09.00	12.00	15.00	18.00	21.00	24.00	Mean	Max.	Min.
May	1967	0.15	0.16	0.16	0.21	0.21	0.19	0.18	0.17	0.18	0.47	0.01
	1968	0.17	0.18	0.18	0.20	0.22	0.21	0.18	0.18	0.19	0.63	0.02
	1969	0.19	0.16	0.17	0.20	0.22	0.20	0.21	0.18	0.19	0.73	−0.01
	1970	0.14	0.13	0.15	0.19	0.20	0.19	0.14	0.13	0.16	0.59	0.00
	Mean	0.16	0.16	0.17	0.20	0.21	0.20	0.18	0.17	0.18	0.61	0.01
Jun.	1967	0.18	0.18	0.20	0.22	0.24	0.21	0.18	0.18	0.20	0.71	0.03
	1968	0.16	0.17	0.18	0.25	0.28	0.22	0.19	0.17	0.20	0.58	0.03
	1969	0.17	0.18	0.18	0.20	0.22	0.21	0.17	0.17	0.19	0.73	−0.18
	1970	0.17	0.18	0.21	0.23	0.27	0.26	0.18	0.17	0.21	0.66	0.01
	Mean	0.17	0.18	0.19	0.23	0.25	0.23	0.18	0.17	0.20	0.67	−0.03
Jul.	1967	0.23	0.21	0.20	0.25	0.27	0.27	0.21	0.20	0.23	0.70	0.02
	1968	0.17	0.20	0.19	0.26	0.27	0.20	0.15	0.16	0.20	0.95	0.00
	1969	0.15	0.16	0.15	0.22	0.26	0.23	0.18	0.16	0.19	0.62	−0.02
	1970	0.17	0.17	0.19	0.21	0.23	0.17	0.15	0.16	0.18	0.69	0.01
	Mean	0.18	0.19	0.18	0.24	0.26	0.22	0.17	0.17	0.20	0.74	0.00
Aug.	1967	0.15	0.17	0.18	0.19	0.23	0.19	0.16	0.13	0.18	0.76	0.03
	1968	0.12	0.12	0.12	0.17	0.19	0.16	0.13	0.13	0.14	0.64	−0.02
	1969	0.14	0.15	0.13	0.16	0.20	0.18	0.16	0.13	0.16	0.60	0.01
	1970	0.16	0.16	0.16	0.18	0.20	0.17	0.15	0.16	0.17	0.63	0.00
	Mean	0.14	0.15	0.15	0.18	0.21	0.18	0.15	0.14	0.16	0.66	0.01
Sept.	1967	0.11	0.11	0.11	0.12	0.12	0.09	0.08	0.08	0.11	0.63	0.00
	1968	0.10	0.11	0.11	0.15	0.14	0.10	0.10	0.10	0.11	0.46	−0.04
	1969	0.11	0.11	0.11	0.13	0.15	0.10	0.09	0.10	0.11	0.37	0.01
	1970	0.13	0.13	0.15	0.20	0.20	0.13	0.13	0.12	0.15	0.61	−0.01
	Mean	0.11	0.12	0.12	0.15	0.15	0.11	0.10	0.10	0.12	0.52	−0.01
Oct.	1967	0.08	0.07	0.07	0.08	0.08	0.06	0.06	0.08	0.07	0.45	−0.03
	1968	0.04	0.04	0.05	0.06	0.07	0.04	0.04	0.04	0.05	0.34	−0.05
	1969	0.06	0.06	0.06	0.08	0.09	0.06	0.06	0.05	0.07	0.43	0.00
	1970	0.05	0.05	0.05	0.06	0.06	0.05	0.05	0.05	0.06	0.43	−0.05
	Mean	0.06	0.06	0.06	0.07	0.08	0.05	0.05	0.05	0.06	0.41	−0.03

evaporation low, in spite of the high amount of radiation, are represented by *d*. The increased evaporation at Podersdorf in the afternoon is caused by a rise of the water temperature due to extreme shallowness.

6.1.3 *Heat balance method*

The heat balance method calculates the energy consumed for evaporation as the remaining unknown, where all other components of the surface energy balance are known:

$$SB + B + H + LV = 0, \qquad (2)$$

SB being net radiation, B the change of the heat content of the lake, H the

Table 6.6. Evaporation totals for Neusiedlersee estimated using empirical evaporation formulae ($V = (a+bu^c)(e_0-e_z)$; V_1 with $c = 1$, V_2 with $c = 0.80$) and using the heat budget method (V_3) off Podersdorf, 1968 and 1969, in mm

		V_1	V_2	V_3			V_1	V_2	V_3
Jan.	1968	14.2	12.7	—	Jul.	1968	170.6	158.3	163.0
	1969	13.4	13.6	—		1969	143.0	142.9	155.2
Febr.	1968	5.5	4.8	—	Aug.	1968	118.1	110.6	118.0
	1969	15.5	15.4	—		1969	115.9	115.2	110.3
Mar.	1968	77.4	70.2	—	Sep.	1968	85.4	79.8	74.7
	1969	21.9	20.8	—		1969	84.6	84.3	78.1
Apr.	1968	103.9	96.5	109.0	Oct.	1968	40.2	37.9	36.8
	1969	75.6	72.7	77.8		1969	57.3	57.1	56.9
May	1968	133.5	123.6	148.8	Nov.	1968	31.0	29.3	—
	1969	164.3	165.1	163.2		1969	28.9	31.0	—
Jun.	1968	150.4	140.8	151.9	Dec.	1968	9.6	9.3	—
	1969	131.8	133.6	143.8		1969	18.7	18.5	—
Year	1968	939.8	873.8						
	1969	870.9	870.2						

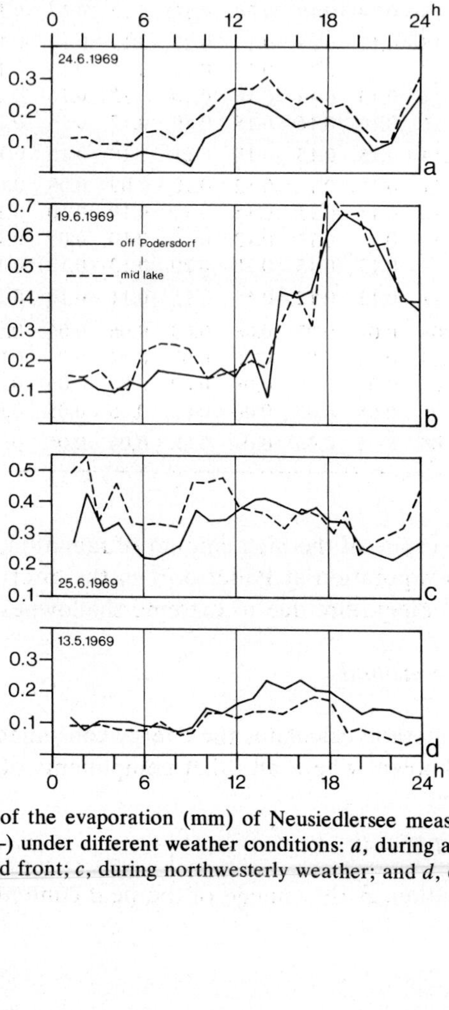

Fig. 6.4. Daily course of the evaporation (mm) of Neusiedlersee measured at mid-lake (---) and off Podersdorf (—) under different weather conditions: *a*, during a southerly situation; *b*, during passage of a cold front; *c*, during northwesterly weather; and *d*, during a period of high pressure.

heat exchange with the atmosphere (sensible heat flux) and $LV = E$ the latent heat flux. The four quantities in equation (2) are generally given in cal per unit of area and unit of time.

This energy budget equation neglects certain components of the heat balance, such as energy transport through precipitation or advective energy supply, for instance through affluents. Of the 4 essential components in (2) SB and B can be estimated fairly exactly. Whereas SB can be measured directly, B must be evaluated indirectly by means of the temperature distribution in the body of water studied. In the case of Neusiedlersee the tautrochrone method was used for the open lake, and because of its shallowness the heat conduction method was also applied to measure the non-negligible energy flow through the bottom of the lake. The tautrochrone method allows the calculation of the changes in the heat capacity of a body of water from its vertical temperature distribution and from the temporal variation of the temperature field pattern. In the heat conduction method the heat flow through the bottom is assumed to be proportional to the vertical gradient of the bottom temperature, the thermal conductivity acting as a factor of proportionality. The tautrochrone method has proved to be very useful particularly in the uppermost strata of a body of water, provided that the temporal change of temperature exceeds the accuracy of measurement. A combination of both methods constitutes a highly efficient approach in estimating the heat flow on the bottom of Neusiedlersee. While the components SB and B are fairly easy to specify this cannot be said of H and E: Sverdrup (Geiger 1961) worked out a method in which the vertical turbulent fluxes of E and H are assumed to be proportional to the specific gradients of their properties i.e. temperature, vapour pressure (exchange coefficient hypothesis):

$$H_K = c_p \rho \, K_H \, \mathrm{d}T/\mathrm{d}z \tag{3a}$$

$$E_K = L\rho \, K_E \, \mathrm{d}q/\mathrm{d}z \tag{3b}$$

where c_p is the specific heat at constant pressure, ρ the atmospheric density, K_H and K_E exchange coefficients of sensible and latent heat flux, $\mathrm{d}T/\mathrm{d}z$ and $\mathrm{d}q/\mathrm{d}z$ the temperature and humidity gradients respectively. A combination of Sverdrup's equations (3a) and (3b) results in

$$H_W = \frac{SB + B}{1 + \frac{L}{c_p} \cdot \frac{K_E}{K_H} \cdot \frac{\mathrm{d}q}{\mathrm{d}z} \cdot \frac{\mathrm{d}T}{\mathrm{d}z}} \tag{4a}$$

$$E_W = \frac{SB + B}{1 + \frac{c_p}{L} \cdot \frac{K_H}{K_E} \cdot \frac{\mathrm{d}T}{\mathrm{d}z} \cdot \frac{\mathrm{d}q}{\mathrm{d}z}} \tag{4b}$$

Provided that in the denominator of (4b), $K_E = K_H$, which seems justified according to the studies in Dobesch (1973) the so-called Bowens ratio is obtained, which is proportional to H/E and constitutes a simplification of (4a) and (4b). This method provides a fair estimation of the sensible and latent heat flux, as long as the temporal changes in water temperature exceed the accuracy of measurement and as long as the Bowens ratio remains measurable by means of defined differences in temperature and humidity between the water and the air. This means that in the case of Neusiedlersee the heat balance method is therefore only of limited applicability in the cold season. Furthermore, its application is only recommended for daily and pentadic means of the data, in order to rule out advective influences via the term B representing change of heat content in (2).

6.1.4 *Aerodynamic profile methods*

The equations (3a) and (3b) present another possibility of evaluating sensible and latent heat fluxes. According to Monin and Obuchow (1958) a definition of the universal functions for the vertical distribution of the wind speed (φ_M), temperature (φ_H) and atmospheric humidity (φ_E) always depends on the vertical coordinate z and a stability length L_v. This is based on considerations of similarities applied to the flow field, the latter characterized by external parameters. From these parameters, scales characteristic of the flow field can be given in terms of velocity (u_*), length (L_v), temperature (T_*), and specific humidity (q_*). Based on these considerations the exchange coefficient (K_M) can be described as follows:

$$K_M(z) = \frac{u_* k z}{\varphi_M} \tag{5}$$

The function φ_M is obtained from the respective gradient measurements

$$\frac{du}{dz} = \frac{u_*}{kz} \varphi_M \, (z/L_v)$$

where k is the Karman constant. Analogous statements can be made for K_H and K_E. In the case of Neusiedlersee φ_M was $(1 - 12.3z/L_v)^{-0.16}$ and $\varphi_H \approx \varphi_E$ was $(1 - 10.8z/L_v)^{-0.44}$ (Dobesch 1973). From these values and the known specific gradients the sensible heat flux (H_K) and the latent heat flux (E_K) could be calculated according to (3a) and (3b).

Another method for ascertaining the sensible and latent heat in adiabatic cases, that is, where $\varphi_M = 1$, starts from the logarithmic profile of wind, temperature and humidity in the form of equation (6). Integration between the limits z_1 and z_0 results in:

$$H_{a_l} = - c_p k^2 \rho \, \frac{T_1 - T_0}{ln z_1 - ln z_0} \, \frac{u_1}{ln z_1 - ln z_0} \tag{7a}$$

$$E_{a_1} = - k^2 \rho \frac{0.623}{p} \, \frac{e_1 - e_0}{ln z_z - ln z_1} \, \frac{u_1}{ln z_1 - ln z_0} \tag{7b}$$

T_0 is the temperature of the surface being evaporated, p the atmospheric pressure and z_0 the parameter of roughness. An additional adiabatic formula can be derived, which dispenses both with the surface temperature

T_0, which is difficult to measure, and with the parameter of roughness z_0: it only requires values from two profile measurement levels, z_1 and z_2:

$$H_{a_2} = -c_p k^2 \rho \frac{T_2 - T_1}{lnz_2 - lnz_1} \frac{u_2 - u_1}{lnz_2 - lnz_1} \quad (8a)$$

$$E_{a_2} = -k^2 \rho \frac{0.623}{p} \frac{e_2 - e_1}{lnz_2 - lnz_1} \frac{u_2 - u_1}{lnz_2 - lnz_1} \quad (8b)$$

In spite of their above mentioned limitation to a neutral or almost neutral stratification the two latter methods are suitable for an estimation of the sensible and latent heat fluxes, as a neutral stratification tends to occur quite frequently above water.

Results from evaluations of H and E according to (7a), (7b), (8a) and (8b) will be discussed together with results from other methods (Table 6.7) in the following section.

6.2 Comparison of methods

Table 6.7 comprises the monthly averages of the daily curves of latent and sensible heat fluxes calculated according to the various methods discussed

Table 6.7. Monthly means of the daily course of H and E values in 1/10 cal cm^{-2}h^{-1} calculated by various methods (mid-lake, July 1969).

Hour	1	2	3	4	5	6	7	8	9	10	11	12
Hw (4a)	16	18	22	21	21	21	16	12	9	2	−3	−7
H_K (3a)	15	15	17	21	23	22	18	11	8	5	−2	−7
H_{a_1} (7a)	15	15	17	21	23	20	16	11	8	2	−2	−5
H_{a_2} (8a)	12	11	12	17	14	16	17	12	10	4	−2	−5
E_w (4b)	81	86	92	77	78	85	79	89	98	107	109	120
E_K (3b)	80	73	75	82	84	82	78	77	72	72	79	94
E_{a_1} (7b)	88	81	87	93	94	90	85	82	80	80	83	96
E_{a_2} (8b)	83	80	94	89	82	85	92	88	88	92	94	113
E_e (1)	89	82	85	91	91	89	82	81	82	92	106	125

Hour	13	14	15	16	17	18	19	20	21	22	23	24	Monthly sum
Hw (4a)	−13	−13	−12	−14	−15	−15	−5	−1	5	7	11	13	2,968
H_K (3a)	−12	−10	−13	−13	−13	−14	−5	−1	3	6	11	13	3,115
H_{a_1} (7a)	−9	−11	−13	−12	−11	−11	−5	−1	3	5	10	11	3,010
H_{a_2} (8a)	−13	−17	−21	−22	−20	−16	−8	−2	2	5	8	10	750
E_w (4b)	139	141	138	139	129	125	119	103	101	92	87	88	77,548
E_K (3b)	97	99	109	107	94	99	94	82	81	79	81	80	63,597
E_{a_1} (7b)	109	114	119	116	99	101	94	80	84	83	86	89	68,622
E_{a_2} (8b)	115	120	135	134	130	131	117	106	104	97	89	87	75,820
E_e (1)	138	143	145	147	130	129	123	104	102	100	96	93	78,895

above (Dobesch 1974). Since the sensible heat values are very low, they show only slight differences. The main energy turnover of a body of water is due to its latent heat. It is striking that by day the mean evaporation values differ by 15 to 20 per cent on an average, whereas by night the difference amounts to only about 5 per cent. In a few individual cases these differences may, however, be considerably greater. In such cases the E_w and E_e, apart from E_{a2}, produce about 1,000 cal cm^{-2} month^{-1} more, which corresponds to an additional evaporation of 2 cm of water per month. All the values gained from daytime profile measurements are below the E_W and E_e values, except

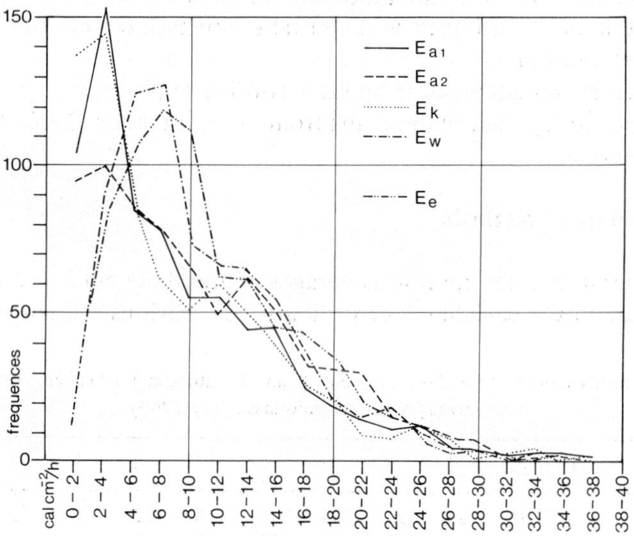

Fig. 6.5. The absolute frequencies of the evaporation values for E_{a1}, E_{a2}, E_k, E_w and E_e, mid-lake.

for E_{a2} which adjusts itself in the late afternoon to the values ascertained empirically or by the heat balance method. For all E, apart from E_W, which reaches its maximum at the time of day when irradiation is at its highest, between noon and one p.m., the peak values are recorded between three and four p.m. corresponding to the onset of the highest saturation deficit. The frequency distribution of the E values according to Fig. 6.5 shows a higher absolute frequency for the E values calculated from profile methods at intervals of 2 to 4 cal cm^{-2} h^{-1}, while E_W and E_e reach their maximum frequency at interval of 6 to 8 cal cm^{-2} h^{-1}.

These striking differences are due to the fact that both the empirical and the heat balance method, even for periods with little energy turnover, result in higher E values. This is due amongst other things to their high degree of dependence upon wind speed.

References

Dobesch, H. 1973. Das Wind-, Temperatur- und Feuchteprofil über einer freien Wasserfläche. Arch. Met. Geoph. Biokl., Ser. A, 22: 47–70.

Dobesch, H. 1974. Die numerische Bestimmung der Transporte fühlbarer und latenter Wärme mittels verschiedener Methoden über einer freien Wasserfläche. Arch. Met. Geoph. Biokl., Ser. A, 23: 263–284.

Geiger, R. 1961. Das Klima der bodennahmen Luftschicht. Vieweg & Sohn, Braunschweig.

Kopf, F. 1964. Die wahren Ausmaße des Neusiedler Sees. Österr. Wasserwirtschaft 16.

Monin, A. A. & Obuchow, A. M. 1958. Fundamentale Gesetzmäßigkeiten der turbulenten Vermischung in der bodennahen Schicht der Atmosphäre. Statistische Theorie der Turbulenz 199. Akademie Verlag, Berlin.

Neuwirth, F. 1971. Ergebnisse von vergleichenden Messungen mit Verdunstungswannen im Gebiet des Neusiedler Sees. Arch. Met. Geoph. Biokl., Ser. A, 20: 361–382.

Neuwirth, F. 1973. Experiences with evaporation pans at a shallow steppe-lake in Austria. Proceedings of the Int. Symp. on the Hydrology of Lakes, Helsinki IAHS Publication 109: 290–297.

Neuwirth, F. 1974. Über die Brauchbarkeit empirischer Verdunstungsformeln dargestellt am Beispiel des Neusiedler Sees nach Beobachtungen in Seemitte und in Ufernähe. Arch. Met. Geoph. Biokl., Ser. B, 22: 233–246.

Richter, D. 1969. Ein Beitrag zur Bestimmung der Verdunstung von freien Wasserflächen, dargestellt am Beispiel des Stechlinsees. Abh.d.MHD d. DDR 11(88).

Sutton, O. G. 1953. Micrometeorology. McGraw-Hill, New York.

Szestay, K. 1958. Orientierungsangaben über die Verdunstung von freien Wasserflächen in Ungarn. Vizügyi Közlemények, Budapest.

World Meteorological Organization. 1966. Measurement and estimation of evaporation and evapotranspiration. WMO Technical Note 83.

7 Water balance

H. Dobesch and F. Neuwirth

7.1 Introduction

Before the water economy of a lake can be satisfactorily dealt with a thorough knowledge of its water budget is indispensable. In this respect, too, Neusiedlersee is unique. Nearly all of the factors of significance for the water budget of the lake are on a different scale from those connected with the other lakes of Austria. A natural input of water via small gulleys is either lacking or only poorly developed. The terrain is so flat that a difference in elevation of only 10 cm can be of greater consquence here than a difference of 100 m in a mountainous district.

In order to protect the lake from its two main sources of danger, on the one hand, the climate, which keeps the lake in a perpetual state of oscillation between the extremes of desiccation and flooding, and the thoughtless destruction inflicted by man himself on the other, it is essential that the water relations of the lake be thoroughly and exactly understood.

Certain climatic situations in the past have led to enormous fluctuations in the water levels of the lake, at times even resulting in its complete desiccation.

7.2 A historical survey of important events connected with the water level of the lake

From the seventeenth century onwards more or less reliable qualitative records have been kept. According to Bendefy (1973) particularly high or low water levels were found as follows:

1616	Water level sank by 1.6 m
1638–1640	Lake completely dried out
1668	Very high water level
1676–1677	Very low water level
1683 1693	Large portions of the lake dried out and were then used for crop farming
1728	Again, rapid loss of water, large numbers of fish died
1738	It was possible to wade across the lake from Rust
1740–1741	Water level rose rapidly; dams had to be built
1742–1756 1765–1767 1770–1790	Extensive areas of the surrounding pastures and cultivated land were flooded
1786	Water depth of lake was 2 m
1807–1812	Water level dropped considerably
1813–1830	Water level rose rapidly, up to a maximum of 2 m
1836–1838	Rapid water losses

1855–1865	Further rapid losses of water
1866–1870	Lake completely dried out. Buildings erected on former lake bottom
1881–1882	Rapid refilling of lake with a steady high water level
1892–1895	Considerable losses of water
1902–1905	Water level again rising
1905–1907	Only parts of the lake covered with water

The consequences of the last period of desiccation for the surrounding countryside have been very vividly described, as the following excerpts show:

'With the final disappearance of the water [of the lake] the already severe effects of the drought for the agriculture of the district has been still further enhanced. For example, the dew, so beneficial to the plants, is reduced. The otherwise so luxuriant reed growth has disappeared, to be replaced by poor grassland. Fishing came to an end some years ago, when due to evaporation, the salt concentration reached a level detrimental to the fish.

In addition to the threat represented by the soft patches in the mud there was the danger involved by the wind, which inevitably brought with it dust that was stirred up in thick clouds above the lake. It was feared in the entire region as far as the Leitha mountains for its ill effects on the mucous membranes of eyes and respiratory organs. The dense clouds of dust visible for miles in windy weather, consisted mainly of the salts from the dried lake surface.

When we had advanced to a distance of about 400 *Klafter* from the shore the salt deposits were of such proportions that as far as the eye could see the ground appeared to be covered with freshly fallen snow. The deception was perfected by the close resemblance borne by the crystalline crust to a blanket of snow' (H. Moser 1866, our translation).

7.3 Special problems connected with determining the water budget

The water balance of a lake can be described by a so-called water-balance equation, which incorporates active and passive factors as follows:

$$Z_0+Z_u+N_s-V-A = 0 \qquad (1)$$

where Z_0 is surface inflow, Z_u is subterranean inflow, N_s precipitation falling on the lake surface, V evaporation and A drainage. Averaged over a large number of years (1) must hold otherwise the lake would either dry out (deficit) or would have to possess a continuous natural outflow (surplus).

In order to obtain the two principal values, N_s and V, it is essential to know the exact dimensions of the lake. An accurate surface survey was carried out from the Austrian side in 1963 (Kopf 1964) and by the Hungarians in 1967. This was a particularly tedious and exacting undertaking, in the course of which it became clear that the lake is extremely sensitive to any change in water level, as reflected in its area and contents (see Fig. 4.3). As an example an increase of only 50 cm in water level, from 114.98 m[1] to 115.46 m would result in an increase of 89 km^2 in the surface area of the lake, and an increase of 124 million m^3 in its contents (Kopf 1974). The consequence for the water balance is shown by the example in Table 7.1, taken from Szestay (1963). The large relative increase in evaporation and precipitation with increasing

1 Above the level of the Adriatic. All water levels are expressed in this way.

Table 7.1. Mean values for the water balance of Neusiedlersee, assuming different surface areas (Szestay 1963).

Assumed surface area of the lake, F_s (Km^2)	Catchment area of the lake, F_E (km^2)	Inflow Z (m^3/sec)	Precipitation on lake surface, N, uniform rate of flow (m^3/sec)	Evaporation from lake surface, V	Drainage (amount withdrawn), $A = Z+N-V$ (m^3/sec)	Water balance values $W = Z+N = V+A$ (m^3/sec)	$\frac{V}{V+A}$	$\frac{V}{N+Z}$	Remarks
0	1300	2.46	0	0	2.46	2.46	0	0	flowing water (not a lake)
50	1250	2.37	1.12	1.42	2.07	3.49	0.41	0.32	
100	1200	2.28	2.25	2.85	1.68	4.53	0.63	0.49	lake with natural flow
200	1100	2.10	4.50	5.70	0.90	6.60	0.86	0.68	
280	1020	1.96	6.30	8.00	0.26	8.26	0.97	0.76	
315	985	1.80	7.18	8.98	0	8.98	1.00	0.80	closed lake
400	900	1.73	9.00	11.40	−0.67	10.73	1.03	0.84	lake fed artificially

Basic values used in the calculation:
Precipitation: $N = 710$ mm
Evaporation: $V = 900$ mm
Runoff coefficient: $a = 8.5\%$
$F = F_s+F_E = 1300\ km^2$

surface area is obvious. Although the lake survey provided the basis for a reasonably accurate determination of the water balance the problem was by no means solved. In considering the individual terms in (1) it has to be borne in mind that surface inflow (Z_0) is based solely upon one measurement from the Wulka, and can only be roughly estimated for the remaining portions of the catchment area. The runoff coefficient in the lake region amounts to only 8–10 per cent.

Subterranean inflow (Z_u) – groundwater inflow – is only little known (see Chapter 2). In the past it has been taken as the remainder in the water balance equation, which itself is rather unreliable. The measurement of this parameter therefore forms the subject of current studies. Precipitation onto the lake surface (N_s) can be fairly accurately measured, but seems in the past to have been overestimated (Neuwirth 1976). The extensive investigations within the framework of the IHD have undoubtedly been an invaluable aid in determining evaporation from the lake (Neuwirth 1974; Mahringer 1966; Dobesch & Neuwirth 1974) so that this, too, can be considered as safely established.

Runoff, A, of the lake on the Hungarian side via the Einser canal could only be roughly estimated on the basis of incomplete water level records, and from a few measurements. Here again, much more satisfactory material has been gathered in recent years. In summarizing it can be said that, according to present knowledge, the greatest remaining unknown quantity is subterranean inflow. Appreciable contributions to the solution of this problem have been provided by a qualitative determination of groundwater inflow into the lake, and by a hydrological description of the catchment area of the lake (Gattinger 1975; and Chapter 2).

As already mentioned an important problem connected with the water relations of the lake in the past was presented by its drainage. The Einser canal was constructed (1909–1911) with the aim of draining the lake and converting the area into agricultural land. This was prevented, however, by silting up of the canal due to heavy storms, although the lake nevertheless suffered continual losses, lowering its water level by an average of about 50 cm. In this way large areas of the lake acquired the depth of 80 cm and less which proved to be particularly conducive to the growth of reed. The areas covered by reed increased from 62 km² in 1872 to 198 km² in 1965 (see Chapter 20). It can be assumed that within a relatively short space of time the lake would have been emptied (Table 7.2) if drainage had been allowed to continue unchecked, although no measures were taken until 1965. As a result

Table 7.2. Area covered by *Phragmites* since 1872, in km² (Kopf 1974).

	1872	1901	1923	1937	1957	1965	1965(%)
Austrian part	42	69	78	98	107	118	48.1
Hungarian part	20	29	71	65	70	80	86.1
Total	62	98	149	163	177	198	58.5

of an Austro-Hungarian treaty it was possible in 1965 to enforce the 'Regulations for the use of the sluice of the Einser canal', with quite obvious success. These regulations lay down the conditions for opening and closing the sluice, using the amount of precipitation and actual water level as the chief criteria. If the mean annual total precipitation for the last three calendar years is between 570 and 679 mm, water may only be withdrawn if the lake surface level exceeds 115.40 cm. If the mean annual total is between 670 and 750 mm, water may only be withdrawn from the lake if its surface level is above 115.20 cm. If the mean annual precipitation is less than 570 mm then water may only be withdrawn if the water level exceeds 115.50 cm. If the mean annual total precipitation exceeds 750 mm, or if it exceeds 750 mm before the end of the third calendar year, the sluice has to be opened as soon as the water reaches 115.40 cm and kept open until the water surface has subsided to 115.20 cm. The volume of water flowing through the sluices may not exceed a total of 30 million m³ within three successive calendar years. Adherence to these regulations has apparently led to stabilization of the water level of the lake, as Fig. 4.2 shows (Kopf 1967, 1974).

7.4 Quantitative data concerning the water budget of the lake

The above regulations have to be taken into consideration if, despite the difficulties mentioned in the foregoing section, an attempt is to be made to construct an average water balance for Neusiedlersee. Assuming a more or less reliable figure of 650 mm for mean precipitation, a mean evaporation of 900 mm, and a surface drainage of 20 million m³ per year, then according to Kopf (1974) the following equation applies (in million m³ per year) for the average water level between 1930 and 1965 (114.98 m):

$$\begin{array}{ccccccccccc} Z_0 & + & Z_u & + & N_s & - & V & - & A & = & 0 \\ 67 & & 8 & & 144 & & 199 & & 20 & & \end{array}$$

where Z_u is calculated as the remaining quantity. In the absence of any records it will never be known what the average water flow was before 1964. If a figure of, for example, 60 million m³ is assumed for output then the water balance equation is as follows:

$$\begin{array}{ccccccccccc} Z_0 & + & Z_u & + & N_s & - & V & - & A & = & 0 \\ 67 & & 48 & & 144 & & 199 & & 60 & & \end{array}$$

so the subterranean inflows must have been six times as large if the equation is to balance. For the mean water level from 1965–1973 (115.45 m) with withdrawal of 10 million m³ as stipulated by the regulations, the equation is as follows:

$$\begin{array}{ccccccccccc} Z_0 & + & Z_u & + & N_s & - & V & - & A & = & 0 \\ 65 & & 8 & & 162 & & 225 & & 10 & & \end{array}$$

This applies to a water surface area of 250 km² and a water volume of 223 million m³, Z_u being determined as the remaining quantity, thus still involving a considerable degree of uncertainty. The Austro-Hungarian lake commission therefore undertook to establish an accurate water balance for Neusiedlersee within the next few years. This is to be carried out independently by the Austrians and Hungarians although relevant data will be exchanged. A team of experts is to work out the new water balance on the Austrian side: the groundwater inflow (and outflow) will be established by two different methods: (1) by direct calculation from precipitation, evaporation, seepage, surface drainage, measurements of direction and speed of flow etc., employing a mathematical model; and (2) as has so far been the practice, as the remainder in the water balance equation, the individual components of which, especially the outflow drainage A, will be measured with greater accuracy.

It is anticipated that when the project, which will run for several years, is completed, the water balance will have been established with such accuracy as to allow a correct interpretation of all processes and events involved and that the water relations of the lake will have been adequately documented.

References

Bendefy, L. 1973. Relation existing between the periodic fluctuation of the water levels of the Hungarian lakes and solar activity. Proc. Int. Symp. on the Hydrology of Lakes, Helsinki. IAHS Publication 109: 109–114.

Dobesch, H. & Neuwirth, F. 1974. Übersicht über die Ergebnisse aus den hydrometeorologischen Untersuchungen im Gebiet des Neusiedler Sees im Rahmen der Internationalen Hydrologischen Dekade 1966–1974. Wetter und Leben 26: 151–156.

Gattinger, T. 1975. Das hydrogeologische Einzugsgebiet des Neusiedlersees. Verh. Geolog. B. Anst. 4: 331–346.

Kopf, F. 1963. Wasserwirtschaftliche Probleme des Neusiedler Sees und des Seewinkels. Österr. Wasserwirtschaft 15: 190–203.

Kopf. F. 1964. Die wahren Ausmaße des Neusiedler Sees. Österr. Wasserwirtschaft 16.

Kopf, F. 1967. Die Rettung des Neusiedler Sees. Österr. Wasserwirtschaft 19: 139–151.

Kopf, F. 1974. Der neue Wasserhaushalt des Neusiedler Sees. Österr. Wasserwirtschaft 26: 169–180.

Mahringer, W. 1966. Über die Einrichtung meteorologischer Stationen zur Bestimmung der Verdunstung des Neusiedlersees. Wetter und Leben 18: 223–229.

Mahringer, W. 1970. Verdunstungsstudien am Neusiedler See. Arch. Met. Geoph. Biokl., Ser. B, 18: 1–20.

Masch, Moser I. & Hecke, ca. 1866. Die Neusiedler Seemulde im Jahre 1865. Mitt. k.u.k. Geol. Ges., Vienna.

Moser, I. 1866. Der abgetrocknete Boden des Neusiedler See's. Jahrb. Geol. Reichsanst., Vienna.

Neuwirth, F. 1974. Über die Brauchbarkeit empirischer Verdunstungsformeln dargestellt am Beispiel des Neusiedler Sees nach Beobachtungen in Seemitte und in Ufernähe. Arch. Met. Geoph. Biokl., Ser. B, 22: 233–246.

Neuwirth, F. 1976. Niederschlagsverhältnisse im Gebiet des Neusiedler Sees. Wetter und Leben 28: 166–177.

Szestay, K. 1963. Beiträge zu den hydrologischen Grundlagen der Seewasserstandsregulierung. Österr. Wasserwirtschaft 15: 16–22.

8 Currents

M. Jungwirth

The frequent and well-developed currents are primarily the result of wind action. The comparatively small inflows have practically no effect upon the currents in the open lake. Even the Wulka, which plays a considerable role in the water budget is in this respect of little significance since it has to flow for 4.5 km through the western reed belt before reaching the open lake.

Currents due to Coriolis forces and barometric pressure differences between the northern and southern parts of the lake, or caused by seiches, have scarcely been investigated at all (Roth-Fuchs 1929). Such currents can achieve velocities of up to several metres per second and are chiefly encountered in the narrows between the northern and southern parts of the lake, south of a line joining Illmitz and Mörbisch.

These relatively strong currents with their very specific pattern of distribution have considerable consequences from the limnological point of view. One of these is their influence on the lake bottom and on the sediments. In zones of stronger water movement material is eroded and carried off. This fine material, known as the characteristic lake turbidity, is deposited or sedimented in calmer water, mainly in the lakeward regions of the reed, but also in the macrophyte zone bordering the northern and western reed belt (see Chapter 23). As a result of the highly specific distribution of the superficial lake sediments, as described in Chapter 13, the submersed vegetation and the benthic fauna also exhibit a significant pattern of distribution (see Chapters 19, 23, 24).

The importance of the lake currents was early recognized. Various authors (Sauerzopf 1959, measurements made under ice; Kopf 1966; own investigations under ice) have attempted to map and measure the currents.

The size of the lake and thus the large number of simultaneous measurements required renders such a project both difficult and expensive. With these problems in mind a current simulation model of Neusiedlersee was constructed in connection with the Man and Biosphere project (Scale 1:10,000). Preliminary experiments with the water-filled model were carried out in a wind tunnel. The currents were rendered visible with the aid of grains of $KMnO_4$ and wood shavings, and the patterns obtained under a variety of wind conditions were photographed.

Although the experiments are incomplete and tests in media of different specific weights have still to be performed the results so far obtained will be discussed briefly. Regardless of the direction of the wind, i.e. whether from the NW or SE, the main current in the northern part of the lake runs in a clockwise direction. A few smaller current systems, flowing in the opposite direction, are mainly due to the presence of reed stands and usually occur in bays.

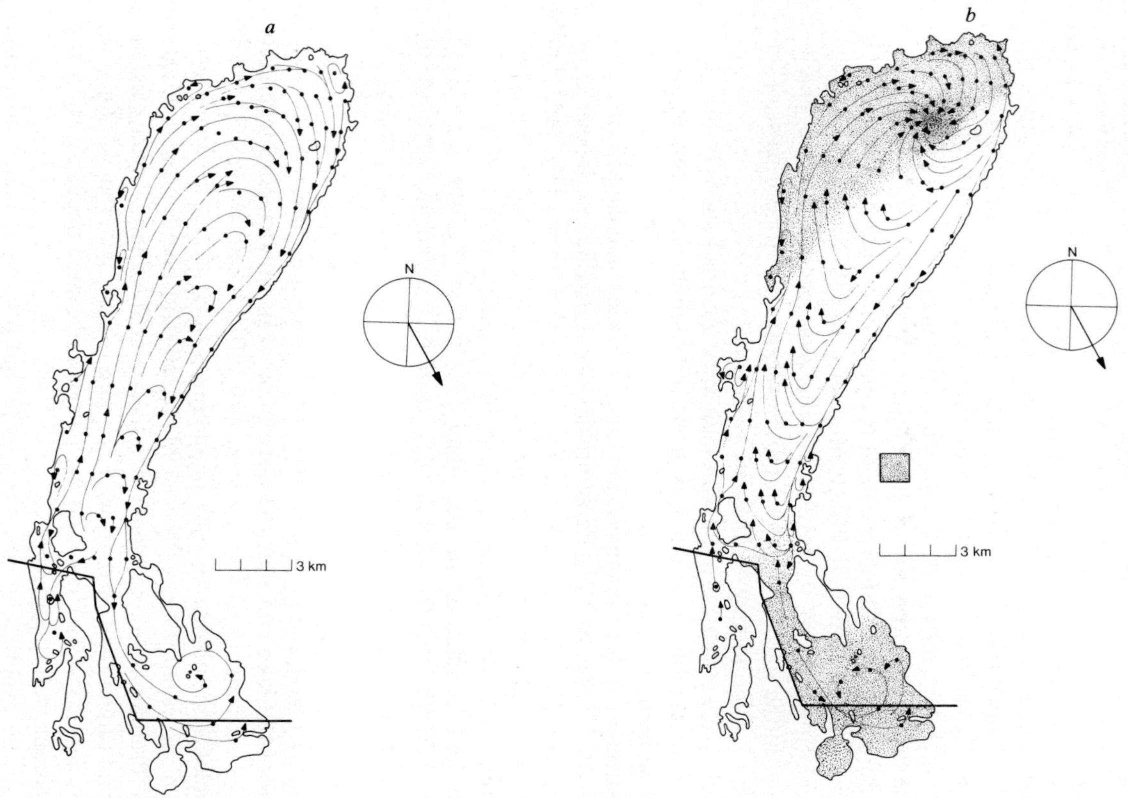

Fig. 8.1. Surface, *a*, and bottom, *b*, currents of Neusiedlersee (current simulation model) with winds blowing from NW. Regions with relatively little water movement are shown up in the model by the accumulation of wood shavings. (▦)

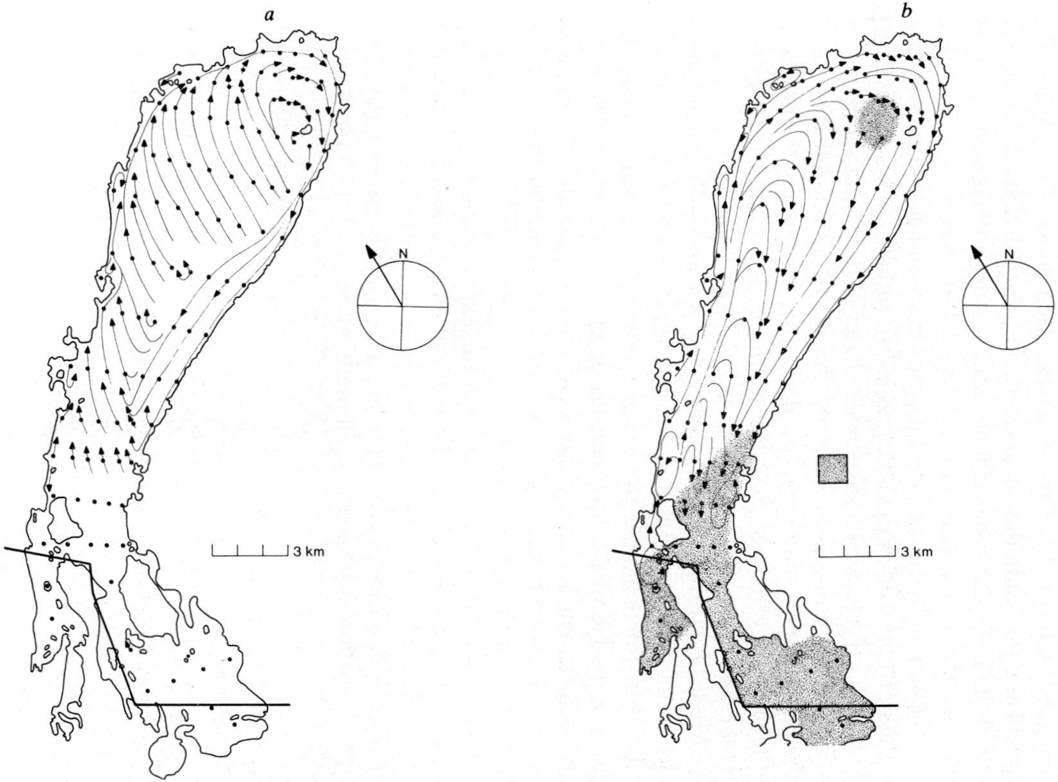

Fig. 8.2. Surface, *a*, and bottom, *b*, currents of Neusiedlersee (current simulation model) with winds blowing from SE. Regions with relatively little water movement are shown up in the model by the accumulation of wood shavings. (▩)

The strongest, or rather the swiftest currents run along the eastern shore in a southerly direction and may have contributed towards the formation of the narrow sand zone (see Chapter 8). Off Illmitz this current turns westwards and finally heads north near Mörbisch, turning to the east in the northern part of the lake. The broader, northward current running along the western reed belt moves more slowly than this relatively narrow current flowing along the eastern shore.

It is striking that the two main and opposing wind directions, NW and SE, result in practically the same flow characteristics. The currents differ merely in the width of the individual components and to a slight degree with respect to their turning points. Regions with relatively little water movement are shown up in the model by the accumulation of the wood shavings. They are situated at the northern end of the lake and in its southern portion. The latter acts as a kind of trap into which suspended material drifts and sediments in the calm water. The observations made on the model agree well with those made on the lake itself. In the regions bordering the reed belt in the northern part of the lake, and even more so in the southern part, a thick layer of soft mud covers the bottom and the water is very shallow.

Surface currents measured in the model (Figs. 8.1 and 8.2) depend upon the direction of the wind as well as upon the configuration of the shore line and the reed belt. For example the current running northwards along the western shore in the lee of the reed belt flows undisturbed when NW winds are blowing, and the surface and bottom currents run parallel. Conversely, when the wind is blowing from the SE, the same holds true for the southerly current running along the eastern shore.

In conclusion, it should be mentioned that the experiments made with the aid of the model largely agree with the measurements made by Sauerzopf (1959) in the lake itself. Discrepancies such as the opposing current system near Podersdorf, as described by Kopf (1966), are probably attributable to the fact that field measurements were not all made at one and the same time.

References

Kopf, F. 1966. Strömungsmessungen im Neusiedlersee (Österr. Teil). Unpublished report.
Sauerzopf, F. 1959. Wasserbewegungen im Neusiedlersee. In: Landschaft Neusiedlersee. Wiss. Arb. Bgld. 23: 51.
Roth-Fuchs, G. 1929. Beiträge zum Problem 'Neusiedler See'. Mitt. Geogr. Ges. Wien 72: 47.

9 The hydrochemical problem

F. Berger and F. Neuhuber

Alpine lakes act in a hydrochemical sense as reservoirs for their input, most of which enters on the surface. The ionic relationships of the incoming waters are shifted by limnochemical processes, to a considerable degree in the case of meromictic lakes, and to a lesser extent in lakes with total circulation. A comparison of the chemical characteristics of the surface inflows and of the lake water of Neusiedlersee reveals very striking differences in every respect.

Preliminary observations suggested that the lake is also fed by subterranean sources, as indicated by the holes occurring in the ice cover. The final demonstration was provided by extensive infrared photography from the air and by temperature measurements (see Chapter 2) over large areas.

In the following, the hydrochemical differences between the visible inflow and the lake water itself are used as a basis for stoichiometric deductions concerning the chemical characteristics of the incoming groundwater. From the wealth of hydrochemical data already available and from the present authors' own investigations between 1957 and 1977 those values have been chosen which are best suited to a quantitative treatment and appear to be characteristic of recurrent conditions in the lake.

The use of uniform terminology for all data employed is important, and some definitions that have proved to be of particular value are given below:

1. The concentration is given exclusively in, or converted to, milliequivalents per litre (meq l^{-1}).

2. The total concentration is taken to be the sum of the cations $Na^{1+}+K^{1+}+Mg^{2+}+Ca^{2+}$. The sum of the anions, $Cl^{1-}+SO_4^{2-}+$ alkalinity, equals the sum of the cations. Other components are ignored since they are present in such negligible quantities that they have no effect on the ion balance.

3. The alkalinity A (in meq l^{-1}) is given by:

$$A\times10^{-3} = [HCO_3]+2\,[CO_3^{2-}]+[OH^-]$$

The brackets indicate gram-mol per litre. It should be borne in mind that the alkalinity, sometimes termed 'carbonate hardness', need not necessarily be a measure either of hardness or of alkalinity, but is merely a nonspecific indication of the number of equivalents of bases bound to weak acids. Any metal ion can fulfil the function of the base, and in natural waters the acids may be carbonic acid, silicic acid and sometimes boric acid.

4. The sum of the alkaline earths Ca and Mg is denoted by E in the following (total alkali earths). The expression 'total hardness' has been avoided to prevent confusion with the 'degree of hardness' thus: E = meq l^{-1} (Ca+Mg).

5. The $(E-A)$ value: in a pure solution of carbonates of Ca and Mg the alkalinity is equal to E, and the difference $E-A = 0$. Any increase ($E-A$ positive) indicates that Ca or Mg or both are present in the water in the form of salts of stronger acids, as chlorides, sulphates, or nitrates (mineral acid hardness). If $E-A$ is negative, however, this means that alkali salts (K, Na carbonates or silicates) are present. Since sodium is easily the dominant ion and silicates are converted to carbonates by the ubiquitous following definition for an approximate estimate: a negative $(E-A)$ value, if its sign is changed, is equivalent to the soda content.

$E-A$ is a most useful value in limnochemistry and hydrogeology, and has the additional advantage of being simple to determine. Limnochemical processes are dominated by the CO_2 consuming assimilation of aquatic plants and by CO_2-producing reactions in the mud. The equilibrium system existing between air carbon dioxide and calcium carbonate is kept in a dynamic state by change of temperature and irradiation. Whenever the solubility product of $CaCO_3$ (10^{-8}) is exceeded, 'biogenic' precipitation of calcium takes place. In addition, the aerogenic precipitation of calcium, which is dependent upon the CO_2 equilibrium between water and air, also plays a role. The important point is that E and A are both diminished by exactly the same amount when $CaCO_3$ is precipitated so that $E-A$ remains unchanged. The same applies for the reverse situation which can arise when $CaCO_3$ is dissolved from the mud or from incrustations on aquatic plants. E and A are again changed by the same amount and in the same sense so that once more $E-A$ remains unaltered and can thus be regarded as an invariable in limnological processes.

Strictly speaking, the above only holds true under aerobic conditions. In anaerobic regions a sulphate reduction, for example, is possible and the sulphide formed raises the alkalinity, A increases, E remains constant and thus $E-A$ decreases. In water containing H_2S, far down in meromictic lakes, such a change in $E-A$ can be observed with increasing depth.

Apart from this, changes in $E-A$ indicate the inflow of water with a different $E-A$. The $E-A$ that results according to the laws of mixing also makes it possible to deduce the $E-A$ of the unknown inflow.

The hydrochemical components of Neusiedlersee appear to remain fairly constant, with certain long-term fluctuations which should probably be interpreted as being due to differing conditions of equilibrium. A quantitative evaluation can best be made by selecting examples from the mass of data available, a method which experience has proved to be justifiable.

9.1 Surface inflow and the lake water

Fig. 9.1 shows the main ionic components of the few inflows and of the lake water itself. The values have been selected from data obtained over the past 20 years. The Ca:Mg ratio in the inflows is almost 'dolomitic', i.e. 1:1, whereas the Mg values in the lake water are nine to ten times those of calcium. The total concentrations in Wulka and Kroisbach are very similar, at about 10 meq l^{-1}, whereas the Purbach, with about 4 to 5, has the lowest

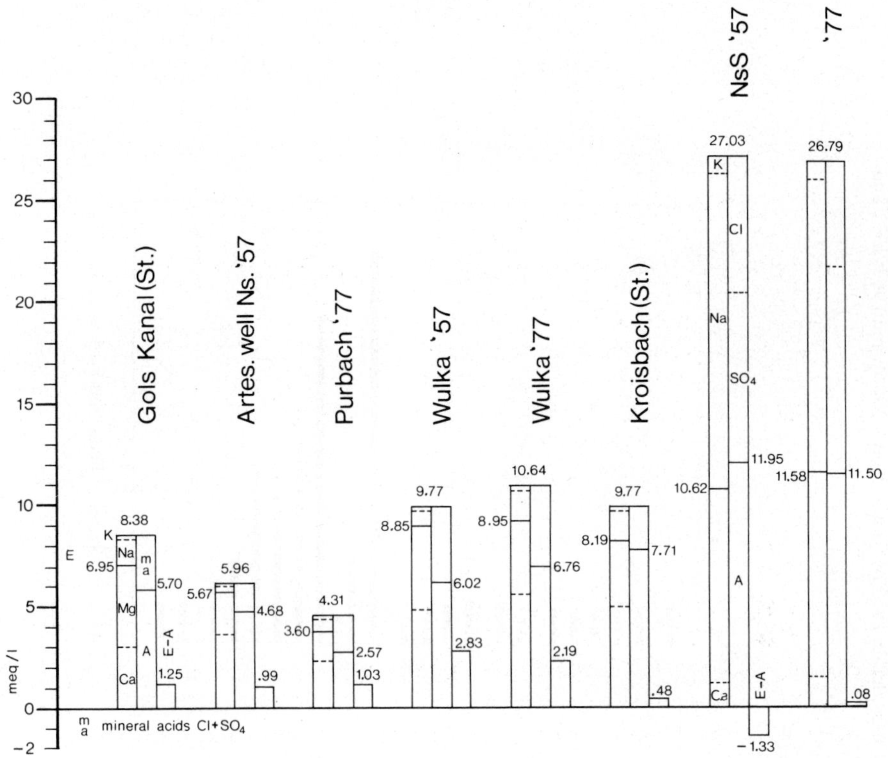

Fig. 9.1. Surface inflows and lake water: main constituents, A = alkalinity, E = Ca+Mg (Stehlik 1972).

concentration. The water in the artesian well near Neusiedl is a so-called alkaline earth-bicarbonate water containing very little Na and K. All inflows exhibit a positive $E-A$.

Total concentrations in the lake water off Rust were found to be strikingly similar after an interval of 20 years, and even the alkalinity had changed only slightly. The $E-A$ values, however, were negative in 1957, not only in Rust but in the entire area, and the lake contained 1–2 meq 1^{-1} of soda. Twenty years later E and A were almost equal and the $E-A$ was only just positive. On account of the large amount of CO_3 ions the Ca content is low, whilst the ion present in the largest quantities is sodium (up to 15).

Before considering the various changes undergone by inflowing water before it becomes 'lake water' some earlier data and some results of the present authors' own investigations made between 1967 and 1971 and in 1977 will be presented (Fig. 9.2).

The series starts with the analysis of Sigmund and Würtzler from year 1830 (cited in Emszt 1904) and ends with the data from 1977 shown in Fig. 9.2. The total concentration of 25–30 meq l^{-1} observed in 1830 was also found in the fifties and again in the seventies (Fig. 9.3). Evaporation and water exchange oscillate around this dynamic equilibrium. In the older data the soda contents of 12 and 11 meq 1^{-1} obtained by calculation are surprisingly

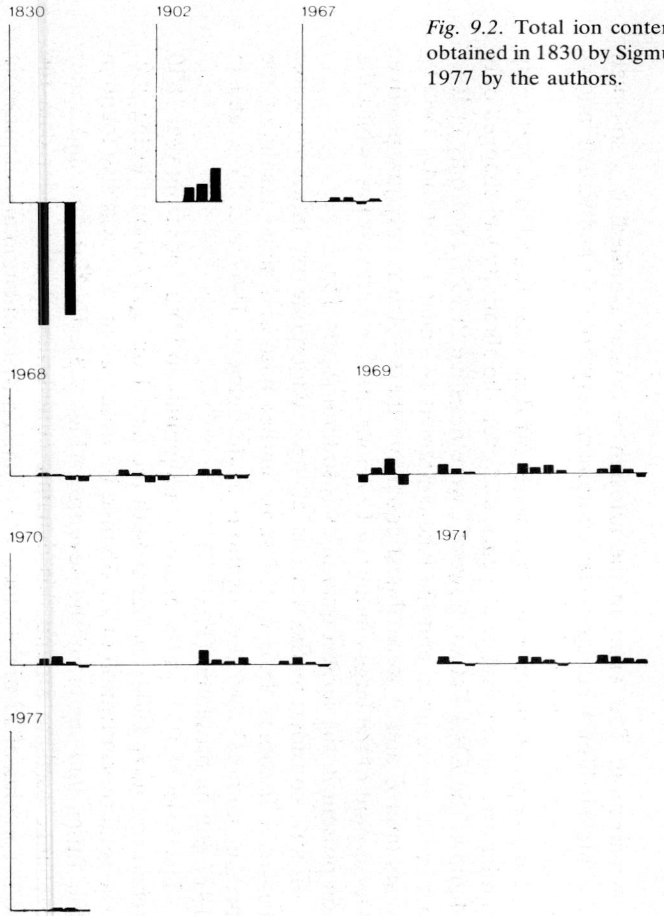

Fig. 9.2. Total ion content and $E-A$ value. Some examples recalculated from analytical data obtained in 1830 by Sigmund and Würtzler, in 1902 by Emszt (Emszt 1904), and 1967–1971 and 1977 by the authors.

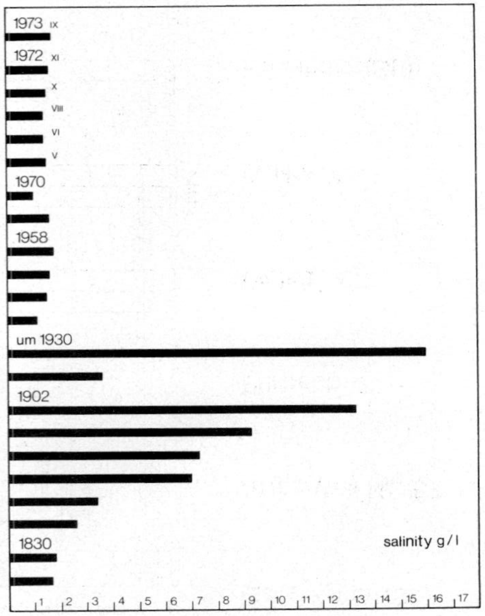

Fig. 9.3. Salinity. Data obtained as in Fig. 9.2.

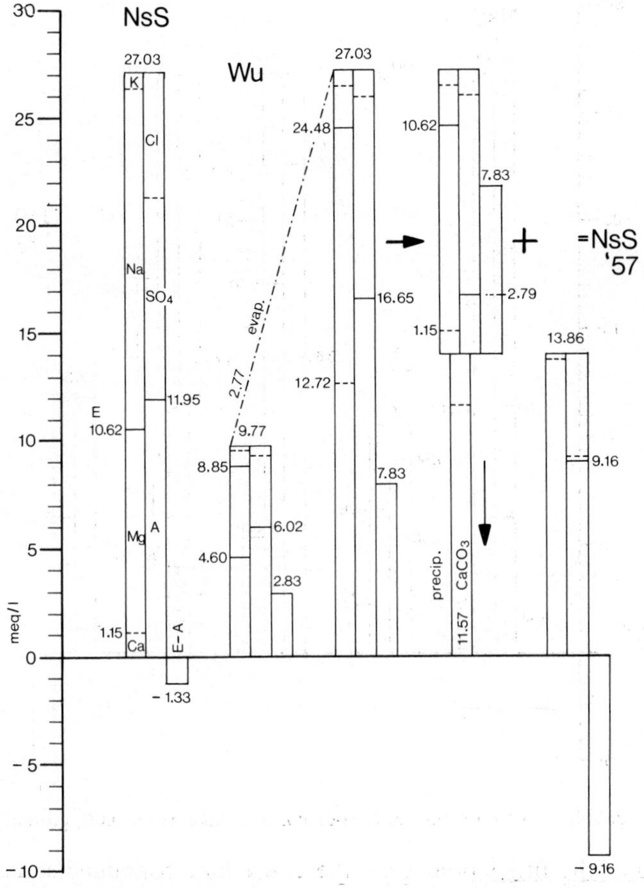

Fig. 9.4. Stoichiometric pattern for conversion of Wulka river water, *Wu*, into lake water, *NsS* (1957): concentration by evaporation, precipitation, addition of 'calculated' groundwater, and final concentration (not in the graph). Soda content: lake water 1.33, groundwater 9.16.

high, although much higher soda contents occur in the salt pans of the Seewinkel and in interstitial water of the lake mud.

When, at the turn of the century, the water level of the lake was again very low, according to Emszt (1904), the conditions in the lake were remarkably different. The sites from which samples were taken between Rust and the southernmost part of the lake were no longer connected by water. This is apparent from the fact that the total concentrations measured differed by as much as 1:5 and even reached 200 meq l^{-1}. In all samples (three are shown in the diagram) $E-A$ was as high as 26 meq l^{-1}, and soda was nowhere present. These values were in the main due to the predominant role played by the evaporation residues of surface inflows.

From 1957 to 1959 Stehlik (1972) carried out thorough and precise investigations, some of which were continued up to 1970 and included studies on inflows. Data from our own sporadic observations, made independently and in ignorance of Stehlik's work, are in complete agreement with his results.

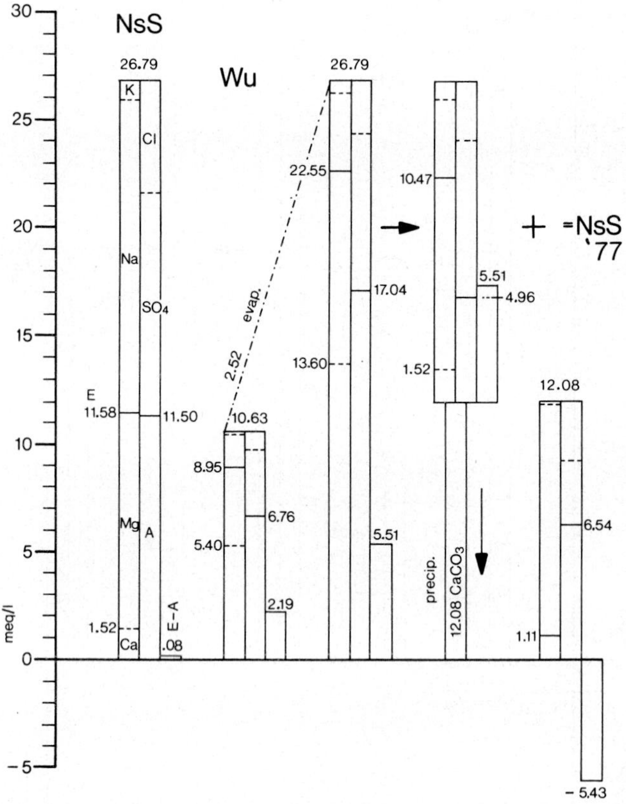

Fig. 9.5. As Fig. 9.4, with data from 1977. Soda content: lake water 0.00, groundwater 5.43.

At the end of the fifties practically the entire lake contained soda, although ten years later it was only detectable in a few places.

Of the investigations carried out regularly from 1967 to 1971 the results from only four sampling sites (from north to south and distributed over the seasons) are presented in the graph. Wherever Na and K values are available total concentrations have also been given. The latter fluctuate between 16 and 24 meq l^{-1}, that is, several units lower than in 1957 and 1977. A distinct increase is detectable from north to south, and the $E-A$ values suggest a tendency towards soda content in the same direction.

Apart from periods of desiccation the total concentration in the lake water seems to oscillate between two positions, the higher of these between 26 and 30 meq l^{-1} and the lower at about 16–24 meq l^{-1}. It is not clear whether the higher value is connected with a larger soda content. In 1977 the total concentration again attained the values observed in 1957, although in contrast to the situation in the latter year the $E-A$ value was not negative. In the intervening years, particularly towards the end of the sixties, $E-A$ was mainly positive.

So much for the empirical data, the question that now has to be considered is how the water from the perpetually inflowing streams and rivulets is altered to attain the very different composition of the lake water and how much of

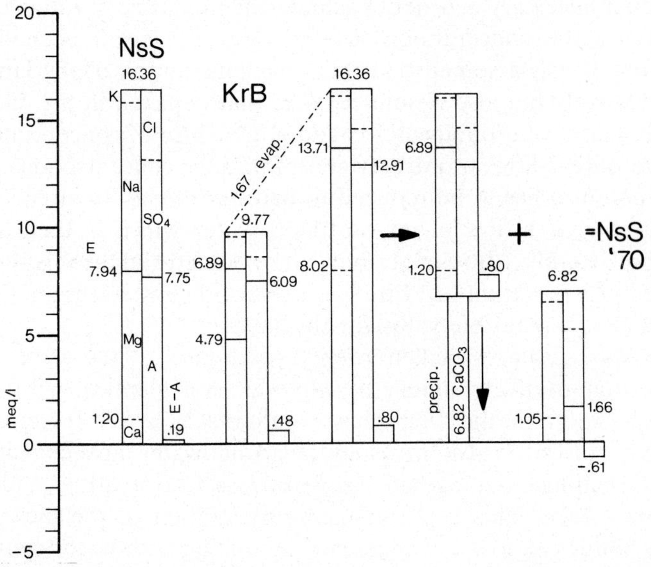

Fig. 9.6. As Fig. 9.4, but for Kroisbach (*KrB*), the southernmost inflow, and lake, *NsS* (1970), calculated from analytical data given by Stehlik (1972). The lake was in a state of rather low concentration with no soda content. Groundwater 0.61 soda.

which salts are steadily added in order to effect this continuous transformation.

Inflowing water is subjected in the lake to:

(1) evaporation;
(2) aerogenic and biogenic decalcification;
(3) addition of groundwater of a composition such that, after mixing, the exact ionic relationships of lake water are achieved.

In nature these three processes proceed simultaneously. In theory, however, they can be separated and considered individually. Three examples are presented: Wulka and lake, 1957 (Fig. 9.4); Wulka and lake, 1977 (Fig. 9.5); Kroisbach and lake, 1970 (Fig. 9.6). From Fig. 9.4 it can be seen that the total concentration is 27.03 meq l^{-1} in the lake water and 9.77 meq l^{-1} in the Wulka. The Wulka water is concentrated by evaporation until the lake concentration of 27.03 is reached, that is 2.77 times. All concentrations are increased by this factor; the solubility product of $CaCO_3$ is well exceeded and precipitation ensues. This proceeds until the E value of the lake water is attained, and this is 13.86 meq lower than that of the concentrated Wulka water. In solution remain 1.15 units of Ca: of the 12.72 meq l^{-1} Ca only 11.57 are precipitated, and the difference between this figure and 13.86 is covered by $MgCO_3$, precipitating at the same time. This process of precipitation leaves the $E-A$ value unchanged. The next step after precipitation is the addition of groundwater to the Wulka water, as shown in the last column. This groundwater contains almost solely NaCl and 9.16 meq l^{-1} sodium carbonate. Of this, 7.83 meq l^{-1} are needed in order to adjust the positive $E-A$ to the value of -1.33 found in the lake water. In fact a second phase of evaporation is necessary in order to achieve the final concentration of the lake water because the concentrated Wulka water has again been diluted by groundwater. If it is assumed that the ionic constitution of the latter agrees not only relatively but also absolutely with that depicted in the diagram this means that 1 litre of groundwater is added to 1 litre of concentrated Wulka water to produce 2 litres of mixed water of half the concentration of the lake water. It therefore has to be reduced to half its volume. In all, 2.77 litres of Wulka water give 1 litre of concentrated water which is then altered by precipitation. To this, 1 litre of groundwater is added and the mixture reduced once more by evaporation to 1 litre. Concentration is therefore in the ratio of 3.77:1, which is a plausible value for this lake.

Fig. 9.5 shows samples taken twenty years later at the same sites. The results are quantitatively similar to, although not identical with, the older data. Evaporation in the first phase amounts to 2.52, the precipitation consists solely of $CaCO_3$ (no Mg) and the groundwater must contribute more chloride and sulphate but less soda (nevertheless 5.43 meq l^{-1}), and the result is a soda-free lake. This is a convincing illustration of the fact that soda-containing water can lose its 'character' by mixing with water containing no soda, by reason of a switch from negative to positive $E-A$ values.

Fig. 9.6 also shows an example from the south of the lake, at the entry of the Kroisbach (analysis data from Stehlik 1972). A soda-free lake water results despite the fact that the groundwater component must contribute 1.66 meq soda for every litre of Kroisbach water, as well as sodium sulphate and sodium chloride, in order for the composition of the lake water to be achieved. $E-A$ remains positive and the total concentration following the precipitation of 6.82 meq of $CaCO_3$ is 16.36.

These examples were chosen to illustrate the fact that the surface inflow is continually modified by large and calculable quantities of salts entering the lake in the groundwater. The latter must be of the rare 'soda' type from which sodium carbonate and other salts would crystallize out on evaporation.

It should be borne in mind here that in earlier publications salts were listed which were in practice of no real significance since the solubility products had not been taken into consideration. From his data, for example, Emszt (1904) constructed a salt mixture 'in the usual manner', in which sodium bicarbonate appears, although stoichiometrically calcium and magnesium salts predominate and in fact conversion would occur in solution and upon evaporation. Fig. 9.2 shows that the water samples analysed by Emszt were actually soda free.

Although the hydrology of the Neusiedlersee area has to a large extent been elucidated in recent years (Gattinger 1974a, 1974b, 1975), the underlying hydrochemical reactions and the quantitiative relationships between the ionic ratios in inflows, lake water and infiltrating groundwater, were long the subject of mere speculation. Although from the north groundwater containing soda cannot be expected, it may well arise mainly from the southwest and also from the Seewinkel in the east, of which the groundwater movement is known. The latter region contributes about 25 million m^3 of water annually.

The prevailing west winds and the rare but strong east winds help to mix the water of the lake along its east-westerly axis. Almost all analyses along the north-south axis reveal a distinct gradient in $E-A$ values, already seen in Fig. 9.2. A study of 640 hydrochemical determinations from the years 1967 to 1971, made over the entire Austrian portion of the lake, reveals 60 cases with a clear soda content. Forty-six of these samples were taken from sites south of the line joining Mörbisch and Illmitz.

Diffuse groundwater input becomes obvious under the ice, in the absence of wind-induced water movements. This was the case in the winter of 1958 when a number of samples taken between Rust and Mörbisch revealed a quite definite stratification within 10–30 cm of water depth. There was no noticeable increase in soda content with depth. However, it did increase towards the south, just as in the ice-free lake. The ice covering in February facilitated sampling from the entire lake with the exception of the Hungarian part. The water was found to be free of soda in 82 places, and only in 2 sampling sites in the south was a small amount detectable.

Conditions of extreme stagnation prevail in the interstitial water of the lake mud so that steep gradients develop. Schroll & Krachsberger (1977) carried out mineralogical and chemical investigations on cores of 15 cm length from three horizons (15–30, 60–75 and 120–135 cm) and from 5 sites in a transverse profile off Rust, as well as from one site near the entry of the Wulka. Interstitial water was extracted from 100 g of mud with 1 litre of distilled water and analysed. The total concentrations and $E-A$ values calculated from the water content and extraction values are of considerable interest and are shown in Fig. 9.7.

1. As compared with the lake water (18 meq l^{-1}), the interstitial water is highly concentrated, the values rising to about 10 times in the open lake and to about 30 times in the reed zone with the highest concentration in the east (520 meq l^{-1}), leading here to the salt soils of the Seewinkel.

2. The soda content also exhibits a gradient in the same direction. Core II from the reed near the lake is soda free with the exception of the lowest horizon. Core I from the reed near the canal of Rust exhibits highly negative $E-A$ values in all horizons. Unless this effect is the result of a concentration resulting from the poor exchange conditions in the reed zone, an infiltration of soda-containing groundwater might be suspected since in 1977 the open canal of Rust was found to contain soda near the lake shore, although there was none in the open lake itself. But investigations in 1967–1971 show that the distribution of $E-A$ in the lake must be influenced by soda addition from the southwestern region.

3. Core VI near the entry of the Wulka into the lake has a 'normal' concentration profile. The horizons apparently are subject to the influence of the groundwater flow accompanying the Wulka.

4. The cores from III, IV and V indicate the enormous concentration gradient starting at the Seewinkel shore. The soda solution only 20 centimetres below the mud surface exceeds 100 times the quantity found in the lake. In spite of the high soda content the transport of soda into the lake water must be much lower than in the southwestern part of the lake.

5. In almost all core profiles a certain degree of concentration is seen in the

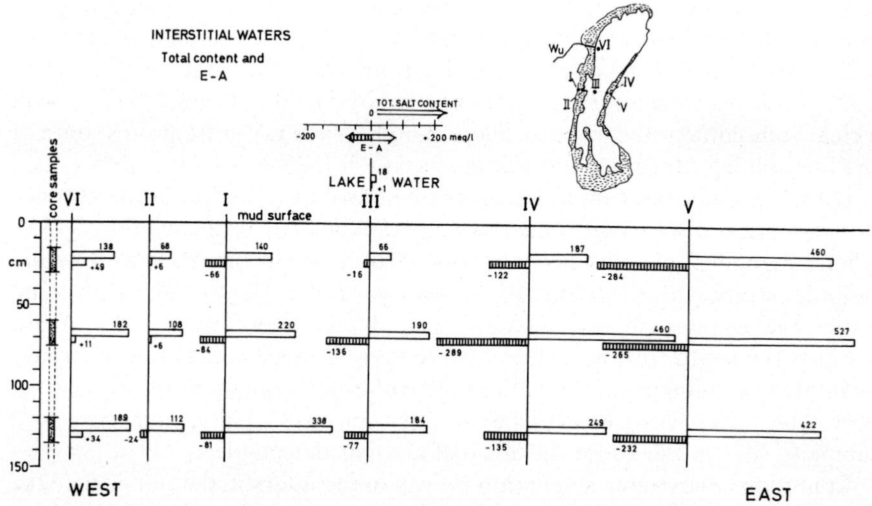

Fig. 9.7. Interstitial waters, total ion content and $E-A$, calculated from analytical data given by Schroll and Krachsberger (1977). Cross-section eastwards from Rust: I–V. Core VI was drilled near the mouth of Wulka river. The total ion content of the interstitial water samples is up to thirtyfold higher than in the supernatant lake water (at Core III). The soda content shows remarkable gradients from east to west, and from 70 cm depth in the mud upward to the mud surface.

middle horizons (60–75 cm). This may result from the greater horizontal infiltration at this level. Furthermore, the washing-out effect of the turbulent lake water is less here than in the upper horizons.

Hydrochemically Neusiedlersee can be considered as a gigantic shallow mixing pan into which normal mountain water flows from the north and west whilst from the southwest, south and east soda-containing groundwater infiltrates via the lake bottom. The inflowing 'normal' mountain water, which can also enter the lake as groundwater, is continually transformed into Neusiedlersee water by means of diffusion and exchange. Although the ionic relationships in the lake undergo changes at longer intervals a characteristic picture is repeatedly reestablished.

The salts of the lake water originate therefore from low-concentrated mountain water with a relative high silicate content and from high concentrated water ascending from marine and brackish sediments. The ionic composition of these high-concentrated waters differs very much depending on the horizon from which they come. An explanation of the soda fed to the lake can be founded in the occurrence of silicates, particularly feldspars, with a high proportion of alkali (plagioclases) exposed to hydrothermal breakdown. Alkali silicate solutions are also produced by weathering of volcanic rocks and tuffs. Ubiquitous carbon dioxide converts the alkali silicate solutions into alkali carbonate and free silicic acid. The increase of the soda content in the south leads to the supposition of a stronger soda addition in the south district of the lake. This is verified by chemical analysis of spring waters, with very high soda concentrations, running from the Hungarian side into the lake (for example the spring of Hegyköi; Pásztó).

New data are continually being added to the wealth of results already available on springs, wells and groundwaters. Their interpretation will help to elucidate the origin of and paths taken by the soda-containing groundwaters, the presence of which is in fact 'labelled' by the negative $E-A$ values.

References

Berger, F. 1971. Zur hydrochemischen Charakterisierung von Sodagewässern. Sitz. Ber. Öst. Akad. Wiss., math. -nat. Kl. 179: 171–181.

Emszt, K. 1904. 2. Mitteilung aus dem chem. Labor der agro-geologischen Aufnahmsabt. der kgl. ungar. Geol. Anstalt. Jahresber. kgl. ung. Geolog. Anstalt 212–224.

Fritsch, V. & Tauber, A. F. 1959. Die Mineralwässer des Neusiedler Seegebietes. Wiss. Arb. Bgld. 21: 34–39.

Gattinger, T. 1974a. Das Werden des Seeraumes. In: Löffler, H. (ed.), Der Neusiedlersee. Molden, Wien.

Gattinger, T. 1974b. Das Kernproblem des Sees. In: Löffler, H. (ed.) Der Neusiedlersee. Molden, Wien.

Gattinger, T. 1975. Das hydrogeologische Einzugsgebiet des Neusiedlersees. Verh. Geolog. B. anst. 4: 331–346.

Knie, K. 1959. Über den Chemismus der Wässer im Seewinkel und des Neusiedlersees. Wiss. Arb. Bgld. 23: 65–68.

Löffler, H. 1959. Zur Limnologie, Entomostraken- und Rotatorienfauna des Seewinkelgebietes (Burgenland, Österreich). Sitz. Ber. Öst. Akad. Wiss., math.-nat. Kl. 168: 315–362.

Löffler, H. 1974. Der Neusiedlersee. Naturgeschichte eines Steppensees. Molden, Wien.

Neuhuber, F. 1971. Ein Beitrag zum Chemismus des Neusiedlersees. Sitz. Ber. Öst. Akad. Wiss., math.-nat. Kl. 179: 225–231.

Neuhuber, F. & Schroll, E. 1974. Chemie, Salz- und Nährstoffhaushalt des Sees. In: Löffler, H. (ed.), Der Neusiedlersee. Molden, Wien.

Schroll, E. 1959. Zur Geochemie und Genese der Wässer des Neusiedler Seegebietes. Wiss. Arb. Bgld. 23: 55–64.

Schroll, E. & Krachsberger, H. 1977. Beitrag zur Kenntnis des Chemismus der Porengewässer des Neusiedlerseeschlammes. Biol. Forsch. Anst. Bgld. 24: 35–62.

Stehlik, A. 1972. Chemische Topographie des Neusiedlersees. Sitz. Ber. Öst. Akad. Wiss., math. nat. Kl. 180: 217–278.

Stehlik, A. 1975. Chemische Untersuchungsergebnisse vom Neusiedlersee aus den Jahren 1971–1974. Biol. Forsch. Anst. Bgld. 13: 86–99.

0 Phosphorus and nitrogen

F. Neuhuber, H. Brossmann and P. Zahradnik

10.1 Introduction

Hydrochemical investigations on Neusiedlersee have been made over a great number of years. However, at the beginning of this investigation in 1970 there were no published data on nutrient conditions of the lake.

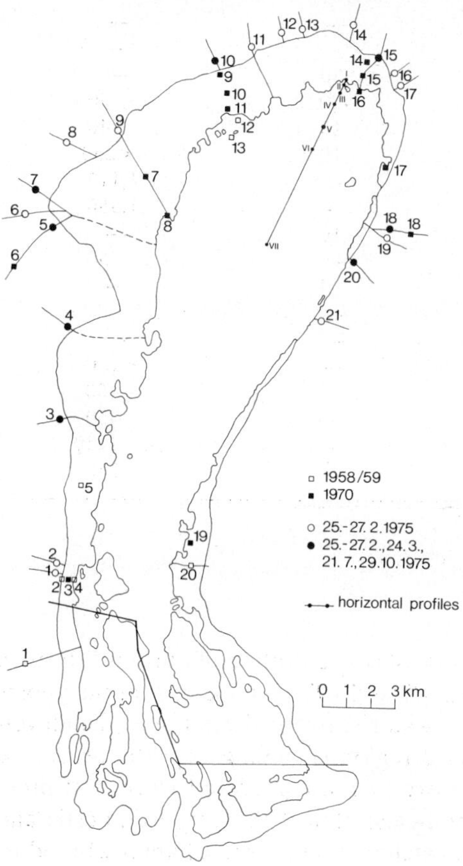

Fig. 10.1. Map of Neusiedlersee showing sampling stations (□ ■ for Table 10.1, ○ ● for Tables 10.3 and 10.4, —●—●— for Fig. 10.3.

According to data published by Stehlik (1972) (Fig. 10.1, Table 10.1) the reed belt receives considerable quantities of nutrients from the inflows (Fig. 10.1, Table 10.1). Within the reed belt itself the concentration of nutrients is rather high and drops gradually towards the open lake. As a consequence the

Table 10.1. Stehlik's analysis (1972) from twenty sampling points (Fig. 10.1).

Open lake	PO_4^{2-} (μg P l^{-1})	NO_3^- (μg N l^{-1})	NH_4^+ (μg N l^{-1})
1958/1959	0.5–3.0	20–30 100 200	40–120
1970	1.5	50	90–190
Inputs reed belt, bays			
1 I	190	3,160	1,400
2 C	140	—	390
3 C	—	—	630
4 C	—	—	230
5 R	70	—	390
6 I	480	3,390	80
7 C	10	740	310
8 C	10	<230	50
9 C	—	1,113	230
10 C	—	1,350	1,320
11 C	10	—	—
12 B	—	50	90–190
13 B	4	—	—
14 R	—	—	230
15 R	—	—	160
16 B	—	<110	120
17 R	—	450	50
18 C	30	450	80
19 R	—	680, 1130	60, 310
20 C	—	<230	200

I: natural inflows
C: channels
R: reed belt water
B: bay water

open lake has the lowest concentration of nutrients. From a comparison of the data for the years 1959 and 1970 it appears that the phosphorus and nitrogen content of the lake water scarcely altered over the eleven years.

The Neusiedlersee region is gaining importance as a recreation area, and certain parts have even been considered worthy of preservation as natural reserves. But nonetheless almost the entire domestic and industrial sewage from the northern catchment area (see Chapter 2) used to flow into the reed belt or directly into the lake itself. Sewage plants constructed recently should reduce the nutrient content in the lake. Work carried out over the past few years will show to what extent the natural balance is threatened by human activity.

Investigations included measurements of phosphorus fractions, Kjeldahl nitrogen, nitrate and nitrite in the open water, phosphorus, Kjeldahl nitrogen and some determinations of organic carbon and amino acids in the uppermost sediment layers, as well as phosphorus, nitrate and nitrite in the inflows. After

Table 10.2. Summary of the methods used.

	Water	Sediment
Sampling	Water from ½ m depth	Each 1 cm of the upper layer
Sampling tools	Ruttner sampler	Gilson sampler
Treatments	Filtration: membrane filters (Sartorius) 0.45μ	Drying cupboard (105°C), pulverization
Analysis	Unfiltered water: total phosphorus (P_t) filtered water: soluble phosphorus (P_s) orthophosphate (P_o) nitrate nitrite	Weighed sediment: total phosphorus Kjeldahl-nitrogen (Kj-N)
Methods	Usual procedure of EAWAG (1974)	Golterman (1969)

preliminary investigations in 1970/1971, which involved only a small portion of the lake, the programme for 1972/1973 covered almost the entire Austrian part of the lake. Only the Wulka was studied at first, but in 1975 an investigation covered almost all inflows and canals. Subsequently, repeated analyses were carried out on selected inflows. The methods employed are summarized in Table 10.2.

10.2 The inflows

As shown in Table 10.3 the water entering the lake brings a considerable load of phosphorus, nitrate and nitrite. The highest concentrations can be detected in the canals that receive influents from sewage plants, but due to the small flow rates, the calculated loads are rather low. If the inflows are arranged numerically with respect to their flow rates the order of flow rate corresponds on the whole with the order of the load carried into the lake (Table 10.4).

For the inflows where the flow rate is small, differences in this rate seem to be of little significance for loading changes. The order appears to depend more upon the actual concentrations present in Table 10.4, for example inflow 3 has high and inflow 7 low concentrations, while inflow 4 has a greater N-load.

The main part of the phosphorus load comprises dissolved phosphorus, P_s, with a highly fluctuating orthophosphate, P_o, portion (P_s being 80–100 per cent, in inflow 19, 70–95 per cent, and in inflow 5, 50–80 per cent). The percentages of P_o and particulate phosphorus, P_p, in the total load tend to increase with the flow rate. Since the latter is not only determined by precipitations but also by uncontrolled addition of waste water the values show large deviations. The same holds for the N/P ratio (N from $\Sigma(NO_3+NO_2)$: P_o, P_s or the total phosphorus content, P_t respectively), which also shows a tendency to increase with greater water flow on account of the readier solubility of nitrogen compounds and enhanced ammonium oxidation.

Table 10.3. Phosphorus, nitrate and nitrite concentrations (mg, P N 1^{-1}) and load s(mg P, N sec^{-1}) of the inflows sampled before running into the reed belt (sampling stations in Fig, 10.1).

	Concentration					Load				
Station	P_t	P_s	P_o	NO_3	NO_2	P_t	P_s	P_o	NO_3	NO_2
1	6.9	6.85	0.7	4.9	0.4	24	24	3	18	1.4
2	13.0	13.0	4.0	0.4	0.01	65	65	20.2	2	0.03
3[x o]	3.3–11.3	3.3–10.5	1.4–6.0	2.0–13.0	0.01–0.55	47– 150	46– 142	10–105	21– 390	0.4– 9
4[x o]	3.6– 5.7	3.4– 5.4	0.7–3.3	1.4–22.4	0.02–1.04	32– 178	29– 169	6–145	13–1,120	0.2– 52
5[o]	0.7– 3.1	0.5– 1.5	0.3–1.4	3.0– 6.6	0.05–0.12	414–1,950	330–1,095	73–860	660–7,560	12–518
6	–	–	1.3	7.6	0.01	–	–	3	15	0.01
7[o]	1.6– 2.4	1.5– 2.4	0.2–1.2	2.9– 5.1	0.06–0.35	27–45	26–42	3–16	40–109	1–5
8	–	–	0.1	4.7	0.01	–	–	0.2	7	0.01
9	4.0	3.7	3.0	19.5	0.60	40	37	30	195	6
10[x o]	4.4–12.4	3.5–10.8	1.9–7.9	3.2–11.9	0.26–2.65	53–74	44–64	19–47	29–149	2–26
11[x]	–	–	1.2	9.5	0.02	–	–	47	380	0.7
12	9.1	7.5	2.3	8.2	0.66	59	48	15	53	4
13	7.1	6.8	4.8	2.1	0.07	14	14	10	4	0.1
14	4.9	4.4	1.1	13.8	0.07	135	121	31	379	2
15[o]	0.4–1.7	0.2–1.5	0.2–0.8	1.8–25.0	0.07–0.85	35–315	14–252	8–103	161–4,700	5–155
16	–	–	6.7	0.7	0.09	–	–	3.4	0.4	0.05
17[x]	–	–	3.0	7.0	0.13	–	–	60	140	3
18	0.1	–	0.1	6.6	0.01	2	–	1	165	0.1
19[o]	0.2–0.9	0.2–0.9	0.2–0.5	0.5–15.0	0.01–0.20	38–312	28–254	18–188	139–7,800	1–88
20[x o]	1.6–5.3	1.5–5.2	1.0–5.0	0.4–25.0	0.09–0.16	22–240	22–232	15–141	5– 500	2–5
21	–	–	0.1	2.0	0.01	–	–	2	60	0.3

[x] Canals receiving effluents of treatment plants, sampling dates 25–27 February 1975.
[o] 25–27 February, 24 March, 21 July, 29 October 1975.

Table 10.4. Inflow stations arranged according to flow rates and loads.

Flow rate	Load P_t	P_s	$N(NO_3+NO_2)$
5	5	5	5
19	19	19	19
15	15	15	15
20	3	3	3
7	20	20	4
4	4	4	20
3	10	10	7
10	7	7	10

The ratio $N:P_s$ calculated from mean loads (geometric means) shows that effluents from sewage plants, with the exception of inflow 15, carry higher phosphorus loads ($N:P_s$ =1–1.5) than the other inflows ($N:P_s$ = 2.5–7.0). In accordance with the high proportion of P_s the P_s/P_p ratio in these effluents is also high ($P_s:P_p$ = 15–35). When the P_s/P_p ratio is low, that of N_{NO_3}: N_{NO_2} is also small and this may indicate that the sewage plant (inflows 15 and 10) is overloaded.

The low P_s/P_p ratio (6.3) and high N_{NO_3}: N_{NO_2} ratio (77) of the Golser canal (inflow 19) that flows for about 10 km through intensively cultivated agricultural land, reflect the influence of manuring. The Wulka (inflow 5), the only natural inflow on the Austrian side, has the lowest P_s/P_p ratio (2.5). Its N/P_s ratio of 4.8 suggests that domestic and industrial waste plays a greater role in the loading than in the Golser canal. Investigations carried out in 1972 showed that the highest loading occurs at the time of year when the sugar beet is processed (higher P_t content and low proportion of P_o).

An attempt to measure the decrease in P_o following entry into the reed belt revealed a decrease of 75 $\mu g\ l^{-1}$ over a distance of 500 m, the curve for the decrease becoming progressively steeper. It can therefore be assumed that the interior of the reed belt has a considerable capacity to take up P_o.

10.3 The uppermost sediment layers

10.3.1 *Phosphorus*

The mean values obtained for the total inorganic and organic phosphorus compounds in the uppermost sediment layers (0.55 parts per thousand)[1] are lower than those reported for rock, the figure for basic rock being 0.7 to 1.3 parts per thousand (Golterman 1975). Sixty-three per cent of the samples gave values between 0.5 and 0.6 parts per thousand (see Fig. 10.2).

The phosphorus content is related to the distribution of hard and soft bottom, which in turn is due to the effects of wind. Soft mud, which is formed by abrasion of the lake bottom and by biogenic precipitation of the carbonates of alkaline earth (Geyer & Mann 1939), is gradually carried to the reed

1 Dry weight, as are all such figures in this chapter.

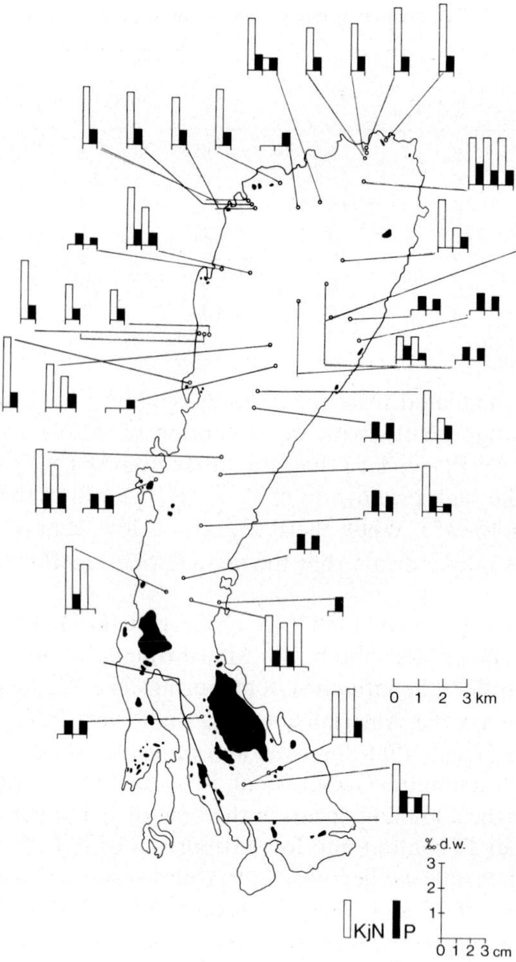

Fig. 10.2. Phosphorus and nitrogen content of the uppermost sediment layers (1972).

belt. Sedimentation can then proceed undisturbed (Tauber 1959). The shore zone is bordered by a thick layer of soft mud. In the centre of the northern part of the lake, a hard bottom predominates, although it may be covered sometimes and in places by a very thin layer of soft mud. Hard and soft muds appear to form a mosaic in the middle of the lake and in the southern part the lake bottom is covered by a thick layer of soft mud (see Chapters 13 and 23).

The phosphorus content of the uppermost soft mud layer amounts to approximately the highest values cited above (0.61 parts per thousand), whereas in the upper hard bottom and lower soft mud layer the values are at about the lower limits given above (0.49 parts per thousand). Higher figures were found for the upper soft mud layer of Neusiedl, Rust, in the southern part of the lake, and off Podersdorf (0.70 parts per thousand). Lower values were found in the region where the Wulka enters the lake (0.43 parts per thousand). In the latter region as well as in the central northern part of the lake a large number of samples from the lower soft mud layer and the hard bottom gave low phosphorus values (0.20–0.38 parts per thousand). In view of the physico-chemical properties of the sediment and of the lake water, the high content of inorganic colloids, the additional precipitation of alkaline earth-carbonates (Wieden 1959), pH (8.5–9), alkalinity (9.5–10 meq l^{-1}), and finally the mixing of water sediment (inorganic sediment forms the largest portion, by weight, of the suspended matter) it seems likely that, apart from the binding of phosphorus by organisms, inorganic phosphorus fixation may also take place. This fixation may be due to precipitation, anion exchange, or adsorption (Hepher 1958).

A series of temperature experiments in which it could be demonstrated by means of P^{32} (as PO_4 in HCl) that phosphorus uptake increased with rising temperature, may indicate the participation of an inorganic process in phosphorus elimination. Laboratory experiments with lake water carried out by Gunatilaka in 1976 (personal communication) showed that the seston is capable of removing all available phosphates in solution. The efficiency of the process seems to be governed by the quantity of seston in suspension and by pH. An attempt was made to quantify the phosphate uptake by abioseston and bioseston; the methods used included inactivation of the bioseston by chemical and physical methods. Experiments with tracers showed a close agreement with the results obtained from those above, and also showed that the inorganic uptake by seston was nearly five times the organic incorporation by bacteria and algae.

In spite of the increase of phosphorus in the water from 1972 to 1975 the phosphorus content of the sediment off shore was maintained, an indication that considerable transport of sediment into the reed belt occurs. This also provides favourable conditions for mineralization.

Investigations concerning the sedimentation of suspended matter in the reed belt using sedimentation cylinders revealed that the larger part is deposited within the first 100 m (Table 10.5). A very small portion may be

Table 10.5. Sedimentation in reed belt.

Exposure time of sedimentation cylinders	0 m		50 m		100 m		150 m	
Days	Wet (cm)	Dry (kg m^{-2})	Wet (cm)	Dry (kg m^{-2})	Wet (cm)	Dry (kg m^{-2})	Wet (cm)	Dry (kg m^{-2})
12	2.7	5.1	0.5	0.9	1	1.8	0.2	0.3
28	5.6	10.2	1.1	3.2	1.2	1.9	–	–
43	8.5	20.4	5.2	9.9	0.64	1.5	–	–
215 (67 ice)	37	–	29	–	0.5	–	0.4	–

transported as far as 150 m into the reed. The individual values given here differ greatly due to the dependence of sediment transport on the direction and strength of the wind. The phosphorus content of such sediments increases with distance from the open lake, the values being similar in magnitude to the upper values of the lake sediments (0.68–0.85 parts per thousand) although they may even rise considerably higher due to organisms and detritus (1.25 parts per thousand). A determination of the exchangeable phosphorus according to the procedure of Egner-Riehm (Schlichting & Blume 1966) showed that approximately 25 per cent of the sediment phosphorus can only be set free by anion exchange. The proportion of exchangeable phosphorus is slightly higher in the reed sediment than in that bordering the reed belt to the lakeward side.

10.3.2. *Kjeldahl nitrogen*

Kjeldahl nitrogen (Kj-N) of the uppermost sediment layers shows a mean content of 1.74 parts per thousand (Fig. 10.2). The region bordering the reed belt has a higher concentration (2.03 parts per thousand) than the open lake (1.61 parts per thousand). Also the uppermost sediment layer contains more (2.03 parts per thousand) than the lower layer (1.22 parts per thousand). This is in agreement with the normal lake situation, where the Kj-N decreases towards the open lake from the shore and also with the depth.

The same holds for the Kj-N/P ratio, which amounts to 4–5 in the uppermost sediment layer near the reed belt but drops in the open lake to 2–4, and sometimes even as low as 1, in the lower layers of the open lake.

Investigations have shown that amino acids account for 28 per cent of the Kj-N. According to Golterman (1975) this means that organisms account for 0.7 per cent, with a nitrogen content of 0.08 per cent (amino acids: 10 per cent, other organic N-compounds: 3 per cent of living material) and a carbon content of 0.4–0.5 per cent (C/N of living material: 5–6). Humus estimations revealed an organic carbon content of 3–5 per cent – Wieden (1959) obtained 2.2–5.0 per cent and Weisser (1970) 3.0–3.5 per cent. Thus about one tenth of the organic carbon and two thirds of the Kj-N are due to organisms. By far the larger portion of the carbon is present as humus, with a very small nitrogen content. The phosphorus content due to organisms is also very small (0.007 per cent, P = 1 per cent of living material, C/P: 60), amounting to only one eighth of the sediment phosphorus.

From the relatively insignificant amounts of phosphorus, Kj-N and carbon in the sediment and the small quantities of Kj-N-containing compounds not deriving from organisms, it seems likely that an almost complete mineralization of the organic compounds takes place, a process favoured by the continual removal of the uppermost sediment layers. The continuous movement of suspended material which only ceases with sedimentation in the reed belt, prevents the accumulation of nutrients in the sediment of the lake.

10.4 The lake water

10.4.1 *Phosphorus compounds*

The total phosphorus content, P_t, as well as its individual components exhibit considerable fluctuations in space and time. In the main, these fluctuations are the result of processes that depend upon the wind, such as the continual changes in the amount of matter suspended in the open lake, and the mixing of reed and lake water.

Such processes are clearly recognizable in horizontal profiles drawn from the edge of the reed out into the open lake (see I–VIII in Fig. 10.1, and Fig. 10.3). With increasing distance from the reed belt, and usually within a few

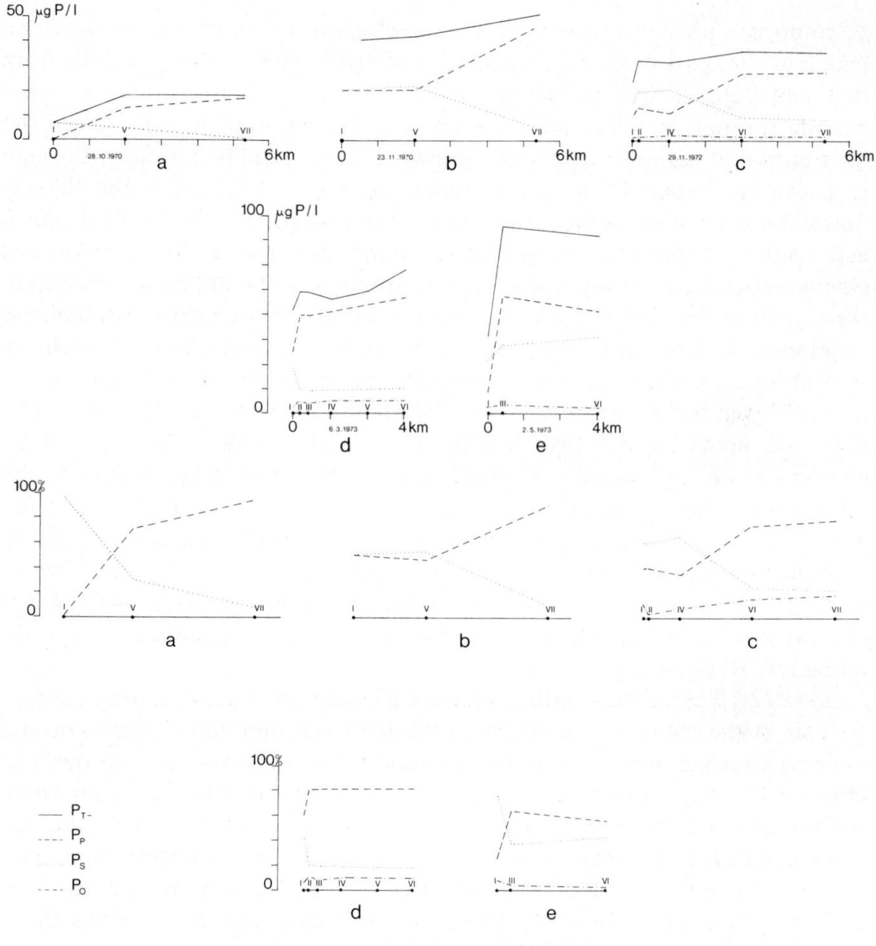

Fig. 10.3. Different forms of phosphorus and their portion expressed in percentages of total phosphorus, in horizontal profiles (see also Fig. 10.1).

hundred metres, P_t rises substantially due to an increase in P_p, whilst P_s and hydrolysable phosphorus, P_h (the latter calculated from the difference between P_s and P_o), decrease in the same direction. P_o, however, fluctuates only slightly and usually increases somewhat towards the open lake. The lake water contains a larger proportion of P_p and a smaller proportion of P_s than the reed water. As will be shown later, the higher P_s content in certain parts of the reed is largely due to P_s influx from the inflows emptying into or flowing through the reed belt. It can be assumed that the sewage from summer houses along the outer fringe of the reed accounts at times for a significant addition of P_s, although this is difficult to measure. The influence of tributaries on the P_s content of the lake is shown by a sample of lake water taken off Podersdorf at the same time as the sample of Fig. 10.2 (P_p as point VII, P_s:130 μg l^{-1}).

10.4.1.1 The distribution of dissolved phosphorus compounds

P_s comprises phosphorus compounds involved in reactions within the open lake plus that part of dissolved phosphorus drifting out to the open lake from the reed belt, as well as the phosphorus compounds from inflows either passing through the reed belt or emptying directly into the lake. Since the greater part of the inflowing water empties into the reed belt or flows through it, it can be assumed that it undergoes a process of filtration and that its nutrient composition undergoes additional changes as a result of biochemical action in the course of its delayed access to the open lake. If the extent of such changes depends not only upon the nutrient load of the inflow but also upon the retention time in the reed zone, that is, on the ability of the reed soil and vegetation to store water and nutrients, then the reed channels ought to exhibit higher concentrations than reed areas into which the inflowing water soaks. This in fact seems to be shown by Stehlik's (1972) data (Table 10.1).

In the areas of the lake where waste water enters, the P_s and P_o concentrations are usually higher than in the open lake, although the influence of the inflows is very variable. This shows up clearly under ice cover: in two series taken under ice (3 February and 13 March 1971) the P_o concentrations were below 2 μg l^{-1} in the open lake; they were slightly higher in the marginal zone of the reed uninfluenced by water inflow (4–7 μg l^{-1}) and attained their highest values in the region of sewage inflow (Neusiedl: 62 μg l^{-1}; Rust: 40 μg l^{-1}).

In 1972/1973, sewage influx was occasionally detectable during ice-free periods in the shore zones off Neusiedl, Breitenbrunn and Podersdorf, and under favourable wind conditions it could be followed far out into the lake (Fig. 10.4). The greatest influx of sewage can be observed in the region where the Wulka enters the lake, although here the reed belt is at its widest and the Wulka, which is not channelled, splits up into many branches soon after entering the reed belt and comes in close contact with the reed ground. In 1975 the concentrations of P_s and P_o in the entire lake were several times larger than those for 1972/1973 (Fig. 10.5). Peaks in influx of sewage were measured on the western shore (Breitenbrunn to Oggau) at the time when the water level was at its maximum in spring. The effect of canals from Neusiedl, Gols and Podersdorf does not show up so strongly in these particular years due to high concentrations in the northern part of the lake. As Fig. 10.4 shows, the part of the lake south of the Oggau-Hölle profile occasionally contains lower concentrations than the northern part of the lake at the same time. Sewage entering in the drainage canals from Rust and Mörbisch in this region of the lake can sometimes be detected.

The mean values for P_s and P_o obtained at sampling points near the shore

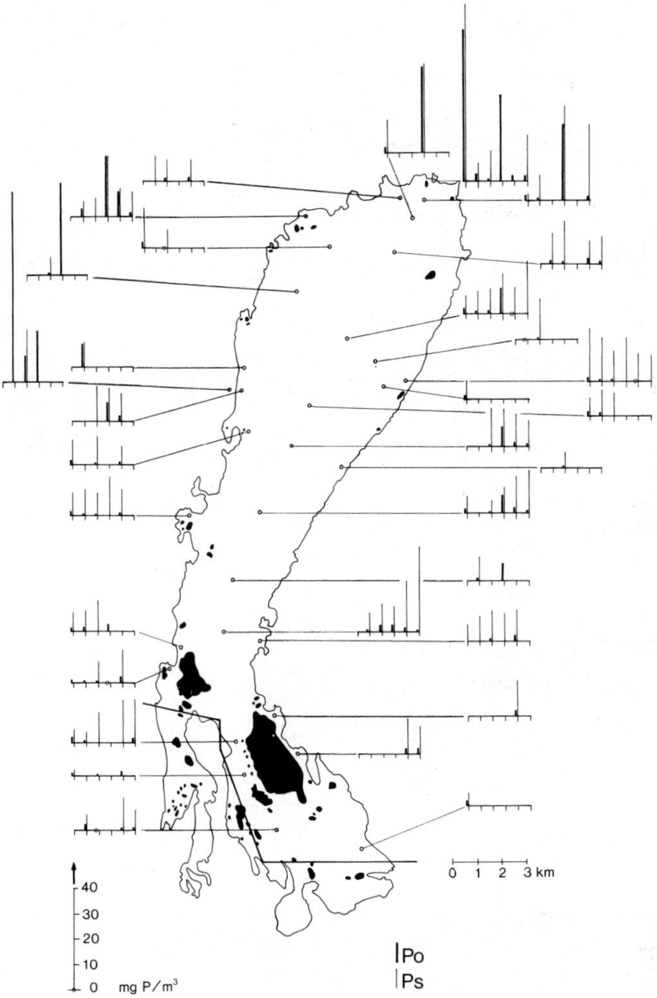

Fig. 10.4. Local distribution of dissolved phosphorus and orthophosphate in samples taken 2 Feb., 19 May, 22 Jun., 24 Aug., 6 Oct., 1972, and 5 Sep. 1973.

are therefore as a rule higher than the mean values from points in the lake itself (Table 10.6) and are also subject to greater fluctuations.

With the large increase in dissolved phosphorus compounds found very recently the ratio of the dissolved fractions to one another has also changed (1972/1973: P_h:P_o>6; 1975: P_h:P_o usually 1–3).

10.4.1.2 The distribution of particulate phosphorus

Particulate phosphorus is made up of the phosphorus compounds from plankton, from organisms stirred up from the bottom by the wind, e.g. bacteria (see Dokulil 1975), and of phosphorus-containing mineral particles.

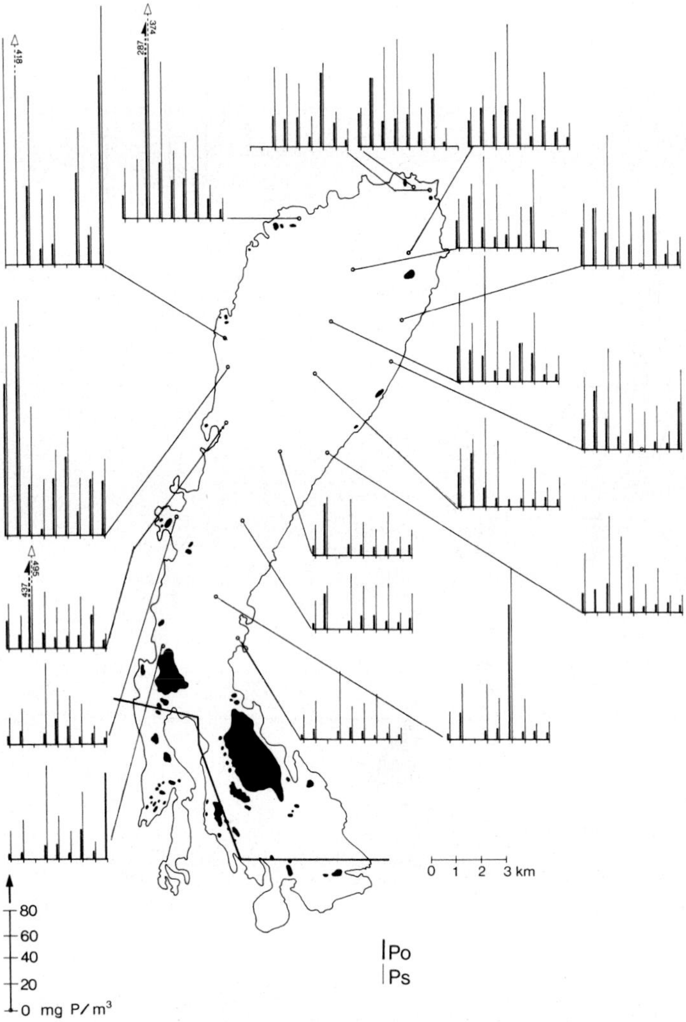

Fig. 10.5. Local distribution of dissolved phosphorus and orthophosphate in samples taken 12 Mar., 1 Apr., 23 Apr., 3 Jun., 27 Jun., 28 Jul., 3 Sep., 30 Sep., and 5 Nov. 1975.

A comparison of the actual P_p concentration with the calculated percentage P_p in the suspension (Fig. 10.6) reveals two parameters as being responsible for the considerable fluctuations in concentration of P_p (for data on solid material see Chapter 15).

1. *Wind* A relation exists between P_p concentration and the amount of solid material in suspension (Table 10.7). A change in the wind situation rapidly leads to a change in P_p concentration. As a rule, the wind conditions up to ten hours prior to sampling govern the P_p content found, but under extreme conditions an effect can last for up to three days although the wind has dropped.

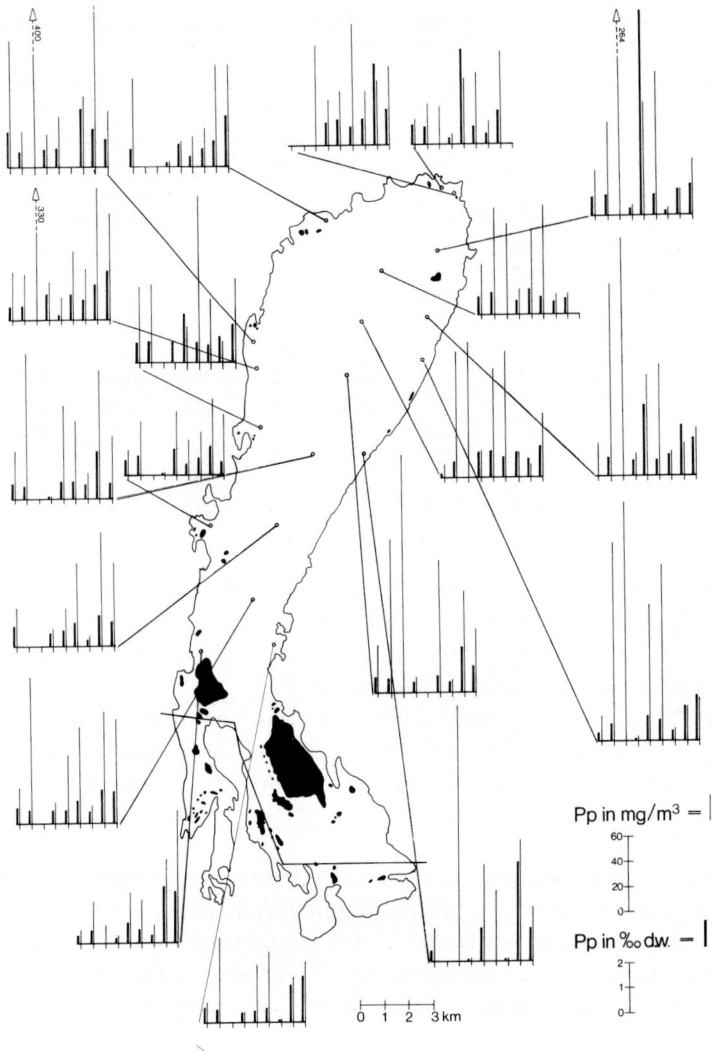

Fig. 10.6. Local distribution of particulate phosphorus in samples taken as in Fig. 10.5.

As Fig. 10.7 shows, the distribution of the P_p ($\mu g\ l^{-1}$) corresponds approximately to that of the solid material and is also mainly influenced by the wind. P_p concentration increases in the wind direction and in proportion to its force. Thus in the horizontal profiles shown (Fig. 10.2) it is lower in shore regions than in the open lake due to the prevailing northwest wind.

P_p (parts per thousand, dry weight) usually exhibits a distributional pattern similar to that of P_p ($\mu g\ l^{-1}$). The percentage P_p in the solid material usually increases with increasing quantities of the latter (*a*, *b*, and *c* in Fig. 10.7).

If no such rise in P_p (parts per thousand, dry weight) is detectable (*c* in Fig. 10.7) it can be assumed that mixing of the water occurred or that the influence of the factors causing a rise in P_p content was smaller. The constant wind speed of 10–20 km h^{-1} during sampling suggests mixing in this case.

Table 10.6. Geometric mean values from the sampling stations near the reed belt and the middle part of the lake (1975), d.w. = dry weight.

	Inshore					Midlake				
Date	P_s	P_o	P_p (parts per 1,000 d.w.)	NO_3	NO_2	P_s	P_o	P_p (parts per 1,000 d.w.)	NO_3	NO_2
	(μg P l^{-1})			(μg N l^{-1})		(μg P l^{-1})			(μg N l^{-1})	
12 Mar.	49	15	0.58	156	5.5	34	13	0.54	167	3.6
1 Apr.	56	20	0.63	–	–	47	31	0.63	–	–
21 Apr.	126	50	–	–	–	80	18	–	–	–
3 Jun.	64	12	0.28	182	1.1	52	9	0.36	185	0.5
27 Jun.	41	17	0.91	199	1.3	29	9	1.14	237	0.4
28 Jul.	34	7	0.61	239	1.4	34	12	0.75	226	0.7
3 Sep.	50	17	0.48	223	2.6	42	11	0.42	212	2.2
30 Sep.	20	6	1.63	78	2.0	14	8	1.13	101	1.9
5 Nov.	24	6	1.45	179	1.2	20	11	1.01	160	0.8

When the wind drops a process of settling can be observed. This leads to a rise in P_p (parts per thousand, dry weight) which indicates that coarse particles with a lower P_p content settle first and the smaller particles with a higher content of P_p remain in suspension.

The settling process is seen first in the part of the lake in the lee of the wind (*e* in Fig. 10.7) and may also be enhanced by water drifting in from the reed belt. If the process of settling continues for longer periods the area with lower P_p (μg 1^{-1}) increases (*d* in Fig. 10.7).

2. *Biomass* Seasonal alterations in total numbers of organisms are expressed in a very variable P_p content of the solid material. Three maxima can be seen for 1975 (April, June, September), the value rising over the course of the year. The course taken by the percentage P_p is the opposite of that taken by P_s (Fig. 10.8). A comparison of shore values and lake values reveals almost

Table 10.7. Relation between P_p (μg l^{-1}), and solid material in water and wind.

	Wind direction	Speed (km h^{-1})	P_p (μg l^{-1})	Solid material (mg l^{-1})
1972/1973	NW	1.6	29	–
	NW	4.5	14.2	–
	N	7.5	17.9	–
	SE	17.3	74.1	–
	SSE	22.0	77.9	–
1975	NW	2.4	9.5	30.3
	SE	4.5	14.3	29.3
	NW	5.0	57.5	67.5
	SE	9.8	55.2	44.4
	NW	10.0	72.3	122.0
	NW	18.2	157.3	–

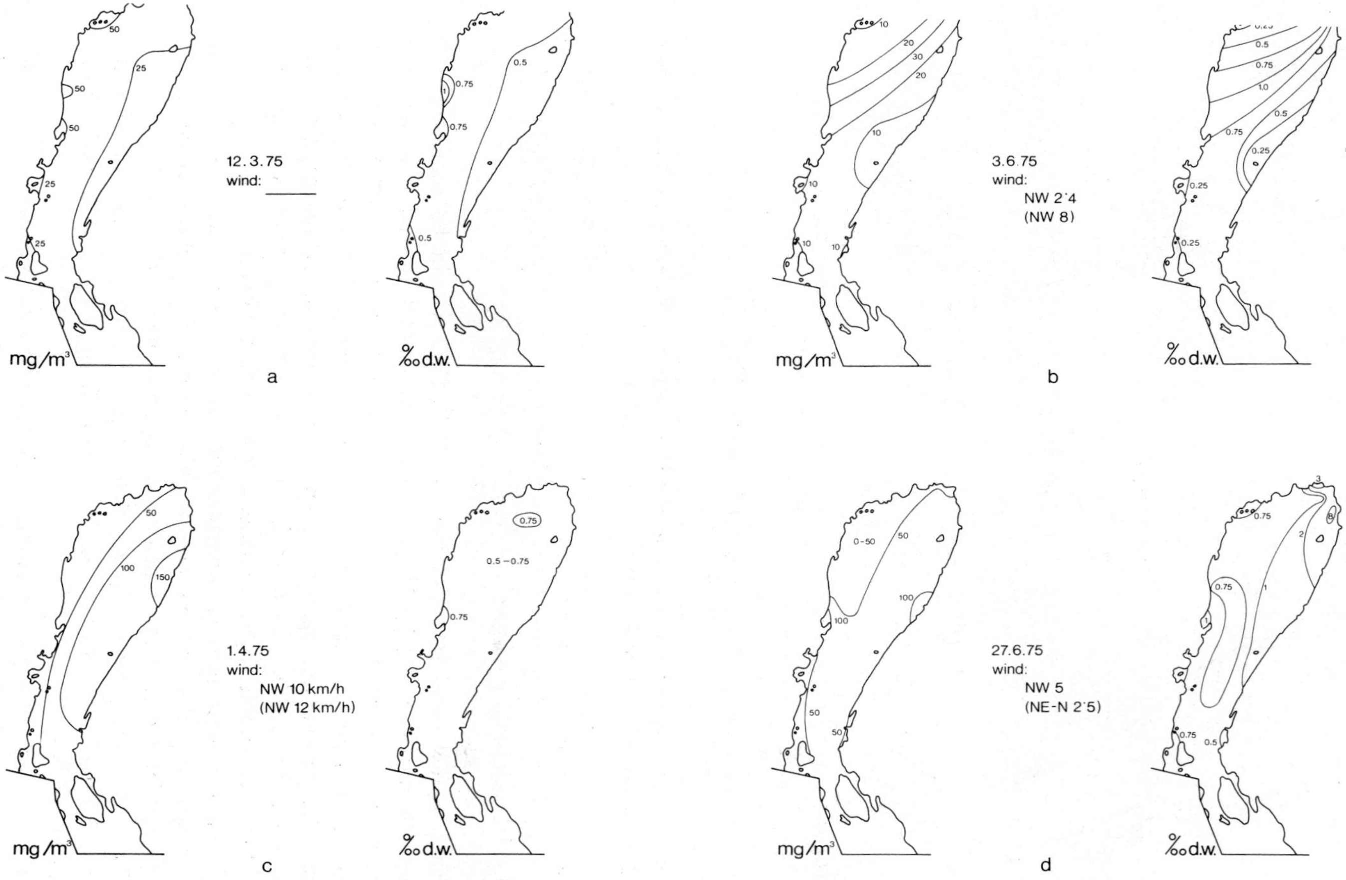

Fig. 10.7. Local distribution of particulate phosphorus expressed in mg m^{-3} and in parts per thousand of suspended material on different sampling dates.

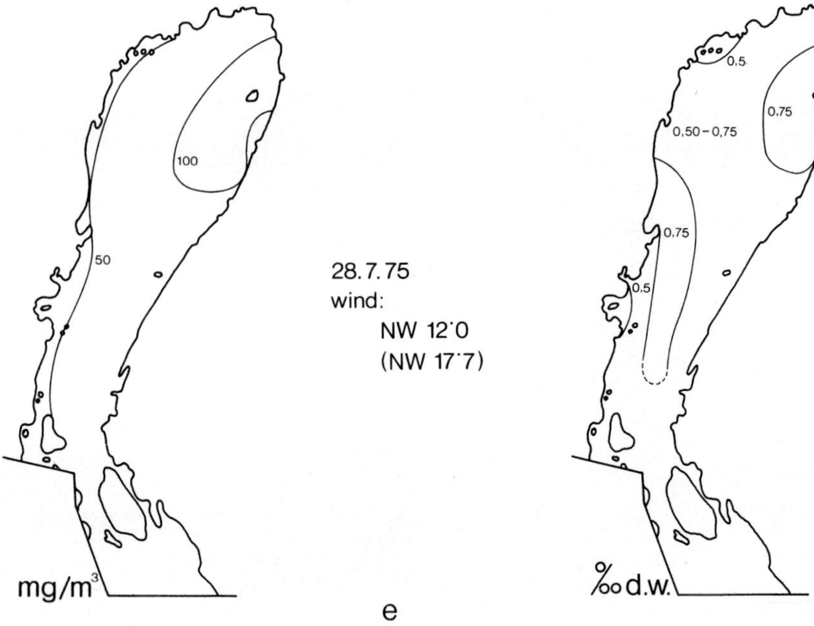

Fig. 10.7.—continued.

identical figures in spring, whereas the P_p is higher in the open lake in summer and in the shore region in autumn (Table 10.6).

10.4.2 *Nitrogen compounds*

10.4.2.1 Kjeldahl nitrogen

Although few data are available for 1970/1971 it can be seen that the Kjeldahl nitrogen (Kj-N) shows smaller horizontal fluctuations than the phosphorus compounds. The somewhat higher values recorded from the shore region may be attributable to a higher ammonium nitrogen content. Since the particulate organic nitrogen compounds in the macrophyte zone bordering the reed belt are also high, the same explanation can be assumed to hold for the soluble organic nitrogen-containing compounds.

Temporal differences in concentration are also slight, except beneath the ice, where the values of dissolved Kjeldahl nitrogen stay almost constant and a substantial decrease in particulate nitrogen is observed.

The concentrations of particulate organic nitrogen lie between 0.3 and 0.5 mg N l^{-1} (total Kj-N: 1.0–1.2 mg N l^{-1}; dissolved Kj-N: 0.7–0.9 mg N l^{-1}).

If it is assumed that the particulate nitrogen originates exclusively in living material, then the dry weights and the carbon and phosphorus contents of the organism calculated from these particulate nitrogen values (according to Golterman 1975) are considerably higher than the carbon and phosphorus

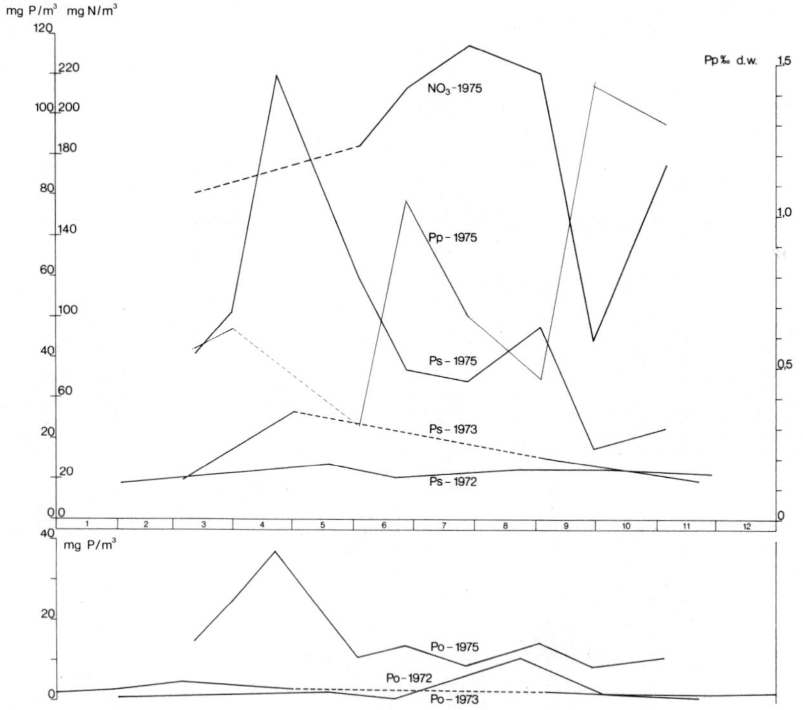

Fig. 10.8. Annual course of phosphorus and nitrate in 1972, 1973 and 1975 (geometrical mean values).

values actually measured. It can therefore be supposed that only a small amount of the particulate nitrogen is present in organisms.

10.4.2.2 Nitrate and nitrite

Nitrate content (Fig. 10.9), like that of phosphorus, is subject to considerable spatial and temporal fluctuations. Since our own results only extend back as far as 1975 we can merely conclude from a comparison of our data with those of Stehlik (Table 10.1) that the nitrate concentration in the lake has also increased in recent years.

The nitrate content is usually higher in the open lake than in the shore region (Table 10.6). As investigations still in progress show, the ammonium content behaves in the opposite manner, so that with increasing distance from the reed belt it can be assumed that this rise in nitrate is caused by oxidation.

If a higher nitrate concentration is encountered in the shore region it can be assumed that this is due to an increased inflow of sewage water. This shows up particularly well during periods with precipitation (Fig. 10.10). Since the region with higher nitrate concentration coincides with one of higher P_p (parts per thousand, dry weight), it can be assumed that drifting of water from the reed, and thus also of inflows, occurs.

An influence of P_p on the nitrate content can be detected at the time of the

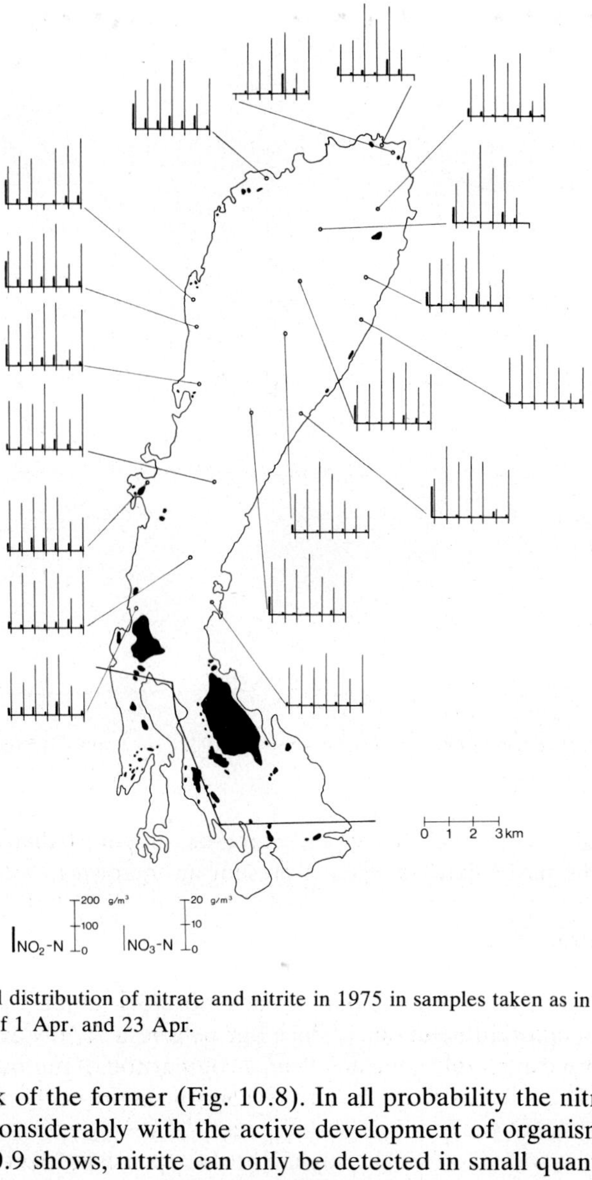

Fig. 10.9. Local distribution of nitrate and nitrite in 1975 in samples taken as in Fig. 10.5, with the exception of 1 Apr. and 23 Apr.

autumn peak of the former (Fig. 10.8). In all probability the nitrate content drops very considerably with the active development of organisms.

As Fig. 10.9 shows, nitrite can only be detected in small quantities due to the favourable oxygen conditions prevailing in the lake. Here too, somewhat higher values are found in the shore region due to drifting reed water.

10.4.3 *Summary of the nutrient conditions in Neusiedlersee*

During the years covered by this investigation an increase of soluble phosphorus loading is detectable in Neusiedlersee. In the absence of sufficient evidence an increase in soluble nitrogen compounds can only be suspected. Point sources of phosphorous are mainly responsible for this rise, in that they considerably disrupt the natural balance that had become established between the reed zone and the lake (input of soluble nutrients from the reed zone into the lake, transport of particulate nutrients from the lake into the reed). As could be seen for 1975 (Fig. 10.5) P_s also influences the P_p content of the lake. Since no data on solid material are available from the years 1970–1972 an increase of the P_p can only be deduced from plankton studies (see Chapters 17 and 22).

As phosphorus studies have shown a similar change has not yet commenced in the upper superficial sediment since the favourable conditions in the lake

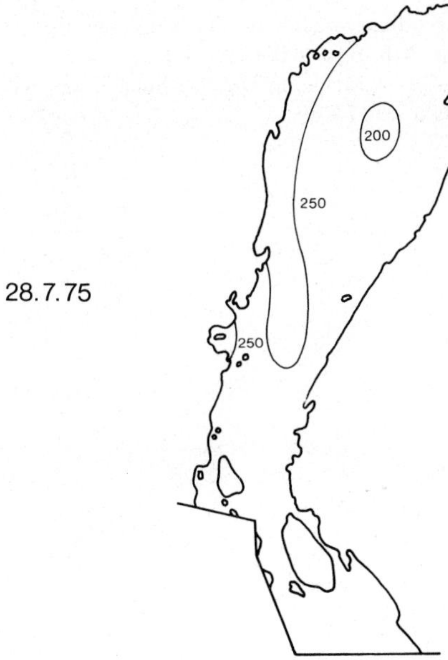

Fig. 10.10. Distribution of nitrate showing the inflow from the northwest (g m^{-3}).

and continuous translocation of sediment into the reed belt prevent the accumulation of nutrients. But it is only a matter of time before this increase of nutrients can be detected in the sediment, if the input of nutrients is not reduced noticeably.

The reed belt serves only to a limited extent as a depository for the particulate nutrient compounds of the open lake and for the highly loaded inflow.

References

Dokulil, M. 1975. Bacteria in the water and mud of Neusiedlersee (Austria). Symp. Biol. Hung. 15: 135–140.

Eidgenössisches Department des Inneren 1974. Vorläufige Empfehlungen über die regelmäßige Untersuchung der schweizerischen Oberflächengewässer. Bern.

Geyer, F. & Mann, H. 1939. Limnologische und fischereibiologische Untersuchungen am ungarischen Teil des Fertö. Arb. Ungar. Biol. Forschungsinst. Tihany 9: 64–193.

Golterman, H. L. 1969. Methods for chemical analysis of fresh waters. IBP Handbook 8. Blackwell, Oxford.

Golterman, H. L. 1975. Physiological limnology. Development in Water Science 2. Elsevier, Amsterdam.
Hepher, B. 1958. On the dynamics of phosphorus added to fishponds in Israel. Limnol. Oceanogr. 3: 84–100.
Neuhuber, F. 1978. Die Phosphorsituation des Neusiedlersees. Österr. Wasserwirtschaft 30, 5/6: 94–99.
Schlichting, E. & Blume, H.-P. 1966. Bodenkundliches Praktikum. Paul Parey, Hamburg.
Stehlik, A. 1972. Chemische Topographie des Neusiedler Sees. Sitz. Ber Öst. Akad. Wiss., math.-nat. Kl. 180 (8–10): 217–278.
Tauber, F. 1959. Trübung und Sedimentverfrachtung im Neusiedlersee. Wiss. Arb. Bgld. 23: 81–88.
Weisser, P. 1970. Die Vegetationsverhältnisse am Neusiedler See: pflanzensozilogische und ökologische Studien. Wiss. Arb. Bgld. 45: 1–83.
Wieden, P. 1959. Sediment-petrographische Untersuchungen des Schlammes vom Neusiedler See (Bgld). Wiss. Arb. Bgld. 23: 73–80.

1 Oxygen conditions

F. Neuhuber and L. Hammer

Trophogenic and tropholytic processes in Neusiedlersee proceed in a body of water which, due to morphology and wind conditions, exhibits scarcely any vertical oxygen gradient. On the other hand, very distinct horizontal variations in oxygen content occur, due to the presence of the two sub-biotopes, the open lake and the reed belt, each with its own ratio of oxygen-producing and oxygen-consuming processes. The transitional zone linking these two sub-biotopes coincided with the former macrophyte belt (see Chapter 19) adjoining the reed belt on its lakeward fringe, although its position depends upon the wind direction. It can thus shift at times into the reed belt or even be displaced to the open water beyond the former macrophyte belt. At the same time wind conditions determine the intensity with which the transitional zone is developed. As Table 11.1 shows, the oxygen content of the water of the open lake is near to saturation level or even slightly above it (88–105 per cent). Values measured in the bays are usually similar to those found at points remote from the shore. Only in the Hungarian part of the lake is the influence of reed water observable in places remote from the reed fringe. Due to breakdown processes the oxygen content of the reed water is lower than in the lake and is subject to much greater fluctuations (0–70 per cent). Values recorded for the open lake zone in front of the reed fringe (0–1 km) clearly reflect the alternating effects of reed and lake water in addition to the macrophytes (24–125 per cent).

The biological oxygen demand also rises considerably from the open lake towards the reed belt (lake: 0.02–0.7 mg O_2 l^{-1} day^{-1}; shore zone: 0.35–1.7; reed: 1.1–2.7).

The large degree of turbulence to which the open lake is usually exposed is enough to ensure oxygen saturation throughout the entire body of water, so that only on the rare occasions when a longer period of calm weather occurs is a weak vertical oxygen gradient observable (Stehlik 1972). Thus biological processes influencing the oxygen content of the open water are masked by the physical processes.

The significance of these biological processes for the oxygen content of the lake can be seen from the yearly course of oxygen saturation (Fig. 11.1). From April until June the lake is highly supersaturated (measured in the bay of Neusiedl) which indicates intensive assimilation. According to Zakovsek (1961) phytoplankton assimilation is the chief cause of this supersaturation.

The somewhat lower values measured at the same time close to the bottom are thought to be due to greater oxygen consumption in the sediments.

In summer and autumn the saturation values are much lower (80–100 per cent), because oxidation is accelerated as a result of warming up of the water and the uppermost sediment layers.

Table 11.1. The oxygen data of Neusiedlersee.

Observer	Date	Location			O_2(mg l^{-1})	% saturation	BOD_1 (mgO_2 l^{-1} day^{-1})
Zakovsek	21 Sep. 1951	bay	N	Neusiedl	10.82	104	0.42
Stehlik	23 Jul. 1959	centre	M	Illmitz	8.3	93	
		shore	M	Hölle	8.05	92	
		shore	N	Neusiedl	7.75	92	
	4 Nov. 1959	centre	M	Illmitz	12.0	102	
		reed	M	Illmitz	8.4	70	
	13 Nov. 1959	centre	N	Breitenbrunn	14.2	105	
		shore	N	Neusiedl	12.8	100	
		canal	N	Breitenbrunn	2.5	20	
		reed	N	Neusiedl	3.6	30	
		reed	N	Neusiedl	2.4	20	
		reed	N	Neusiedl	1.7	14	
		reed	N	Neusiedl	0	0	
Schmid	29 Aug. 1968	reed	N	Neusiedl	3.2	36	2.32
		shore	N	Neusiedl	6.16	68	1.17
	30 Aug. 1968	shore	N	Neusiedl	6.20	68	1.20
	31 Aug. 1968	reed	N	Neusiedl	1.49	18	2.16
		shore	N	Neusiedl	2.00	24	
	1 Sep. 1968	reed	N	Neusiedl	2.64	30	1.10
		shore	N	Neusiedl	5.41	58	
	19 Sep. 1968	shore	N	Neusiedl	6.80	66	1.73
	9 Oct. 1968	reed	N	Neusiedl	4.46	39	2.67
		shore	N	Neusiedl	6.74	67	1.06
	13 Oct. 1968	reed	N	Neusiedl	1.96	39	
		shore	N	Neusiedl	6.85	66	2.35
	27 Oct. 1968	reed	N	Neusiedl	3.85	32	
		hore	N	Neusiedl	11.87	98	0.54
	28 Oct. 1968	reed	N	Neusiedl	4.27	40	
		shore	N	Neusiedl	12.56	115	1.39
	3 Nov. 1968	reed	N	Neusiedl	5.63	52	1.38
		shore	N	Neusiedl	9.94	90	0.48
	10 Nov. 1968	reed	N	Neusiedl	2.24	22	2.22
		shore	N	Neusiedl	10.86	98	
Pásztó	13 Jun. 1970	bay	S	Fertőrákos	8.05		
	8 Jul. 1970	bay	S	Fertőrákos	4.7		
		bay	S	Rucás	5.8		
		bay	S	Madárvárta	5.71		
		bay	S	Hegykői	4.41		

Observer	Date	Location			O_2(mg l^{-1})	% saturation	BOD_1 (mgO_2 l^{-1} day^{-1})
Pásztó	28 Jul. 1970	bay	S	Fertőrákos	8.80	102	
	25 Sep. 1970	bay	S	Rucás	10.88		
	25 Nov. 1970	bay	S	Fertőrákos	11.3	88	
		bay	S	Madárvárta	12.1	94	
Dokulil	21 Jul. 1971	bay	N	Neusiedl	9.4	96	1.06
	10 Aug. 1971	bay	N	Neusiedl	8.83	104	0.74
		centre	N	Neusiedl	8.19	95	0.34
	25 Aug. 1971	bay	N	Neusiedl	9.52	104	0.50
		centre	N	Neusiedl	9.25	100	0.25
	23 Sep. 1971	reed	M	Rust	4.90	50	
	4 Mar. 1972	shore	N	Neusiedl	13.32		1.27
		bay	N	Neusiedl	15.76		
		2 km from shore	N	Neusiedl	15.73		0.69
	4 May 1972	shore	N	Neusiedl	8.73	83	0.47
		bay	N	Neusiedl	10.82	102	0.27
		2 km from shore	N	Neusiedl	10.66	98	0.23
Pesendorfer & Stehlik	2 Feb. 1974	mean value from 3 sample stations in lake centre			12.8	99	0.9 BOD_2
Neuhuber	13 Mar. 1975	shore	N	Weiden	10.65	93	0.35
		shore	N	Golser canal	10.71	95	0.49
		shore	M	Illmitz	10.22	92	
		centre	N		10.63	93	0.40

Parts of the lake: N, northern; M, middle; and S, southern.

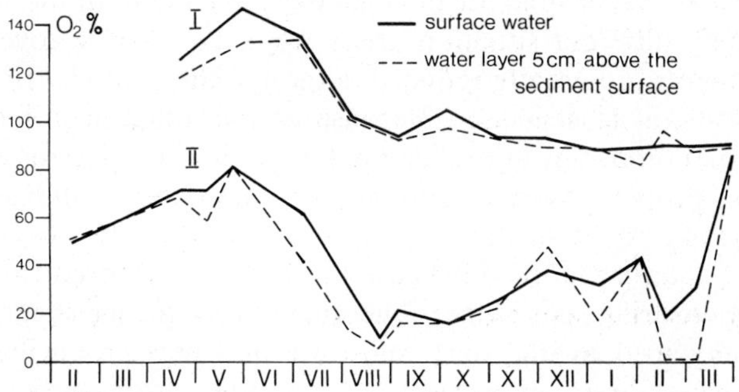

Fig. 11.1. Oxygen content of Neusiedlersee (Zakovsek 1961) I: bay of Neusiedl; II: 'Rohrlacke' within the reed belt near Neusiedl.

In the reed belt (measured in the 'Rohrlacke' near Neusiedl) the ratio of oxygen production to oxygen consumption is displaced in favour of the latter, and supersaturation is never attained. Nevertheless, here too, a higher oxygen content can be observed in spring than in summer and autumn. A vertical gradient of oxygen content is rarely seen. Continuous horizontal water drift seems to provide better mixing than would be expected in such a biotope.

The daily course of oxygen content is different in reed and lake water (Fig. 11.2). Whereas in the reed water the daily rhythmicity in oxygen content is well developed, it is barely detectable in the vicinity of the reed fringe. The diurnal rhythmicity seen in the reed water results from a combination of very intense consumption processes and the considerable powers of assimilation of the macrophytes, particularly *Utricularia vulgaris* (see Chapter 21) and epiphytic algae. In this semiaquatic biotope with its dense vegetation and excessive quantities of organic breakdown products, consuming processes greatly affect the oxygen content of the water. Their influence is at a maximum in July and August due to warming-up, and results in totally anaerobic conditions in the reed water at night.

The daily increase in oxygen content of the water, which usually occurs between 10,00 and 16,00 h reduced due to the shading effect of the reed plants and the increased intensity of respiration with rising temperature.

In the summer months oxygen production lags behind consumption. This agrees with the results of studies carried out by Maier (1973) and Draxler (1973) on *Utricularia*. They found that growth of *Utricularia vulgaris* is on the whole completed by the end of June, and with increasing age of the shoots oxygen consumption gradually exceeds production.

The oxygen content of reed and lake water is influenced by water exchange between the two sub-biotopes, and thus depends upon wind conditions. Studies so far carried out have shown that the reed water has a far greater influence on the oxygen content of the water of the shore zone of the open lake than vice versa. Northwest winds blowing at measuring stations in the southeast of the lake (Illmitz) lead to outdrift of reed water (6 July 1977), a fact that can only be explained by currents in the lake (see Chapter 8) and possibly also seiches, the lake water being diverted along the shore, whilst water is expelled from the reed. Under northwesterly conditions lake water is apparently partly responsible for the high oxygen content of the reed water.

A completely different situation arises when the lake is covered by ice. Water movements are greatly reduced, although the elasticity of the ice can lead to currents, at least locally. Because of the different influence of the wind, ice formation begins in the reed belt with brief ice periods and spreads out to the lake proper. Given appropriate air temperatures this can occur very rapidly on account of the shallowness of the water. The ice is always thicker in the open lake than in the reed because the latter is sheltered from the wind and the snow covering lasts longer. The thickness of the ice cover in the reed is further influenced by the dark substrate and heat-producing biological processes. The addition of warmer water, either from surface inflow or subterranean sources (for example in the shore region north of Podersdorf, the bay of Breitenbrunn, and the reed belt near Rust), can in places lead to a reduction in the thickness of the ice covering in the reed belt and the open lake.

When the lake is completely frozen over settlement of seston takes place. Oxygen conditions, on the other hand, scarcely alter in the part of the open lake remote from the reed fringe (about 1 km from the edge of the reed), as long as enough light for assimilation can penetrate the ice cover (Table 11.2). The lack of a vertical oxygen gradient in this region of the lake, and the frequently higher saturation values in the bottom water of the zone in front of the reed fringe, are the result of the well-developed bottom algal populations (Prosser, personal communication). When the reed water is covered by ice and snow, a vertical oxygen gradient immediately develops (Zakovsek 1961: surface water 30 per cent, bottom water 7 per cent oxygen saturation, after one day of ice cover) and results within a short period of time in a completely anaerobic state in the reed water.

In the lake region in front of the reed fringe (0–1 km) the oxygen content increases with depth and with distance from the reed fringe, but decreases with duration of ice cover. The possible reasons for this will now be discussed.

1. Intensity of assimilatory activity is reflected in a daily rhythmicity in oxygen saturation in both surface and bottom water (Fig. 11.3). The maximum and minimum oxygen saturation of surface and bottom water at the sampling sites (with the exception of the 50 m point) differ only very slightly. This means that either assimilation in the two layers is equal and the consumption processes in the sediment scarcely have any effect on the oxygen content of the bottom water, or the assimilation of the bottom algae is greater than that of the planktonic algae and the consumption processes have a greater influence on the oxygen content of the bottom water. Due to the high organic content of the sediment in the shore zone (see Chapter 24) the latter possibility is the more likely.

2. Decrease in the oxygen content at the various sampling points during longer periods of ice cover (Table 11.2, Winter 1968/1969) is much more pronounced in the surface water than nearer the bottom. Thus it can be assumed that apart from local consumption processes there is also an inflow of warmer reed water, although this still has to be confirmed. The oxygen content of the surface water might then be more affected by the accumulation of the warmer reed water at the surface.

Oxygen conditions also change in connection with long-term algal development. Thus the greater nutrient supply in recent years has led to active algal growth in the entire body of water (see Chapter 17), even under ice. Although no oxygen measurements are available it can be assumed that the oxygen content of the surface water has also increased. The higher oxygen content of the open lake near the reed fringe in the winters of 1970 and 1971 (Table 11.2) indicates changes in assimilatory activity.

At times when the water level is extremely low and the ice cover at the same time lasts particularly long, the open lake becomes anaerobic. Such a situation occurred in 1929 and had catastrophic effects on the lake's fishery (Varga & Mika 1937). The present high nutrient content warrants the assumption that an extreme situation of this nature might nowadays develop more readily.

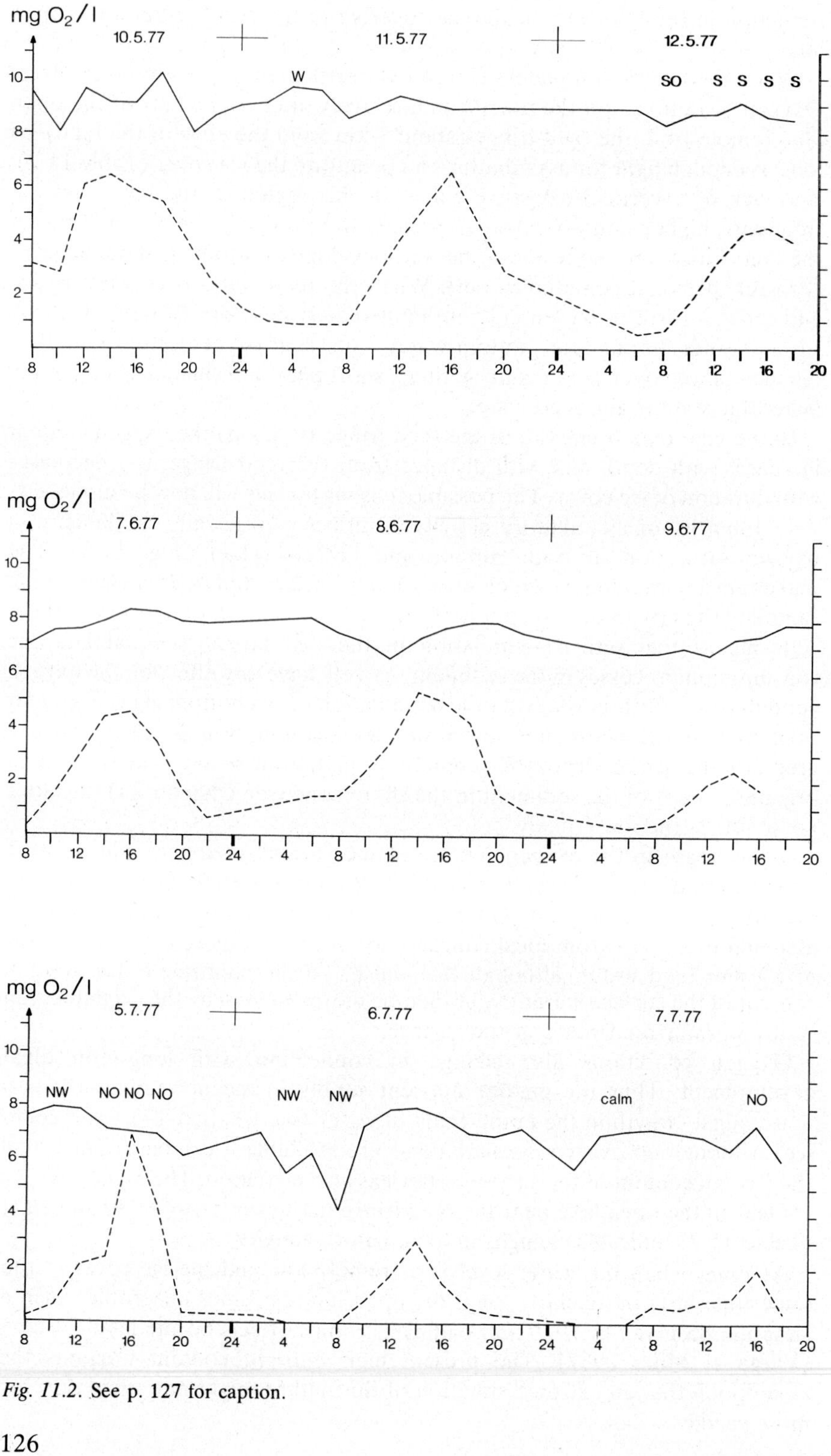
mg O2/l
10.5.77
11.5.77
12.5.77
W
SO
S S S S
mg O2/l
7.6.77
8.6.77
9.6.77
mg O2/l
5.7.77
6.7.77
7.7.77
NW
NO NO NO
NW NW
calm
NO

Fig. 11.2. See p. 127 for caption.

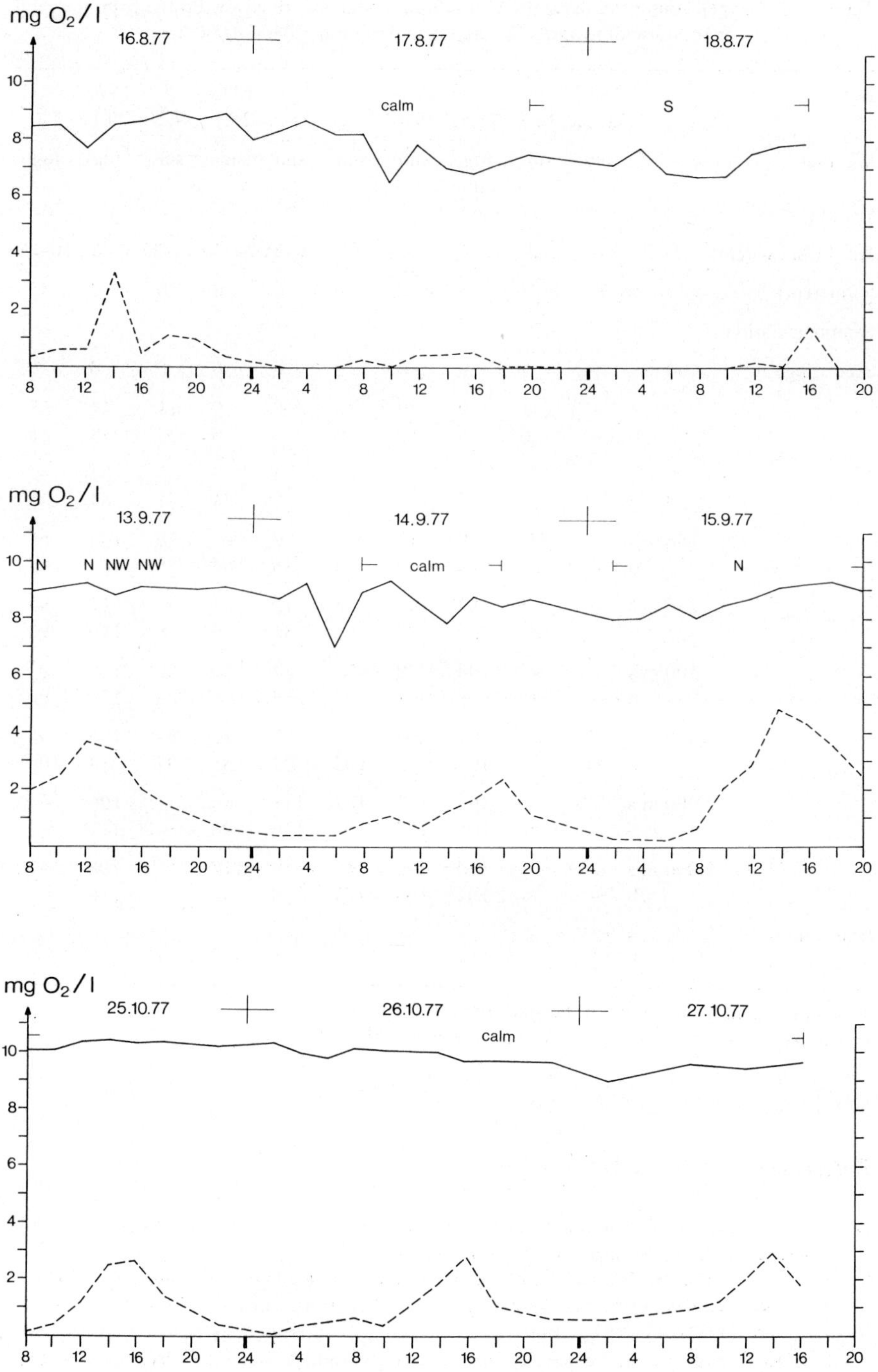

Fig. 11.2. Daily course of the oxygen content of Neusiedlersee near Illmitz. Sampling stations: ——: jetty of Illmitz; – – –: reed water 20 m inside the reed belt. Sampling depth: 15 cm below the water surface.

Table 11.2. Oxygen content in percentage saturation under the ice cover. Profile from the reed near Neusiedl towards the lake. Sampling time: 0800–1000 hrs.

Date		1968			1969			1970		1971	
		21.12.	12.1	15.1.	19.1.	7.2.	26.2.	27.1.	9.2.	20.1.	3.2.
Weather		sun	sun	fog, rain-fall	sun	sun	sun	sun	sun	fog	fog
Ice thickness (cm)		–	25	15–25	22–25	25	20–25	25–30	15–25	30	10–15
Snow (cm)		0	8–10	5	2	0	0	0	0	0	5
Sampling station											
reed fringe	100 m	–	0	0	0	0	0	0	0	0	0
	0 m s	–	1	0	0	0	0	0	11	44	68
	b	–	0	0	0	0	0	0	39	44	88
	50 m s	–	–	–	–	–	4	–	73	82	–
	b	–	–	–	–	–	7	24	75	84	–
	150 m s	–	55	48	18	0	9	–	78	100	86
	b	–	59	43	37	23	36	64	76	103	85
	250 m s	–	63	63	42	36	12	77	87	111	93
	b	–	56	59	62	66	38	79	78	110	90
	500 m s	–	–	98	78	31	24	80	86	115	97
	b	–	–	95	91	67	56	80	83	119	96
	1,000 m s	102	–	107	106	103	99	82	87	106	98
	b	110	–	108	104	105	93	85	87	114	100
	2,000 m s	–	–	105	103	102	111	90	–	106	–
	b	–	–	105	102	106	116	92	–	111	–
	3,000 m s	–	–	110	92	105	110	110	–	108	–
	b	–	–	101	104	107	109	95	–	116	–
lake centre	s	105	–	–	–	–	–	–	–	–	–
	b	116	–	–	–	–	–	–	–	–	–

s: sample taken from 0.3–0.6 m beneath the water surface
b: sample taken from 0–0.3 m above the sediment surface

References (*continued on p. 130*).

Draxler, G. 1973. Gaswechselmessungen an *Utricularia vulgaris*. In: H. Ellenberg (ed.), Ökosystemforschung. Springer, Berlin, pp. 103–107.

Geyer, F. & Mann, H. 1939. Limnologische und fischereibiologische Untersuchungen am ungarischen Teil des Fertö. Arb. Ungar. Biol. Forschungsinst. Tihany 9: 64–193.

Jorga, W. & Weise, G. 1977. Biomasseentwicklung submerser Makrophyten in langsam fließenden Gewässern in Beziehung zum Sauerstoffhaushalt. Int. Rev. Hydrobiol. 62(2): 209–234.

Maier, R. 1973. Produktions- und Pigmentanalyse an *Utricularia vulgaris* L. In: H. Ellenberg (ed.), Ökosystemforschung. Springer, Berlin, pp. 87–101.

Maucha, R. 1930. Sauerstoffschichtung und Seentypenlehre. Verh. Int. Ver. Limnol. 5: 75–101.

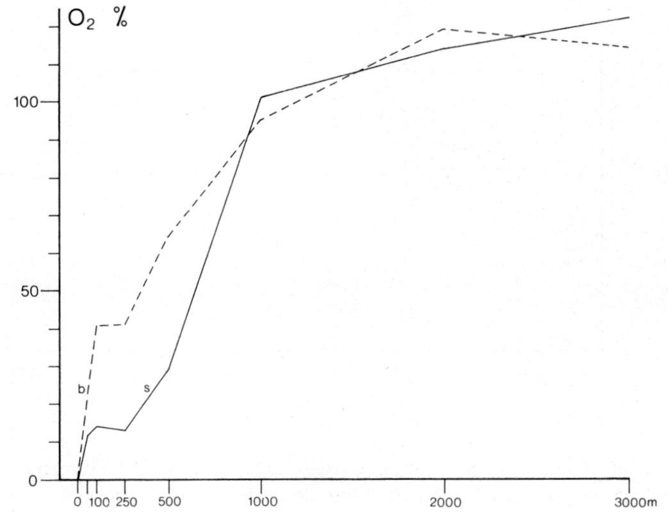
O2 %
100
50
0
b
s
0 100 250 500 1000 2000 3000m

A

Fig. 11.3. A. Caption p. 130.

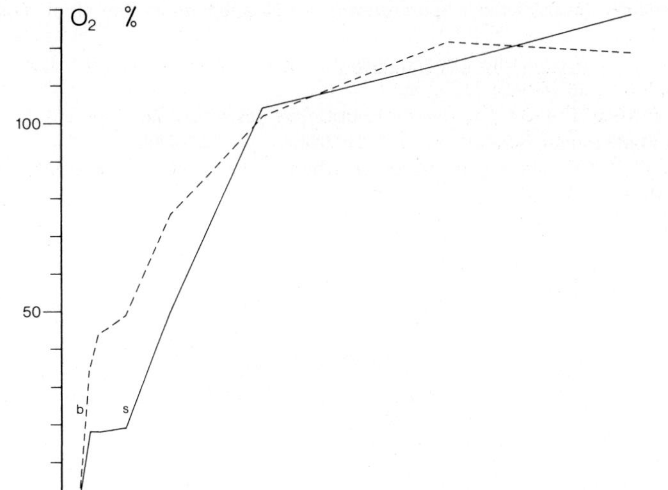
O2 %
100
50
0
b
s
0 100 250 500 1000 2000 3000m

B

Fig. 11.3. B. Caption p. 130.

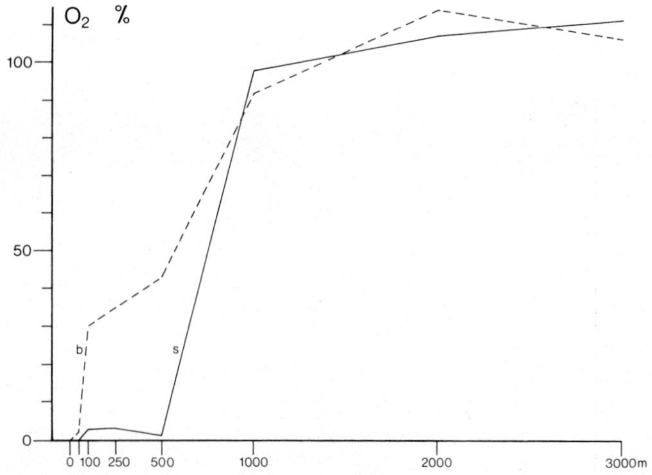

C

Fig. 11.3. Daily course of the oxygen content under ice on 26 February 11,00–13,00, 17,00–19,00 hrs (A, B) and 27 February 1969, 06,00–08,00 hrs (C) from a horizontal profile near Neusiedl, *s* = surface, *b* = bottom.

Pásztó, P. 1976. A Fertő-tó vizminőség. – In: L. Aujesky, S. Somogyi & F. Schilling (eds.), A Fertő-táj Monográfiáját előkészítő to adatgyűjtemény. Volume 2: Természeti adottságok: Kiegészîto a Fertö-táj hidroszférája és vízgazdálkodása c. kötethez. Budapest, pp. 2–153.

Pesendorfer, H. & Stehlik, A. 1975. Organochloropesticid-Rückstände und andere Wasserinhaltsstoffe (einschließlich Spurenstoffe) des Neusiedlersees. Ber. Biol. Forschungsinst. Bgld. 3: 171.

Stehlik, A. 1972. Chemische Topographie des Neusiedlersees. Sitz. Ber. Öst. Akad. Wiss., math.-nat. Kl. 180 (8–10): 217–278.

Varga, L. & Mika, F. 1937. Die jüngste Katastrophe des Neusiedler Sees und ihre Auswirkung auf den Fischbestand. Sonderber. Arch. Hydrobiol. 31: 527–546.

Zakovsek, G. 1961. Jahreszyklische Untersuchungen am Zooplankton des Neusiedler Sees. Wiss. Arb. Bgld. 27: 85.

12 Sediments

A. Preisinger

The first description of the sediments occurring in Neusiedlersee was the result of extensive investigations by Moser of samples taken from the lake bed following its complete evaporation and drying out in 1865 (Moser 1866). Although the water level was very low in 1891, 1900, 1922, 1933 and 1949 (Sauerzopf 1959), the lake bed has not again been completely dry since 1870. Nearly 100 years elapsed before publication of further studies on the lake's sediments (Tauber and Wieden 1959; Wieden 1959; Schroll and Wieden 1960; Löffler 1971; Blohm 1974). It is evident from these above studies and from personal investigations that the sediments forming the present bed of Neusiedlersee are neither physically nor mineralogically homogeneous. Furthermore, the general rule that the grain size of marine and lacustrine sediments decreases with increasing distance from shore does not apply to Neusiedlersee.

12.1 Sediment samples

Core samples taken by ramming or drilling showed striking macroscopic colour variations, widely differing water contents, and mainly quantitatively, but also qualitatively, different solid components with grain sizes varying from sand, 2 mm–0.063 mm, to silt, 0.063 mm–2 μm, to clay, $<2\ \mu$m (Fig. 12.1). Quartz pebbles in the size range of fine to coarse gravels are sporadically present in the sediments. Gravel banks are also found in the present lake area – one example of which is the gravel island between Illmitz and Rust – whose gravels are probably related to those of the Seewinkel (Tauber 1959). The formation of the upper layers in the southern part of the Seewinkel seems to have resulted from fluvial deposition of the Raab-Rabnitz system during the Würm ice age (Fuchs 1974).

The solid components of the sediment samples can be divided into two main groups (Table 12.1): the allochthonous (or detrital) components which originated outside the lake and entered it via aeolian or fluvial transport, and the autochthonous components which formed in the lake itself either through the flora and fauna or through the combined effect of photosynthesis and supersaturation of the lake, with salts produced by the process of evaporation. The relationship salt ecology – water ecology – sediment formation is discussed elsewhere (Preisinger 1979).

The physical characteristics, the grain size distribution and the various compositions of the solid components permit the differentiation of the following sediment layers: soft mud, mud, compact mud, and compact sediments, which are sometimes interrupted by thin sand layers (Table 12.2). The sequence of these layers is generally chronological, with the most recent

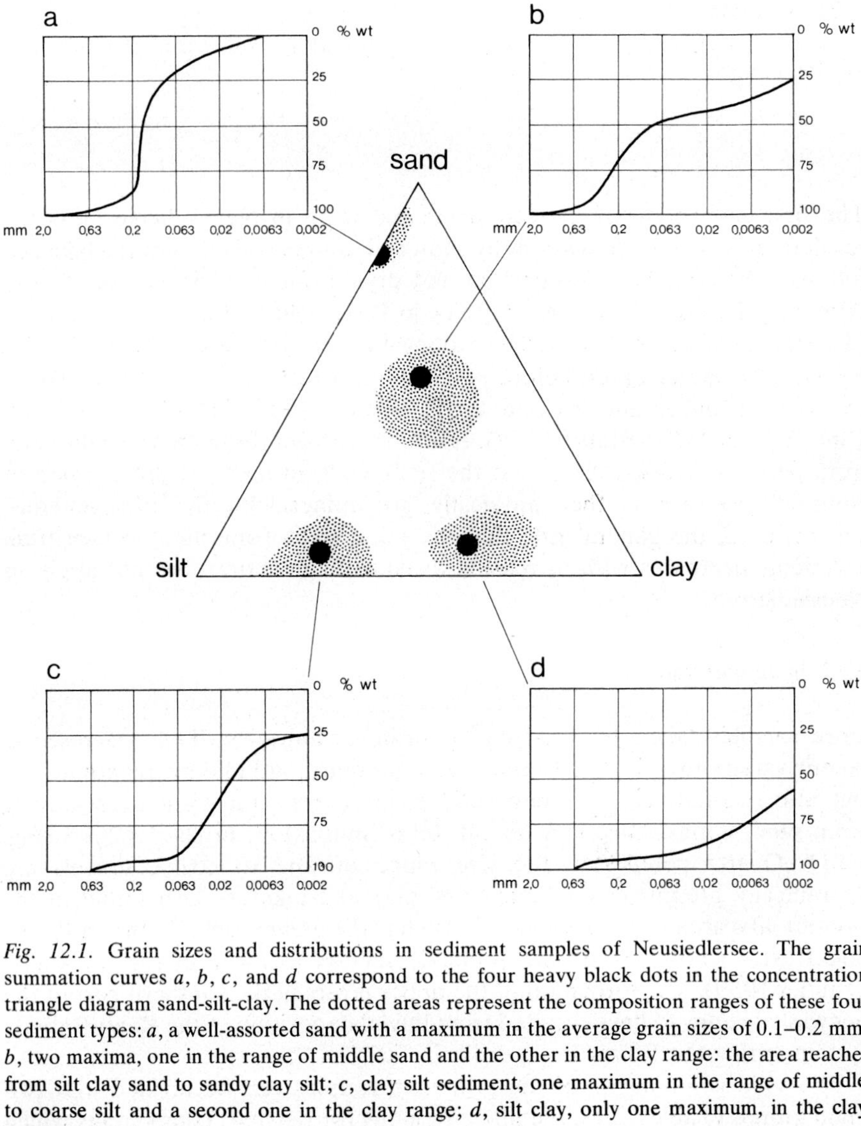

Fig. 12.1. Grain sizes and distributions in sediment samples of Neusiedlersee. The grain summation curves *a*, *b*, *c*, and *d* correspond to the four heavy black dots in the concentration triangle diagram sand-silt-clay. The dotted areas represent the composition ranges of these four sediment types: *a*, a well-assorted sand with a maximum in the average grain sizes of 0.1–0.2 mm; *b*, two maxima, one in the range of middle sand and the other in the clay range: the area reaches from silt clay sand to sandy clay silt; *c*, clay silt sediment, one maximum in the range of middle to coarse silt and a second one in the clay range; *d*, silt clay, only one maximum, in the clay range.

at the top. The thickness of the layers or the absence of individual ones depended on temporal and/or local events.

12.2 Origin of the sediment layers

The present sediments covering the bed of Neusiedlersee in the area from Neusiedl to the Hungarian border are composed of three chronologically distinguishable groups of layers.

(1) Layers which originated before the formation of the recent Neusiedlersee, that is, before the sinking of the present lake bed area towards the end

Table 12.1. Solid components of Neusiedlersee sediments.

Allochthonous components			Autochthonous components		
Grain size	Amount in %wt	Minerals	Amount in %wt	Substances	Particle size
sand & silt	15–40 5–30 >5	dolomite calcite quartz	variable	bacteria algae plant materials zoobenthos shells (calcite) diatoms (opal)	>2 μm
	>5	mica, 14 Å-chlorite, oligoclase, K-feldspar, *garnet*	<1	pyrite	
clay	30–40	illite, 14 Å-chlorite, *montmorillonite*	~60 1–8	Mg-calcite,[a] protodolomite[b] organic substances	<2 μm
	~100	calcite, montmorillonite, illite, 14Å-chlorite			

Infrequent minerals are in italics.

[a] Calcite containing $MgCO_3$ (<25 mol %) with nearly statistical distribution of the Mg and Ca ions in the calcite lattice (Grünberg & Preisinger 1969).

[b] Dolomite with an excess of $CaCO_3$ (<15 mol %) and only partial ordering of the Mg and Ca ions (Graf & Goldsmith 1956).

of the Würm ice age (Küpper 1957; Riedl 1964): these layers are thus characterized by the absence of autochthonously formed minerals such as Mg-calcite and protodolomite. Their composition is that of the compact sediment in Table 12.2 which corresponds to the deeper layers found by Moser (1866).

(2) Layers which originated after formation of the lake bed but before 1865: these layers thus contain the characteristic autochthonously formed minerals such as Mg-calcite and the protodolomite diagenetically formed from them (*D* in Fig. 12.1). Their composition is that of the compact mud in Table 12.2. The water content of these layers is low due to their drying out in the five years from 1865 to 1870. Otherwise their composition is that of the uppermost mud layers of the lake bed observed by Moser (1866).

(3) Layers which originated after 1870 and are characterized by the presence of autochthonously formed minerals such as Mg-calcites and protodolomites (*B* and *C* in Fig. 12.2): they correspond to the soft mud and other mud layers in Table 12.2.

Table 12.2. Sediment layers of Neusiedlersee.

Time	Layer	Colour	Water content (%)	Grain size	Minerals	Thickness (cm)
	soft mud	olive-black	>60	silt clay, type *d*	Mg-calcite, protodolomite, calcite, dolomite, mica, quartz, illite, 14Å-chlorite, oligoclase, *pyrite*, *montmorillonite*	open lake: 0–20 reed area: 40–100
After 1870	mud	light olive-grey	30–60	clay silt, type *c*	dolomite, calcite, quartz, mica, Mg-calcite, protodolomite, 14Å-chlorite, illite, oligoclase, *K-feldspar*, *pyrite*	0–40
		olive-grey	30–60	silt clay, sandy silt clay, type *d*, type *b*	dolomite, quartz, calcite, mica, Mg-calcite, protodolomite, 14Å-chlorite, illite, oligoclase, K-feldspar, *pyrite*	0–25
Prior to 1865 but after bed formation	compact mud	light olive-grey	<25	clay silt, type *c*	calcite, dolomite, quartz, mica, protodolomite, 14Å-chlorite, illite, oligoclase, K-feldspar, *Mg-calcite*, *pyrite*	0–40
			<25	sandy clay silt, type *b*	dolomite, quartz, calcite, mica, protodolomite, oligoclase, 14Å-chlorite, illite, montmorillonite, K-feldspar, *Mg-calcite*, *pyrite*	0–100 in the southern part of the lake
After bed formation	sand	white-grey	<15	sand, type *a*	quartz, dolomite, *calcite*, *oligoclase*, *mica*	0.2–5
Prior to bed formation	compact sediment	greenish grey	<25	sandy clay silt, and quartz pebbles	dolomite, calcite, quartz, oligoclase, mica, K-feldspar, 14Å-chlorite, montmorillonite, illite	100–700

Infrequent minerals are given in italics.

The allochthonous components of all these layers carried to Neusiedlersee reflect the mineralogical composition of the Leitha mountains, the Wulka basin, and the hills of Rust. The characteristic feature of the recent Neusiedlersee sediments is the presence of autochthonously formed minerals.

The Mg-calcite and protodolomite crystals are rhombohedra with an edge length of about 0.5 μm. The chemical composition of the Mg-calcite varies

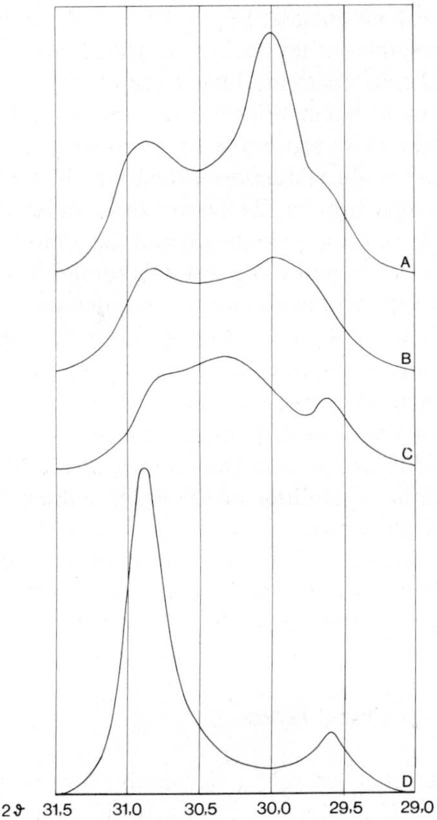

Fig. 12.2 The 2θ values ($Cu_{K\alpha}$-radiation) of the (014)-reflection of the Mg-calcites and protodolomites: *A*, turbidity of the lake water after calm weather (September 1975). Mg-calcites (2θ = 30.00°, 2B = 0.45°) correspond to a range of 15–21 mol % $MgCO_3$ with a maximum at 18 mol% ($Ca_{0.82}$ $Mg_{0.18}$ CO_3). Protodolomites (2θ = 30.85°, 2B = 0.45°) correspond to a mean composition of $Ca_{1.05}$ $Mg_{0.95}$ $(CO_3)_2$; *B*, <1 μm fraction of the mud layer from a drilled core sample from the lake bed; *C*, <1 μm fraction of the same mud layer only a few cm deeper; *D*, <1 μm fraction of the compact mud layer underlying the mud layer. My thanks are due to Doz. Dr. J. Kurzweil for the preparation of the <1 μm fraction.

with the Mg/Ca ratio in the lake water; the greater the Mg/Ca ratio, the greater the Mg content in the Mg-calcite. The ratio is at its minimum in spring and its maximum in autumn (Knie 1959). Thus the Mg-calcite, which is formed in connection with plant assimilation (macrophytes and microphytes), has a lower Mg content in spring than in autumn. The strong seasonal dependence of photosynthesis is also reflected in the amount of Mg-calcite formed. Crystal nucleation takes place on the surfaces of submersed macrophytes (*Potamogeton pectinatus* L. and *Myriophyllum spicatum* L.) which occur in the open lake off the reed areas (Varga 1931; Schiemer & Weisser 1972) and on the surfaces of microphytes. When the chemical composition of the crystal nuclei falls within the instability range of Mg-calcite (>25 mol % $MgCO_3$), as is the case, for instance, for the macrophytes in Neusiedlersee, the growing Mg-calcite crystallites will within a very short time be protodolomite inside and Mg-calcite outside. In Fig. 12.2, *B*, *C*, and *D* show the <1 μm fraction of different sediment layers from a drilled core sample: *B* and *C* are from a mud layer, *D* a compact mud layer sample. It is obvious here that the protodolomite is formed diagenetically from the Mg-calcite. The protodolomite growth is a very slow process with a growth rate of several hundred angstroms per 1,000 years (Peterson 1966) which is why the particles of sample *D* (0.5–0.6 μm) are hardly larger than those of *B* and *C* (about 0.5 μm). Due to their extremely small size and the almost perpetual motion of the lake water when free from ice, the crystals remain suspended in the water for a long time and are responsible for the greater part of its turbidity. This part of the turbidity is shown by *A* in Fig. 12.2: the higher peak indicates Mg-calcites with a mean composition of $Ca_{0.82}\ Mg_{0.18}\ CO_3$ the lower one Mg-calcites with about 47.5 mol % $MgCO_3$, that is, protodolomites with a mean composition of $Ca_{1.05}Mg_{0.95}(CO_3)_2$. The protodolomite component of the lake water turbidity arises from the interior of the Mg-calcites as well as from the protodolomite crystallites of the older sediments which are stirred up by the motion of the water.

In addition to the Mg-calcite and protodolomite already mentioned, some calcite originates from ostracod shells. Pyrite, which is produced predominantly in the mud by bacterial action, is also present.

12.3 Location of the sediment layers

In 1866 Moser noted that the salts ('zickstaub') formed by evaporation and efflorescence during the drying up of the lake were for the most part blown away. The remaining lake bed sediments, now the 'compact mud' layers, but which then, so soon after the evaporation of the lake, still contained considerable water, were not uniformly distributed over the lake bed. This applies even more so to the present lake bed. The regional distribution of the layers 'soft mud' (water content >60 per cent by weight) and 'mud' (water content 30–60 per cent) in the uppermost sediment layers (the first centimetre) of the lake bed is shown in Fig. 13.1.

The regional differences may result from the following local conditions:

(1) Water level changes. Alterations in the tributaries and outlets of the lake, such as the regulation of the Rabnitz in 1558 (Sauerzopf 1959), the construction of a surface drainage canal, the 'Einserkanal', in 1910, which brought about an extreme overgrowth of reeds (Weisser 1973), and the new regulation, 1965, which maintains the water level near 115.5 m. Seasonal and annual fluctuations in the water level in the flat lake bed area lead to shore

displacements and to islands at low water levels and hence to grain separation by size on their shore zones and, at high water levels, to sediment deposition, especially in the southern to southeastern part of the lake.

(2) Wind influences. The prevailing NW wind or the less frequent SE wind (Steinhauser 1970) effect an aeolian transport of minerals as well as sediment transport, possibly by clockwise currents in the northern part of the lake when ice free, and grading due to the southward shear on the water.

(3) Biological changes. The rapid expansion of the reed zone over half of the lake area after 1910 resulted in an accumulation of the fine sediments ('soft mud'), especially in the northern and northwestern part of the lake, and a reduced influx of the coarse silt fraction from the lake's tributaries. The macrophyte belt off the reed zone not only accumulates the fine sediments but together with the microphytes is responsible for the formation of Mg-calcite and protodolomite, which extracts more Ca than Mg from the lake water and accounts for its relatively high Mg content.

References

Blohm, M. 1974. Sedimentpetrographische Untersuchungen am Neusiedler See/Österreich. Dissertation, University of Heidelberg.

Fuchs, W. 1974. Bericht über Exkursionen in die Oststeiermark, in das südliche Burgenland und nach Westungarn zur Klärung der Herkunft der Seewinkelschotter. Verh. Geol. Bundesanst. Wien 4: 118–121.

Geological Society of America 1963. Rock color chart. Geological Society of America, New York.

Graf, D. L. & Goldsmith, J. R. 1956. Some hydrothermal syntheses of dolomite and protodolomite. J. Geol. 64: 173–186.

Grünberg, W. & Preisinger, A. 1969. Magnesium-hältiger Calcit in tierischen Harnkonkrementen. Naturwiss. 56: 518–519.

Knie, K. 1959. Über den Chemismus der Wässer im Seewinkel und des Neusiedler Sees. Wiss. Arb. Bgld. 23: 65–68.

Küpper, H. 1957. Erläuterungen zur geol. Karte Mattersburg-Deutschkreuz. Geol. B.-A., Vienna.

Löffler, H. 1971. Beitrag zur Kenntnis der Neusiedler See-Sedimente. Sitz. Ber. Öst. Akad. Wiss., math.-nat. Kl. 179: 313–318.

Moser, I. 1866. Der abgetrocknete Boden des Neusiedler See's.-Jahrb. Geol. Reichsanst. 16, 338–344.

Peterson, M. N. A. 1966. Growth of dolomite crystals. Amer. J. Sci. 264: 257–272.

Preisinger, A., 1979. Sedimentbildung im Neusiedler See.

Riedl, H. 1964. Erläuterungen zur morphologischen Karte der eiszeitlichen Flächensysteme im Flußgebiet der Wulka und an der Südostabdachung des Leithagebirges. Wiss. Arb. Bgld. 31: 175–195.

Sauerzopf, F. 1959. Die Wasserstandsschwankungen des Sees. Wiss. Arb. Bgld. 23, 92–101.

Schiemer, F. & Weisser, P. 1972. Zur Verteilung der submersen Makrophyten in der schilffreien Zone des Neusiedler Sees. Sitz. Ber. Öst. Akad. Wiss., math.-nat. Kl. 180: 87–97.

Schroll, E. & Wieden, P. 1960. Eine rezente Bildung von Dolomit im Schlamm des Neusiedler Sees. Tschermaks Min. Petrogr. Mitt. 3 (7): 286–289.

Steinhauser, F. 1970. Kleinklimatische Untersuchungen der Windverhältnisse am Neusiedlersee. Part one: Die Windrichtungen. Időjárás 74: 76–88.

Tauber, A. F., 1959. Geologische Stratigraphie und Geschichte des Neusiedlerseegebietes. Wiss. Arb. Bgld. 23: 18–24.

Tauber, A. F. & Wieden, P. 1959. Zur Sedimentschichtfolge im Neusiedlersee. Wiss. Arb. Bgld. 23: 68–73.

Varga, L. 1931. Interessante Formationen von *Potamogeton pectinatus* L. im Fertö (Neusiedler See). Arb. Ungar. Biol. Forschungsinst. Tihany 4: 349–355.
Weisser, P. 1973. Die Verschilfung des Neusiedler Sees. Umschau 73: 440–441.
Wieden, P. 1959. Sediment-petrographische Untersuchung des Schlammes vom Neusiedler See (Bgld). Wiss. Arb. Bgld. 23: 73–80.

3 The superficial sediments: their characterization and distribution

M. Jungwirth

The sediments of the present lake are characterized by autochthonously formed minerals, Mg-calcite and protodolomite, as well as by differences in water content. The superficial sediments usually have a higher water content (>60 per cent: soft mud; 30–60 per cent: mud); only few areas are composed of compact mud with <25 per cent water content (see Chapter 12).

For further characterization of the superficial sediments, samples of the uppermost centimetre were analysed for water loss at 105°C and ignition loss of dried substance at 450°C. For this purpose sediment cores of the uppermost 10 centimetres were taken with plexiglass tubes of 5 cm diameter at 94 positions (Fig. 13.1).

From these cores subsamples of the first centimetre were taken in glass tubes 1 cm in diameter. Water content, determined as difference between wet weight and dry weight (105°C for 24 hours), is given as percentage of wet weight. The ignition loss, determined as the difference between dry weight and ash weight at 450°C for 24 hours, is given as percentage of dry weight. It was quantitatively proven by X-ray diffraction that at 450°C for 24 hours no calcite, Mg-calcite, protodolomite or dolomite was destroyed due to loss of CO_2.

As can be seen from Fig. 13.1 the highest values of water content (up to 90 per cent) and ignition loss (up to 10 per cent) occur within bays or channels of the northwestern reed belt. It should be remarked that the high values of up to 30 per cent ignition loss found for the reed belt itself by Farahad & Nopp (1966) cannot be compared because they were determined at 600°C for 8 hours. At such an ash temperature protodolomite and Mg-calcite, which are highly concentrated in the soft mud of the reed belt, decompose through loss of CO_2.

Fig. 13.2 shows the water contents and ignition losses of the first centimetre of the superficial sediment along the profile Golserkanal–Breitenbrunn. The water contents, which range from 35 to 83 per cent, parallel the ignition losses, which range from 0.9 to 6.6 per cent. The area of soft mud with water contents >60 per cent also shows the greatest ignition losses and corresponds to the macrophyte zone described by Schiemer and Weisser (1972). Only within this macrophyte zone can a high content of detritus from submerged vegetation be observed. Between positions 6 and 7 there is an inhomogeneous mixture of mud, compact mud and sand so that the ignition loss is slight (1.4–2.0 per cent). The same can be observed at position 1.

As can be seen from Fig. 13.1, soft mud (water content >60 per cent) is

*(Identified by C. Jelinek and A. Preisinger).

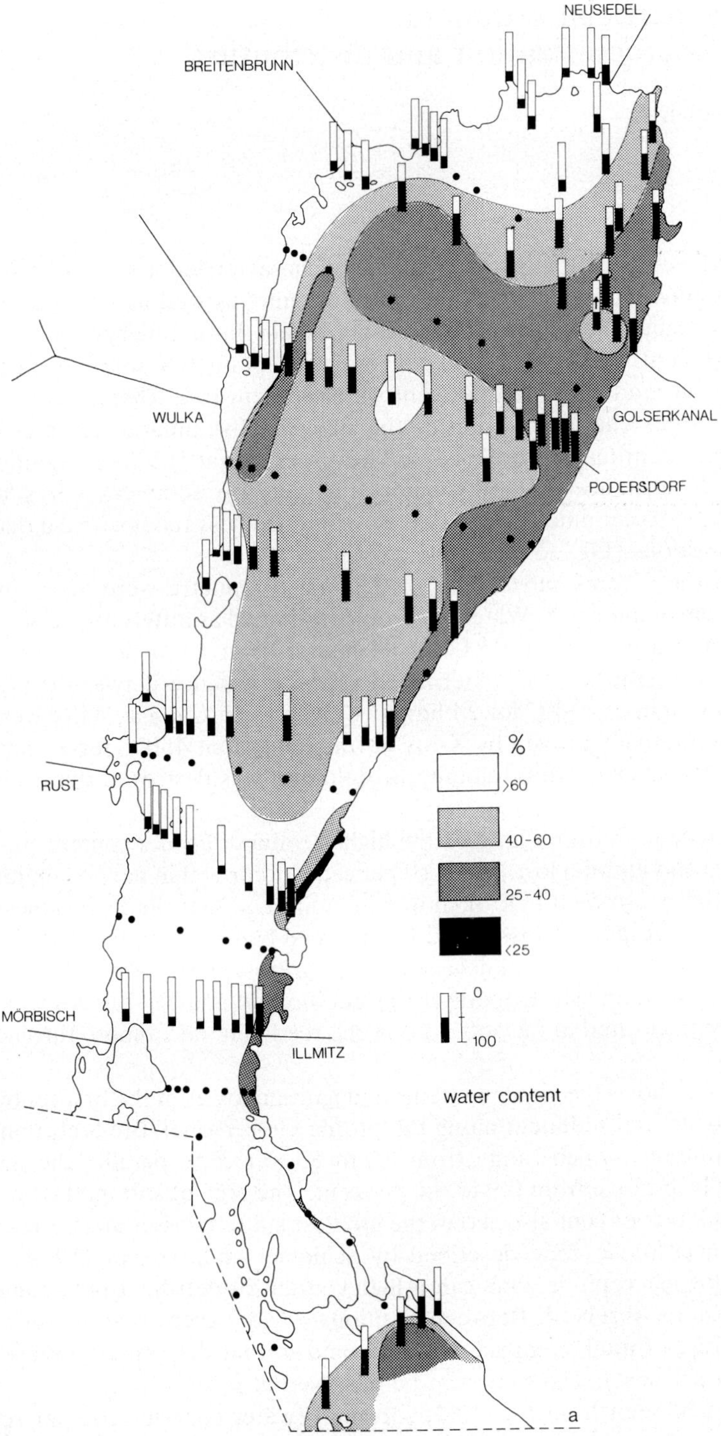
NEUSIEDEL
BREITENBRUNN
WULKA
GOLSERKANAL
PODERSDORF
RUST
%
>60
40-60
25-40
<25
0
100
water content
MÖRBISCH
ILLMITZ
a

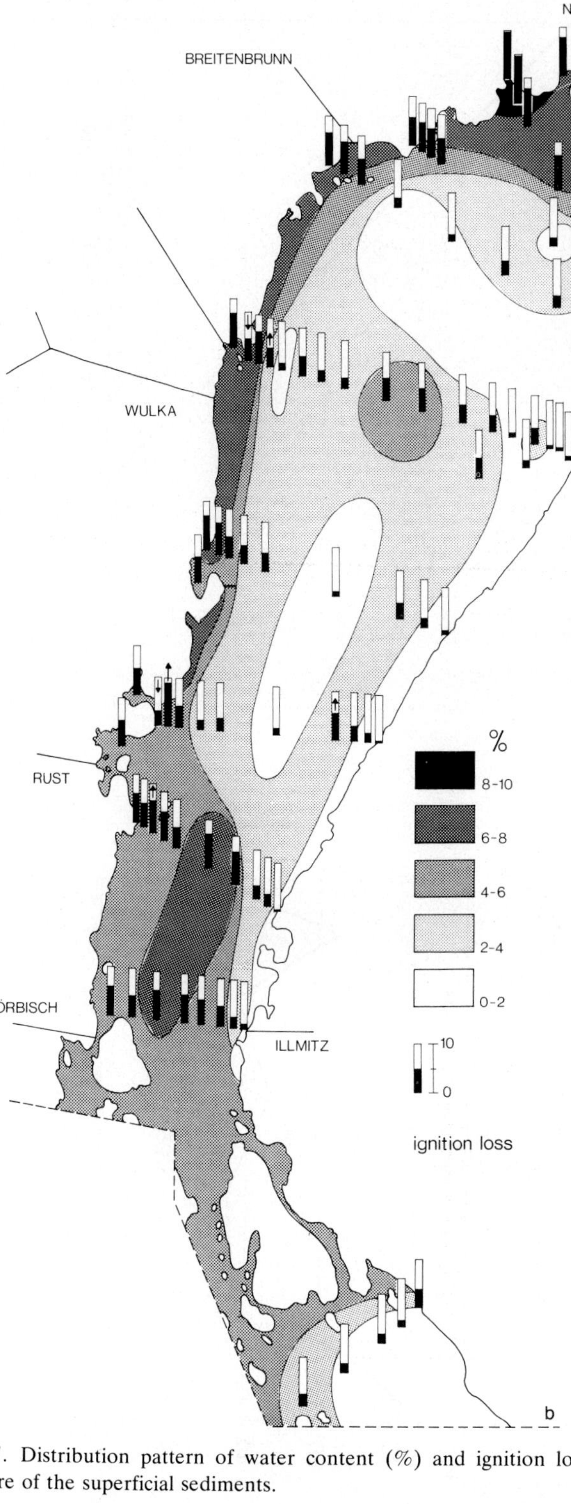

Fig. 13.1. Distribution pattern of water content (%) and ignition loss (%) within the first centimetre of the superficial sediments.

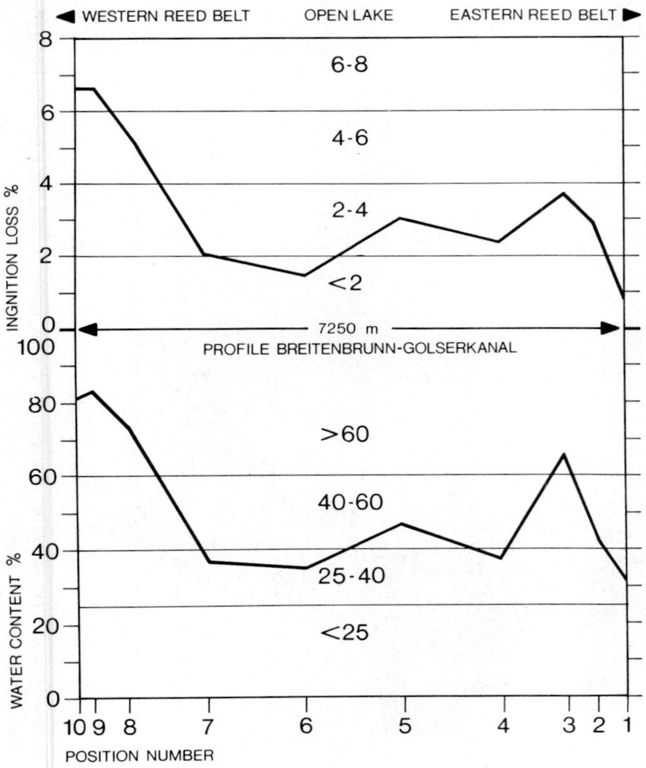

Fig. 13.2. Water contents (%) and ignition losses (%) along the profile Breitenbrunn–Golserkanal (compare Table 13.1).

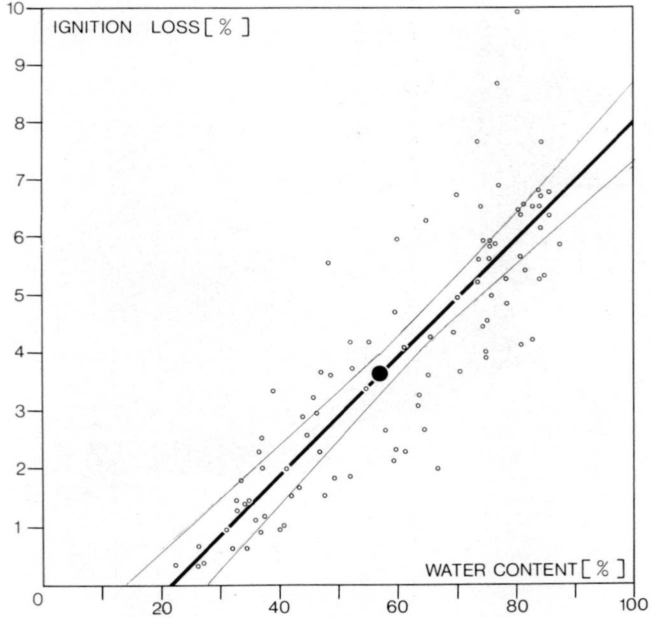

Fig. 13.3. Water content/ignition loss regression of the top centimetre at 94 positions (see Fig. 13.1) and its 99 per cent confidence limits.

$y = bx + a$, $n = 94$, $b = 0.104$, $a = -2.24$

concentrated in the northern and western parts of the lake in front of the reed belt as well as in the part south of Rust. In the eastern shore zone the bottom is formed by compact mud and sand with low water contents and ignition losses. The central part of the northern lake basin is dominated by compact or soft mud. Local combinations with sandy material result in very low ignition losses. Fig. 13.3 shows the water content/ignition loss regression and its 99 per cent confidence limits for the uppermost centimetre (compare Table 13.1).

Table 13.1. Water contents and ignition losses of the top centimetre at ten positions along the profile Golserkanal–Breitenbrunn (see Fig. 13.2).

Position	1	2	3	4	5	6	7	8	9	10
Water content (%)	31.0	44.1	64.7	36.5	46.2	34.7	37.0	73.2	82.8	80.9
Ignition loss (%)	0.9	2.9	3.6	2.3	2.9	1.4	2.0	5.3	6.6	6.6

The recent distribution pattern of the sediments is largely due to strong currents which result from the shallowness of the lake and the frequent wind action. Above all the prevailing winds from northwest induce strong clockwise currents, which, due to their sorting effect, may represent the major reason for the local occurrence of compact sediments or sands along the eastern shore. Fine and soft material eroded in that region is carried off and deposited in areas with little water movement, such as the reed belt itself, the northern and western macrophyte zone and the southern part of the lake. Here, fine inorganic material from all over the lake and organic matter produced mainly by submerged vegetation is accumulated and trapped (Jungwirth 1977).

References

Farahad, A. Z. & Nopp, H., 1966. Über die Bodenatmung im Schilfgürtel des Neusiedler Sees. Sitz. Ber. Öst. Akad. Wiss., math.-nat. Kl. 175: 237–255.

Jungwirth, M. 1978. Ein Beitrag zur Beziehung Strömung – Sedimentbeschaffenheit – Bodenfauna des Neusiedlersees. Biol. Forschungsinst. Bgld. Ber. 29, 52–59.

Schiemer, F. & Weisser, P. 1972. Zur Verteilung der submersen Makrophyten in der schilffreien Zone des Neusiedler Sees. Sitz. Ber. Öst. Akad. Wiss., math.-nat. Kl. 180: 87–97.

4 Electron microscopic investigations of the mud sediments

E. Schroll

14.1 Introduction

The following observations supplement the more modern and comprehensive investigations of Chapter 12 (see also Blohm 1974). Wieden (1959) and Schroll and Wieden (1960) first described the occurrence of protodolomite in the mud sediments of this lake. They published the first electron micrograph of the mud and suspected the diagenetical alteration of protodolomite into dolomite. One of the objects of this supplementary study was to provide better electron microscopic documentation of the carbonate mineralization.

The samples were taken from a 150 cm core, obtained in 1969 by drilling in the middle of the lake in the east-west profile Rust-Untere Wiese (Core 3 in Schroll & Krachsberger 1977).

The sample (No. 3/1) consists of soft, dark-grey mud from a depth of 0–15 cm containing Mg-calcites, muscovite, illite and quartz. The second sample (No. 3/5) was selected from the brownish-grey silt from 60–75 cm. The mineral composition is characterized by dolomite, little calcite, feldspars, muscovite and 14 Å-chlorite.

In conformity with Preisinger (Chapter 12) the following minerals have been observed in the lake sediments:

Carbonates: Mg-calcite (low-Mg-calcite), protodolomite (high-Mg-calcite), calcite and dolomite;

Silicates: opal, quartz, 14 Å-chlorite, illite, mica, montmorillonite, oligoclase and alkali-feldspar;

Other minerals: pyrite.

The occurrence of Mg-calcites is more or less restricted to the soft and compacted mud layers containing 30 per cent or more water. The sandy sediments contain dolomite and calcite only. The Mg-calcites disappear first.

14.2 Investigations using scanning electron microscopy (SEM)

Scanning electron micrographs of low-Mg-calcite from Lake Balaton (Hungary) have been published by Müller (1969). Imperfectly formed crystallites of Mg-calcite in the range of 1 μm form aggregations of more than 10 μm.

The micrographs of the soft mud (Fig. 14.1) of Neusiedlersee reveal larger particles, mostly sheets of phyllosilicates (illite or mica) of 10 μm or more, embedded in aggregations of small particles (1 μm or less). Similar aggregations to those seen by Müller (1969) were observed. In addition, a mixture of fine particles is found, consisting of carbonates and silicates. This can be seen from an X-ray analysis diagram (Fig. 14.2) taken from an agglomeration of

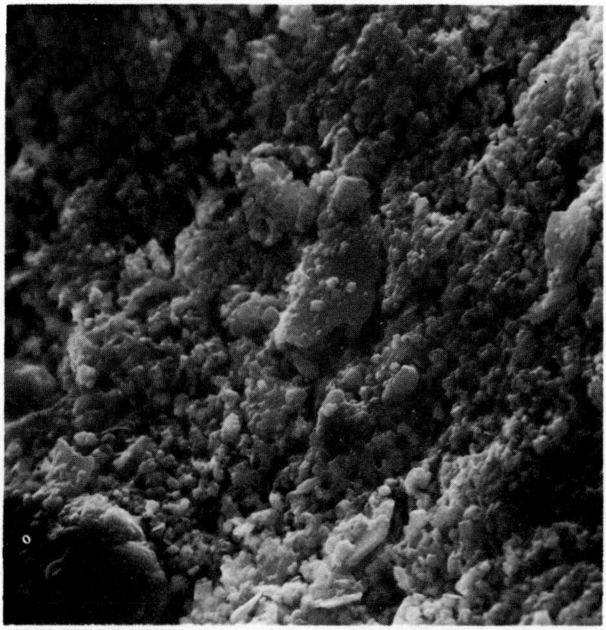

Fig. 14.1. Scanning electron micrograph of soft mud of core 3/5–10 (magnification 1,000×).

fine particles. It is not easy to differentiate carbonates and silicates on the basis of morphology alone.

Scanning of the distribution of the main elements Si, Al and K, and Mg gives an idea of the distribution of the mineral phases. The largest crystals are sheet-like silicates. However, the pattern of silicon covers a larger area than

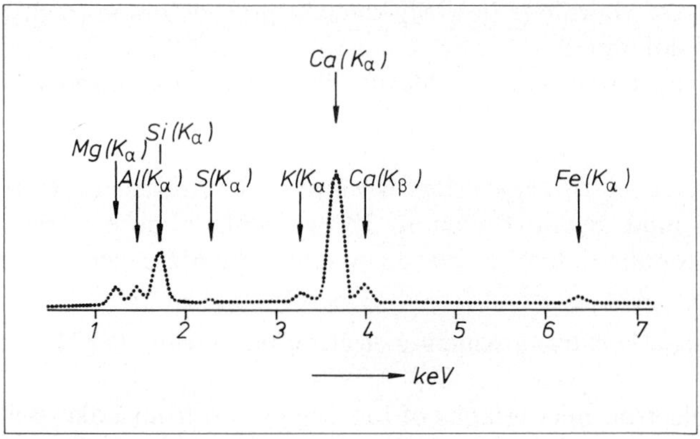

Fig. 14.2. X-ray diagram of the soft mud.

the K-Al pattern. Since no small quartz grains, opal or diatom debris are present, calcium and magnesium do not stand out very distinctly.

14.3 Investigations using transmission electron microscopy (TEM)

Pictures of the soft mud zone are compared with those of the deeper layers (Figs. 14.3, 14.4).

The soft mud is characterized by the presence of organic materials (plant debris and so on). Silicate sheets, black crystallites (0.1 μm or less) and their aggregations are found. The sample of the solidified sediment, however, consists of larger silicate particles and crystallites (0.1 up to 1 μm) of apparently rhombohedral habit.

Most instructive pictures are obtained by applying the replication technique, in which a carbon film is deposited using vacuum evaporation. The

Fig. 14.3. Micrographs of soft mud of core 3/0–15 cm. Carbon replica, stripped by removing the carbonates in 2nHCl, showing mainly biogenic debris and carbonate particles (magnification 15,000×).

replica is then stripped by removing the mineral particles in acids, carbonates in hydrochloric acid (2nHCl), or silicates in hydrofluoric acid (1:1). The morphological properties of the minerals are conserved by the carbon replica. This method can probably be developed further for diagnostic purposes.

Following treatment with hydrochloric acid the carbonates of the soft mud zone are shown to form crystallite aggregations without sharp borders. Well-formed rhombohedra are rarely observed. However, the rhombohedral habit can be recognized on the surfaces.

In contrast to these observations, the carbonate particles of deeper layers are characterized by sharp edges, rhombohedral crystallites and distinct growth phenomena (Fig. 14.3, 14.4).

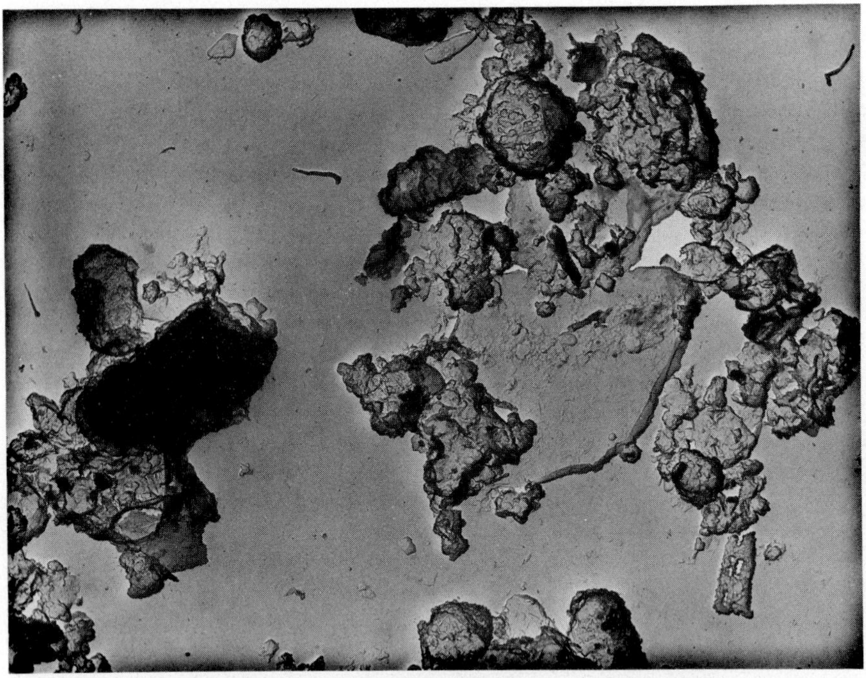

Fig. 14.4. Micrograph of the Fig. 14.3 sample. Carbon replica, stripped by removing both the silicates and carbonates, showing the sheets of mica and Mg-calcites, partly ball-like (magnification 7,500×).

14.4 Conclusion

In conclusion it can be said that although the carbonate particles investigated cannot be recognized, the observations conform with the results of Müller et al. (1972) who suggested that high-Mg-calcite (protodolomite) is transformed into dolomite during diagenesis. The low-Mg-calcite is less stable and disappears first.

The observed distribution of the Mg/Ca ratios in the pore solutions of the lake sediments, especially of the zone of compact mud, would support this idea. Further, the morphology of the allochthonic carbonates, calcite and dolomite, supplied by fluvial and aeolic transport deserve further study.

References

Blohm, M. 1974. Sedimentpetrographische Untersuchungen am Neusiedler See/Österreich. Dissertation, University of Heidelberg.

Müller, G., Irion, G. & Förstner, U. 1972. Formation and diagenesis of inorganic Ca-Mg-carbonates in lacustrine environment. Naturwiss. 59: 158–164.

Müller G., 1969. Sedimentbildung im Plattensee, Ungarn. Naturwiss. 56: 606–615.

Schroll, E. & Wieden, P. 1960. Eine rezente Bildung von Dolomit im Schlamm des Neusiedler Sees. Tschermaks Min. Petr. Mitt. 3(7): 286–289.
Schroll, E. & Krachsberger, H. 1977. Beitrag zur Kenntnis des Chemismus der Porenwässer des Neusiedlersees. Biol. Forsch. Anst. Bgld. 24: 35–62.
Wieden, P. 1959. Sediment-petrographische Untersuchung des Schlammes vom Neusiedler See (Bgld). Wiss. Arb. Bgld. 23: 73–80.

15 Optical properties, colour and turbidity

M. Dokulil

15.1 Surface incident radiation

Insolation of the lake area varies little in consecutive years (Table 15.1) and is usually higher than in other parts of Austria (Chapter 5). The maximum

Table 15.1. Total annual incoming radiation (TIR), albedo and the deviation of the radiation integral from the long-term average.

	TIR (Kcal m^{-2})	Deviation from the average (%)	Albedo (%)
1967[a]	1,021,040	+5.0	14.6
1968	962,389	−1.0	11.3
1969	979,707	+0.8	12.8
1970	987,684	+1.6	16.1
1971	992,180	+2.0	14.4
1972	930,186	−4.2	
1973	~1,010,487[b]	~+4.0	

[a] According to Mahringer (1969)
[b] Extrapolated because 45 days were missing; total recorded was 899,580.

deviation from the average of this region was five per cent during the period 1967 to 1973.

Variation of total incoming radiation (TIR), albedo and wind speed irrespective of direction are shown in Fig. 15.1. Surface radiation loss is unusually high (mean 13.8 per cent) compared with that of other bodies of water on account of considerable back-scattering due to turbidity. The latter results from inorganic particles stirred up from the sediment by wind action. Turbidity and wind speed account for more than 50 per cent of the observed variance in the percentage albedo (r = 0.71, F (4.13): 3,260, $p \leqslant 0.05$).

15.2 Underwater light attenuation

Light measurements were performed with a selenium photocell and Schott filters RG_2, VG_9, BG_{12} and $BG_{12}+UG_1$, with optical maxima at 650, 540, 460 and 390 nm respectively. Total percentage energy flux, E, was calculated according to Vollenweider (1969).

Examples of light penetration are illustrated in Fig. 15.2. Red light is the most penetrating spectral component in the reed water, closely followed by green due to the high content of humic compounds (D in Fig. 15.2). In the lake water under the ice, C, green light replaces the red on account of

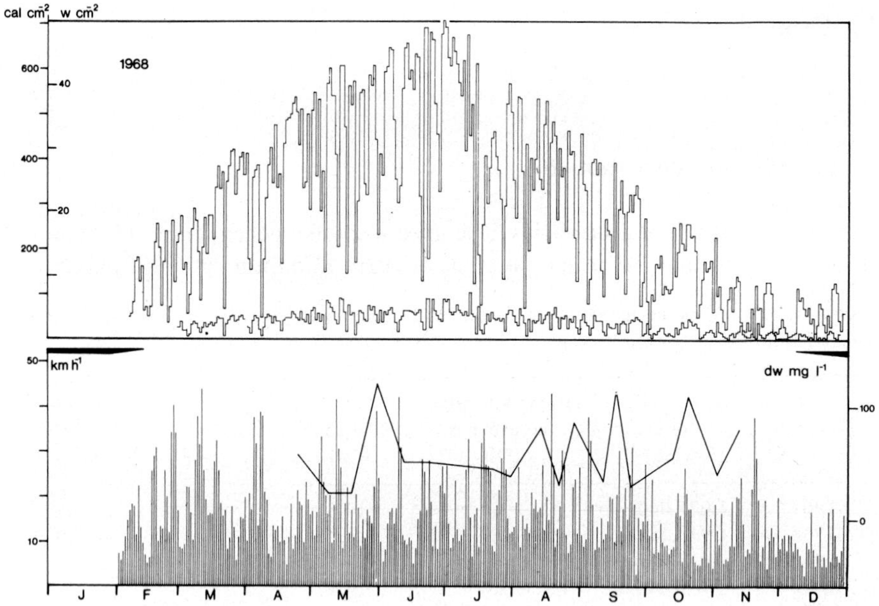
cal cm^{-2} w cm^{-2}
1968
600
400
200
40
20
50
km h^{-1}
10
dw mg l^{-1}
100
0
J F M A M J J A S O N D

Fig. 15.1. Caption on p. 154.

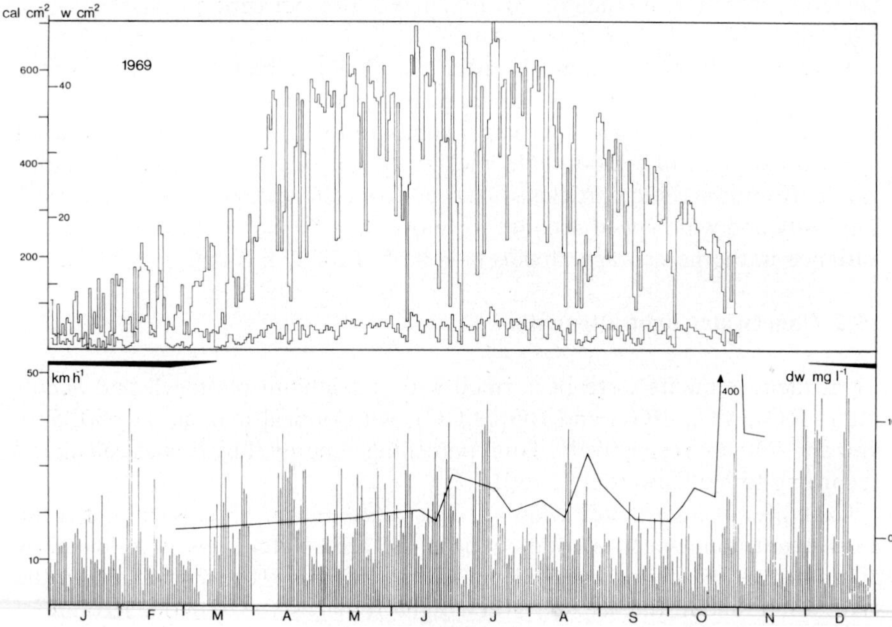
cal cm^{-2} w cm^{-2}
1969
600
400
200
40
20
50
km h^{-1}
10
dw mg l^{-1}
400
100
0
J F M A M J J A S O N D

cal cm⁻² w cm⁻²
1970
km h⁻¹
dw mg l⁻¹
J F M A M J J A S O N D

Fig. 15.1. Caption on p. 154.

cal cm⁻² w cm⁻²
1971
km h⁻¹
378
dw mg l⁻¹
J F M A M J J A S O N D

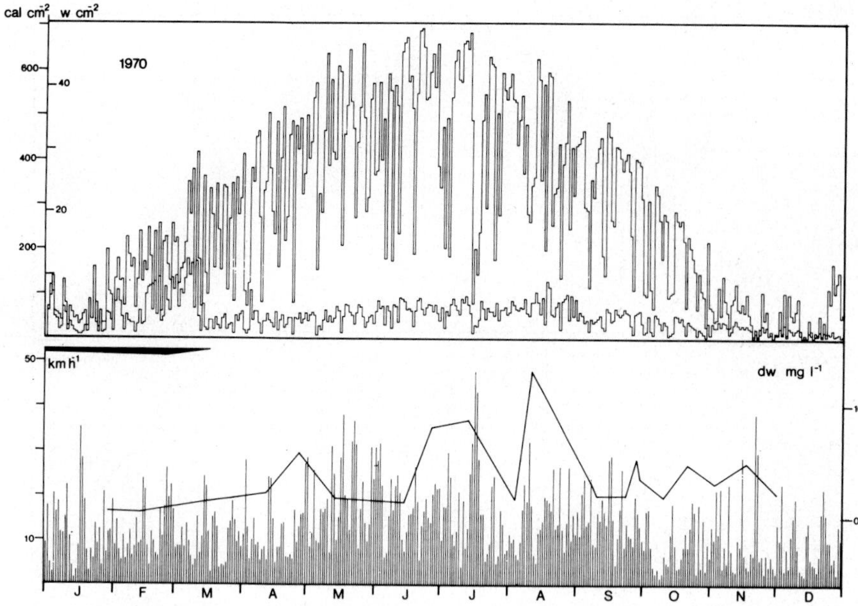

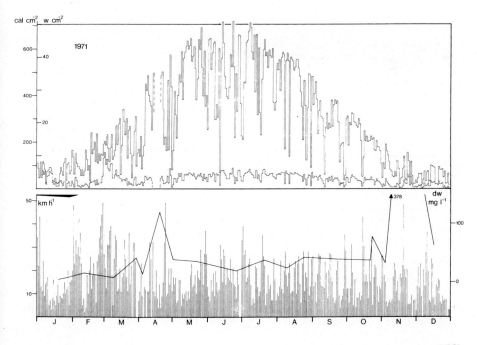

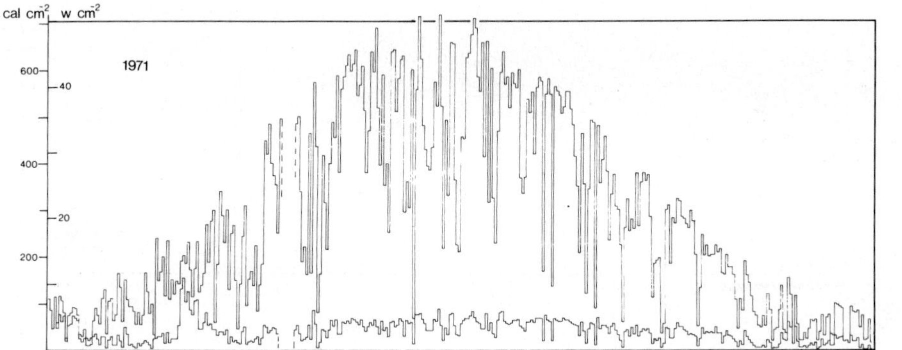

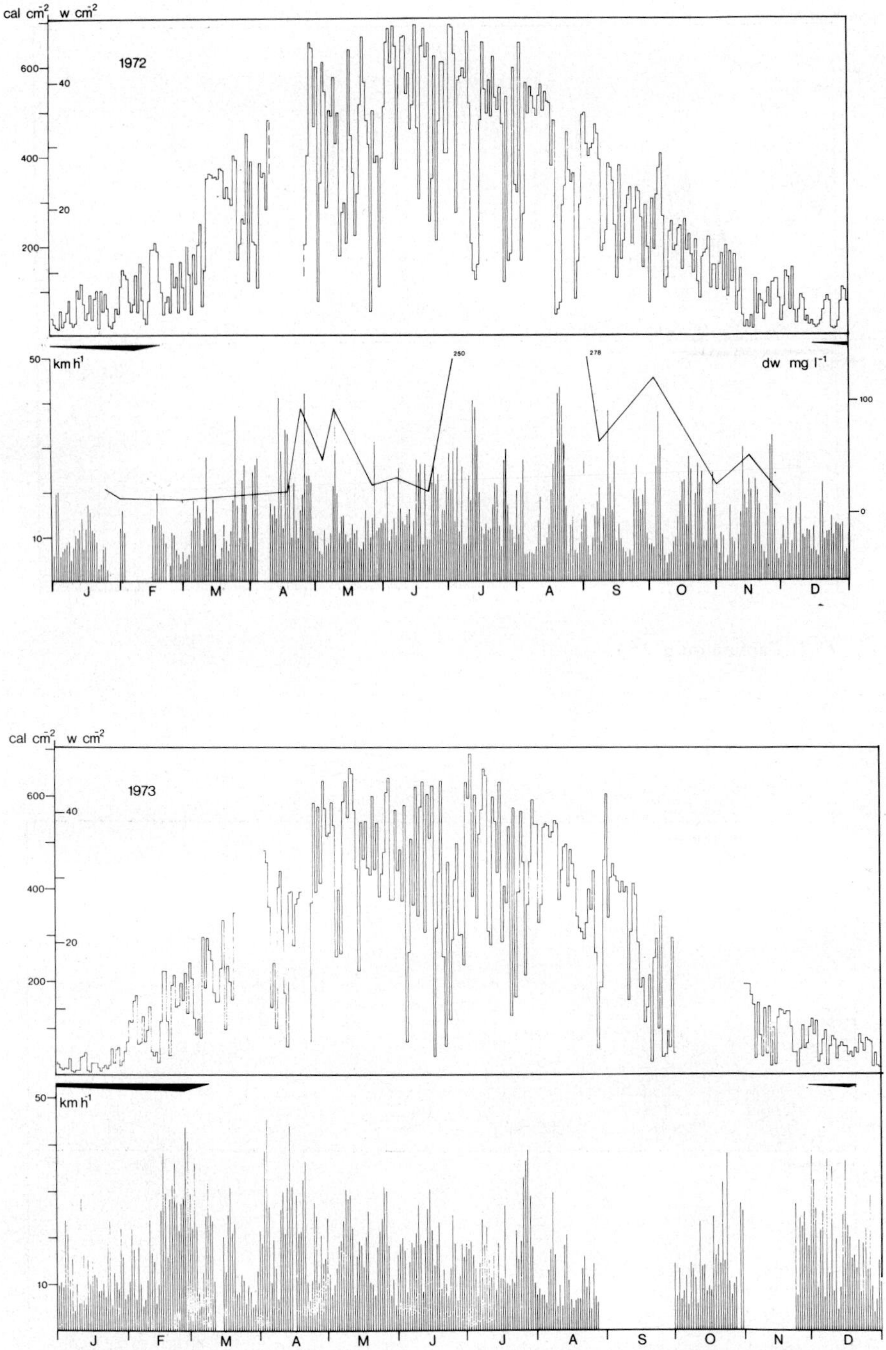

Fig. 15.1. Daily variation of total incoming radiation (TIR) for the years 1968 to 1973, albedo and wind speed irrespective of direction, with turbidity measured as mg l^{-1} dry weight.

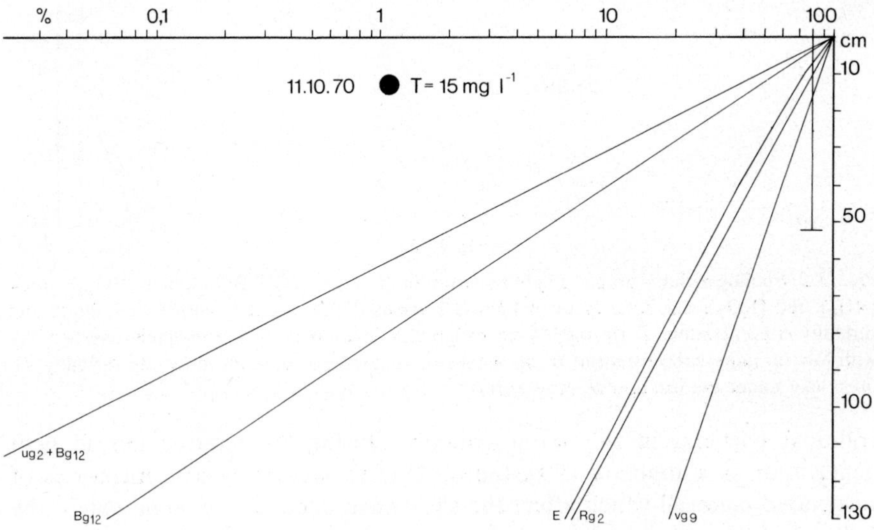

Fig. 15.2a. (Legend, see p. 156)

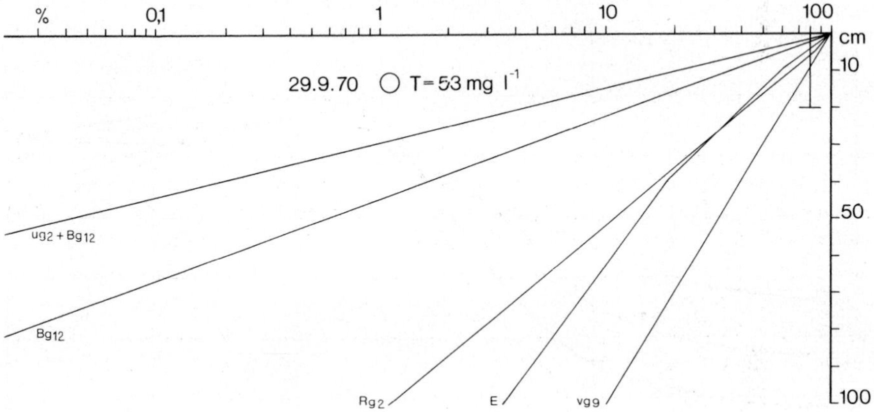

Fig. 15.2b. (Legend, see p. 156)

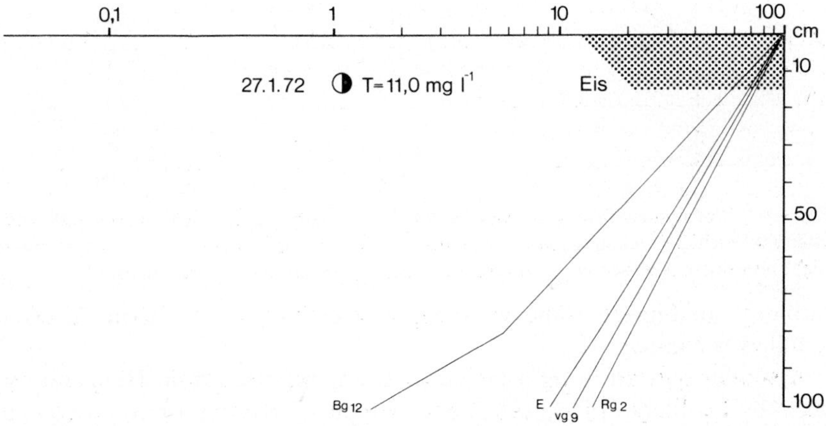

Fig. 15.2c. (Legend, see p. 156)

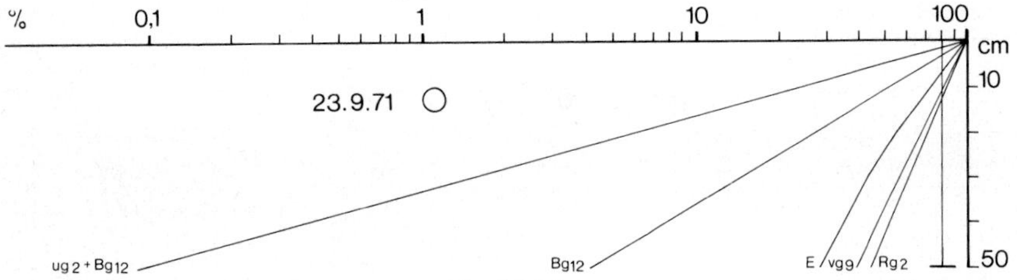

Fig. 15.2d.

Fig. 15.2. Similogarithmic graph of light penetration of violet (UG_2+BG_{12}), blue (BG_{12}), green (VG_9), red (RG_2), and total incident light (*E*) versus depth, together with Secchi depth and turbidity concentration, *T*, (in mg l^{-1} dry weight): *A*, open lake on a completely overcast day with low turbidity concentration; *B*, open lake on a cloudless day with moderate turbidity; *C*, open lake under the ice; and *D*, reed water.

colloidal particles in the water column. During the ice-free period light penetration is a function of turbidity. Due to increasing concentrations of suspended material which affect the shortwave band and to some extent the red end of the spectrum green is the dominating wavelength (*A* and *B*). The blue spectral block is strongly absorbed in any case due to water colour or

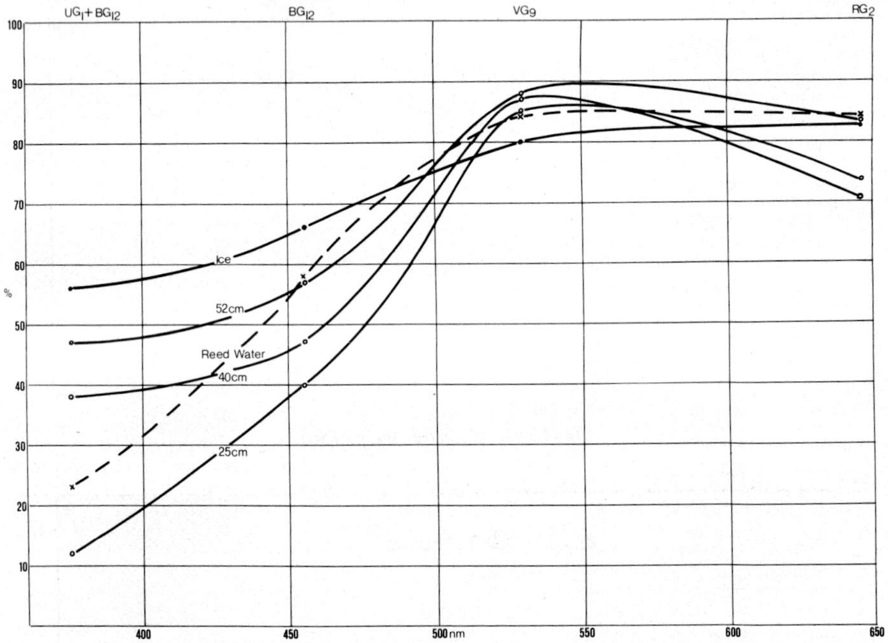

Fig. 15.3. Percentage light transmission for 10 cm path-length versus wavelength and the different Secchi disc readings indicated on the curves, showing wavelength selectivity of turbidity. Light transmission under ice and in the reed water (broken line) is also indicated.

turbidity. In Fig. 15.3 the wavelength selectivity for different Secchi disc readings is compared.

A similar pattern of light penetration was reported from Hungarian soda lakes by Dvihally (1958, 1961) and is typical of turbid or coloured waters

(Sauberer 1945; Schmolinsky 1954; Vollenweider 1960, 1961; Talling 1965, Talling et al. 1973).

Light attenuation is compared with earlier measurements (see Fig. 15.4).

According to Sauberer (1952) and Panosch (1973) the highest transparency of the lake water is in the orange spectral region, which was not

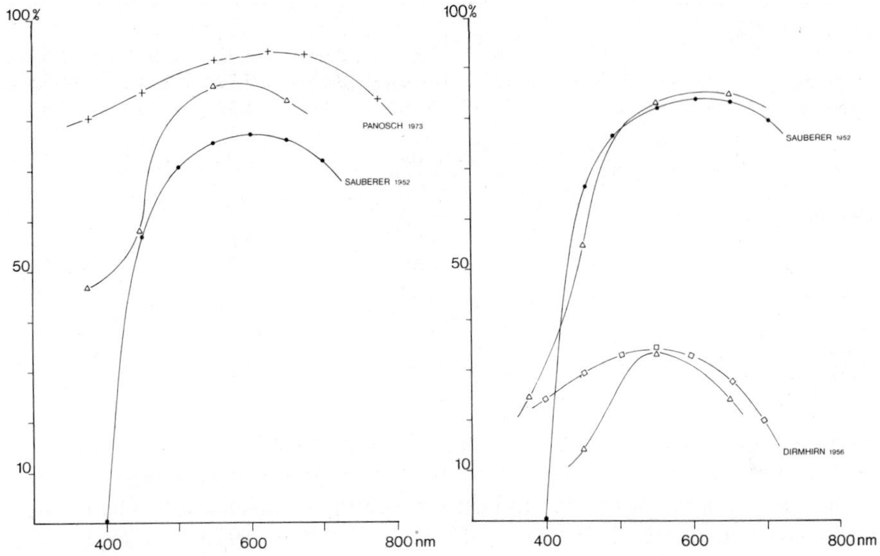

Fig. 15.4. Comparison of percentage light transmission found in this investigation with data of Sauberer (1952), Dirmhirn (1956), and Panosch (1973). Left: open lake in summer; upper right: reed water; lower right: beneath ice cover.

included in this study. The large difference at the shortwave band between the data of Sauberer and those of Panosch and the present author might be due partly to instrumentation and partly to the lower water level and therefore greater turbidity in the early fifties. Deviations from Panosch's data can be reduced to the fact that his readings were only taken on calm, cloudless days.

Except at 400 nm the two penetration curves from the reed water resemble each other (*B* in Fig. 15.4). Light measurements made below 35 cm of ice as reported by Dirmhirn (1956) showed a considerable blue light transmission which could not be confirmed recently (*B*). This discrepancy might be explained by differences in turbidity concentration and by the fact that Dirmhirn made only one measurement close to the reed fringe.

The idea of Sauberer and Ruttner (1941) of characterizing the optical properties of lakes by a code number is inapplicable to Neusiedlersee. Since percentage transmission per metre at 400, 500 and 600 nm is often considerably below ten per cent the code becomes meaningless when rounded off to tens.

The vertical extinction coefficients, ε_v^λ, calculated according to Vollenweider (1955), are collected in Table 15.2. The mean total energy coefficient (E_{b+g+r}) from this study is very close to the single observation made by

Table 15.2. Vertical extinction coefficients (ε_v^λ, ln m^{-1}) of different wavebands measured in Neusiedlersee by various authors. Values of Panosch and Dokulil are means. Data of Sauberer and Dirmhirn are from single observations.

	nm						
	450	500	550	600	650	700	$E_{(b+g+r)}$
Sauberer (1952)							
relative calm	5.87	3.44	2.89	2.68	2.72	3.29	3.83
stormy	20.54	19.02	18.34	17.15	17.72	19.66	18.87
reed water	4.08	2.51	2.06	1.80	1.76	2.14	2.63
Panosch (1973)	1.88		1.06	0.96	0.99	2.63	1.31
Dokulil (this study)							
open lake	6.57		1.97		2.91		3.31
reed water	6.05		1.85		1.55		2.58
lake ice-covered	4.61		2.62		3.27		3.50
Dirmhirn (1956)							
lake ice-covered	3.54	3.16	3.08	3.17	3.64	4.74	3.42

Sauberer on a relatively calm day in 1951. Considerable differences can be observed in specific wave bands, especially in the green spectral region, which at present has a smaller ε_v due to better green light transmission. The highest mean coefficient recorded during the years 1968 to 1973 was 13.45 (see Table 15.3), whereas a mean vertical extinction coefficient of 18.87 has been

Table 15.3. Relationship between Secchi depth (z_{SD}), total vertical extinction coefficient (ε_v^E), depth of euphotic zone (z_{eu}) and relative total light intensity at z_{SD} ($I_{z_{SD}}$). Values in italics are graphical extrapolations. Indices of light penetration (IP) and optical depth notation (OC) for z_{eu} are also given.

z_{SD} (m)	ε_v^E (mm^{-1})	$z_{SD} \cdot \varepsilon_v^E$	z_{eu} (m)	z_{eu}/z_{SD}	$I_{z_{SD}}$ (%)	IP	OD
0.40	3.38	1.35	1.80	4.50	28.5	0.30	8.8
0.45	1.98	0.89	*2.50*	5.55	42	0.50	7.1
0.52	2.08	1.08	*2.00*	3.84	30	0.48	6.0
0.47	2.31	1.09	1.80	3.83	33	0.43	5.8
0.58	2.32	1.35	*2.00*	3.46	27	0.43	6.7
0.28	3.81	1.07	1.30	4.64	40	0.26	7.1
0.08	13.45	1.08	0.37	4.63	39	0.07	7.2
0.30	2.85	0.86	1.80	6.00	43	0.35	7.4
0.33	3.03	1.00	1.60	4.85	37	0.33	7.0
0.10	10.60	1.06	0.46	4.60	37	0.09	7.0
0.15	7.05	1.06	0.69	4.60	38	0.14	7.0
0.20	5.40	1.08	0.91	4.55	36	0.19	7.1
0.25	4.09	1.02	1.14	4.56	32	0.24	6.7
Mean		1.08		4.59	35.7	0.29	6.99
Standard deviation		±0.14		±0.67	±5.2	±0.14	±0.71

calculated from Sauberer's data on a stormy day. The extreme values for the different ε_v^λ on this day may possibly find a similar explanation to that outlined previously.

Figures for the reed water and under the ice cover are in much better agreement with earlier observations. Better green light transmission and less blue are observed in both cases. The effect at the shortwave band is assumed to result from a higher concentration of colloidal particles.

The applicability of the standard attenuation concept, proposed by Vollenweider (1961), was tested in Fig. 15.5 by plotting ε_v^E against corresponding ε_v^λ.

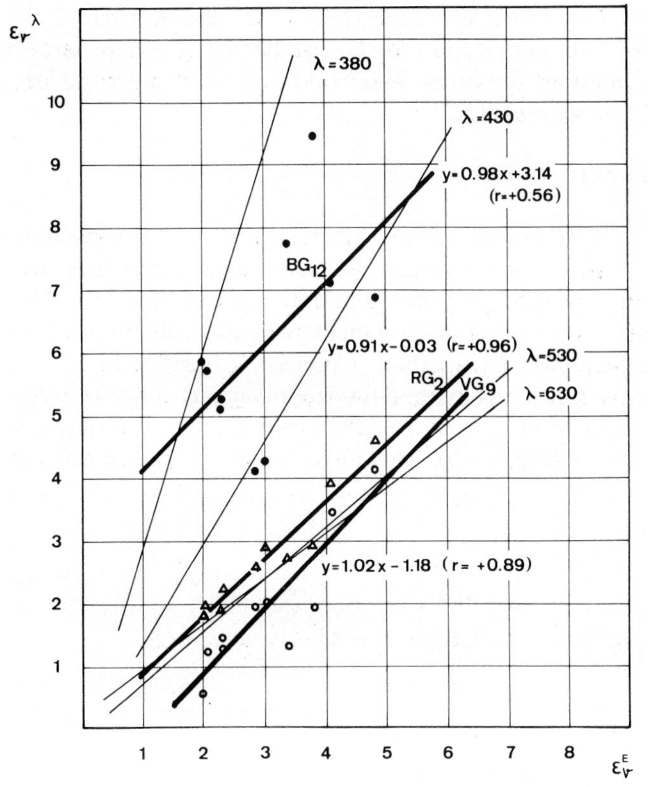

Fig. 15.5. Specific vertical extinction coefficient (ε_v^λ) versus corresponding total vertical extinction coefficient (ε_v^E) for three different wavebands. Calculated regression lines are given together with correlation coefficients. Standard attenuation curves (thin lines) proposed by Vollenweider (1961) for comparison.

Whereas values for the green and red spectral region fit the standard lines relatively well, blue light extinction does not fit at all. Although light attenuation in Neusiedlersee is primarily a function of scattering, Vollenweider's concept has to be precluded because variable light absorption in the blue leads to a large variation of data points.

Scattered light calculated as a percentage of incoming radiation, according to Withney (1938), is 28.1, 45.1, and 49.4 per cent for blue, green and red respectively.

Indices of light penetration, IP = $(\varepsilon_v^E)^{-1}$, and optical depth, OD = $z\varepsilon_{min}$ $(\ln 2)^{-1}$ introduced by Talling (1965) are included in Table 15.3. Each unit of optical depth corresponds to a halving of light intensity. The minimum vertical extinction coefficient is replaced here by ε_v^E. Talling also related optical depth to euphotic zone (z_{eu}), which is defined as the depth of one per cent of subsurface radiation. In Neusiedlersee z_{eu} corresponds on an average to 6.99 units OD and consequently

$$z_{eu} \simeq 4.8\ (\varepsilon_v^E)^{-1}$$

as compared to a factor of 3.7 established by Talling (1965) for East African lakes, and 4.2 by Balon & Coche (1974) for Lake Kariba.

Indices of light penetration for Neusiedlersee generally agree with Secchi depth. Calculations for other water bodies suggest a greater deviation with increasing transparency.

15.3 Secchi disc visibility

Transparency, measured with a white disc of 25 cm diameter (average reflection 84 per cent, no viewer used), varies tremendously between 8 and 80 cm during the year, depending entirely on turbidity (Fig. 15.6, vertical). Measurements are in good agreement with data obtained by Kopf (1967) using a slightly different method. Comparing our Secchi disc readings and turbidity concentrations with data from various authors using the same technique both in marine and fresh waters reveals a perfect correlation of both parameters (Fig. 15.6). This implies that the Secchi disc can be widely

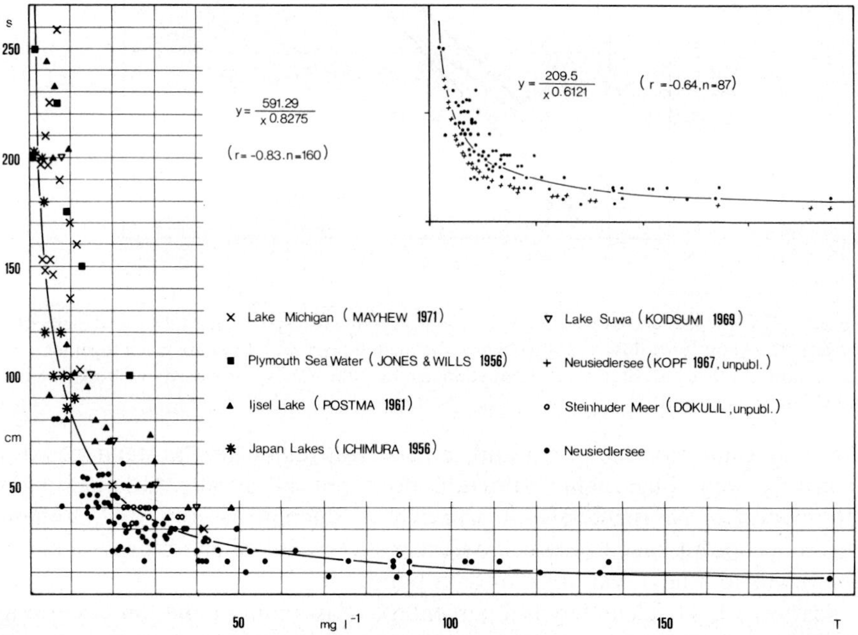

Fig. 15.6. Calculated regression curves of Secchi depth (cm) versus inorganic turbidity (mg l^{-1} dry weight). Upper right: Neusiedlersee alone.

employed to predict turbidity. Application of this correlation to other water bodies is possible only if grain sizes are approximately the same, since particle size influences Secchi disc visibility (Postma 1961).

Thirteen Secchi disc readings have been systematically combined with simultaneous photometer measurements of total light penetration (Table 15.3). The mean factor relating Secchi depth to the total vertical extinction coefficient as calculated from these data amounted to 1.08. The range of this factor (0.86 to 1.35) corresponds with those quoted by or calculated from different authors (Table 15.4). On an average z_{SD} equals the depth at which

Table 15.4. Factors relating Secchi depth to euphotic zone, the mean vertical extinction coefficient and incident light left at z_{SD} for a variety of bodies of water.

Lake	z_{eu}/z_{SD}	% irr. at z_{SD}	$\varepsilon_v^E z_{SD}$	Reference
Torneträsk	1.3	2.5	3.64	Rhode (1969)
Zürichsee	1.6	8.0	2.90	Rhode (1969)
Bodensee (Lake Constance)	1.7	7.0	2.75	Rhode (1969)
Lago Maggiore	1.7	6.3	1.76	Rhode (1969)
Lac Leman (Lake Geneva)	1.9	10.0	2.42	Rhode (1969)
Lake Erken	2.0	8.0	3.02	Rhode (1969) and Vollenweider (1960)
Lago Mergozzo	2.0	5.0	3.26	Vollenweider (1960)
Ossiachersee	2.0	13.0	2.31	Rhode (1969)
Lago di Garda	2.1	20.0	2.21	Rhode (1969)
Gr. Plöner See	2.2	6.0	2.09	Rhode (1969)
Mariut	2.2	10.0	2.83	Vollenweider (1960)
Wörthersee	2.4	15.0	1.90	Rhode (1969)
Lago Varese	2.4	20.0	1.93	Rhode (1969)
Nine Japanese lakes	2.4	15.0	1.90	Ichimura (1956)
Finstertaler See	2.5	16.7	1.91	Tilzer (1972)
Vombsjön	2.5	13.8	2.21	Gelin (1975)
Oceans	2.5		0.85	Strickland (1958)
Lake Chad	2.7	17.0	1.40	Lemoalle (1973)
George's Bank	3.0			Riley (1941)
Goleta Bay	3.1	22.6	1.57	Holmes (1970)
Lago Maggiore	3.5	15.0	1.30	Vollenweider (1960)
Lake Kariba	3.6	22.7	1.96	Balon & Coche (1974)
Lago Tarfala	3.8	26.0	1.22	Rhode (1969)
Great Slave Lake	4.3			Rawson (1950)
Neusiedlersee	4.6	35.7	1.08	Dokulil, this study
Lake Erie	5.0			Verduin (1956)

35.7 per cent of incident light is found. The general order of magnitude was assumed to be about 15 per cent (Vollenweider 1969), but from Table 15.4 it appears that it ranges at least from 2.5 per cent up to 35 per cent. A general tendency of increasing equivalent light with decreasing Secchi depth can be discerned.

On this basis depth of visibility can be used to estimate the euphotic zone. In Neusiedlersee Secchi depth has to be multiplied by 4.6 on an average to obtain the depth of the euphotic zone. The range (3.46 to 6.00) in Table 15.3

exceeds the extreme values quoted by different authors for marine and fresh waters (Table 15.4).

15.4 Colour of the lake

The apparent colour of the open lake water varies between greenish, grey and white according to weather conditions and turbidity concentration. Water inside the reed belt is invariably coloured brown by humic substances.

To determine the true colour and to detect humic compounds in the open lake, measurements were made on membrane-filtered water in the spectrophotometer at 430 nm (Table 15.5). The highest values were recorded in

Table 15.5. Ranges of water colour in 1969 characterized by the extinction $\times 10^{-3}$ at 430 nm of membrane-filtered lake water using distilled water as a blank.

Month	Water from the reed belt	Submerged macrophyte zone	Open lake
Jan.		15	
Feb.		23	
Mar.		25	
Apr.		10	
May	21–22	7–15	7
Jun.	18–23	12–18	8–12
Jul.	12–18	12	10–14
Aug.	22–30	10–13	8–12
Sep.	19–32	9–20	9
Oct.	10–36	5–20	10–12
Nov.	19–20	18–20	10
Dec.	17	25	8
Year	10–36	5–25	7–14

the reed water, especially in autumn when humic substances are liberated as a result of foliage loss of *Phragmites*.

Dependent upon wind direction trails of brown water sometimes extend out into the lake. This is demonstrated in Fig. 15.7 using measurements made in a bay sheltered from west and north by *Phragmites*. The highest colour concentrations occur when the wind blows from the north, whereas turbidity is highest and colour lowest with southerly winds.

15.5 Turbidity

The concentration of total suspended inorganic particles was measured gravimetrically on membrane filters (pore size 0.2 μm) using the method of Banse et al. (1963). Dry weight analysis of turbidity proved to be the most reliable technique (Dokulil 1976), in agreement with results of Duckrow & Everhart (1971). The concentration of the particles stirred up from the

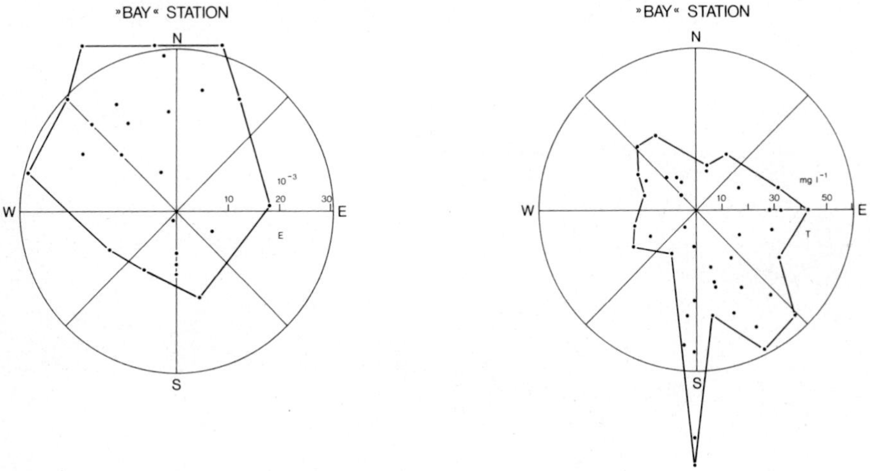

Fig. 15.7. Colour (extinction $\times 10^{-3}$) and turbidity (mg l^{-1} dry weight) for different wind directions at a 'bay station' sheltered from west and north by *Phragmites*.

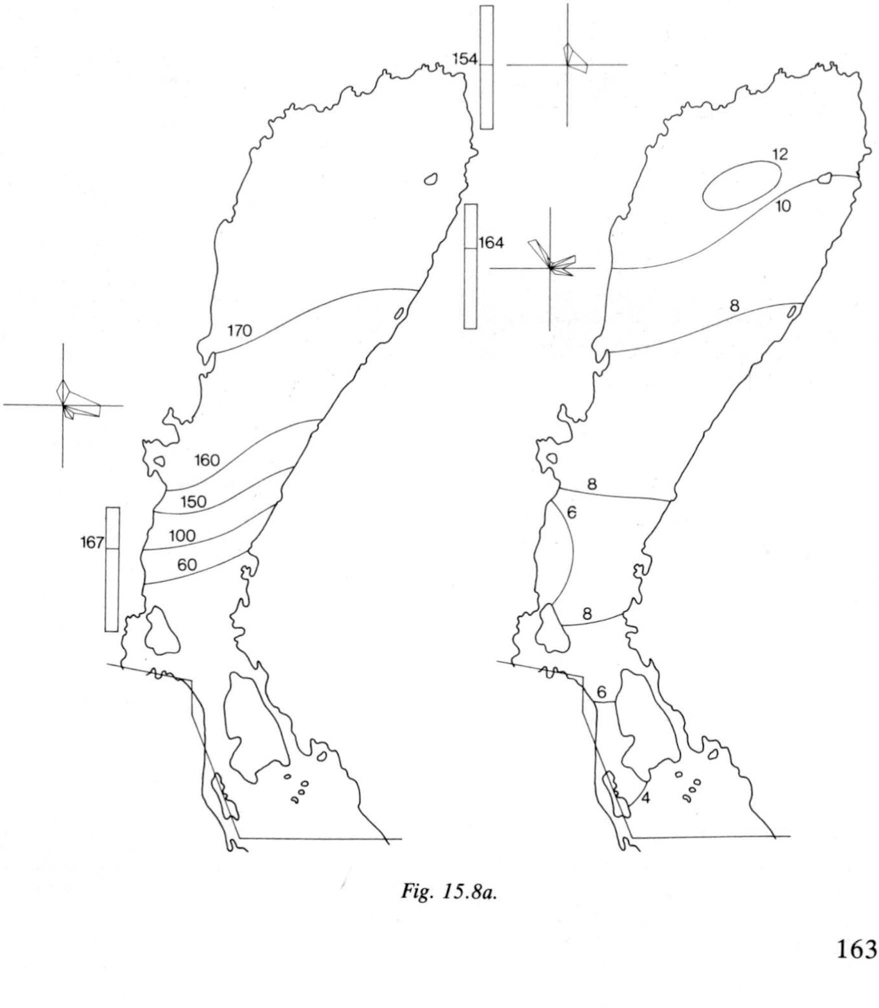

Fig. 15.8a.

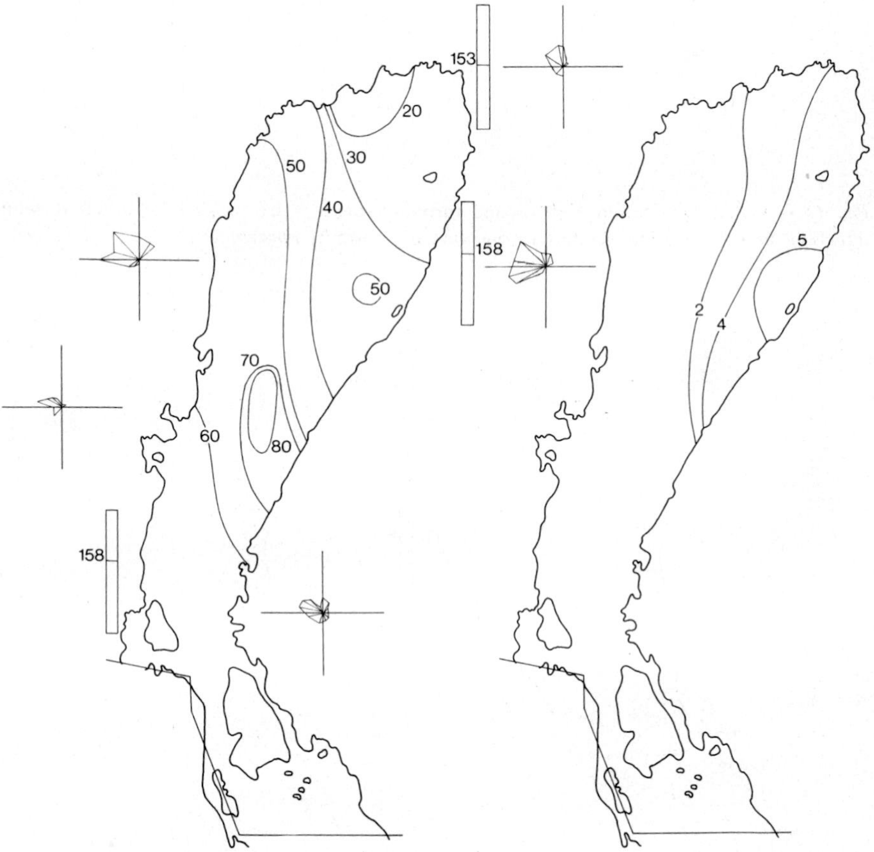

153
20
30
50
40
50
158
70
60
80
158
5
2
4

Fig. 15.8b.

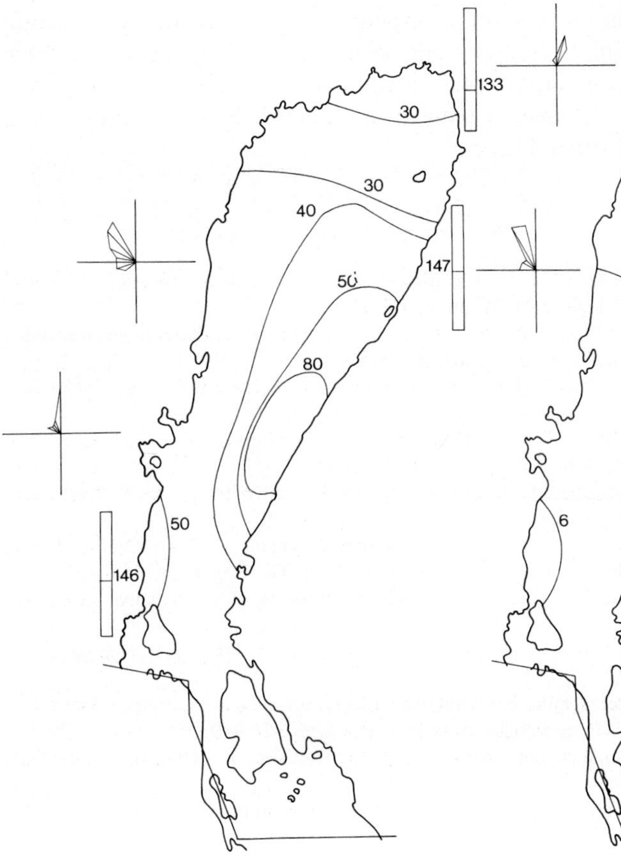

Fig. 15.8c.

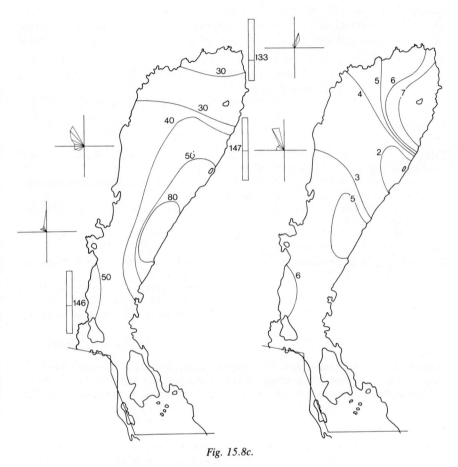

Fig. 15.8. Horizontal distribution of turbidity (left, mg l^{-1} dry weight) and chlorophyll *a* (right, $\mu g\ l^{-1}$) in Neusiedlersee. Contour lines of the lake exclude the reed belt and the Hungarian part. Wind speed and direction for the 24 hours prior to sampling are given together with water level for different stations. *A*: 2 May 1971; *B*: 22 May 1971; *C*: 22 May 1971.

bottom by wind action (Sauberer 1953) ranges from less than 10 mg l^{-1} to more than 500 (Fig. 15.5). Values of less than 10–15 mg l^{-1} are usually found under the ice due to sedimentation and in summer following very rare prolonged periods of calm weather (compare wind speeds in Fig. 15.1, Steinhauser 1970a, 1970b). Typical horizontal distributions of turbidity (Fig. 15.8) result from the prevailing wind directions (NW and SE). Additional examples of this type are given by Dokulil (1974a, 1974b, 1975). From the graphs it is evident that the amount of suspended material is largely a function of wind action and can to some extent be predicted from speed and fetch alone (Dokulil i.p.). Horizontal water movements associated with wind stress also lead to a distinct horizontal pattern in chlorophyll concentration (Fig. 15.8) which will be discussed later.

References

Balon, E. K. & Coche, A. G. 1974. Lake Kariba: a man-made tropical ecosystem in central Africa. Monographiae biologicae 24. W. Junk, The Hague.

Banse, K., Falls, C. F. & Hobson, L. A. 1963. A gravimetric method for determining suspended matter using Millipore filters. Deep Sea Res. 10: 639–642.

Dirmhirn, I. 1956. Über eine Beobachtung der Struktur der Eisdecke auf dem Neusiedler-See. Wetter und Leben 8: 73–75.

Dokulil, M. 1974a. Der Neusiedlersee. (Österreich). -Ber. Naturhist. Ges. Hannover 118, 205–211.

Dokulil, M. 1974b. Die Seetrübe und ihre Bedeutung. In: H. Löffler (ed.), Der Neusiedlersee: Naturgeschichte eines Steppensees. Molden, Vienna, pp. 52–54.

Dokulil, M. 1975. Horizontal- und Vertikalgradienten in einem Flachsee (Neusiedlersee, Österreich). In: Verh. Ges. Ökologie, Wien 1975. W. Junk, The Hague, pp. 177–187.

Dokulil, M., i.p. Light-turbidity-wind relations in a shallow lake: a predictive model. In preparation.

Duckrow, R. M. & Everhart, W. H. 1971. Turbidity measurements. Trans. Am. Fish. Soc. 100: 682–690.

Dvihally, Z. T. 1958. Untersuchung der selektiven Lichtabsorption in Natrongewässern vom Gesichtspunkt der Produktionsbiologie. Acta Biol. Hung. 8: 347–359.

Dvihally, Z. T. 1961. Seasonal changes in the optical characteristics of a Hungarian sodic lake. Hydrobiologia 17: 193–204.

Gelin, C. 1975. Nutrients, biomass and primary productivity of nannoplankton in eutrophic Lake Vombsjön, Sweden. Oikos 26: 121–139.

Holmes, R. W. 1970. The secchi disc in turbid coastal waters. Limnol. & Oceanogr. 15: 688–694.

Ichimura, S. 1956. On the ecological meaning of transparency for the production of matter in phytoplankton communities of lakes. Bot. Mag. 69: 219–226. Tokyo.

Jones, D. & Wills, M. S. 1956. The attenuation of light in sea and estuarine waters in relation to the concentration of suspended solid matter. J. Mar. Biol. Ass. U.K. 35: 431–444.

Koidsumi, K., Nagasawa, T. & Kawashima, S. 1969. Quantitative relations between concentrations of solids in suspension and the transparency or the turbidity of waters. Japanese J. Limnol. 30: 125–138.

Kopf, F. 1967. Schlammbewegung im Neusiedlersee. Unpublished report.

Lemoalle, J. 1973. L'Energie lumineuse et l'activité photosynthétique du phytoplancton dans le Lac Tchad. Orstom sér. Hydrobiol. 7: 95–116.

Mahringer, W. 1969. Der Strahlungshaushalt des Neusiedler Sees im Jahre 1967. Arch. Met. Geoph. Biokl., Ser. B, 17: 51–72.

Mayhew, M. C. 1971. Seston in Lake Michigan and Green Bay. In: R. R. Howmiller & A. M. Beeton (eds.), Center for Great Lakes studies special report 13. Center for Great Lakes Studies, Milwaukee, pp. 38–44.

Panosch, K. 1973. Das Lichtklima des Neusiedlersees. Dissertation, University of Vienna.

Postma, H. 1961. Suspended matter and secchi disc visibility in coastal waters. Neth. J. Sea. Res. 1: 359–390.

Rawson, D. S. 1950. The physical limnology of Great Slave Lake. J. Fish. Res. Bd. Canada 8: 1–66.

Rhode, W. 1969. Standard correlations between pelagic photosynthesis and light. In: C. R. Goldman (ed.), Primary productivity in aquatic environments. University of California Press, Berkeley, pp. 365–381.

Riley, G. A. 1941. Plankton studies IV: George's Bank. Bull. Bingham Oceanogr. Col. 7.

Sauberer, F. 1945. Beiträge zur Kenntnis der optischen Eigenschaften der Kärntner Seen. Arch. Hydrobiol. 41: 258–314.

Sauberer, F. 1952. Über das Licht im Neusiedlersee. Wetter und Leben 4: 12–15.

Sauberer, F. 1953. Der Windeinfluß auf die Trübung des Neusiedlersees. Wetter und Leben 5: 200–203.

Sauberer, F. & Ruttner, F. 1941. Die Strahlungsverhältnisse der Binnengewässer. Akademie, Leipzig.

Schmolinsky, F. 1954. Einige Ergebnisse vergleichender Lichtmessungen an Seen des Hochschwarzwaldes und der Schweiz. Arch. Hydrobiol. Suppl. 20: 615–632.

Steinhauser, F. (1970a): Kleinklimatische Untersuchung der Windverhältnisse am Neusiedlersee. Part one: Die Windrichtungen. Időjárás 74: 76–88.

Steinhauser, F. 1970b. Kleinklimatische Untersuchung der Windverhältnisse am Neusiedlersee. Part two: Die Windstärken. Időjárás 74: 324–345.

Strickland, J. D. H. 1958. Solar radiation penetrating the ocean: a review of requirements, data and methods of measurement, with particular reference to photosynthetic productivity. J. Fish. Res. Bd. Canada 15: 453–493.

Talling, J. F. 1965. The photosynthetic activity of phytoplankton in east African lakes. Int. Rev. Ges. Hydrobiol. 50: 1–32.

Talling, J. F., et al. 1973. The upper limit of photosynthetic productivity by phytoplankton: evidence from Ethiopian soda lakes. Freshwat. Biol. 3: 53–76.

Tilzer, M. 1972. Dynamik und Produktivität von Phytoplankton und pelagischen Bakterien in einem Hochgebirgssee (Vorderer Finstertaler See, Österreich). Arch. Hydrobiol. Suppl. 40: 201–273.

Verduin, J. 1965. Primary production in lakes. Limnol. & Oceanogr. 1: 85–91.

Vollenweider, R. A. 1955. Ein Nomogramm zur Bestimmung des Transmissionkoeffizienten sowie einige Bemerkungen zur Methode seiner Berechnung in der Limnologie. Schweiz. Z. Hydrol. 17: 205–216.

Vollenweider, R. A. 1960. Beiträge zur Kenntnis optischer Eigenschaften der Gewässer und Primärproduktion. Mem. Ist. Ital. Idrobiol. 12: 201–244.

Vollenweider, R. A. 1961. Photometric studies in inland waters I: Relation existing in the spectral extinction of light in water. Mem. Ist. Ital. Idrobiol. 13: 87–113.

Vollenweider, R. A. 1969. The methodology of light measurements. In: R. A. Vollenweider (ed.), A manual on methods for measuring primary production in aquatic environments. IBP Handbook 12. Blackwell, Oxford, pp. 158–171.

Withney, L. V. (1938). Transmission of solar energy and scattering produced by suspensions in lake waters. Tr. Wisc. Acad. 31: 201.

Biology, ecology and production

6 The algal vegetation of Neusiedlersee

E. Kusel-Fetzmann

16.1 Introduction

Apart from the considerable fluctuations in the salt concentration of Neusiedlersee resulting from its changeable water level, long-term alterations are detectable in the proportion of the individual ions. Formerly, Neusiedlersee was regarded as a soda lake, but recent research has shown that this is nowadays true only of the southern part, whereas Ca, Mg and SO_4 have greatly increased in the northern section (Neuhuber 1971). A drop in the water level can bring about the drainage of large marginal areas and salt concentrations may rise rapidly. Heavy rains, on the other hand, can lead to a decrease in salinity. In locations of this nature the algae must therefore be able to endure high osmotic stresses in addition to possessing ion-specific resistance. Since algae present in brackish waters on the sea-shore are exposed to similar factors it is not in the least surprising to find algae of marine littoral origin thriving in Neusiedlersee (e.g. *Bacillaria paradoxa, Gyrosigma macrum, Enteromorpha intestinalis*). Many species found in the lake also exist in other saline inland waters (they will be discussed in more detail in connection with diatoms). A specific soda species is one that is also encountered in other soda waters, e.g. *Surirella peisonis*, which has been found in Guru Göl in Iran (Löffler 1956) and which was termed a 'brackish water species bound to soda waters by Hustedt 1959, and Cholnoky 1968.

A further component of the algal flora is formed by the resistant eutrophic species that otherwise inhabit fresh or slightly polluted waters. Schmid (1973) established the origin of a number of diatoms on the basis of their physiological behaviour. Other groups of algae have not yet been studied from this point of view.

In general, it can be said that this unique algal flora has not so far been adequately studied. No exhaustive records of the algal groups are available and even in those groups that have been investigated, unrecognized elements are certainly present and some species may gradually have vanished and been replaced by others.

The only data on algae found in Neusiedlersee before the desiccation of 1868 are Grunow's (1860–63) lists. Following his investigations on Lake Balaton in 1912 Pantocsek studied the diatoms of Neusiedlersee. Further publications concerning these organisms are those of Hustedt (1959) and Schmid (1973). The dinoflagellates, cryptomonades and *Euglena* were investigated by Schiller (1955, 1956, 1957); Kusel-Fetzmann (1974) has published on the Chlorophyceae. Loub (1955) covered all groups of algae in an autumn study, and the development and sexual processes of a few selected species were observed by Geitler (1969, 1970 a, b), and Tschermak-Woess (1972, 1973 a, b).

The following survey of species from all classes of algae is based upon the above publications and some additional studies, as well as on personal collections made during the course of a number of years. The latter, however, are not yet completely evaluated and will be described in greater detail in individual publications. In general, species that are only found sporadically or that have been described on the basis of unique finds (e.g. by Schiller) are not included. Interested readers are advised to refer to the relevant literature quoted.

16.2 Algal communities in the lake

Whereas in most lakes the pelagic, benthic and littoral communities can be clearly distinguished, a sharp distinction is impossible in the case of Neusiedlersee. The open lake and the reed belt, which is 3 km wide in some places and itself contains various smaller habitats, form two large biotopes associated with completely different environmental conditions.

Due to the shallowness of the lake the open water is constantly mixed by wind and currents. Fine mud stirred up from the bottom produces a permanent turbidity which gives the water a grey and opaque appearance. In the reed belt, on the other hand, currents are arrested, particles settle and humic substances originating from the decomposition of organic matter give a yellowish-brown colour to the water.

Since the biocenoses of Neusiedlersee have not yet been investigated in great detail, an attempt will be made in the following pages to outline some of their more fundamental aspects.

16.3 Phytoplankton

The almost permanent turbulence is responsible for the fact that, in addition to the small genuine forms of plankton (*Nitzschia acicularioides*, flagellates, *Euglena acus*, *Sphaerocystis*, *Monoraphidium*, *Botryococcus braunii* etc.) one invariably encounters the larger, sturdy forms that originate in the benthos or from the periphyton on the macrophytes. *Surirella peisonis* and *Campylodiscus clypeus* in particular, as well as species of *Gyrosigma*, *Rhopalodia* and *Navicula*, are characteristic components of almost any plankton sample whether it be taken near the reed belt, in the middle of the lake or off the reedless eastern shore near Podersdorf. That *Surirella* and *Campylodiscus*, however, cannot propagate when freely suspended, but require contact with the substrate, was suggested by the preliminary experiments of Schmid (1973); see also Harder & Witsch (1942).

A few species generally predominate in the plankton samples, but these vary greatly from year to year, and completely new species may become dominant. Whereas *Monoraphidium* (= *Ankistrodesmus falcatus var. spirilliformis*) dominated in the fifties (according to Ruttner-Kolisko and Ruttner 1959), with up to 3 million cells per litre, it later receded and only in 1972 exceeded these values again. In the years 1968–70 Dokulil found *Oocystis lacustris* (never before found in the lake, and *Cyclotella comta* as prevalent species, while in 1970 especially in the southern part of the lake, *Botryococcus braunii* was observed in bloom and was present in all plankton samples. Although *Pediastrum duplex* was on no occasion found by Loub (1955) it predominated in 1975 along with *Sphaerocystis* and *Planktosphaeria*, which have been found in the lake only since this date. In 1976 another new alga reached great numbers: *Lobocystis dichotoma*, formerly not found in Austria. Huge masses of a colourless bluegreen alga, *Lampropedia hyalina*, occur in the plankton of the macrophytic zone near the reed belt (up to 21 million cells per litre) in August according to Dokulil (1971).

The reasons underlying such enormous changes in the spectrum of species are largely unknown. In addition to chemical changes such as are caused by fluctuations in the water level (Neuhuber 1971; Stehlik 1972) new species may be imported by passive dispersal.

Plankton communities also develop in the calmer and deeper water of the larger canals or reedless pools within the reed belt. Small flagellates, bluegreen algae, diatoms and Chlorophyceae can even attain population densities five times those prevailing in the open lake (Loub 1955). Schiller (1955, 1956, 1957) based his studies of the flagellates chiefly on the plankton of the canal of Rust and he reported numerous new dinoflagellates, cryptomonades and eugleninae.

16.4 Phytomicrobenthos

Benthic biocenoses can only develop in the open lake in winter under the ice cover: they include, for example, *Gyrosigma spencerii, G. macrum, Cymatopleura solea, Surirella peisonis, Amphiprora costata, Nitzschia-* and *Navicula* spp., as well as *Dictyosphaerium* and species of *Scenedesmus, Euglena* and *Phacus.*

An algal layer can only develop in the mud in calm bays and between the reed islands. Thus Loub (1955) found *Surirella peisonis*, *Campylodiscus clypeus*, *Nitzschia sigmoidea*, *Cymatopleura solea* and *Pediastrum boryanum* to be prevalent in the muddy bays near Jois and Winden. A sample taken from the bay off Neusiedl in September 1975 was dominated by *Oscillatoria* in addition to *Gyrosigma macrum*, *Bacillaria paradoxa*, *Euglena tripteris* and *Euglena subehrenbergii.* If such epipelic algae become detached from the bottom by gas bubbles they drift to the water surface and their constituents may easily enter the plankton. Similar algal communities are found on the bottom of the canal-like regions between large reed stands in the souther part of Neusiedlersee just off the Hungarian border.

In the canals and clearings in the reed belt where the water is clear, olive-brown algal mats generally cover the mud. The latter is rich in organic substances and often turns into black sapropel just below the surface. Such ‘epipelic’ algal communities comprise the greatest numbers of species of all the biocenoses of Neusiedlersee. The species composition, and particularly the ratio of predominances, often varies over short distances and changes in the course of a few weeks. Whether a reed canal opens into the lake or runs parallel to the reed margin is reflected in its species composition. Species of the open lake penetrate far into the former, whereas the different ecological conditions offered by the latter account for the presence of other forms. Due to greater evaporation in the shallow marginal areas the salt concentration is higher than in the open lake. At the inland end of the canal of Rust the salt concentration is higher than that at the point where it enters the lake. Consumption processes in the mud influence oxygen conditions particularly in the water strata nearest the bottom (see chapters 11, 26.5). A certain amount of hydrogen sulphide is formed, and this gas can only be tolerated by a few species (the diatoms *Navicula oblonga* and *N. radiosa*, apart from bluegreen algae and sulphur bacteria).

The mud below deeper water is covered mainly by such diatoms as *Cymatopleura solea*, *Gyrosigma spencerii*, *Nitzschia lorenziana*, *N. tryblionella*, *Anomoeoneis sphaerophora*, *Caloneis amphisbaena*, *Navicula cuspidata*, *N. pygmaea* and a host of others. In some places light brown flakes some centimetres in diameter are formed, consisting purely of *Bacillaria paradoxa*. For many years *Cylindrotheca gracilis* was found in small areas near Neusiedl and Podersdorf. The former habitat has now been filled in, but in 1975, a fresh large-scale occurrence was observed at the reed margin only a few hundred metres distance. For the time being its existence seems to be secured.

As the water level sinks or the temperature rises the bluegreen algae and sulphur bacteria increase in number, lending a greenish or blackish appearance to the coverings.

In all of these algal associations striking numbers of *Euglena* occur, specimens as large as 200–300 μm being no rarity (*Euglena ehrenbergii*, *Eu. subehrenbergii*, *Eu. oxyuris*).

16.5 Periphyton

An important contribution to the algal vegetation of Neusiedlersee is made by the periphyton on the wide variety of substances surrounding the open water.

In the zone of submersed macrophytes (*Potamogeton pectinatus* and *Myriophyllum spicatum*), where the waves are already subdued to some extent, algae mixed with grey mud form a thick covering on the plants with the consequence that the photosynthesis of the latter may be substantially hindered (Weisser 1970). The following species were found by Loub (1955) in such periphyton communities: *Rhopalodia gibba*, *Navicula radiosa*, *Mastogloia smithii*, *Gyrosigma*, *Pleurosigma* and the sessile forms *Rhoicosphenia curvata* and *Gomphonema olivaceum*. In addition *Closterium* sp., *Cosmarium biretum*, *Staurastrum*, *Euglena*, *Campylodiscus* and *Surirella peisonis*, *Gyrosigma macrum* and the gelatinous tubes of *Cymbella lacustris* are frequently encountered today.

In addition to *Phragmites communis* numerous other submerged aquatic plants, particularly *Utricularia vulgaris*, are found in the reed belt, especially at the edge of open pools and channels. Loose threads of algae harbouring a high proportion of free-living forms are able to develop right down to the bottom of the clear water. The reed stems are hung with thread-like *Oedogonium*, *Tribonema*, *Mougeotia*, *Spirogyra* etc., which themselves bear tufts of *Synedra affinis, Gomphonema, Cymbella* and *Rhoicosphenia* (on gelatinous stalks) and *Epithemia* species attached to the surface of other algae. In between, flagellates and diatoms move about freely, accompanied by Chlorococcales, Desmidiales, the Xanthophycea *Ophiocytium* and the Crysophycea *Phaeothamnium.* In addition, the drift shoots of *Utricularia* harbour Oscillatoriae, *Spirulina*, *Lyngbya* and often *Bacillaria paradoxa.* Not only are the old, dying part of the shoots colonized by bluegreen algae, but by *Beggiatoa* and other sulphur bacteria, by colourless flagellates (*Anthophysa*) and ciliates as well, all of which contribute to the rapid decomposition of the shoots.

In recent years, these periphyton biocenoses have exhibited marked changes in species composition. In the years 1963/64 for example, *Utricularia* was never overgrown with *Bacillaria* in the reed belt of Rust, although from 1972 onwards it dominates in every sample taken. On the other hand, *Amphipleura pellucida* has always been plentiful near Rust, but totally absent in Neusiedl until found in 1975 growing in larger quantities on *Utricularia* in the reed canal near the lake museum.

'Aufwuchs' found on the jetties, the supports of the bathing-huts and concrete structures for example, off Mörbisch, Rust, Neusiedl or Podersdorf are very different from those encountered in the macrophyte zone and in the reed belt. The former permanent type of substrate favours colonization by more stable biocenoses, e.g. with *Calothrix parietina* (Loub 1955). *Cladophora glomerata* and *Stigeoclonium* are frequent in summer, but in places exposed to battering by waves, and especially in winter, thick gelatinous accumulations of diatoms develop. Besides *Gomphonema* (chiefly *G. olivaceum var. calcarea*) and *Cymbella* (species sessile on gelatinous stalks) some particularly characteristic forms inhabit branched gelatinous tubes. Apart from a sporadic occurrence of *Cymbella prostrata*, also found, for example, in the littoral of Lunzer Untersee, the most frequently encountered of these forms are *Cymbella lacustris* and *Nitzschia filiformis.* Isolated occurrences of *Cymbella lacustris* in the bottom mud of Neusiedlersee were already reported by Hustedt in 1959, but his studies involved only preserved specimens treated with acids. He therefore failed to observe their strange habits inside branched gelatinous tubes that form muddy tufts several centimetres in length. *Nitzschia filiformis* develops particularly tough and smooth sheaths in which the cells can slide freely to and fro.

According to Drum (1969) and Schmid (1973) the ecological significance of such gelatinous tubes (e.g. in the diatoms of the watt or *Amphipleura rutilans* occurring in Van See, a lake extremely rich in soda; Gessner, 1959) lies in the creation of a milieu largely independent of the chemistry of the surroundings. The diatoms are thus able to exist in extreme locations (with high or strongly fluctuating salt concentrations) although lacking a particularly high plasmatic tolerance. It is interesting that *Cymbella lacustris*, which otherwise does not develop gelatinous tubes (Hustedt 1930) does so, and to a striking extent, only in Neusiedlersee.

16.6 Periodic changes in the algal vegetation

In the course of the year Neusiedlersee can develop large extremes of temperature. In late autumn the shallow lake cools down faster than deeper waters. In winter it may be covered by ice and the water may even freeze down to the bottom mud, especially in the shallow marginal areas in the reed belt; anaerobic conditions may destroy (see chapter 11) most of the algae. In one such sample only *Navicula oblonga* and *Navicula radiosa* were found alive, while *Gyrosigma*, *Nitzschia* and others were dead.

Some algae, however, are even favoured by low temperatures and can survive freezing. Schiller (1954, 1955) found a rich protophyte vegetation of Volvocales, Tetrasporales, Cryptomonades, *Euglena* and Gymnodiniae under the ice and slushy snow in the canal of Rust. Although some individuals were frozen they began to swim about again 30–90 minutes after thawing. In winter, *Gymnodinium pascheri*, a species of widespread occurrence, is also frequent in Neusiedlersee.

As a rule, the diatoms in the mud coverings of the reed belt occur in greater numbers in winter, whereas the bluegreen algae increase in quantity later on. Some species of bluegreen algae are better developed in early spring than later in the year, or an alternation of species takes place.

High temperatures are rapidly attained in the lake during the warm season, the average in June being 22°C (maximum 29.1°C) at a depth of 25 cm. Temperatures of this magnitude favour the growth of such Chlorophyceae as flourish in warmth (e.g. masses of *Pediastrum duplex* in September 1975 at 20°C). In addition, *Gomphonema* and *Cymbella* species, as well as *Bacillaria paradoxa*, are found in considerable numbers in summer.

Short-term aperiodic fluctuations within the reed belt are chiefly the result of currents caused by the wind. North-westerlies, for example, cause large patches of the reed to be left dry near Neusiedl, whilst the quickly-rising turbid water resulting from south-east winds washes away large numbers of algae. A period of calm is necessary before the algal communities can re-establish themselves.

16.7 Systematic enumeration

16.7.1 *Cyanophyta*

The Cyanophyceae are an important algae group in Neusiedlersee forming part of the plankton, but they are even more important in the epipelic algae communities and in the boundary growths on reed and *Utricularia.* This complicated algae group has not yet been separately studied. Geitler (1959, 1970) described some interesting species of the lake, which are enumerated here, together with the results of personal collections and the most important species described by Loub (1955). Dokulil (1971) dealt with the apochlorotical *Lampropedia hyalina* Schroeter, which occurs in large numbers (maximum value in August: 21 million cells/litre) in the plankton of the reed belt and of the macrophyte zone throughout the year.

	Plankton	Benthos	Periphyton
Lampropedia hyalina Schroeter	+	−	−
Microcystis pulverea (Wood) Migula	+	+	−
Coelosphaerium kuetzingianum Nägeli	−	+	+
Chroococcus minutus (Kuetz.) Nägeli	+	−	−
Chroococcus turgidus (Kuetz.) Nägeli	−	+	+
Merismopedia glauca (Ehr.) Nägeli	−	+	+
Merismopedia punctata Meyen	+	−	−
Chamaesiphon subaequalis Geitler	−	−	+
Calothrix parietina (Näg.) Thuret	−	−	+
Nostoc planctonicum W. Poretzky and Tschernow	−	+	−
Anabaena flos-aquae (Lyngby) Breb.	+	+	−
Anabaena viguieri Denis et Fremy	+	−	+
Aphanizomenon gracile Lemm.	+	−	−
Pseudanabaena catenata Laut. (Plate 16.1, Fig. 6)	+	+	+
Spirulina jenneri (Stitzenb.) Geitler	+	+	+
Spirulina major Kütz. (Plate 16.1, Fig. 5)	+	+	−
Spirulina raphidioides Geitler	+	−	−
Spirulina subsalsa Oerst. (Plate 16.1, Fig. 9)	−	+	−
Spirulina subtilissima Kütz.	−	+	−
Oscillatoria amphibia Ag.	−	+	−
Oscillatoria brevis Kütz.	−	+	−
Oscillatoria chalybaea Mertens (Plate 16.1, Fig. 3)	−	+	−
Oscillatoria chlorina Kütz. (Plate 16.1, Fig. 4)	−	+	−
Oscillatoria lauterbornii Schmidle	−	+	−
Oscillatoria limosa Ag. (Plate 16.1, Fig. 1)	−	+	−
Oscillatoria okeni (Ag.) Gomont	−	+	−
Oscillatoria princeps Vaucher	−	+	−
Oscillatoria pseudoacutussima Geitler	−	+	+
Oscillatoria subtilissima Kütz.	−	+	+
Oscillatoria tenuis Kütz. *var. tergestina* (Kütz.) Rab. (Plate 16.1, Fig. 2)	−	+	−
Lyngbya limnetica Lemm.	+	−	−
Lyngbya hieronymusii Lemm. (Plate 16.1, Fig. 7)	−	+	+
Lyngbya martensiana Menegh.	−	−	+

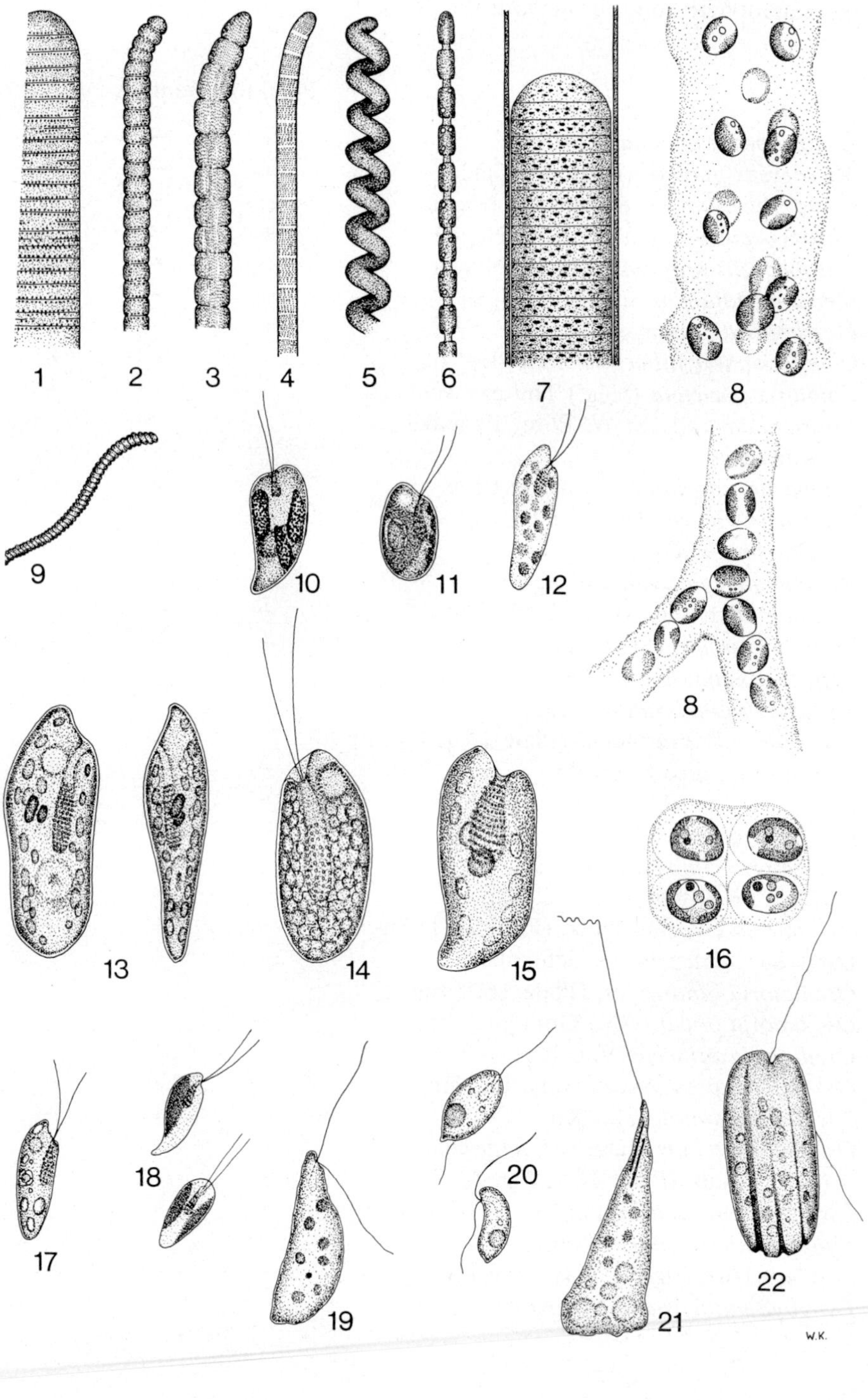
1
2
3
4
5
6
7
8
9
10
11
12
8
13
14
15
16
17
18
19
20
21
22
W.K.

16.7.2 *Flagellata*

Of the various flagellate groups the Cryptomonads, Dinoflagellata and Euglenae have been studied in great detail by Schiller (1952a, 1954, 1955, 1956, 1957). He concentrated mainly on the canal of Rust, where the centrifuged material of a five-year research programme showed 450 new species of many systematic groups. Since quite a number of them appear to be unique or rare finds it is essential to take samples from the entire lake area to prove whether these finds were really characteristic elements of the flora of Neusiedlersee. The following list incorporates data of Schiller, of Loub (1955) and the author's own findings (of *Euglena*, *Phacus*, *Lepocinclis* and *Trachelomonas* in particular).

a) Cryptomonads: The study of Schiller (1957) was focused upon the canal of Rust, although no doubt many of these species occur in other parts of the lake as well. Loub (1955) also specified representatives of this group. Only the forms that occur in larger numbers will be mentioned here.

Cryptochrysis minor Nygaard: Canal of Rust, spring, frequent at temperatures between 14 and 18°C

Rhodomonas tenuis Skuja (Plate 16.1, Fig. 17)

Chroomonas nana Schiller: canal of Rust, spring, frequent at 12–20°C

Chroomonas cor Schiller: canal of Rust, in autumn and winter at 0.5–3°C

Chroomonas caerulea (Geitler) Skuja (Plate 16.1, Fig. 11)

Chroomonas unamacula Schiller (Plate 16.1, Fig. 18)

Chroomonas cyanea Schiller

Cyanomonas curvata Schiller (Plate 16.1, Fig. 10)

Cryptomonas appendiculata Schiller: canal of Rust, frequent in May at 20–22°C

Plate 16.1.

1. *Oscillatoria limosa*
2. *Oscillatoria tenuis*
3. *Oscillatoria chalybaea*
4. *Oscillatoria chlorina*
5. *Spirulina major*
6. *Pseudanabaena catenata*
7. *Lyngbya hieronymusii*
8. *Phaeothamnium confervicola*
9. *Spirulina subsalsa*
10. *Cyanomonas curvata*
11. *Chroomonas caerulea*
12. *Chilomonas* sp.
13. *Cryptomonas platyuris*
14. *Cryptomonas ovata*
15. *Cryptomonas reflexa*
16. *Sarcinochrysis granifera*
17. *Rhodomonas tenuis*
18. *Chroomonas unamacula*
19. *Heteronema* sp.
20. *Anisonema acinus*
21. *Peranema trichophorum*
22. *Entosiphon sulcatum*

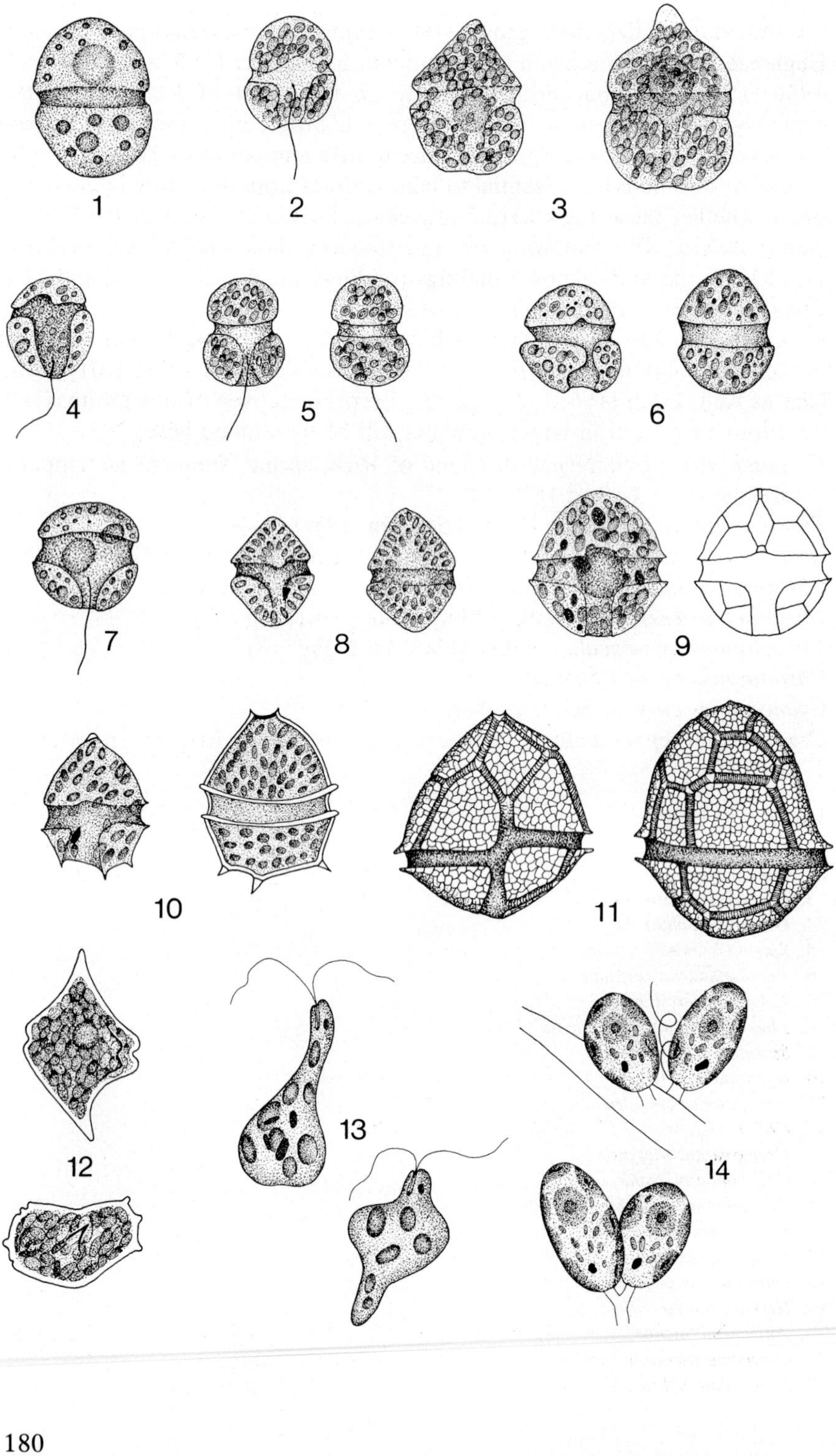
1
2
3
4
5
6
7
8
9
10
11
12
13
14

Cryptomonas erosa Ehrenberg: Neusiedlersee, rare
Cryptomonas komma Schiller: canal of Rust, in January frequent at 0.5–2°C
Cryptomonas marsonii Skuja: Neusiedlersee, canal of Rust, at 12–14°C
Cryptomonas obovata Skuja: canal of Rust, at 10–12°C
Cryptomonas ovata Ehrenberg, et var.: Neusiedlersee, canal of Rust (Plate 16.1, Fig. 14)
Cryptomonas peisonis Schiller: canal of Rust, under ice
Cryptomonas platyuris Skuja: author's own collections in the reed near Neusiedl (Plate 16.1, Fig. 13)
Cryptomonas rostrata Troitz.: canal of Rust, average frequency, at 15–18°C
Cryptomonas reflexa (Marsson) Skuja (Plate 16.1, Fig. 15)
Cryptomonas violacea Schiller: canal of Rust under ice, chromatophores metallic blue; 19–21 μm long
Colorless *Chilomonas* sp. are occasionally found in the reed belt between the decaying remains of plants (Plate 16.1, Fig. 12)

b) Dinoflagellata: Studied by Schiller in the year 1954 and in great detail in 1955.
Gymnodinium pascheri (Suchland) Schiller is a hibernal form frequent in the whole lake (Plate 16.1, Fig. 2). If the water temperature exceeds 4°C this species turns into *Gymnodinium veris* Lindemann (Plate 16.2, Fig. 3) and finally forms horned cysts. (Plate 16.2, Fig. 12)
Gymnodinium cyaneum Schiller: in the canal of Rust in winter (Plate 16.2, Fig. 5)
Gymnodinium glaucum Schiller: February and March, also under ice, in the canal of Rust
Gymnodinium coronatum Wolosz.: summer form in the canal of Rust
Gymnodinium knollii Schiller: in the canal of Rust in spring, heterotrophic (Plate 16.2, Fig. 7)
Gymnodinium mitratum Schiller: in the cold season
Gymnodinium paradoxiforme Schiller (Plate 16.2, Fig. 1)
Amphidinium bidentatum Schiller: in summer and autumn on the eastern shore of Neusiedlersee, also in the Zicksee (= salt pool) near Podersdorf

Plate 16.2.
1. *Gymnodinium paradoxiforme*
2. *Gymnodinium pascheri*
3. *Gymnodinium pascheri, status veris*
4. *Amphidinium vorax*
5. *Gymnodinium cyaneum*
6. *Glenodinium bieblii*
7. *Gymnodinium knollii*
8. *Glenodinium amphiconicum*
9. *Peridinium hiemale*
10. *Peridinium inconspicuum*
11. *Peridinium cinctum*
12. *Gymnodinium pascheri*, cyste.
13. *Eutreptia viridis*
14. *Colacium cyclopicola*

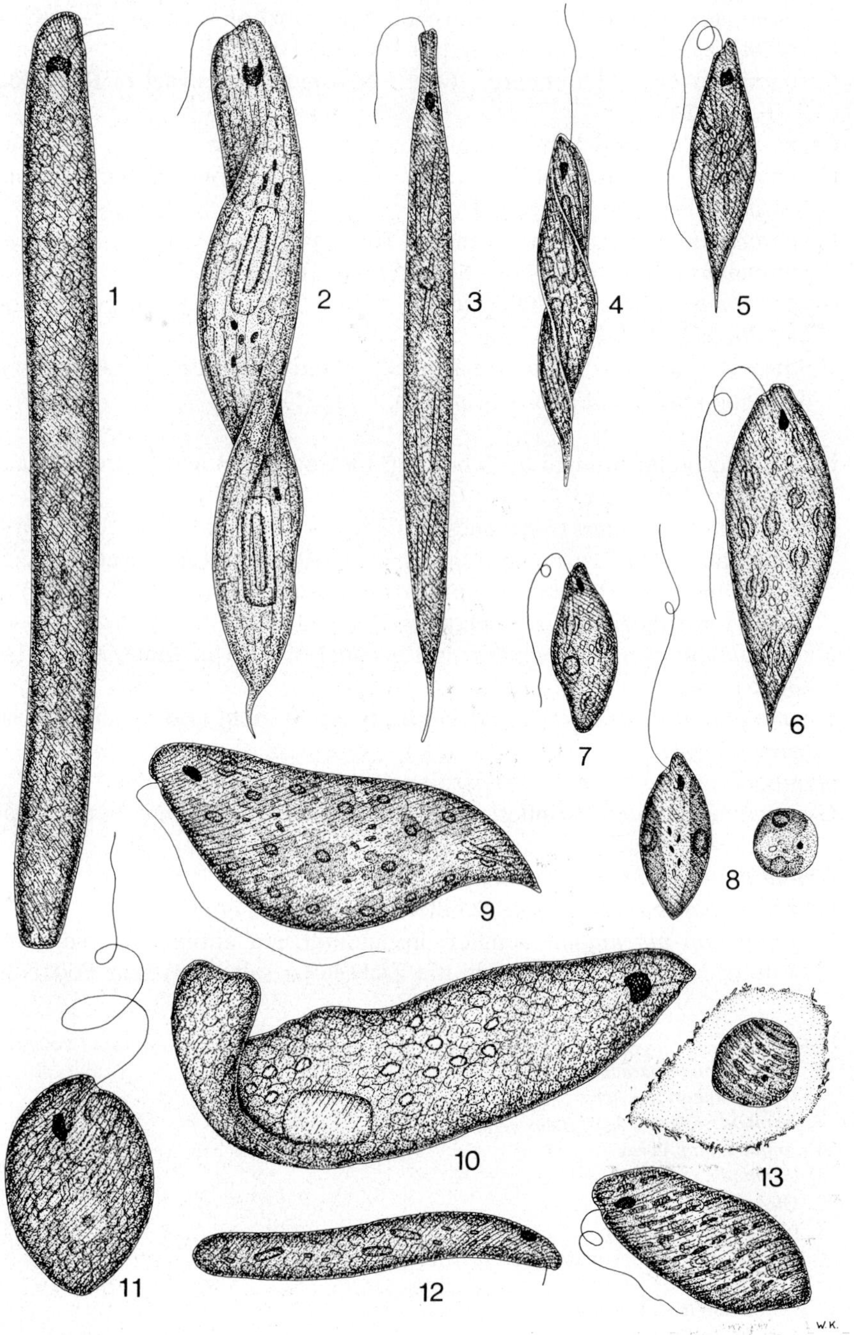

1
2
3
4
5
6
7
8
9
10
11
12
13
W.K.

Amphidinium caerulescens Schiller: in the canal of Rust in February and March, bluish-gray to yellowish
Amphidinium sauerzopfii Schiller: rare
Amphidinium vorax Schiller: canal of Rust (Plate 16.2, Fig. 4)
Massartia hiemalis Schiller: has cyanells, rare, in the canal of Rust under ice
Glenodinium amphiconicum Schiller: canal of Rust, May–October Plate 16.2, Fig. 8
Glenodinium bieblii Schiller: canal of Rust, in Summer (Plate 16.2, Fig. 6)
Glenodinium coronatum Wolosz.: canal of Rust, in summer
Peridinium hiemale Schiller: canal of Rust, winter (Plate 16.2, Fig. 9)
Peridinium inconspicuum Lemmermann: in the open lake in summer (Plate 16.2, Fig. 10)
Peridinium cunningtonii Lemmermann: canal of Rust in summer
Peridinium penardiforme Lindemann: Neusiedlersee in summer
Peridinium palatinum Lauterborn: canal of Rust in winter
Peridinium cinctum (Müller) Ehrenb.: in the reed belt near Neusiedl (Plate II, 11)

c) Euglenophyta: of the 80 different *Euglena* species found by Schiller (1956) 53, chiefly in the canal of Rust, were new. Many of them were described on the basis of sporadic finds without latin diagnoses. A critical re-examination of many of these species is indispensible. Only those species that occur often or in several places will be mentioned here; our own finds are included.
Euglena acus Ehrenb.: in the plankton and in the reed belt, very frequent (Plate 16.3, Fig. 3)
Euglena aestivalis Schiller: frequent in summer, smaller than *Euglena intermedia*
Euglena ehrenbergii Klebs: in the plankton
Euglena gracilis Klebs (Plate 16.3, Fig. 7)
Euglena impleta Schiller: in the canal of Rust in summer and autumn
Euglena intermedia (Klebs) Schmitz *var. klebsii* Lemm.: very frequent in the reed belt (Plate 16.3, Fig. 12)
Euglena limnophila Lemm.: near Rust in autumn, is also present in Lake

Plate 16.3.
1. *Euglena subehrenbergii*
2. *Euglena oxyuris*
3. *Euglena acus*
4. *Euglena tripteris*
5. *Euglena viridis*
6. *Euglena sociabilis*
7. *Euglena gracilis*
8. *Euglena pisciformis*
9. *Euglena velata*
10. *Euglena mobilis = E. ehrenbergii*
11. *Euglena texta var. salina*
12. *Euglena intermedia var. klebsii*
13. *Euglena oblonga*

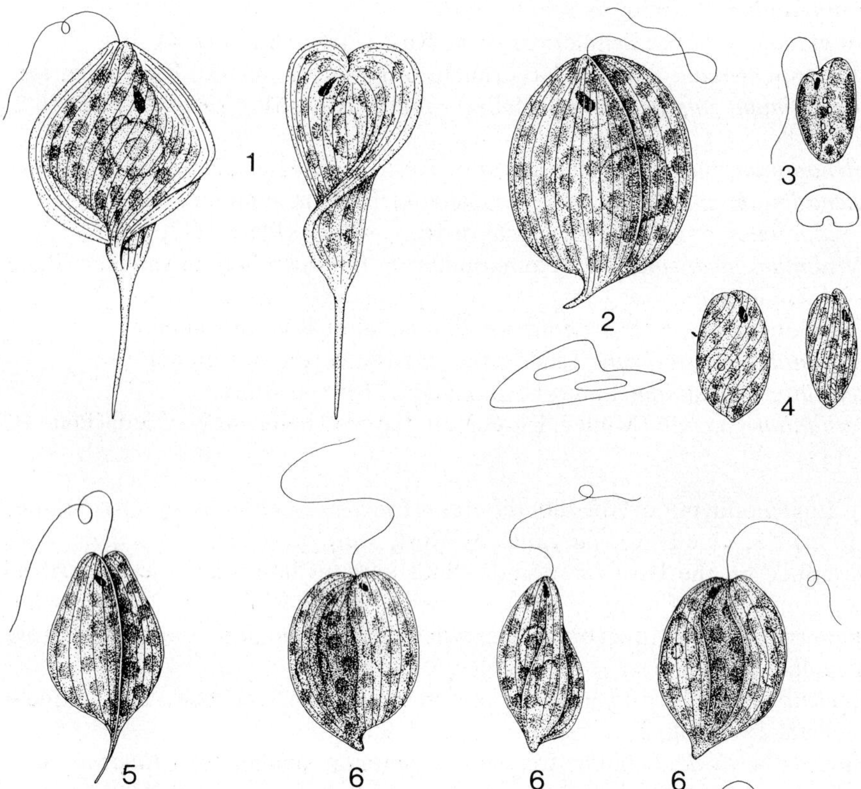
1
2
3
4
5
6
6
6
7
8
9
10

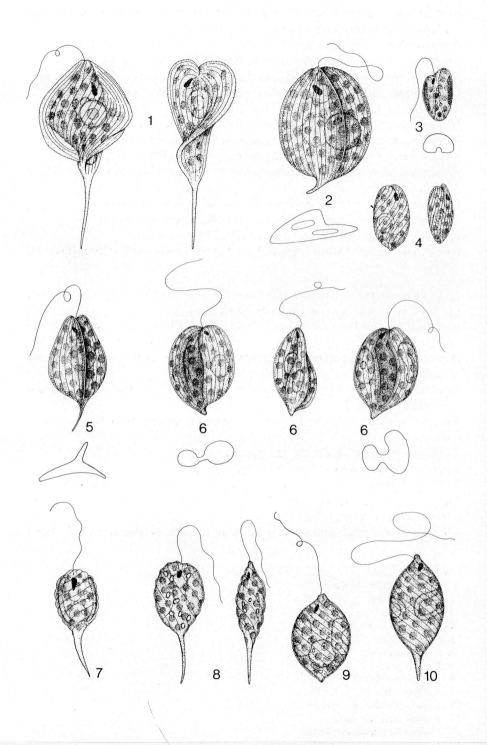

Balaton

Euglena mobilis Schiller: is perhaps identical with *Eu. ehrenbergii* (Plate 16.3, Fig. 10)

Euglena oblonga Schmitz (Plate 16.3, Fig. 13)

Euglena oxyuris Schmarda: very frequent, both in the plankton and in the reed (Plate 16.3, Fig. 2)

Euglena pisciformis Plebs: called *Eu. agilis* Carter by Schiller, with some new varieties (Plate 16.3, Fig. 8)

Euglena proxima Dangeard: in summer till autumn

Euglena repulsans Schiller (1952), in the plankton and in the canal of Rust

Euglena sociabilis Dangeard (Plate 16.3, Fig. 6)

Euglena subehrenbergii Skuja: in the plankton, belongs to the same group as *Eu. ehrenbergii* (Plate 16.3, Fig. 1)

Euglena texta (Duj.) Klebs *var. salina* Fritsch (= *Lepocinclis salina* Fritsch): in a reed pool near Neusiedl, canal of Rust (Plate 16.3, Fig. 11)

Euglena tripteris (Duj.) Klebs: occurs in a large form (100–120 μm) and in a small form (75–85 μm); in the plankton and in the reeds near Neusiedl and Rust (Plate 16.3, Fig. 4)

Euglena velata Klebs (Plate 16.3, Fig. 9)

Euglena viridis Ehrenberg: has larger cells than usual, in the reed belt near Neusiedl and Rust (Plate 16.3, Fig. 5)

Colacium cyclopicola (Gicklhorn) Bourr.: very frequent on *Cyclops* in the plankton in September 1975 (Plate 16.3, Fig. 14)

Eutreptia viridis Perty: highly metabolic, has two flagella (Plate 16.2, Fig. 13)

Anisonema acinus Duj.: like all colourless Eugleninae in the reed belt (Plate 16.1, Fig. 20)

Heteronema sp. (Plate 16.1, Fig. 19)

Peranema trichophorum (Ehrenb.) Stein (Plate 16.1, Fig. 21)

Entosiphon sulcatum (Duj.) Stein, (Plate 16.1, Fig. 22)

Trachelomonas hispida (Perty) Stein em. Defl. (Plate 16.9, Fig. 4)

Trachelomonas volvocina Ehrenb.: among *Utricularia* in the reed belt.

Lepocinclis caudata Da Cunha *var. nasuta* Conrad: near Rust, Breitenbrunn (Plate 16.4, Fig. 10)

Lepocinclis fusciformis (Cater) Lemm. em. Conrad: near Rust (Plate 16.4, Fig. 9)

Lepocinclis ovum (Ehr.) Lemm.

Plate 16.4

1. *Phacus tortus*
2. *Phacus pleuronectes*
3. *Phacus agilis*
4. *Phacus pusillus*
5. *Phacus triqueter*
6. *Phacus curvicauda, different views*
7. *Phacus pyrum*
8. *Phacus pseudonordstedtii*
9. *Lepocinclis fusciformis*
10. *Lepocinclis caudata var. nasuta*

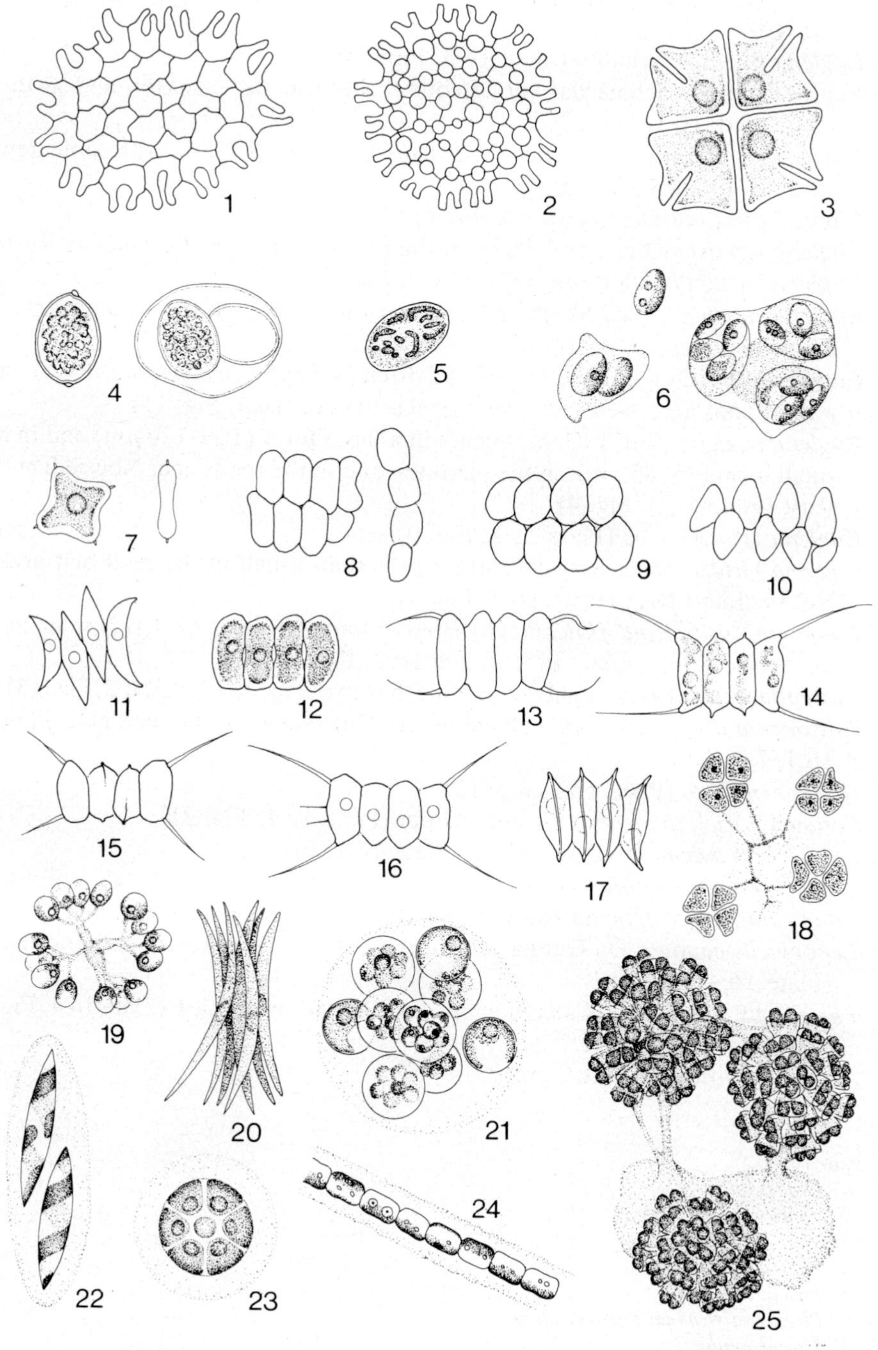
1
2
3
4
5
6
7
8
9
10
11
12
13
14
15
16
17
18
19
20
21
22
23
24
25

Phacus agilis Skuja: Rust, in a test basin planted with reed, canal of Rust, canal near Weiden (Plate 16.4, Fig. 3)
Phacus curvicauda Pochmann: frequent in the reed belt between Neusiedl and Rust (Plate 16.4, Fig. 6)
Phacus elegans Pochmann: Rust, Neusiedl, in the reed belt
Phacus helikoides Pochmann
Phacus inflexus (Kiss.) Pochmann: Rust
Phacus lismorensis Playfair: Rust, Mörbisch
Phacus oscillans Klebs: canal near Weiden and Rust
Phacus platyaulax Pochmann: in the reed belt and the canals near Rust, Breitenbrunn, Neusiedl, Illmitz
Phacus pleuronectes (O.F. M.) Duj.: reed pool near Neusiedl, canal of Rust (Plate 16.4, Fig. 2)
Phacus pseudonordstedtii Pochmann: reed belt (Plate 16.4, Fig. 8)
Phacus pusillus Lemm. (Plate 16.4, Fig. 4)
Phacus pyrum (Ehr.) Stein: reed belt near Neusiedl (Plate 16.4, Fig. 7)
Phacus stokesii Lemm: canal of Rust
Phacus tortus (Lemm.) Skvortz.: reed margin near Neusiedl (Plate 16.1, Fig. 1)
Phacus triqueter (Ehr.) Duj.: (Plate 16.4, Fig. 5)

d) Volvocales: so far no systematic studies have been carried out. Some

Plate 16.5.

1. *Pediastrum boryanum*
2. *Pediastrum duplex*
3. *Pediastrum tetras*
4. *Oocystis solitaria*
5. *Glaucocystis nostochinearum*
6. *Oocystis lacustris*
7. *Tetraedron minimum*
8. *Scenedesmus platydiscus*
9. *Scenedesmus ecornis var. disciformis*
10. *Scenedesmus ovalternus*
11. *Scenedesmus acuminatus*
12. *Scenedesmus balatonicus*
13. *Scenedesmus quadricauda*
14. *Scenedesmus opoliensis*
15. *Scenedesmus armatus*
16. *Scenedesmus spinosus*
17. *Scenedesmus acutiformis*
18. *Crucigenia tetrapedia*
19. *Dictyosphaerium pulchellum*
20. *Ankistrodesmus falcatus*
21. *Sphaerocystis schroeteri*
22. *Closteriospira lemanensis*
23. *Planktosphaeria gelatinosa*
24. *Ulothrix amphigranulata*
25. *Botryococcus braunii*

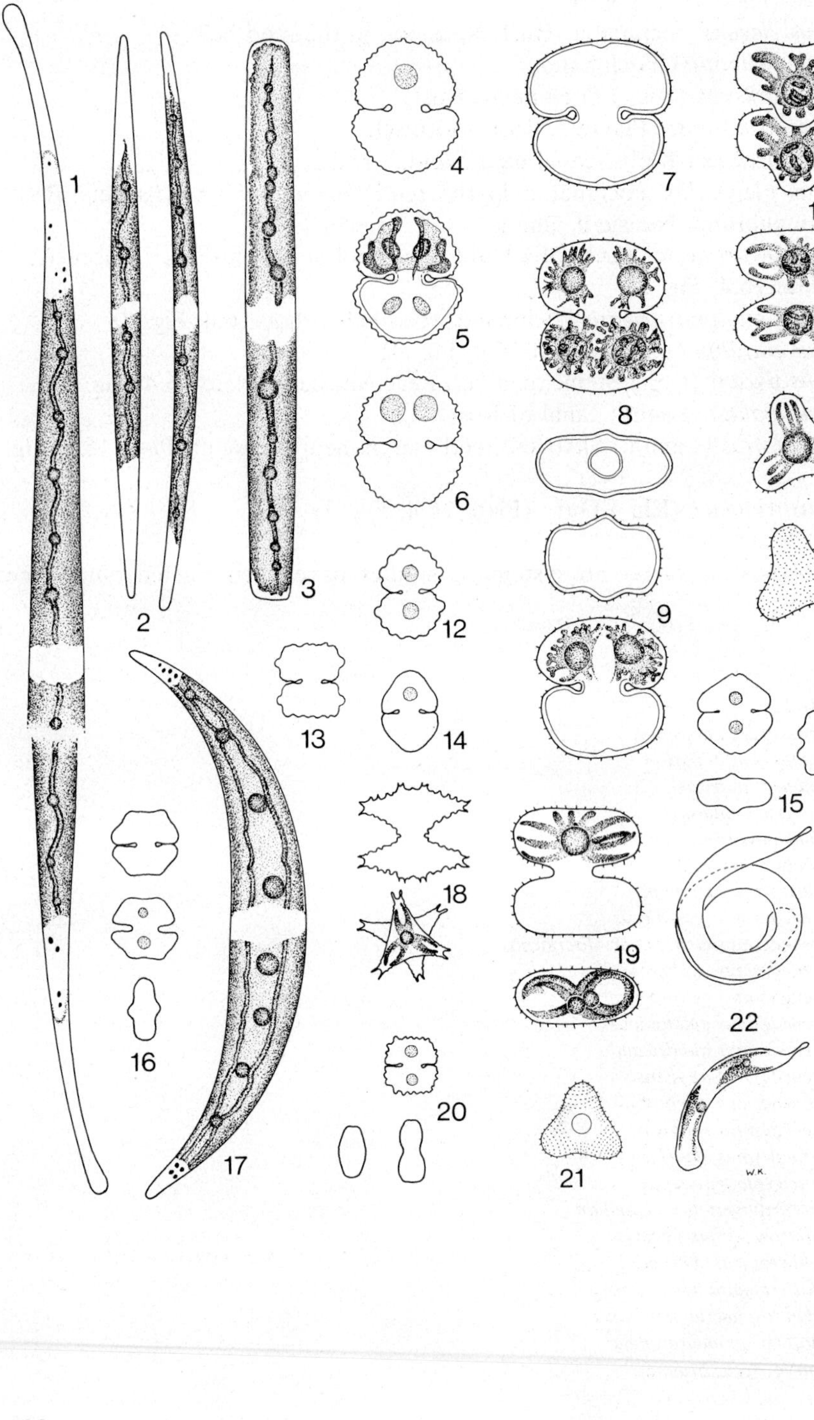

1
2
3
4
5
6
7
8
9
10
11
12
13
14
15
16
17
18
19
20
21
22
W.K.

forms will be mentioned here, in particular those described by Schiller (1954).

Carteria globulosa Pascher: canal of Rust

Chlamydomonas paramucosa Schiller: in the canal of Rust from November to February

Chlorogonium hiemalis Schiller: in the canal of Rust under ice

Chlorogonium maximum Skuja: also in the canal of Rust under ice

Haematococcus pluvialis Flot. em. Wille: in a furrow in the reed near Rust

Pandorina morum (Müller) Bory: in the reed belt near Neusiedl

16.7.3 *Chrysophyta*

Chrysophyceae have not yet been studied systematically in Neusiedlersee, a fact that is certainly primarily due to the difficulty of specifying these generally frail organisms. Nevertheless some excellent studies have been made on individual species: Tschermak-Woess (1972) studied a pure culture of the alga *Sarcinochrysis granifera* (Plate 16.1, Fig. 16) with its individual stages of development and she proved that this species, first found by Mack (1954) in the muddy covers on *Myriophyllum* and described by him as *Chrysocapsa granifera*, in fact belonged to the genus of *Sarcinocrysis*. Geitler (1970) identified another Chrysophycea forming palmelloid threads as *Phaeothamnium confervicola* Lagerh. (Plate 16.1, Fig. 8) (see also Kusel-Fetzmann 1974). He also found the creeping threads of *Apistonema expansum* Geitler on reed leaves. Schiller (1954) encountered *Chromulina biococca* Schiller in samples taken in the canal of Rust from below the ice. In February 1975 a *Mallomonas* sp. occurred in the reed fringe near Rust, and in September it was found between *Utricularia* – *Dinobryon sertularia* E. near

Plate 16.6.

1. *Closterium pronum*
2. *Closterium acutum*
3. *Gonatozygon monotaenium*
4. *Cosmarium punctulatum*
5. *Cosmarium varsoviense*
6. *Cosmarium botrytis*
7. *Cosmarium biretum var. trigibberum*
8. *Cosmarium biretum var. trigibberum*
9. *Cosmarium biretum var. trigibberum*
10. *Staurastrum alternans*
11. *Staurastrum alternans*
12. *Cosmarium subgranatum*
13. *Cosmarium regnellii var. chondrophorum*
14. *Cosmarium granatum*
15. *Cosmarium scopulorum*
16. *Cosmarium praecisum var. suecicum*
17. *Closterium parvulum*
18. *Staurastrum hexacerum*
19. *Staurastrum alternans forma biradiata*
20. *Cosmarium humile*
21. *Staurastrum dilatatum*
22. *Ophiocytium majus*

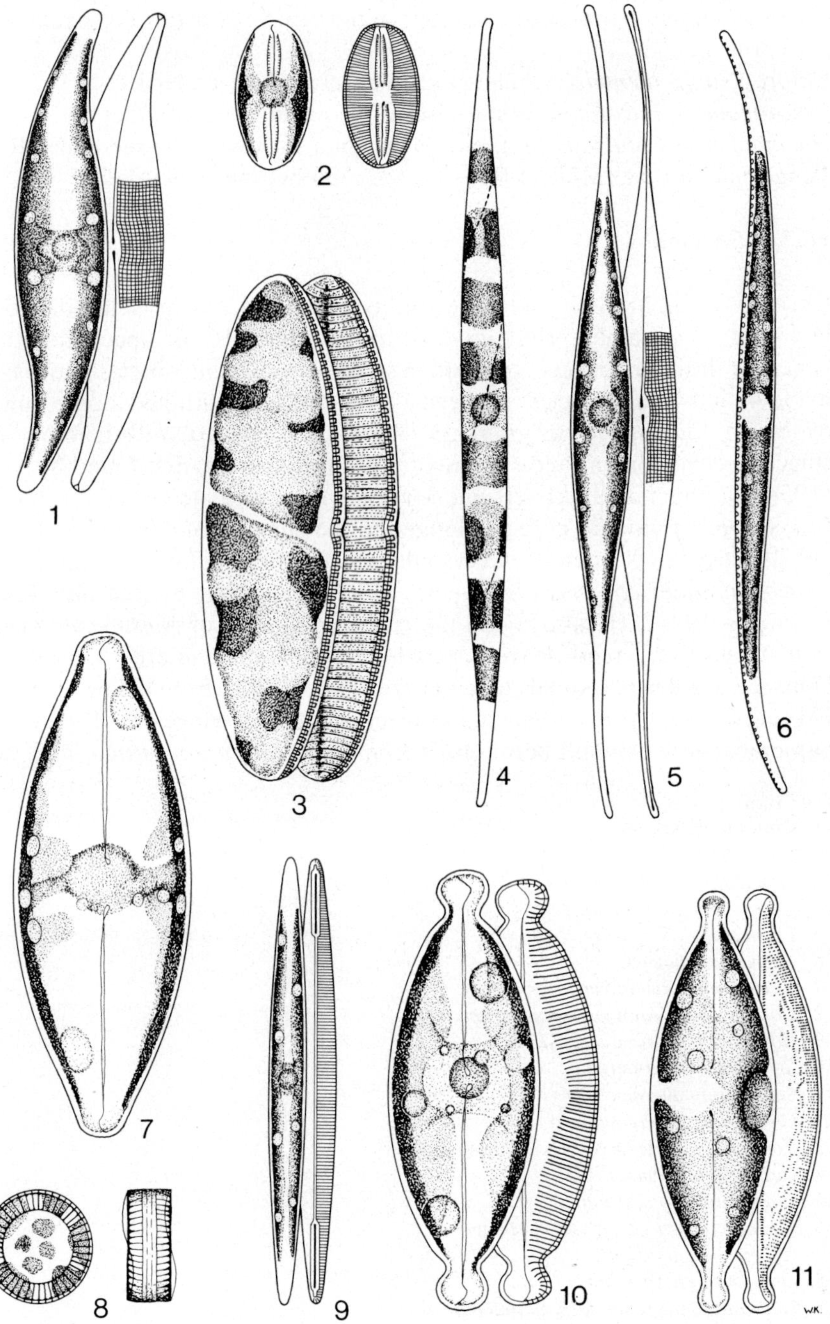
1
2
3
4
5
6
7
8
9
10
11
W.K.

Neusiedl (see also Loub 1955). *Synura uvella* E. was found in a reed pool. In the reed belt near Rust the colourless flagellate *Anthophysa* sp., which occurs in colonies, is quite frequent; iron deposits often give their jelly stalks a brown colour.

16.7.4 *Bacillariophyta*

The diatoms are the best studied algal group of Neusiedlersee and of the saline lakelets in the Seewinkel east of the lake, where the chemical conditions are even more extreme (Legler 1941, Hustedt 1959c). Nevertheless the group has still been insufficiently investigated, partly due to the size of the lake, the variety of its biotopes and the gradual changes to which it is subjected.

Grunow (1860–63) was the first to specify about 50 diatoms living in Neusiedlersee. Many of them had not previously been known and they were given the specific denotation '*peisonis*'. A more exact study of Pantocsek (1912) enumerated 149 species, many of which must be eliminated as synonyms. Hustedt (1959a) estimated the total number in the lake at 150 species, he himself identifying 130. But in recent years some species not yet mentioned by him have been found, such as *Gyrosigma macrum*, which was first reported in 1968 and has since become temporarily prevalent (Kusel-Fetzmann 1973).

Considering the size of the lake this number of algae species is rather small, a fact strongly emphasized by Hustedt (1959b). This is mainly due to the chemistry of the lake and its strong fluctuations. The periods of desiccation also exert a selective effect, even though occurring at long intervals (the last one in 1868), since a new population has to develop. It cannot come from nearby waters of different chemical composition (the similar but smaller salt lakelets are likely to dry up at the same time). Waters containing a similar diatom flora are hundreds of kilometers away. Loub (1955), who himself identified 140 diatom species in Neusiedlersee, drew up comparisons to some such places: the salt waters (sodium chloride) of Oldesloe (Hustedt 1925) contain 134 species, 70 of which also occur in Neusiedlersee. 200 species were reported in the saline area of Sperenberg (Kolbe 1927), 50 per cent of which occur in Neusiedlersee. Woronichin (1926) ascertained 40 diatoms in the saline bitter lakes, 15 of which are also found in Neusiedlersee.

Plate 16.7.
1. *Gyrosigma spencerii*
2. *Navicula pygmaea*
3. *Nitzschia tryblionella*
4. *Cylindrotheca gracilis*
5. *Gyrosigma macrum*
6. *Nitzschia lorenziana*
7. *Caloneis permagna*
8. *Cyclotella meneghiniana*
9. *Amphipleura pellucida*
10. *Caloneis amphisbaena*
11. *Anomoeoneis sphaerophora*

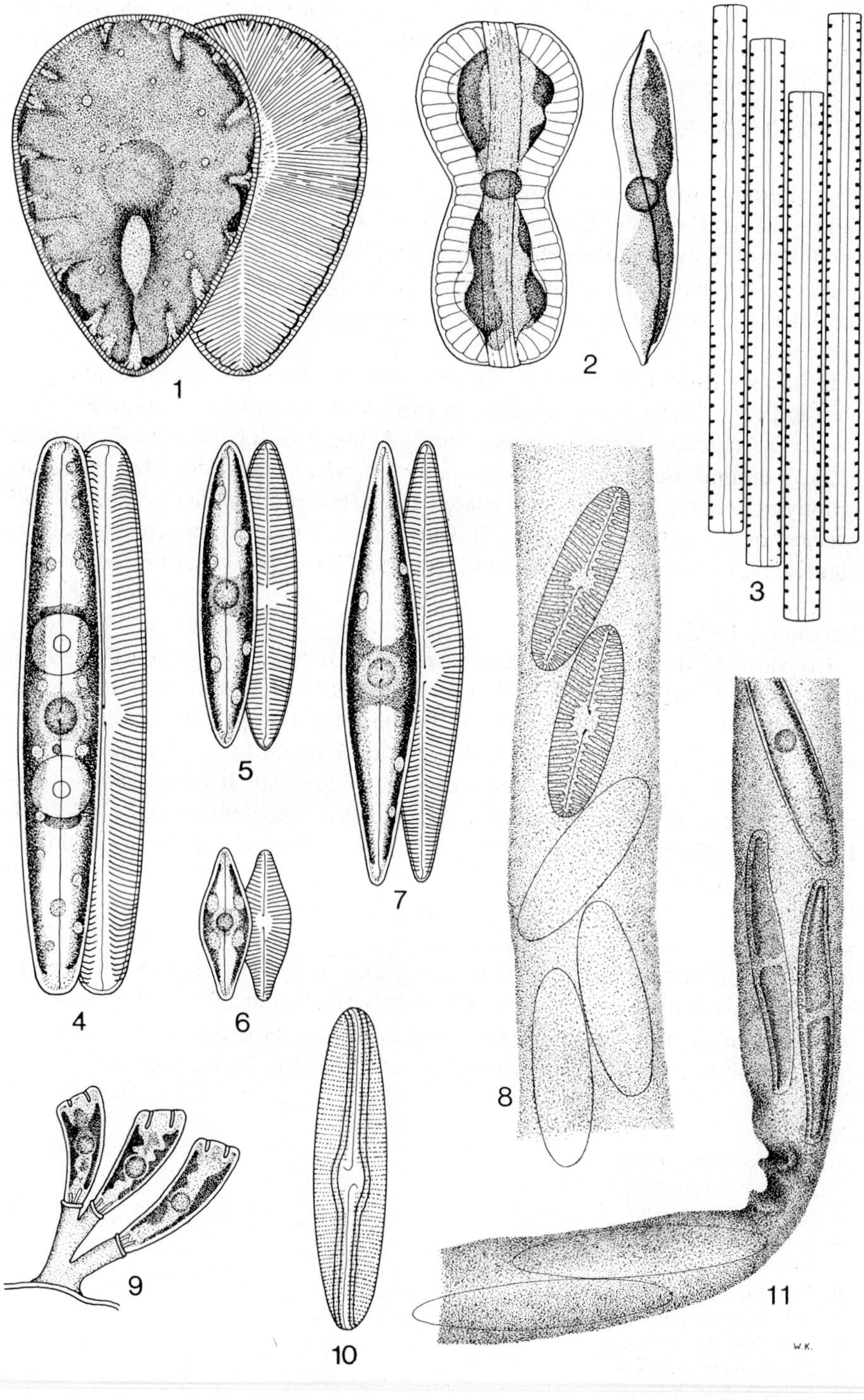
1
2
3
4
5
6
7
8
9
10
11
W.K.

A number of species are – if not endemic – at least limited to natron lakes and are known from Hungary (e.g. Szemes 1959, in the Lake of Szelid) or Iran, (Guru Göl, Löffler 1956). According to Hustedt (1959b) *Amphiprora costata, Caloneis permagna, Gyrosigma peisonis, Surirella peisonis, Scoliotropis peisonis, Campylodiscus clypeus, Navicula halophila*, and *Cylindrotheca gracilis* belong to these species (see also Schmid 1973).

The plankton lacks the *Melosira-*, *Asterionella-*, and *Tabellaria*-species, which are otherwise frequent and very characteristic of lakes. Quite a few species are present in the periphyton, only some of them occurring in masses (e.g. *Gomphonema olivaceum var. calcarea*, *Cymbella pusilla*, *Synedra pulchella*, *Synedra tabulata*).

	Plankton	Benthos	Periphyton	
Melosira varians Ag.	–	+	+	
Cyclotella meneghiniana Kütz. (Plate 16.7, Fig. 8)	+	+	+	
Cyclotella comta (E.) Kütz.	+	+	+	
Stephanodiscus hantzschii Grunow	+	–	–	
Chaetoceros muelleri Lemm.	+	–	–	halophyte
Diatoma elongatum Ag.	–	+	+	
Diatoma vulgare Bory	–	–	+	
Fragilaria capucina Desm.	–	–	+	
Fragilaria crotonensis Kitton	–	+	+	
Synedra acus Kütz.	+	–	+	
Synedra capitata Ehr.	–	–	+	
Synedra pulchella Kütz.	–	+	+	halophyte
Synedra tabulata (Ag.) Kütz. (= *affinis* Kütz.!)	+	+	+	halophyte
Synedra ulna (Nitzsch.) Ehr. + formae	+	+	+	
Eunotia lunaris (E.) Grunow	–	+	+	
Cocconeis placentula Ehr.	–	–	+	
Achnanthes brevipes Ag.	–	+	+	
Achnanthes minutissima Kütz.	+	–	+	
Rhoicosphenia curvata (Kütz.) Grun. Plate 16.8, Fig. 9)	–	–	+	

Plate 16.8.

1. *Surirella peisonis*
2. *Amphiprora costata*
3. *Bacillaria paradoxa*
4. *Navicula oblonga*
5. *Navicula gracilis*
6. *Navicula cryptocephala*
7. *Navicula radiosa*
8. *Cymbella lacustris*
9. *Rhoicosphenia curvata*
10. *Scoliotropis peisonis*
11. *Nitzschia filiformis*

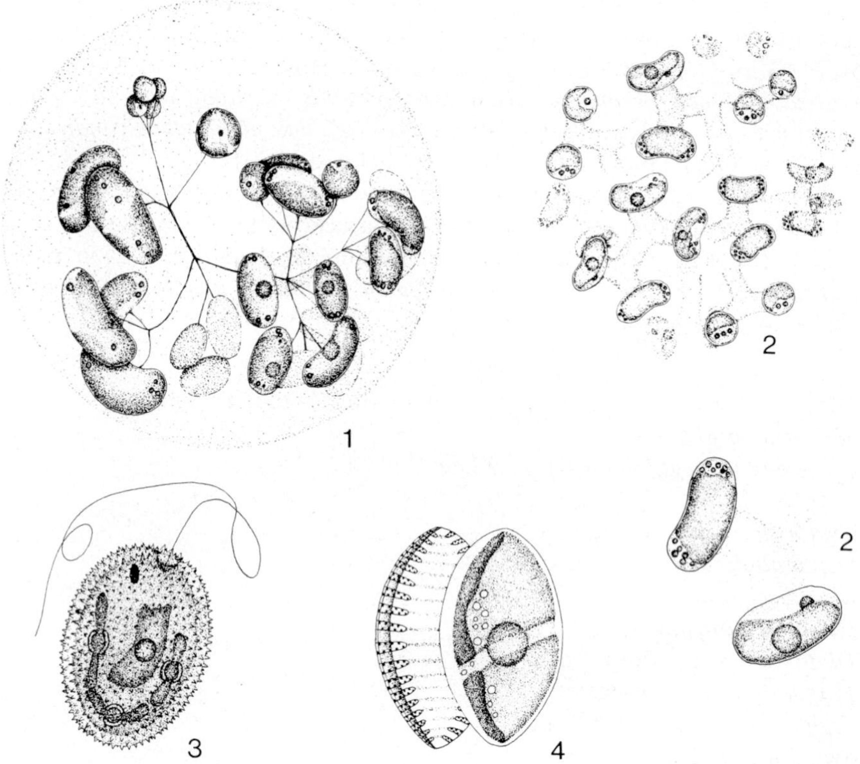

Plate 16.9.
1. *Dictyosphaerium reniforme*
2. *Lobocystis dichotoma*
3. *Trachelomonas hispida*
4. *Nitzschia navicularis*

	Plankton	Benthos	Periphyton	
Mastogloia smithii Thwait.	+	+	+	
Amphipleura pellucida Kütz. (Plate 16.7, Fig. 9)	–	+	+	
Amphipleura rutilans (Trent.) Cl.	–	–	+	halophyte
Stauroneis spicula Hickie	+	+	–	halophyte
Stauroneis wislouchii Poretz, u. Anis.	–	+	–	halophyte
Anomoeoneis costata (Kütz.) Hust.	–	+	–	soda-halophyte
Anomoeoneis sphaerophora (Kütz.) Pfitz. (Plate 16.7, Fig. 11)	–	+	–	halophyte
Navicula cuspidata Kütz.	+	+	–	
Navicula halophila (Grun.) Cleve	+	+	–	soda-halophyte
Navicula halophiloides Hustedt	+	–	–	endemic
Navicula cincta (E.) Ralfs	–	+	–	
Navicula cryptocephala Kütz. (Plate 16.8, Fig. 6)	+	+	+	

	Plankton	Benthos	Periphyton	
Navicula gracilis Ehr. (Plate 16.8, Fig. 5)	–	+	+	
Navicula hungarica Grun. + *var. capitata*	+	+	–	
Navicula oblonga Kütz. Plate 16.8, Fig. 4)	–	+	+	
Navicula radiosa Kütz. (Plate 16.8, Fig. 7)	+	+	+	
Navicula reinhardtii Grun.	–	+	–	
Navicula salinarum Grun.	–	+	–	
Navicula stundlii Hustedt	–	–	+	
Navicula protracta Grun.	+	+	+	
Navicula pygmaea Kütz. (Plate 16.7, Fig. 2)	+	+	–	
Navicula tenuipunctata Hustedt	+	–	–	
Pinnularia kneuckeri Hustedt	–	+	–	
Pinnularia microstauron var. brebissonii (Kütz.) Hustedt	–	+	–	
Pinnularia major Kütz.	–	+	–	
Caloneis amphisbaena (Bory) Cleve (Plate 16.7, Fig. 10)	–	+	+	
Caloneis permagna (Bail.) Cleve (Plate 16.7, Fig. 7)	–	+	+	soda-halophyte
Caloneis bacillum (Grun.) Mereshck.	–	+	–	
Caloneis silicula (E.) Cl. + *var. peisonis*	–	+	–	
C. schumanniana var. biconstricta Grun.	–	+	–	
Gyrosigma acuminatum (Kütz.) Rab.	+	+	–	
Gyrosigma attenuatum (Kütz.) Rabenh.	+	+	+	
Gyrosigma peisonis (Grun.) Hust.	+	+	+	soda-halophyte
Gyrosigma spencerii (W.Sm.) Cl. (Plate 16.7, Fig. 1)	–	+	–	soda-halophyte
Gyrosigma macrum (W. Smith) Cleve (Plate 16.7, Fig. 5)	+	+	–	chloride-halophyte
Pleurosigma elongatum W. Smith	+	–	+	chloride-halophyte
Scoliotropis peisonis (Grun.) Hust. (Plate 16.8, Fig. 10)	–	+	–	soda-halophyte
Amphiprora costata Hustedt (Plate 16.8, Fig. 2)	+	+	+	halophyte
Amphiprora paludosa W. Smith	+	+	+	halophyte
Amphora ovalis Kütz.	+	+	–	
Amphora veneta Kütz.	+	–	–	
Amphora commutata Grun.	–	+	–	
Cymbella prostrata (Berkel.) Cl.	–	–	+	
Cymbella pusilla Grun.	–	+	+	soda-halophyte

	Plankton	Benthos	Periphyton	
Cymbella lacustris (Ag.) CI. (Plate 16.8, Fig. 8)	–	+	+	
Cymbella affinis Kütz.	+	–	+	
Cymbella cistula (Hempr.) Grun.	+	+	+	
Cymbella lanceolata (E.) vH.	–	+	–	
Cymbella aspera (E.) Cl.	–	+	+	
Gomphonema parvulum (Kütz.) Grun.	–	–	+	
Gomphonema constrictum Ehr.	–	+	+	
Gomphonema olivaceum (Lyngbye) Kütz.	–	+	+	
Gomphonema olivaceum var. calcarea Cl.	–	+	+	
Epithemia argus Kütz.	+	+	+	
Epithemia zebra (E.) Kütz.	–	–	+	
Epithemia turgida (E.) Kütz.	+	+	–	
Epithemia sorex Kütz.	+	+	+	
Rhopalodia gibba (E.) O. Müller	+	+	+	
Rhopalodia gibberula (E.) O. Müller	–	+	–	
Cylindrotheca gracilis (Breb.) Grun. (Plate 16.7, Fig. 4)	+	+	–	soda-halophyte
Hantzschia amphioxys (E.) Grun.	–	+	+	
Hantzschia vivax (W. Smith) Hust.	+	+	–	soda-halophyte
Hantzschia spectabilis (E.) Hust.	+	+	–	soda-halophyte
Bacillaria paradoxa Gmelin (Plate 16.8, Fig. 3)	+	+	+	marine-brackish water
Nitzschia tryblionella Hantzsch (Plate 16.7, Fig. 3)		+	+	chloride-halophyte
Nitzschia salinarum Grun.		+	+	halophyte
Nitzschia apiculata (Greg.) Grun.		+	–	halophyte
Nitzschia hungarica Grun.		+	–	halophyte
Nitzschia subamphioxoides Hust.		+	–	halophyte
Nitzschia vitrea Norman		+	–	soda-halophyte
Nitzschia dissipata (Kütz.) Grun.		–	+	
Nitzschia peisonis Pantocsek		+	+	
Nitzschia acicularioides Hustedt		–	–	endemic
Nitzschia admissa Hustedt		+	+	
Nitzschia amphibia Grun.		+	+	
Nitzschia communis Rabenh.		+	+	
Nitzschia frustulum (Kütz.) Grun.		+	–	
Nitzschia jugiformis Hustedt		+	–	
Nitzschia kützingioides Hust.		+	+	endemic
Nitzschia navicularis (Breb.) Grun. (Plate 16.9, Fig. 3)		–	–	halophyte
Nitzschia legleri Hustedt		+	–	halophyte
Nitzschia subcapitellata Hust.		+	–	halophyte

	Plankton	Benthos	Periphyton	
Nitzschia palea (Kütz.) W. Smith	–	+	–	
Nitzschia subtiloides Hustedt	+	+	–	halophyte
Nitzschia geitleri Hustedt	+	+	–	endemic
Nitzschia obtusa W. Smith	–	+	+	halophyte
Nitzschia sigmoidea (E.) W. Smith	+	+	+	
Nitzschia acicularis W. Smith	+	+	–	
Nitzschia lorenziana Grun. *var. subtilis* (Plate 16.7, Fig. 6)	+	+	–	halophyte
Nitzschia filiformis (W. Smith) Hust. (Plate 16.8, Fig. 11)	–	–	+	halophyte
Cymatopleura solea (Breb.) W. Smith	+	+	+	
Surirella ovata Kütz.	+	+	+	
Surirella peisonis Pant. (Plate 16.8, Fig. 1)	+	+	+	soda-halophyte
Surirella tenera Greg.	+	+	+	
Surirella robusta E. *var. splendida* (E.) V. Heur	–	+	–	
Campylodiscus clypeus Ehr. and *var. bicostata* (W.Sm.) Hust.	+	+	+	soda-halophyte

16.7.5 *Chlorophyta*

(a) Chlorococcales: Neusiedlersee contains fairly many species of Chlorococcales (= Protococcales) and they exist in all biotopes (see Kusel-Fetzmann 1974).

Eutetramorus globosus Walton: found by Schiller (1952a) in the canal of Rust
Characium braunii Bruegger: growing on *Cladophora*
Pediastrum duplex Mayen: in the reed belt, on macrophytes and of recent years in great numbers in the plankton; not present before 1965 (Plate 16.5, Fig. 2)
Pediastrum boryanum (Turp.) Menegh.: scattered occurrences in the reed belt (Plate 16.5, Fig. 1)
Pediastrum tetras (Ehr.) Ralfs: reed belt near Rust (Plate 16.5, Fig. 3)
Oocystis lacustris Chodat: plankton, benthos (Plate 16.5, Fig. 6)
Oocystis solitaria Wittrock: plankton, benthos (Plate 16.5, Fig. 4)
Tetraedron minimum (A. Braun) Hansgirg: in the plankton, scattered occurrences in other places (Plate 16.5, Fig. 7)
Scenedesmus species are found in greater numbers in the reed, between *Utricularia*, and in the plankton. They can vary a great deal; most important are:

Scenedesmus acuminatus (Lagerh.) Chodat: (Plate 16.5, Fig. 11)
Scenedesmus acutus Meyen
Scenedesmus acutiformis Schröder (Plate 16.5, Fig. 17)
Scenedesmus armatus Chodat. (Plate 16.5, Fig. 15)
Scenedesmus balatonicus Hortobagy (Plate 16.5, Fig. 12)
Scenedesmus disciformis (Chodat) Fott and Komarek (Plate 16.5, Fig. 9)

Scenedesmus opoliensis Richter (Plate 16.5, Fig. 14)
Scenedesmus ovalternus Chodat (Plate 16.5, Fig. 10)
Scenedesmus platydiscus (G. M. Smith) Chodat (Plate 16.5, Fig. 8)
Scenedesmus quadricauda (Turp.) Bréb. (Plate 16.5, Fig. 13)
Scenedesmus spinosus Chodat (Plate 16.5, Fig. 16)
Botryococcus braunii Kütz: Scattered between *Utricularia*, also in the plankton; in October 1970 even flowering in the southern part of the lake (Plate 16.5, Fig. 25)
Dictyosphaerium pulchellum Wood. (Plate 16.5, Fig. 19)
Dictyosphaerium reniforme: Rare, in the plankton (Plate 16.9, Fig. 2)
Crucigenia tetrapedia (Kirchner) W. and G. S. West (Plate 16.5, Fig. 18)
Ankistrodesmus falcatus (Corda) Ralfs (Plate 16.5, Fig. 20)
Monoraphidium contortum (Thur.) Legnerova = *Ankistrodesmus falcatus var. spirilliformis* G. S. West: Single, generally screw-like twisted cells, in the plankton
Closteriospira lemanensis Rev.: in the pool near the museum of Neusiedl; among *Utricularia* (Plate 16.5, Fig. 22)
Coelastrum microporum Nägeli
Coelastrum proboscideum Bohlin
Planktosphaeria gelatinosa G. M. Smith (Plate 16.5, Fig. 23) and *Sphaerocystis schroeteri* Chodat (Plate 16.5, Fig. 21): larger amounts of these were for the first time found in the plankton off Rust in September 1975.
Lobocystis dichotoma Thompson is new in the plankton of Neusiedlersee. Before 1975 this species had not been recorded from Neusiedlersee or elsewhere in Austria. In 1976, July, it occurred in great quantities in samples off Rust, Breitenbrunn and Neusiedl (Plate 16.9, Fig. 1)
Glaucocystis nostochinearum Itzigs.: with bluegreen cyanelles, in the reed belt (Plate 16.5, Fig. 6).

(b) Filamentous Chlorophyceae: Representatives of the filamentous Chlorophyceae, such as Ulotrichales, Microsporales, Oedogoniales, Chaetophorales, Zygnemales and Chladophorales, are present in various biotopes of Neusiedlersee. They have hardly been examined at all since they tend to be sterile; usually the reproductive stages are necessary for an identification.

In spring floating green threads of *Ulothrix*, *Spirogyra* and *Oedogonium* grown on the reeds on the margins of larger canals or pools, whereas *Stigeoclonium* and *Cladophora* are found on the props and concrete walls of the embankments. Later in the year floating mats of *Mougeotia-*, *Spirogyra-*species and *Rhizoclonium hieroglyphicum* (Kütz.) Stockm. can be found in calm, shallow bays. In the opaque water of the lake sometimes huge *Cladophora* mats grow out from the macrophytes. *Ulothrix amphigranulata* Skuja occurs in short threads among *Utricularia* near Neusiedl (Plate 16.5, Fig. 24). *Enteromorpha intestinalis* Grev., an alga occurring in brackish waters, is periodically found in the reed belt south of Podersdorf.

(c) Desmidiaceae: Although Desmidiaceae are most frequent in acid waters containing few electrolytes, they are found in striking quantities in the plankton, and even more in the canals and pools within the reed belts, where the water is rich in humus. Only the most frequent of the 30 taxa will be mentioned here (see Kusel-Fetzmann 1974).

Gonatozygon monotaenium De Bary: in the mud covering *Utricularia* near Rust (Plate 16.6, Fig. 3)
Closterium acutum Bréb.: in the periphyton (Plate 16.6, Fig. 2)
Closterium pronum Bréb.: in the reed belt only and not in the plankton as in other lakes (Plate 16.6, Fig. 1)
Closterium parvulum Nägeli: The most frequent Desmidiacea in the benthos, in the periphyton and on macrophytes (Plate 16.6, Fig. 17)
Closterium dianae Ehr. var. *pseudodianae* (Roy) Krieger
Closterium lanceolatum Kütz.
Cosmarium biretum Bréb. *var trigibberum* Norst.: This characteristic *Cosmarium* is one of the most salient algae in Neusiedlersee in Chlorophyceae mats, growing on *Potamogeton pectinatus*, but also in the Oberer Stinker and in Darscho Lacke – two salt lakelets of a considerably higher concentration in the Seewinkel (Plate 16.6, Fig. 7–9)
Cosmarium botrytis Menegh. (Plate 16.6, Fig. 6)
Cosmarium granatum Bréb. (Plate 16.6, Fig. 14)
Cosmarium humile (Gay) Nordst.: very frequent among *Utricularia* (Plate 16.6, Fig. 20)
Cosmarium impressulum Elfv. *var. crenulatum* (Näg.) Krieger
Cosmarium meneghinii Bréb.
Cosmarium punctulatum (Bréb. (Plate 16.6, Fig. 4)
Cosmarium praecisum Borge *var. suecicum* (Borge) Krieger (Plate 16.6, Fig. 16)
Cosmarium cf. regnellii Wille *var. chondrophorum* Skuja (Plate 16.6, Fig. 13)
Cosmarium scopulorum Borge: very frequent in the lake at all seasons (Plate 16.6, Fig. 15)
Cosmarium subgranatum (Nordst.) Lütken (Plate 16.6, Fig. 12)
Cosmarium varsoviense Racib. (Plate 16.6, Fig. 5)
Staurastrum alternans Bréb. (Plate 16.6, Figs. 10, 11)
Staurastrum alternans fo. biradiata: a *Cosmarium* unlike any known species was often found in the reed belt near Rust. Since its cell wall and its chloroplast had a structure similar to that of the three-radiate cells of *Staurastrum alternans*, it was therefore placed here in this list (Plate 16.6, Fig. 19)
Staurastrum dilatatum Ehr. (Plate 16.6, Fig. 21)
Staurastrum hexacerum (Ehr.) Wittr.: the most frequent *Staurastrum* in Neusiedlersee (Plate 16.6, Fig. 18)

16.7.6 *Xantophyta*

The literature on algae does not contain any data on this group. Our own collections showed that *Ophiocytium majus* Nägeli (Plate 16.6, Fig. 22) was very frequent in the muddy coverings of *Utricularia*, whereas *Mischococcus confervicola* Näg. was extremely rare. Thready *Tribonema* species are also found in the periphyton on reeds among other algae and Weisser (1970) encountered *Vaucheria* sp. in the reed belt.

16.7.7 *Charophyta*

The occurrence of Characeae in Neusiedlersee was studied by Weisser (1970) in the course of his sociological and ecological investigations in the reed belt. *Chara ceratophylla* is the only species to be found in the bay of Rust down to one metre. The other species are generally present in shallow pools, in old fairways in the reed and on the landward margin of the reed belt, joining *Bolboschoenus* (Oggau, Weiden). *Chara crinita* inhabits places of stronger salinity such as the eastern lake shore toward Zitzmannsdorfer Wiese and near Stinker See.

Chara ceratophylla Wallr.
Chara crinita Wallr.
Chara delicatula Ag.
Chara foetida A. Braun and forma *paragymnophylla* Migula
Chara fragilis Desvaux
Chara intermedia A. Braun
Chara tenuispina A. Braun

16.7.8 *Rhodophyta*

On agar cultures from samples of algae taken in the bay of Illmitz, Geitler (1970) found besides *Porphyridium aerugineum* Geitler a grey Bangiale, which he called *Porphyridium griseum* Geitler. He also found various *Chantransia*-like creeping thalli, but not a single adult *Batrachospermum*. Up to now no other Rhodophyceae have been found in Neusiedlersee.

References

Cholnoky, B. J. v., 1968. Die Ökologie der Diatomeen in Binnengewässern.–Verlag J. Cramer, Lehre, 1–699.

Dokulil, M. 1971. Über das Vorkommen von *Lampropedia hyalina* Schroeter, einer apochlorotischen Cyanophycee, im Neusiedlersee.– Arch. Hydrobiol, 69: 405–409.

Drum, R. W., 1969. Light and electron microscope observations on the tube-dwelling diatom *Amphipleura rutilans* (Trentep.) Cl.–J. Phycol–5: 21–26.

Geitler, L., 1959. *Spirulina raphidioides* n. sp. und Bemerkungen über ähnliche Planktonalgen.– Öst. Bot. Z. 106: 133–137.

Geitler, L., 1969. Die Auxosporenbildung von *Nitzschia amphibia*.–Öst. Bot. Z. 117: 404–410.

Geitler, L., 1970a. Beiträge zur epiphytischen Algenflora des Neusiedler Sees.– Öst. Bot. Z. 118: 17–29.

Geitler, L. 1970b. Die Entstehung der Innenschalen von *Amphiprora paludosa* unter acytokinetischer Mitose.– Österr. Bot. Z. 118: 591–596.

Geitler, L., 1972. Sippen von *Gomphonema parvulum*, Paarungsverhalten und Variabilität pennater Diatomeen.– Öst. Bot. Z. 120: 252–268.

Gessner, F., 1959. Hydrobotanik II. Hochschulbücher f. Biologie.– Veb, Deutsch. Verlag d. Wiss., Berlin, 1–701.
Grunow, A., 1860. Über neue oder ungenügend gekannte Algen.– Verh. k. k. zool. Bot. Ges. Wien 10: 503–582.
Grunow, A., 1862. Die österreichischen Diatomaceen nebst Anschluß einiger neuer Arten von anderen Lokalitäten und einer kritischen Übersicht der bisher bekannten Gattungen.– Verh. K. K. zool. bot. Ges. Wien 12: 315–472, 545–588.
Grunow, A., 1863. Über einige neue und ungenügend bekannte Arten und Gattungen von Diatomaceen.– Verh. k. k. zoo. bot. Ges Wien, 13: 137–162.
Harder, R. und v. Witsch, H. 1942. Über Massenkulturen von Diatomeen.– Ber. Deutsch. Bot. Ges. 60: 146–152.
Hustedt, Fr., 1925. Bacillariales aus den Salzgewässern bei Oldesloe in Holstein.– Mitt. Geogr. Gesell. Nat. hist. Museum Lübeck, 30: 84.
Hustedt, Fr., 1930. Bacillariophyta (Diatomeae) in: Die Süßwasserflora Mitteleuropas, Hrsg. A. Pascher, Fischer Verlag, Jena, Heft 10.
Hustedt, Fr., 1959a. Die Diatomeenflora des Neusiedler Sees im österreichischen Burgenland.– Öst. Bot. Z. 106: 390–430.
Hustedt, Fr., 1959b. Bemerkungen über die Diatomeenflora des Neusiedlersees und des Salzlachengebietes.– In: Landschaft Neusiedlersee. Wiss. Arb. Bgld. 23: 129–133.
Hustedt, Fr., 1959c. Die Diatomeenflora des Salzlackengebietes im österreichischen Burgenland.– Sitz. Ber. Österr. Akad. Wiss., Math. nat. Kl. I, 168: 387–452.
Kolbe, R., 1927. Zur Ökologie, Morphologie und Systematik der Brackwasserdiatomeen. Die Kieselalgen des Sperenberger Salzgebietes.– Pflanzenforschung 7: 1–146.
Kusel-Fetzmann, E., 1973. *Gyrosigma macrum*-neu für den Neusiedler See.–Öst. Bot. Z., 122: 115–120.
Kusel-Fetzmann, E., 1974. Beiträge zur Kenntnis der Algenflora des Neusiedler Sees I.– Sitz. Ber. Österr. Akad. Wiss., Math.– Nat. Kl I, 183: 5–28.
Legler, Fr., 1941. Zur Ökologie der Diatomeen burgenländischer Natrontümpel.– Sitz. Ber. Österr. Akad. Wiss., Math.– nat. Kl. I, 150: 45–72.
Löffler, H., 1956. Ergebnisse der österreichischen Iranexpedition 1949/50. Limnologische Untersuchungen an iranischen Binnengewässern.– Hydrobiologia 8: 3–4.
Loub, W., 1955. Algenbiozönosen des Neusiedler Sees.– Sitz. Ber. Österr. Akad. Wiss., Math.– nat. Kl. I, 164: 81–107.
Mack, Br., 1954. Untersuchungen an Chrysophyceen V–VII.– Öst. Bot. Z. 101: 64–73.
Neuhuber, F., 1971. Ein Beitrag zum Chemismus des Neusiedler Sees.– Sitz. Ber. Österr. Akad. Wiss., Math.– nat. Kl. I, 179: 225–231.
Pantocsek, J., 1912. A fertö tó kovamoszat viránya (Bacillariae Lacus Peisonis) Pozsony (Preßburg) 1–43.
Ruttner-Kolisko, A. und Ruttner, F., 1959. Zusammenfassung und allgemeine Limnologie.– In: Landschaft Neusiedlersee. Wiss. Arb. Bgld. 23: 195–201.
Schiemer, F. und Weisser, P. 1972. Die Verteilung der submersen Makrophyten in der schilffreien Zone des Neusiedler Sees.– Sitz. Ber Österr. Akad. Wiss., Math.– nat. Kl. I, 180: 87–97.
Schiller, J. 1952a. Neue oder wenig bekannte Mikrophyten aus dem Neusiedler See und benachbarter Gebiete.– Österr. Bot. Z. 99: 363–369.
Schiller, J., 1952b. Über die Vermehrung des Paramylons und über Alterserscheinungen bei Eugleninen.– Österr. Bot. Z. 99: 413–420.
Schiller, J., 1954. Über winterliche pflanzliche Bewohner des Wassers, Eises und des darauf-liegenden Schneebreies I.– Österr. Bot. Z. 101: 236–284.
Schiller, J., 1955. Untersuchungen an den planktischen Protophyten des Neusiedler Sees 1950–1954. I. Teil.– Wiss. Arb. Bgld. 9: 5–66.
Schiller, J., 1956. Untersuchungen an den planktischen Protophyten des Neusiedler Sees. II. Teil: Euglenen.– Sitz. Ber. Österr. Akad. Wiss., Math.– nat. Kl., I, 165: 547–583.
Schiller, J., 1957. Untersuchungen an den planktischen Protophyten des Neusiedler Sees 1950–1954. II. Teil.–Wiss. Arb. Bgld. 18: 4–44.
Schmid, A., 1973. Beiträge zur Ökologie einiger Neusiedlersee-Diatomeen, mit besonderer

Berücksichtigung ihrer Salzresistenz.– Dissertation Univ. Wien 192 pp.
Stehlik, A., 1972. Chemische Topographie des Neusiedler Sees.– Sitz. Ber. Österr. Akad. Wiss., Math.– nat. Kl. I. 180: 217–278.
Szemes, G., 1959. Die Bacillariophyceen des Szelider Sees. In: Das Leben des Szelider Sees, Die Binnengewässer Ungarns, Hersg. Dr. Ernö Donászy. I, 251–416.
Tschermak-Woess, E., 1972. Über die Haptophycee *Sarcinochrysis granifera* aus dem Neusiedlersee.– Österr. Bot. Z. 121: 235–255.
Tschermak-Woess, E., 1973a. Über die bisher vergeblich gesuchte Auxosporenbildung von *Diatoma*.–Öst. Bot. Z. 121: 23–27.
Tschermak-Woess, E., 1973b. Die geschlechtliche Fortpflanzung von *Amphipleura rutilans und* das verschiedene Verhalten der Erstlingszellen von Diatomeen in Gallertschläuchen.– Öst. Bot. Z. 122: 21–34.
Weisser, P., 1970. Die Vegetationsverhältnisse am Neusiedler See. Pflanzensoziologische und ökologische Studien.– Wiss. Arb. Bgld. 45: 1–83.
Woronichin, N., 1926. Zur Biologie der bittersalzigen Seen in der Umgebung von Pjatigorsk (nördl. Kaukasus).–Arch. Hydrobiol. 17: 628–643.

7 Seasonal pattern of phytoplankton

M. Dokulil

17.1 Introduction and methods

Although papers concerning the algal flora of Neusiedlersee appeared as long ago as 1860 investigations were restricted to systematic and qualitative aspects of different algal groups from a few sampling stations (see Chapter 16).

Except for some figures given by Ruttner-Kolisko & Ruttner (1959) and information on net-phytoplankton provided by Zakovsek (1961), no quantitative data on the lake's phytoplankton existed. To fill this gap an intensive programme including measurements on algal primary production was initiated during the IBP and has been pursued continuously from 1968 until now with an interruption in 1974.

Samples were routinely taken from up to 20, but at least from three stations including the submerged macrophyte zone and the reed belt until 1972. Detailed information on phytoplankton biomass, production, and on bacterial biomass and production has been published by Dokulil (1973, 1975a). This chapter concentrates on standing crop aspects of the plankton of the open water.

Quantitative analysis was based on the enumeration of cell numbers by counting with the inverted microscope using Utermöhl's technique (Utermöhl 1931, 1958). Lugols-acetic-acid solution with a small amount of formalin was used as preservation fluid. A minimum of 100 individuals of the more important species was counted, or at least a total of 400 in the subsample, thus yielding a maximum error of about ±20% for a single species and less than ±10% for the entire sample (Holmes & Widring 1956, Javornický 1958, Lund et al. 1958, Hobro & Willén 1975).

Cell counts were converted to biomass by calculating mean cell volumes from size data, applying geometric formulae or plastic models. Average dimensions, mean cell volume and formulae are summarized in Table 17.1. In converting to fresh weight a specific gravity of one was assumed. Average cell surface area was also calculated in a similar way (Table 17.1), following suggestions by Paasche (1960) and Bellinger (1974). These two parameters yield the surface area to volume ratios (S/V) listed in Table 17.1. Organic carbon content of the phytoplankton, according to Findenegg (1971), was assumed to be 12% of fresh weight (75% water content: Nalewajko 1966; carbon equal to 50% of ash-free dry weight: Lund 1964), which is slightly higher than the now widely accepted figure of 10% (e.g. Goldman et al. 1968, Berman & Pollingher 1974).

Chlorophyll estimation provided an additional parameter of phytoplankton biomass. The shallowness of the lake and the strong winds of the region cause

Table 17.1. Average dimensions, mean cell volume, mean cell surface area, the surface/volume ratio of the more important algal species of the phytoplankton and the formulas on which the calculations are based.

	Dimensions (μ)			Volume	Formula	Surface	S/V	Formula
	l	w	h	(μ^3)		(μ^2)		
Chroococcus minutus	5.14			71	$\pi/6.1^3$	83	1.17	$1^2.\pi$
Merismopedia punctata	3.37			20	,,	36	1.78	$1^2.\pi$
Cryptomonas ssp.	21.8	13.1	11.79	1,763	$\pi/6.1.w.h$	789	0.45	num. appr.
Rhodomonas lacustris	13.0	8.0		285	$\pi/12.w^2.(w/2 + 1)$	289	1.01	,,
Oocystis lacustris	10.6	5.4		162	$\pi/6.w^2.1$	154	0.95	,,
Oocystis solitaria	13.0	9.0		552	,,	332	0.60	,,
Tetraedron minimum	8.0	8.0	3.0	192	1.w.h	224	1.17	2(lw + lh + wh)
Chodatella subsalsa	7.5	5.4		117	$\pi/6w^2.l$	116	0.99	num. appr.
Crucigenia tetrapedia	3.0	3.0	1.2	11	l.w.h	32.3	2.95	2(lw + lh + wh)
Monoraphidium contortum	38.0	1.6		97	$(w/2)^2.l.\pi$	195	2.01	$\pi.w(l + w/2)$
Sphaerocystis schröteri	5.0			65	$\pi/6.l^3$	79	1.21	$l^2.$
,, (after division)	2.5			8	,,	20	2.45	,,
Scenedesmus ssp.	10.0	3.5		65	$\pi/6.w^2.l$	90	1.38	num. appr.
Cyclotella sp. & o.sp.	13.0		5.5	750	$(1/2)^2.h.\pi$	490	0.65	$\pi l(h + l/2)$
Synedra acus	102.6	5.4		1,500	av. of $(w/2)^2.l.\pi$ and $\pi/6.w^2.l$	1,786	1.19	$\pi.w(l + w/2)$
Synedra ulna & capitata	160.0	6.8		9,000	$(w/2)^2.l.\pi$	3,490	0.38	,,
small *Fragilaria*	21.0	5.2	4.0	425	l.w.h.	428	1.01	2(lw + lh + wh)
Amphora ovalis	19.0			3,500	from model			
Navicula oblonga	61.6	10.3		5,000	mean of speroid & cylinder	1,584	0.32	num. appr.
Amphiprora costata				6,500	from model			
Rhopalodia gibba				12,000	,,			
small *Naviculas*	21.8	5.8		700	$\pi/6.w^2.l$	320	0.46	,,
small *Nitzschias*	10.0	2.8		400	,,	71	0.18	
Surirella peisonis		1,650 μ^2	20.0	33,000	from model after tracing			
Campylodiscus clypeus	57.0		3.0	9,000	from model			
Euglena acus	100.0	7.2		4,071	$(w/2)^2.l.\pi$	1,780	0.43	,,
Euglena oxyuris	104.0	16.0		20,000	,,	4,149	0.21	,,

frequent resuspension of dead and moribund algal material ($\bar{z}$ = 1.3 m) (cf. Chapter 15), occasionally yielding large quantities of pigment degradation products. The method of Lorenzen (1967) was used to correct total chloroplastic pigment estimations for such phaeopigments. Since 1975 chlorophyll b and c have been included in the measurements, using the trichromatic equation recommended by Scor-Unesco (1966). Since the values obtained were usually low, if detectable at all, they will not be further discussed here.

17.2 Species composition

Species are listed in Table 17.2 (comp. chapter 16) and include earlier observations by Loub (1955), Ruttner (unpubl., 1956–58)[1], and Zakovsek (1961). Taxa are arranged according to Fott (1971). Two species of the blue-green algae, namely *Microcystis pulverea* and *Merismopedia punctata*, where invariably encountered whereas *Chroococcus minutus* was not found prior to 1968.

The Chrysophytes were represented by one species only, not found recently. A striking discrepancy exists concerning the centric diatoms which were not recorded among the phytoplankton by Loub (1955), although his data represent the situation in fall when centric diatoms are abundant. Even Hustedt (1959) recorded *Chaetoceros mülleri* as the main component of the plankton. In contrast, *Cyclotella meneghiniana* was the dominant centric species during the observation period, followed by *Stephanodiscus hantzschii* in abundance. *Chaetoceros mülleri* was extremely rare. Veszprémi (1976) reports *Cyclotella stelligera* from the Hungarian part of the lake in samples from the year 1975.

Among the pennate diatoms a variety of epiphytic and epipelic algae contribute occasionally to the phytoplankton when they become detached due to wind-induced wave action. Epiphytic species are more important as 'tychoplankton' during the vegetation period of submerged plants, which is at a time of year when epipelic algae cannot develop to any great extent on account of sediment surface instability and poor light conditions. In late autumn and especially in early spring, benthic algae stirred up from the mud often form a considerable part of the plankton's diatom population since their maximum development takes place under the ice cover (Prosser, unpubl.).

Diatoms appearing regularly in the plankton are various tiny species of the genera *Nitzschia*, *Synedra*, and *Fragilaria*. Two large and heavy species are sometimes abundant, namely *Surirella peisonis* and *Campylodiscus clypeus*. *Gyrosigma peisonis* and *Scolipleura peisonis*, both characteristic for the lake according to Hustedt (1959), seem to have vanished from the plankton.

A possible explanation is the recent change in water chemistry (see Chapter 10) due to a rise of the water level in 1965 in connection with regulation of the artificial outflow ('Einserkanal'). *Gyrosigma macrum*, never recorded before, appeared in large numbers in 1968 (Kusel-Fetzmann 1973) in sediment samples from the littoral and the reed belt. This species, known to be a brackish-water type, rarely enters the phytoplankton.

[1] I gratefully acknowledge the kind permission of Prof. Dr. Ruttner-Kolisko to include this unpublished material in the present study.

Table 17.2. Species list and number of taxa of the phytoplankton of Neusiedlersee compared with earlier observations.

	Zakovsek (1961) 1950–1952	Loub (1955) Sept./Oct. 1952–1953	Ruttner unpubl. 1956–1958	Dokulil 1968–1974
1. Cyanophyceae (blue green algae)				
Microcystis pulverea (Wood) Mig.		+	?[1]	+
Chroococcus minutus (Kütz.) Näg.		–	–	+
Merismopedia punctata Meyen		+	+	+
Anabaena viguierii Denis et Frémy	+	+	–	–
Anabaena flos-aquae (Lyngb.) Bréb.		+	–	+
Spirulina jenneri (Hass.) Kütz.		+	–	+
Lyngbya limnetica Lemm.		–	+	–
Aphanizomenon gracile Lemm.	+	–	+	+
2. Chrysophyceae				
Chrysocapsa sp. Pascher		+	+[3]	–
Synura uvella Ehr.	+	–	–	–
3. Bacillariophyceae (diatoms)				
Cyclotella meneghiniana Katz.	+	–	+	+
Stephanodiscus hantzschii Grun.		–	–	+
Chaetoceros mülleri Lemm.		–	+	+
Diatoma elongatus Aghard.	+	–	–	–
Synedra acus Kütz.	+	+	+	+
Synedra affinis Kütz.		+	–	–
Synedra nana Meister		+	–	–
Synedra ulna (Nitzsch) Ehr.	+	+	–	+
Synedra capitata Ehr.		–	–	+
Fragilaria brevistriata Grun.		–	–	+
Gyrosigma acuminatum (Kütz.) Rabh.		+	+	+
Gyrosigma peisonis (Grun.) Hust.		+	–	–
Gyrosigma attenuatum (Kütz.) Rabh.		+	–	–
Gyrosigma macrum (W. Smith) Griff et Henfr.		–	–	+
Pleurosigma vgl. salinarum Grun.		+	+[2]	–
Navicula protracta Grun.		+	–	–
Navicula radiosa Kütz.		+	+	+
Scoliopleura peisonis Grun.		+	–	–
Amphiprora costata Hust.		–	+	+
Amphiprora paludosa W. Smith		+	–	–
Amphora ovalis Kütz.		–	+	+
Rhopalodia gibba (Ehr.) O. Müll.		+	–	+
Nitzschia palea (Kütz.) W. Smith		+	+	+
Nitzschia sigmoidea (Ehr.) W. Smith		+	–	+
Nitzschia spectabilis (Ehr.) Ralfs		+	–	–
Cymatopleura solea (Bréb.) W. Smith	+	+	+	+
Surirella ovalis Bréb.		+	+	–
Surirella peisonis Pant.	+	+	+	+
Campylodiscus clypeus Ehr.		+	+	+
Campylodiscus clypeus var. bicostata W. Smith	+	+	–	–
Bacillaria paradoxa Gmelin	+	–	–	+

Table 17.2 (*continued*)

	Zakovsek (1961) 1950–1952	Loub (1955) Sept./Oct. 1952–1953	Ruttner unpubl. 1956–1958	Dokulil 1968–1974
4. Dinophyceae (Dinoflagellates)				
Gymnodinium pascheri		+	+	–
Peridinium cinctum (O. F. Müll.) Ehr.		–	+	+
5. Chlorophyceae (green algae)				
Chlamydomonas sp. Ehr.		+	+	+
Pediastrum boryanum (Turp.) Menegh.	+	+	+	+
Pediastrum boryanum var. longicorne Racib.	+	–	–	–
Pediastrum duplex var. reticulatum Meyen		–	–	+
Pediastrum angulosum (Ehr.) Menegh.		+	–	–
Pediastrum glanduliferum Benn.		+	–	–
Oocystis pusilla Hansg.		+	–	–
Oocystis pelagica Lemm.		+	–	–
Oocystis solitaria Wittr.		–	+	+
Oocystis lacustris Chod.		–	+	+
Chodatella subsalsa Lemm.		+	+	+
Tetraedron minimum (A. Br. Hansg.)		–	–	+
Scenedesmus quadricauda (Turp.) Bréb.		+	+	+
Scenedesmus ecornis var. disciformis (Chod.)		–	–	+
Scenedesmus balatonicus Hortob.		–	–	+
Scenedesmus acutiformis Schröd.		–	–	+
Scenedesmus opoliensis Richt.		–	–	+
Crucigenia tetrapedia (Kirchn.) West & West		+	+	+
Dictyosphaerium ehrenbergianum Näg.		+	–	–
Dictyosphaerium pulchellum Wood.		–	–	+
Botryococcus braunii Kütz.	+	+	+	+
Monoraphidium contortum (Thus.) Kom. Leg. (= *Ankistrodesmus falcatus var. spirilliformis*)		–	–	+
Ankistr. minutissimus Korsch		–	–	+
Ankistr. convolutus Corda.		–	+	–
Kirchneriella obesa (W. West) Schmidle		–	+	+
Sphaerocystis schroeteri Chod.		–	+	+
Polyedrium trigonum (= *Tetraedron trigonum*) (Näg.) Honsg.		–	+	+
Closterium pronum Bréb.		–	–	+
Closterium acerosum (Schrank) Ehr.		+	–	–
Closterium leibleinii Kütz		+	–	–
Cosmarium humile (Goy) Nordst.		–	–	+
Cosmarium praecisum var. suecicum (Borge) Krieger & Gerloff.		–	–	+
Cosmarium scopulorum Borge		–	–	+
Cosmarium retusiforme		+	–	–
Planctosphaeria sp.		–	–	+
Lobocystis dichotoma Thompson		–	–	+

Table 17.2 (*continued*)

	Zakovsek (1961) 1950–1952	Loub (1955) Sept./Oct. 1952–1953	Ruttner unpubl. 1956–1958	Dokulil 1968–1974
6. Cryptophyceae				
Cryptomonas erosa Ehrenb.		–	+	+
Cryptomonas sp. Ehr.		–	+	+
Rhodomonas lacustris Pasch, et Rut.		–	+	+
Rhodomonas minuta Skuja.		–	–	+
7. Euglenophyceae				
Euglena acus Ehr.		+	+	+
Euglena oxyuris Schmarda	+	+	+	+
Euglena tripteris (Duj.) Klebs.		+	?	+
Euglena repulsans Schiller		+	–	–
Phacus pleuronectes (O.F.M.) Duj.		+	+	+
Total	15	47	38	59

1. Ruttner claims a 'Sorochloris aggregata' belonging to the group chlorobacteria which is very unlikely because this group needs anoxic conditions. Very likely it is 'Microcystic pulverea' which is similar in shape.
2. Ruttner gives the name Pleurosigma sp. only.
3. Ruttner simply indicates chrysomonads.

The dinoflagellates are common in the littoral (Schiller 1955a) but *Peridinium cinctum* was found in one sample from the open lake in 1968.

Besides the diatoms green algae are the typical plankton group (compare Fig. 17.1 and 17.2). Within the algal class various new species have been observed, thus increasing the total number found from 14 to 26 (see below). Of these species about 8–10 contribute significantly to the total biomass and will be discussed in the next section.

Considerable changes were observed among the genera *Pediastrum*, *Oocystis* and *Scenedesmus*. Most species listed by Loub (1955) have been replaced by others. Special attention must be drawn to *P. duplex var. reticulatum* first recorded by Kusel-Fetzmann (1974), which formed a major component of the phytoplankton in the years 1975 and 1976. The same applies to *Dictyosphaerium reniforme* and *Lobocystis dichotoma* (see Chapter 16). The desmid *Chlosterium pronum*, very seldom found before, bloomed in August 1973 as evidenced by zooplankton net-samples (Herzig, pers. commun.), and disappeared again. All other desmids are usually extremely rare. *Chlamydomonas* sp. was noticed only twice in 8 years.

Finally, two motile groups must be mentioned. Of particular importance are the cryptophytes which were not recorded in the early fifties. Although previously the euglenophytes were present every year at certain times, both groups have vanished from the plankton since 1972/73.

The total number of species recorded increased from 47 to 38 respectively in the years 1952 and 1958, to 59 in the present study. Their distribution among algal classes is listed below:

	1952/53		1956–58		1968–73	
	No	%	No.	%	No.	%
Cyanophyceae	5	10.6	4	10.6	6	10.2
Chrysophyceae	1	2.2	1	2.6	0	0
Bacillariophyceae	21	44.6	12	31.6	19	32.1
Dinophyceae	1	2.2	2	5.3	1	1.7
Chlorophyceae	14	29.8	12	31.6	26	44.0
Cryptophyceae	0	0	3	7.9	3	5.1
Euglenophyceae	5	10.6	4	10.5	4	6.8
Total	47		38		59	

From this list it is obvious that the increase in number of species recorded is solely due to the presence of more green algal species, raising the contribution of this class from 30 to 45%. The phytoplankton assemblage therefore changed in about 20 years from diatom-dominated to green-algal dominated.

Some of the changes observed in the composition of the algal species may be attributable to variations in water chemistry, especially increased nutrient concentration (see Chapter 10), grazing, climatic conditions or the decline of submerged plants since 1972. The latter is particularly likely in the case of the cryptophytes and euglenophytes.

17.3 Seasonal succession of species

Only data from the years 1956–58 and 1968–73 are considered in Fig. 17.1 since samples from 1975 and 76 have not yet been fully analysed. Our discussion will follow, from top to bottom, the species arrangement given in Fig. 17.1, which differs slightly from the list in Table 17.2.

17.3.1 *Blue-green algae*

Chroococcus minutus appeared from June until the end of August in 1968 and 1969 and development continued up to the end of the year in 1970. A winter population which formed in 1971/72 vanished in May 1972, reappearing in summer 1973. The largest population size is associated with water temperatures between 16 and 23°C (cf. Fig. 17.1 and 17.3).

In the period 1956–58 *Chroococcus* was recorded twice by Ruttner, in extremely low numbers. In contrast, *Microcystis pulverea* (1a in Fig. 17.1) at that time had a high biomass which has rarely been found recently.

Merismopedia punctata occasionally formed a small population during the years 1968–70 with a peak at the end of August 1969. The optimal growth temperature range seems to be 14–18°C. in 1958 *Merismopedia* was present from July to October, with a maximum at the beginning of the period.

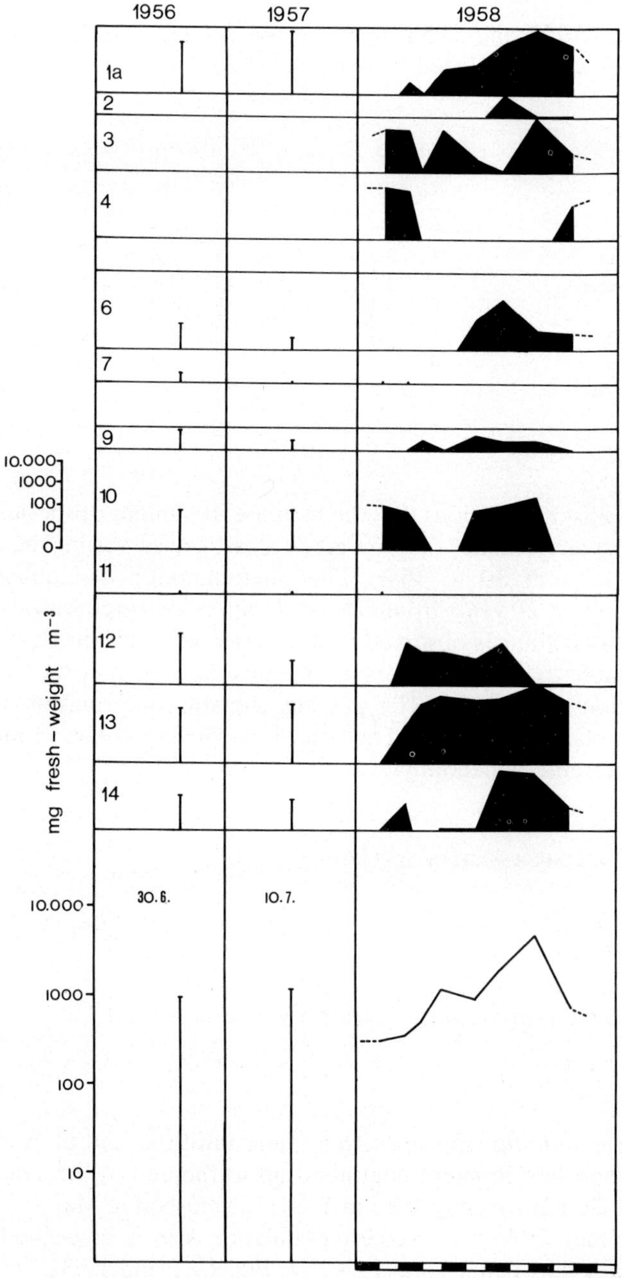

Fig. 17.1. Seasonal succession of phytoplankton in the years 1956–1958 and 1968–1973. Individual algal species as mg fresh weight m^{-3}, total biomass (A) as mg fresh weight per m^3, total cell numbers (B) as 10^9 m^{-3}, and total cell surface area (C) as 10^3 cm^2 m^{-3} (logarithmic plots). 1 = *Chroococcus minutus*; 1a = *Microcystis pulverea*; 2 = *Merismopedia punctata*; 3 = *Cryptomonas* ssp.; 4 = *Rhodomonas lacustris*; 5 = *Chodatella subsalsa*; 6 = *Oocystis lacustris*; 7 = *Crucigenia tetrapedia*; 8 = *Tetraedron minimum*; 9 = *Scenedesmus* ssp.; 10 = *Monoraphidium contortum*; 11 = *Sphaerocystis schröteri*; 12 = centric diatoms; 13 = pennatic diatoms; 14 = *Euglena* ssp.

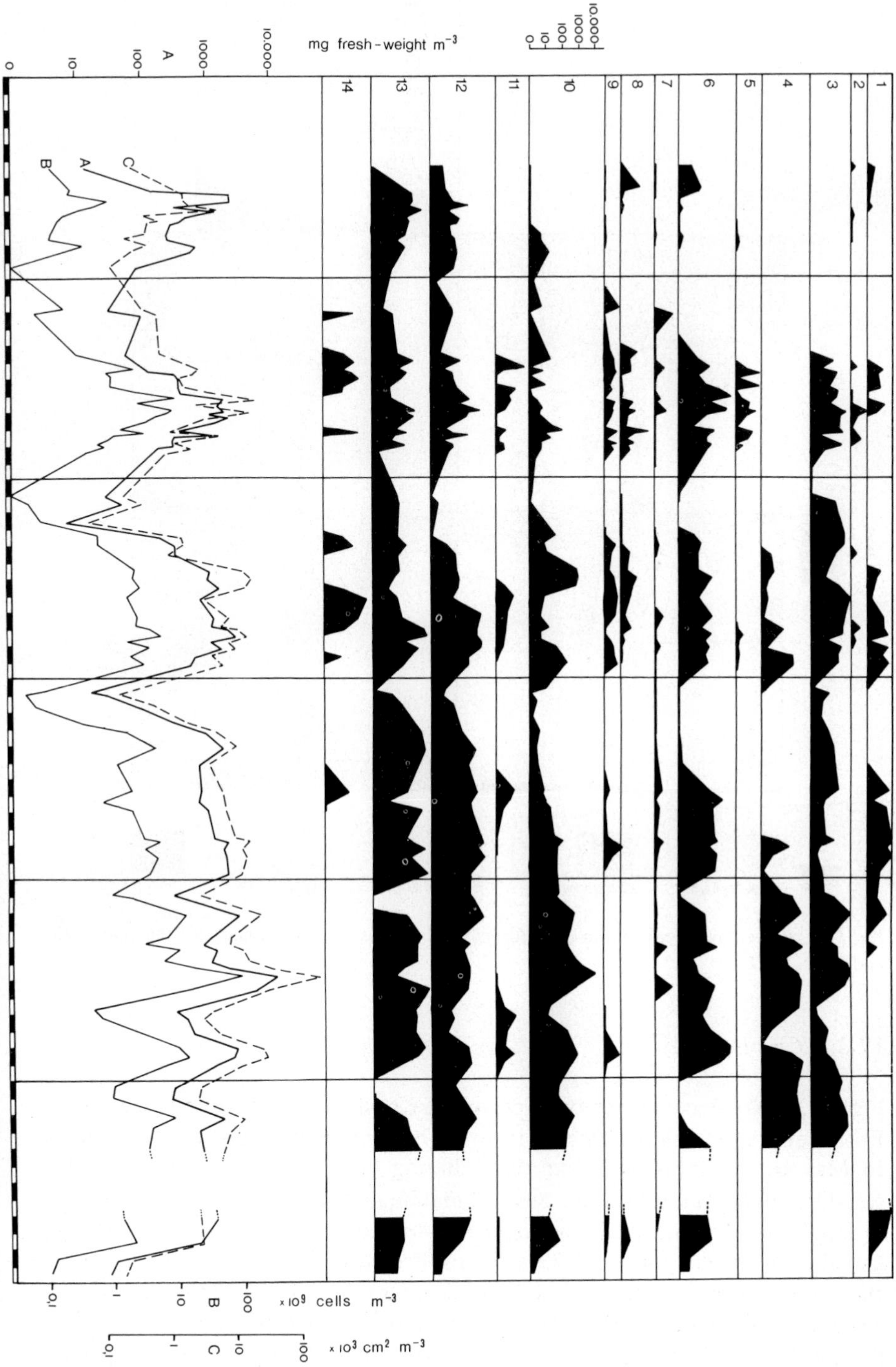
1968
1969
1970
1971
1972
1973
1
2
3
4
5
6
7
8
9
10
11
12
13
14
10.000
1000
100
10
0
mg fresh-weight m⁻³
10.000
1000
100
A
10
0
C
A
B
B
100
10
1
0,1
x 10⁹ cells m⁻³
C
100
10
1
0,1
x 10³ cm² m⁻³

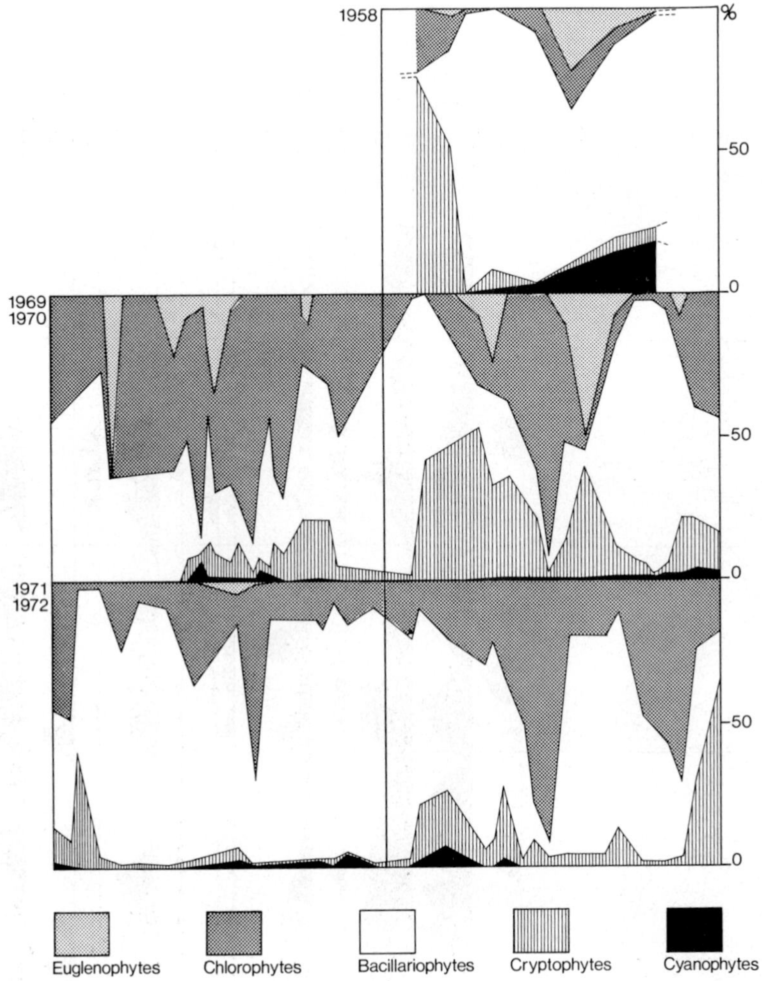

Fig. 17.2. Percentage contribution of different algal groups to total biomass for the years 1958 and 1969–1972.

17.3.2 *Cryptophytes*

Fig. 17.1, No. 3 combines two species of *Cryptomonas* of about equal size. Both were occasionally detected in 1968, never reaching substantial biomass. In May of the following year the standing crop increased and thereafter steadily gained in importance. Several maxima were observed (Fig. 17.1), but the highest values normally occur in July or August. Winter populations are usually low except for 1972/73 as a result of a high autumn biomass and a short freezing period (cf. Fig. 15.1 of Chapter 15). The weather conditions alone (see discussion in the next section), suffice to explain the rapid decline during July and August 1972, also noticeable among the other species. *Cryptomonas*, present all the year round, has a rather wide temperature tolerance, confirming results of Findenegg (1971). The high activity of this genus, mentioned by Findenegg, has to be confirmed by comparison with production measurements. From the cell surface/volume ratio in Table 17.1, a somewhat lower activity is to be expected. Similar low S/V ratios for *Cryptomonas* are reported by Deveaux (1973, p. 44) and Bellinger (1974, p. 159).

Ruttner did not find *Cryptomonas* in 1956 and 1957 but the genus was well documented in the following year. A maximum of 260 mg fresh weight · m^{-3} was reached in early September.

Rhodomonas lacustris, not found during 1956/57, possibly because of insufficient sampling, appeared in March and October 1958. The species reappeared in May 1970 in low numbers, approaching a maximum of 70 mg fr wt. m^{-3} in late November. After a rapid decline only a small fraction of the population survived under the ice. *Rhodomonas* was then undetectable until mid October when a new population developed for a short period. Starting from a small winter stock high biomass figures were reached again in early 1972 (250 mg fr wt. m^{-3}). During this and the following year by the pattern crudely resembles that for *Cryptomonas*. As was pointed out by Hickman (1974) temperature does not limit the growth of *R. lacustris*. From the present study it must be concluded that this is valid for temperatures below 20°C since peaks are always associated with water temperatures equal to or lower than this value. As compared with *Cryptomonas* and other species *Rhodomonas* seemed to be little influenced by turbulent water conditions during July and August 1972. Similar observations were made by Happey (1968, cit. from Hickman 1974). If cell surface/volume ratios are indicative of species activity *Rhodomonas* (S/V = 1.01, Table 17.1) should be more active than *Cryptomonas*.

17.3.3 *Green algae*

This algal class forms a major part of the total phytoplankton biomass (up to 92%, Fig. 17.2). All important species belong to the order Chlorococcales, except in 1974 when *Closterium pronum*, a desmid, bloomed, as was pointed out earlier. Some of the relevant species are known to be very active and to have high S/V ratios (e.g. Bellinger 1974, p. 160). Referring to Table 17.1 *Crucigenia*, *Monoraphidium*, *Sphaerocystis*, *Scenedesmus* and *Tetraedron* have the highest ratios among the species listed. *Oocystis lacustris*, contributing largely to total biomass in terms of fresh weight (Fig. 17.1), is the only species with a low surface/volume ratio. On the other hand, experiments using autoradiographic techniques (Stull et al. 1973) revealed extremely high carbon fixation rates and rapid turnover.

Chodatella subsalsa, which appeared occasionally in very low numbers between 1956 and 1958 was also recorded in 1968–1970 and formed a considerable part of the green algal biomass in 1969 (up to 31 mg fr wt. m^{-3}). Maximum values coincide with temperatures of 13–23°C (see Fig. 17.3). *Chodatella* seems to be a species which is always present in the lake but only capable of forming larger populations in exceptional years. The second

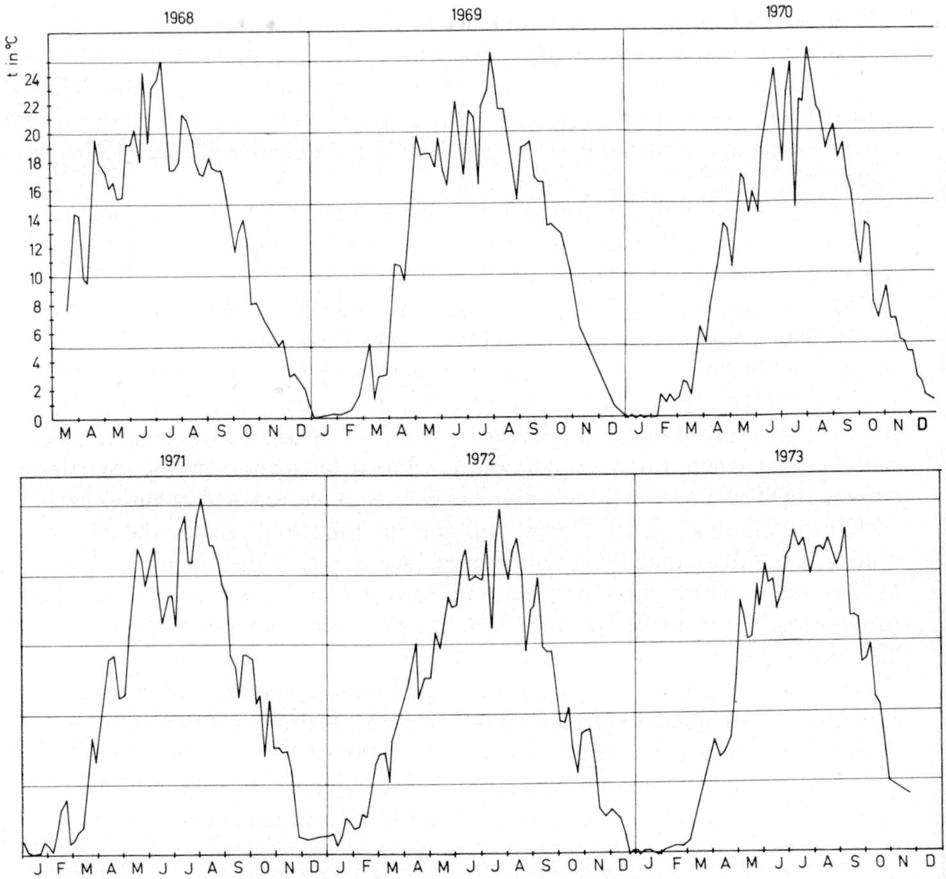

Fig. 17.3. Water temperature as five-day-averages for 1968–1972.

possibility is that *Ch. subsalsa* is not a well defined species but are instead spiny single cells of *Scenedesmus*, an idea which is supported by culture experiments (Overbeck & Stange-Bursche 1965, Swale 1965, Fott 1967, Trainor & Rowland 1968).

Oocystis lacustris was present in the lake throughout the period studied. Population development starts at water temperatures around 10°C. Several peaks of biomass (up to 1,800 mg fr wt m^{-3}) were attained in summer in association with surface temperatures of 19–25°C (see Fig. 17.1 and 17.3). High values were recorded in late autumn in some years at water temperatures below 15°C, possibly due to other favourable factors such as light or nutrients. Low standing crops survived under the ice. The *Oocystis* sp. counted by Ruttner 1956–1958 was very likely *Oocystis lacustris* as evidenced by its cell dimensions. The species appeared in May, attaining a single population maximum of 200 mg in July.

The next species in Fig. 17.1, *Crucigenia tetrapedia*, regularly occurred in the phytoplankton during the ice-free period. Growth started below 5°C. Biomass peaks were at or below 12 mg fresh weight per cubic meter when surface water temperatures were below 16°C. Very low numbers of

Crucigenia appeared from time to time in the years 1956–58.

Tetraedron minimum was not detected by Ruttner 1956–58 but *T.* (=*Polyedrium*) *trigonum* was occasionally found. *Tetraedron* developed a considerable biomass (max. 45 mg fr wt. m^{-3}) in the years 1968–1970, reappearing in 1973 after two years of absence. Its temperature preference for optimum growth is slightly higher than for *Crucigenia* (11–21°C).

Number 9 in Fig. 17.1 is not a single species but represents combined biomass data from at least three *Scenedesmus* species occurring in small numbers. Nevertheless, population size is low and only on one occasion exceeds 10 mg per cubic meter. The same applies to the data from the fifties. Both *Tetraedron* and *Scenedesmus*, like *Chodatella* were never recorded in winter under the ice cover.

Monoraphidium contortum (= *Ankistrodesmus falcatus var. spirilliformis*) seems to be a characteristic species of the phytoplankton assemblage of Neusiedlersee. In 1956–58 *M. contortum* was the dominant green alga and made a large contribution to the biomass (Fig. 17.1 and 17.3). Although second among the green algae in 1968 with respect to biomass, *Monoraphidium* progressively became the most abundant species and attained the exceptionally high biomass of 11 g. m^{-3} in June 1972. Maximum development is attained in summer when temperatures exceed 20° (compare Fig. 17.1 and 17.3). As was mentioned above for *Oocystis*, growth continued in some cases into November and December. Winter populations, usually small, maintained a high biomass in both 1971/72 and 1972/73 with short ice periods (ref. to Fig. 15.1, Chapter 15).

Sphaerocystis schröteri, found on only three occasions and in very small numbers during 1956–58 developed larger populations during late summer and early autumn from 1969 onwards (max. 63 mg fr wt. m^{-3}). Like some other species *S. schröteri* was absent from winter plankton. For significant growth to occur this species seems to need water temperatures above 10°C. Most of the biomass peaks alternate with those of *Chodatella* or *Tetraedron.*

17.3.4 *Diatoms*

The next curve (no. 12 in Fig. 17.1) represents all three centric diatom species. As pointed out in the previous section *Cyclotella meneghiniana* is by far the most important, representing about 90–95 per cent of the biomass. Fresh weight of these species in 1956–58 was less than 200 mg. m^{-3}. In the years 1968–1973 they were present all year round and became increasingly abundant. The main development time is in spring and fall. A peak of 1,600 mg was reached in 1971 and high winter populations were maintained in the following two freezing periods. Stull et al. (1973) in their autoradiographic studies were able to show that for *C. meneghiniana* the renewal times of carbon are comparable to those of green algae and much shorter than for other diatoms.

Since composition of the pennate diatom assemblages is greatly influenced by varying contributions of benthic and epiphytic diatoms, these were combined to give the total pennate biomass (Fig. 17.1, No. 13). The contribution of true plankton forms to these figures varies tremendously.

Except for the last two winter seasons pennate diatoms were continuously present throughout the six years from 1968 to 1973. The main growth periods are spring and autumn, as for the centric diatoms. Biomass figures are usually smaller (max. 2,200 mg fr wt. m^{-3}) than for 1958 when a maximum of 3,200 mg was attained.

17.3.5 *Euglenophytes*

Euglenophytes rarely appeared in 1968 and were insignificant for the biomass. In the years 1969–1971 this algal group (also combined because of large variation in composition) reached summer populations of up to 400 mg fresh weight per cubic meter between July and September. Additional single peaks were observed in spring and fall. Similar seasonal patterns and maximum biomass had been attained in 1958. Submerged macrophytes are supposed to serve as a source for Euglenophyta species, which become detached by wave action and thus appear in the plankton. Since the macrophytes almost completely vanished in 1972 and 1973 euglenophytes disappeared too. Growth does not seem to be limited by temperature because peaks are associated with various temperatures between 5 and 21°C. Euglenophytes are the only group absent from far offshore samples.

17.4 Total biomass

The time course of the total biomass is shown in Fig. 17.1 (lower half A). Cell numbers (Fig. 17.1, B) and total cell surface area (Fig. 17.1, C) are given as additional parameters of phytoplankton population development. In general, all three parameters behave in the same way since most of the important algae belong to approximately the same size class (compare Table 17.1).

After ice break the rapidly rising water temperatures (Fig. 17.3) result in a large population increase which leads to several peaks during the year. Unfavourable light and temperature conditions in November and December reduce algal numbers again to small winter standing crops.

Occasional discrepancies between the three parameters considered are due to the particular algal composition. For instance, on 31 August 1968 fresh weight and surface area were high, whereas cell numbers were low. In this case large species of pennate diatoms, with large cell volumes and small S/V ratios (Table 17.1), contribute substantially to total biomass but are less important in terms of cell numbers. In accordance with the cell numbers, surface area differs from biomass at the end of June 1969 due to the species of euglenophytes, cryptophytes and bacillariophytes with very low S/V ratios which form a large fraction of the phytoplankton assemblage (Fig. 17.2). 30th June 1970 may serve as an example, where surface area behaves differently from the two other parameters on account of a predominance of green algae, especially *Monoraphidium*, all of which have high surface to volume ratios (cf. Fig. 17.1, 17.2 and Table 17.1). The fluctuations observed during the ice-free period are reflections of the wax and wane of algal species or taxonomic groups due to physical conditions such as radiation, wind and seiches which have much more influence in shallow waters than in deep lakes.

The best example of such an effect on the algal population is the tremendous decline of biomass by a factor of more than 30 in July and August 1972, largely caused by unusually bad weather conditions:

No. of days and percentage in parentheses during July and August (62 days) associated with mean wind speeds of

	$25\ km\ h^{-1}$	$30\ km\ h^{-1}$	TIR Kcal cm^{-2} (62 days)$^{-1}$
1969	10 (16 per cent)	8 (13 per cent)	27.9
1970	8 (13 per cent)	4 (6.5 per cent)	26.6
1971	10 (16 per cent)	3 (4.9 per cent)	29.7
1972	16 (26 per cent)	9 (14.5 per cent)	26.4
1973	6 (10 per cent)	3 (4.9 per cent)	27.0

The surface light conditions, together with the strong wind which produces high turbidity, result in weak underwater light intensities, a small euphotic zone (compare Chapter 15) and somewhat lower water temperatures (Fig. 17.3). The particles also have a mechanical effect on the algae, mainly cryptophytes and green algae, particularly *Monoraphidium*. A possible growth inhibition by turbulence as observed by White (1976) on a marine dinoflagellate still has to be proved by laboratory experiments. The breakdown of the population was accompanied by an increase in phaeopigment content (see Fig. 17.5 and the following section), and in PO_4-phosphorus from 0.2 to 15.3 $\mu g\ l^{-1}$. Particulate phosphorus concentration dropped from 32 to 16 μg per litre (see Chapter 10).

High standing crops of phytoplankton were maintained throughout fall and winter 1971/72 due to favourable light and temperature conditions and a short freezing period (compare Fig. 17.1, 17.3, Table 17.3 and Fig. 15.1, of Chapter 15).

The obvious trend towards increasing biomass figures during the years under observation (Fig. 17.1, Table 17.3) clearly reflects a steady increase in eutrophication of the lake (compare P- and N-data, see chapter 10). As a second important factor the general climatic conditions have to be considered, since total incoming radiation, which was below the usual average in 1968, progressively improved from 1969–1971 (up to +2 per cent of the mean; see Table 15.1 of Chapter 15). The year 1972 was extremely unfavourable in this respect (−4.3 per cent) due to a poor summer period as discussed above. Nevertheless a biomass peak occurred in June, with the highest value observed, due to large initial crops in early spring and a better phosphorus supply. Comparatively large populations were maintained in the winters of 1971/72 and 1972/73 because of mild temperatures and the brief ice cover (compare Fig. 15.1, Chapter 15).

Fresh weight calculated from unpublished material of F. Ruttner gave figures around 1,000 mg. m^{-3} for 1956 and 1957 (Fig. 17.1), which is lower than those obtained in 1968–70. In the year 1958 (for which more data are available) biomass was at a maximum in September and comparable to 1972 figures.

In order to obtain a somewhat more general picture of the seasonal biomass change, monthly mean values were calculated from all the available data from

Table 17.3. Integral biomass per season or per year (S) and mean standing crop ($\bar{m}$) per day expressed as mg fresh weight m^{-3} for the years 1968–1973

	mg fresh weight m^{-3}											
	1968		1969		1970		1971		1972		1973	
	S	$\bar{m}$	S	$\bar{m}$	S	$\bar{m}$	S	$\bar{m}$	S	$\bar{m}$	S	$\bar{m}$
Dec.–Feb.			6,009	67	6,334	70	13,305	148	119,245	1,325	99,741	1,108
March–May			7,472	81	27,500	299	97,412	1,059	131,633	1,530	75,995	826
June–Aug.	64,722	819	79,472	873	98,136	1,366	69,443	755	451,118	4,903	(95,364	1,036)
Sept.–Nov.	45,546	501	62,646	688	129,930	1,420	141,480	1,555	130,766	1,437	40,312	443
Total	114,060 (13.6.–31.12.)	671	155,599	426	268,339	735	360,380	987	810,405	2,220	286 267[1]	784

[1] Minimum value because of underestimation of the June–August integral by extrapolation

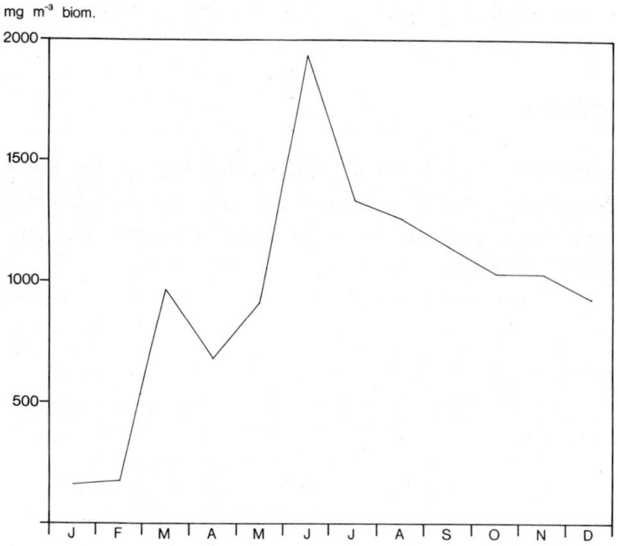

Fig. 17.4. Average seasonal variation of total phytoplankton biomass normalized by computing monthly mean values from all data between 1968–1972.

the years 1968–1973 (Fig. 17.4). On an average, a rapid population increase during and after ice-break culminates in an early spring peak. After a slight slowing down in April a maximum in biomass is reached in June and then slowly declines until the end of year.

The percentage composition of total biomass of different algal classes, discussed in detail above, is represented in Fig. 17.2 for the years 1958 and 1969–72. Diatoms formed the major part of the standing crop for most of 1958, except February and March when cryptophytes contributed more than 50 per cent. Blue-green algae were abundant in the second half of the year, which is of considerable interest since they have been scarce in recent years. In this connection it should be mentioned that preliminary observations have shown that cyanophytes became more frequent again in late 1976.

Chlorophytes and bacillariophytes were the major component in the years 1968–72. Green algae were especially important in summer, peaking mainly in June, whereas diatoms were present in higher percentages during spring and fall. Diatoms contributed the major part (up to 80 per cent) of the total biomass under the ice cover. Fifty per cent of the standing crop was attributable to cryptophytes in spring 1970. They also contributed significantly in early 1971 and 1972 and autumn 1969 and 1970. On the 5 August 1970 motile species (crypto- and euglenophytes) formed about 90 per cent of the biomass, probably due to a combination of calm wind conditions, low turbidity and relatively high radiation (see Fig. 15.1, Chapter 15). Only on one occasion, in March 1969, did euglenophytes constitute more than 50 per cent of the biomass. Some remarkable similarities exist between the years 1958 and 1971. In both years diatoms constituted the major component of the phytoplankton with a high proportion of cryptophytes in spring, peaks of green algae and euglenophytes in summer and of blue-green algae in the second half of the year.

17.5 Chlorophyll-a

Algal pigments were analysed from mid-April 1970 to May 1973 and in 1975 and 1976. Mean chlorophyll-a concentrations of the samples are given in Fig. 17.5 together with phaeopigment content (broken line in Fig. 17.5).

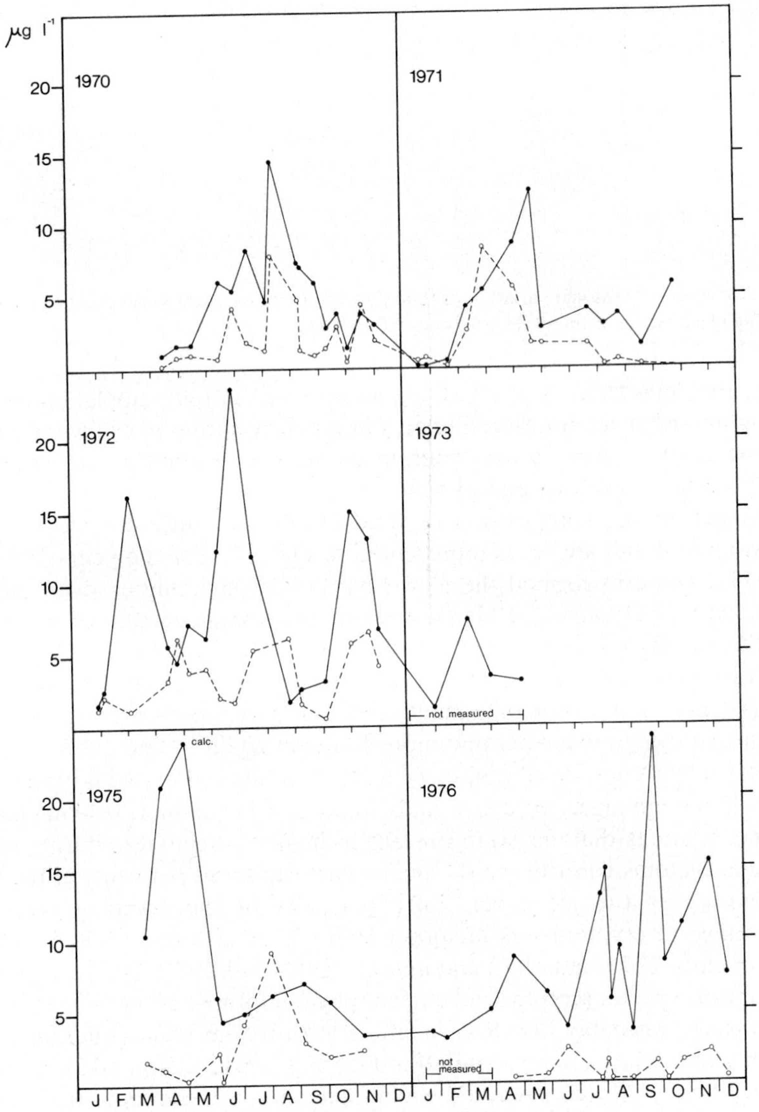

Fig. 17.5. Seasonal variation of chlorophyll-a and phaeopigment (broken line) for the years 1970–1976. Values are averages from 20 sampling points distributed over the entire lake area.

Chlorophyll-a values ranged from 0.1 μg in winter to 14.6 μg per litre in August in the period 1971/72. The latter figure is not represented in the biomass curve of Fig. 17.1 because no counts were made on this day. The remaining part of the curve closely resembles that for the biomass. Phaeopigment concentrations were usually high in these years. The maximum of 8.5 μg per litre by the end of March 1971 was presumably due to resuspension of dead and moribund benthic algae. The latter are killed as a result of sediment instability in the open lake following ice break (Prosser, pers. commun.).

In 1972 and spring 1973 chlorophyll-a is well correlated with biomass whereas phaeopigment follows an inverse course in most cases. The highest pigment content was observed by the end of June 1972 (23.8 μg. 1^{-1}).

Similar mean concentrations were recorded in April 1975 and September 1976. A tendency to increasing pigment content may be deduced from Fig. 17.5, corresponding to the increase in biomass. The trend becomes more obvious if we consider the maximum chlorophyll-a concentrations:

	Aug 1970	May 1971	June 1972	April 1975	Sept 1976
Chl-a (μg 1^{-1})	25	17	26	30	44

This reflects an increasing nutrient input from domestic and agricultural sources (see Chapter 10). As a further consequence littoral samples often differ significantly from those offshore.

17.6 Phytoplankton biomass versus chlorophyll-a

Significant correlations ($p \geqslant 0.001$) are obtained between different parameters of biomass (Table 17.4). Cell counts are closely correlated to surface area and to a lesser extent to chlorophyll-a and fresh weight. Chlorophyll shows the least correlation to algal volumes whereas both individuals and surface area yield the same coefficient.

Table 17.4. Correlation matrix for biomass parameters

Ind.	Vol.	Surf.	Chl.-a	
1	0.85	0.99	0.88	Ind.
	1	0.89	0.85	Vol.
		1	0.88	Surf.
			1	Chl.-a

Paasche (1960) observed the best correlation between cell surface area and cell volume (0.72), a slight connection between cell numbers and surface area (0.50), and no correlation of cell numbers with cell volume (0.16). Since production capacity had the best coefficient with respect to surface area, he concluded that it might be better to use biomass based on surface area instead of volume.

Correlations of fresh weight with chlorophyll-a have been reported by several authors. Phytoplankton of Lake Norrviken (Sweden) revealed coefficients of r = 0.93 and 0.98 dominated by blue-green algae or cryptomonades respectively (Ahlgren 1970). A value of 0.98 was calculated for the Swedish lake Mälaren by Tolstoy (1974) cited according to Willén (1976) who observed a correlation coefficient of 0.71 for Lake Hjälmaren. A significant relationship between biomass and chlorophyll-a was also published for five reservoirs in the GDR (r = 0.71) by Javornický (1974) and by Javornický & Komárková (1973) for the Slapy Reservoir in the ČSSR (r = 0.72).

Chlorophyll content as percentage of fresh weight and carbon: chlorophyll ratios are listed in Table 17.5 together with the major algal component, per cent of total biomass and basic data. Both the lowest and highest contents of chlorophyll-a were observed when diatoms dominated. The range of 0.1–1.3 per cent agrees with observations of various authors on cultural material (0.2–3.0 per cent) from marine and fresh waters (e.g. Krey 1957, Table 1; Strickland 1960, p. 14; Parsons 1961, Table 1; Jørgensen 1964, 1970). A variety of references concerning mean chlorophyll content and maximum – minimum values in various lakes throughout the world have been summarized in Table 17.6. Those lakes dominated by diatoms in general show the same range, with the exception of the maximum figure from Lake Victoria (5.3 per cent) and the 12 per cent value of the Štěchovice Reservoir. The latter is by far the highest content recorded and may be attributable to the large contribution of flagellates or to the accumulation of degradation products since phaeophytin was not determined.

A tendency to decreasing chlorophyll content with increasing biomass can be read off Table 17.5 during diatom-dominated seasons. This confirms observations by Jørgensen (1964) on cultures of *Cyclotella meneghiniana* – a species important in Neusiedlersee as pointed out earlier – and Wright (1959) from field data. A comparison of seasons in which cryptophytes form the major algal component, from Table 17.5, with lakes dominated by this group (Table 17.6) reveals good agreement of the data. Little variation was found among the chlorophyll content of green algae in Neusiedlersee. Most figures are close to the average of 0.53 per cent. For algal cultures of chlorophytes values of 0.27–5.2 per cent are reported in the literature.

Chlorophyll content seems primarily to vary with the time of year, i.e. with light conditions, and appears to be influenced to a lesser extent by species composition and nutrient availability.

Excluding Štěchovice Reservoir an average chlorophyll-a content of 0.88 per cent was calculated for all fresh-water lakes listed in Table 17.6, which is close to the 0.90 per cent for marine mixed populations (for references see Krey 1957 and Hagmaier 1961). There is obviously no difference between the stirring mechanisms in marine and fresh waters affecting the chlorophyll-a concentrations.

The ratio of organic carbon (calc. from fresh weight biomass) to chlorophyll-a varies from 9.5 to 134.7 in Neusiedlersee (Table 17.5). Similar fluctuations were observed in Lake Kinneret (Berman & Pollingher 1974). The changes are not attributable to seasonal variation in phytoplankton

Table 17.5. Biomass, chlorophyll-a, chlorophyll content of biomass and carbon to chlorophyll ratio for Neusiedlersee

Date	B (μgl^{-1})	Chl.a (μgl^{-1})	major algal component	B (%)	Chl.a/ fr. wt. (%)	C/Chl.a ($\mu g/\mu g$)
13. 4.70	196	1.1	Crypt.	53	0.56	22.3
28. 4.70	223	1.7	Crypt. + Bac.	64	0.76	22.8
15. 5.70	335	1.8	Crypt. + Chlor.	72	0.54	23.3
15. 6.70	1,151	6.2	Chlor.	58	0.54	23.2
28. 6.70	1,035	5.6	Chlor.	91	0.54	23.1
15. 7.70	1,548	8.4	,,	42	0.54	23.0
5. 8.70	767	4.8	Crypt. + Eugl.	90	0.62	19.9
9. 9.70	1,290	7.5	Bac.	72	0.58	21.5
29. 9.70	1,705	6.1	,,	95	0.34	34.9
11.10.70	2,817	2.9	,,	92	0.10	121.4
22.10.70	1,150	4.0	,,	93	0.35	36.0
3.11.70	1,619	1.5	,,	88	0.93	134.7
18.11.70	532	4.9	,, + Chlor.	78	0.92	13.6
2.12.70	581	3.1	,, ,,	77	0.53	23.2
20. 1.71	17	0.1	,, ,,	90	0.60	20.0
27. 1.71	30	0.1	Crypt. + Bac.	77	0.33	38.0
20. 2.71	100	0.5	,, ,,	70	0.50	25.0
15. 3.71	325	4.3	Bac.	74	1.32	9.5
2. 5.71	1,880	8.7	,,	90	0.46	27.0
3. 6.71	740	3.8	,,	60	0.51	24.3
21. 7.71	825	4.0	,,	72	0.48	25.8
10. 8.71	748	3.1	Chlor.	70	0.41	30.2
25. 8.71	1,050	3.7	Bac.	86	0.34	35.4
22.10.71	2,085	5.8	,,	79	0.28	44.9
26. 1.72	306	1.7	,,	77	0.56	22.4
3. 2.72	495	2.7	,,	68	0.54	23.0
3. 3.72	3,015	16.3	,,	53	0.54	23.1
13. 4.72	1,061	5.7	,,	64	0.53	23.1
24. 4.72	850	4.6	,,	68	0.54	23.3
4. 5.72	1,351	7.2	,, + Chlor.	70	0.51	23.5
25. 5.72	1,168	6.2	,, ,,	96	0.53	23.6
7. 6.72	2,249	12.4	Chlor.	78	0.55	22.6
22. 6.72	11.817	23.8	,,	90	0.20	62.2
14. 7.72	5,930	12.0	Bac.	75	0.20	61.8
23. 8.72	334	2.0	,,	76	0.60	21.0
7. 9.72	474	2.6	,,	73	0.55	23.1
3.10.72	601	3.2	,, + Chlor.	97	0.53	23.4
30.10.72	2,784	15.0	Chlor.	56	0.56	23.2
15.11.72	2,461	13.0	,,	70	0.52	23.6
29.11.72	1,253	6.0	Crypt. + Bac.	76	0.47	26.0

composition since minimum and maximum values are associated with diatom-dominated assemblages. Correspondence between absolute figures for both lakes was found in particular during seasons dominated by chlorophytes (compare figures in Table 17.6 and 17.5 with Table 4 in Berman & Pollinger 1974).

Table 17.6. Chlorophyll content of fresh-weight biomass and carbon chlorophyll ratio for a variety of lakes.

Lake or Reservoir	Dom. algae	Chl.a % fr.wt.	Range %	C/Chl.a $\mu g\, \mu g^{-1}$	Reference
Loch Neagh, N. Ireland	Cyan	0.35	—	28.6	Gibson et al. (1971, p. 154)
Loch Leven, Scotland	Bac. + Cyan.	0.52	—	24.0	Bailey-Watts (1974, p. 8)
L. Norrviken, Sweden	Cyan.	0.90	0.4–1.2	12.5	Ahlgren (1970, p. 368)
	Crypt.	1.50	0.6–5.7	6.6	
L. Vättern, Sweden	Crypt.	0.73	0.7–0.8	—	calc. by Ahlgren (1970)
	Bac.	0.31	—	—	from Fondén et al. (1968)
L. Hjälmaren, Sweden	Cyan.	1.10	0.1–8.6	—	Willén (1976, p. 68)
L. Mälaren, Sweden	?	0.60	0.5–0.9	—	Tolstoy (1976, p. 48–49) cit. acc. to Willén (1976)
Neuzehnhain R., DDR	Fl. + Bac.	4.2	1.9–9.7	4.7	partly worked up from
Seidenbach R., DDR	Crypt.	0.7	—	17.4	Javornický (1974, Table 4)
Stechlinsee, DDR	Bac.	1.1	0.8–1.5	12.5	
Wummsee, DDR	Bac. + Chl.	1.7	—	7.6	
Slapy R., CSSR	Crypt. + Bac.	1.4	0.6–3.8	19.5	Javornický & Komárková (1973, p. 202, Table 6)
Stechovice R., CSSR	Fl. + Bac.	12.0	—	—	
Vrané R., CSSR	Bac.	296	—	—	Straškraba & Javornický (1973, p. 289, Table 5)
Piburgersee, Austria	Clan. + Chl.	0.53	0.4–2.3	22.4	Rott (1975, p. 66)
Vord. Finstertaler See, Austria	Fl.	100.	0.3–3.2	12.5	calc. from Tilzer & Schwarz (1976, Fig. 5)
Attersee, Austria	Bac.	0.31	0.2–0.5	42.6	calc. from Geipel & Bauer (1976, p. 110, Table 2)
Neusiedlersee Austria	Bac. + Chl.	0.53	0.1–1.3	31.8	this study
L. Kinneret, Israel	Peridin.	0.33	0.1–0.9	111.1	Berman & Pollingher
	Chl.	0.41	0.1–0.9	31.7	(1974, p. 42, Table 4)
L. Victoria, Africa	Bac.	—	0.4–5.3	—	calc. by Ahlgren (1970) from Talling (1965)
Can. Ferry R., USA	Cyan.	0.21	0.1–0.4	—	Wright (1959, Table 1)
Char L., Canada; Meretta L., Canada	Per. + Chrys.	0.32	0.1–0.9	—	Kalff et al. (1972, Fig. 4)
15 ELA-Lakes, Canada	Chrys. + Bac.	0.37	0.2–0.5	—	Schindler & Holmgren (1971, Table 1)
L. Biwa, L. Suwa, L. Yunoko (Japan)		0.36	—	34.4	Mori & Yamamota (1975, p. 384, Fig. 9.2)

High ratios are in general linked to high biomass figures (e.g. 3.11.70 and 22.6.72, Table 17.5), and seem to be typical for slowly growing, nutrient-depleted cells (Thomas & Dodson 1972).

According to Strickland (1971) the C/Chl-a ratio usually varies between 30 and 90 in marine phytoplankton, progressing from eutrophic to oligotrophic. However, the carbon-chlorophyll relationships of most of the lakes listed in Table 17.6 have coefficients considerably lower than 30, suggesting a high degree of eutrophication. Pearl et al. (1976), on the other hand, reported C: Chl ratios between 12.05 and 19.25 for nine North American and New Zealand lakes of different tropic status, including the ultra-oligotrophic L. Tahoe, California (17.24mg C/mg Chl-a). Therefore the question arises whether a basic difference exists between marine and fresh waters with respect to the factors influencing C/Chl-a ratios. More data, based on direct carbon measurement, would be necessary in order to settle this question.

17.7 Horizontal distribution

Vertical distribution of algae in Neusiedlersee is negligible because of the shallowness of the lake and the well mixed water (Dokulil 1975). Even under calm conditions, or beneath the ice, phytoplankton stratification is of minor importance since most of the species are small or have mucilage, which results in low sinking speeds.

Horizontal differences in algal concentration resulting from wind-induced seiches and currents are represented in Fig. 15.8, A–C of Chapter 15. Of course phytoplankton biomass is never uniformly distributed in such a large lake on account of local differences in nutrient availability, currents, insolation and related factors, all complicating the interpretation of horizontal distribution. In general, the highest contents are associated with a down-wind direction due to transportation by trochoidal waves (see Smith & Sinclair 1972). This situation is illustrated in Fig. 15.8, A and B (Chapter 15) for southeast and northwest winds. Fig. 15.8, C reveals a somewhat unclear situation with respect to chlorophyll. Since the wind speed was very low on the day prior to sampling (compare Fig. 15.1 of Chapter 15) pigment distribution may still show little influence of wind stress, whereas turbidity, which is much sooner affected, demonstrates a typical northwest to southeast gradient. Some further examples of horizontal chlorophyll and turbidity distribution in Neusiedlersee are given by Dokulil (1974a, b; 1975). Larger discrepancies in species composition have not yet been found. An enlarged counting programme on this specific question was started in 1975. Results are not yet available.

17.8 Species diversity

Diversity was calculated for the years 1969 and 1972 applying Shannon's formula since this index is now widely used by many ecologists. In general, diversity exhibits the same pattern in both years (Fig. 17.5 and 17.6). Figures were about two bits/individual in spring 1969 when diatoms dominated (compare Fig. 17.2), and reached three by the end of May. After a brief decrease in June due to a predominance of chlorophytes a figure of 3.2–3.3 was reached again as a result of the high contribution of diatoms, crypto- and euglenophytes. A substantial decrease was observed during July and August corresponding with a drop in species number when a few green algae dominated. In late summer the index of diversity oscillates between two and more than three, reflecting percentage species changes (compare Fig. 17.2)

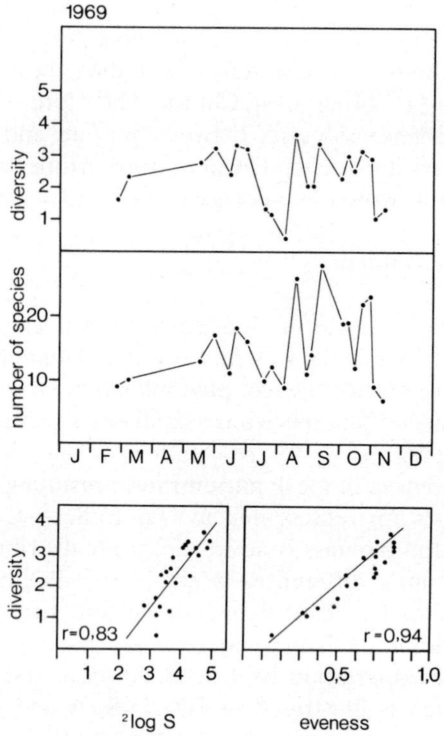

Fig. 17.6. Shannon's diversity index, number of species, and correlations between species diversity and species richness (^{2}log S) as well as species diversity and evenness for the year 1969.

and number of species found (Fig. 17.5).

In 1972 diversity in the open water was about two bits/individual in spring and autumn but never exceeded three (Fig. 17.6). As in 1969 the index dropped below one in summer when *Monoraphidium* dominated (compare Fig. 17.6, 1 and 2). Similar diversity patterns occurred in a bay close to the reed fringe where, however, the number of species, as compared with the open water, was much higher in winter. The diversity index exceeded three on the 3rd March. Many but evenly distributed species seem to be present on the 25th May since the increased number detected is not reflected in the index.

The decrease of diversity during the June/July period was also noticed from Boesels Pond, Ohio, by Winner (1969), from Storhjälmaren, Sweden by Willén (1976) and from Gull Lake, Michigan and Abbot's Pond, U.K. by Moss (1973).

The mean annual diversity is 2.04 in Neusiedlersee which is about the same as observed in Lake Norrviken by Tinnberg (1973) and in the lakes cited above, except Abbot's Pond which has a mean diversity index of 0.97. Higher values (3.03–3.60) were reported from Lake Tuomijärvi by Granberg (1973) and from Trout Lake and Lake Mendota by Sager (1967).

Since diversity is influenced by the number of species and their evenness both factors were tested according to DeJong (1975, p. 225). In the year 1969

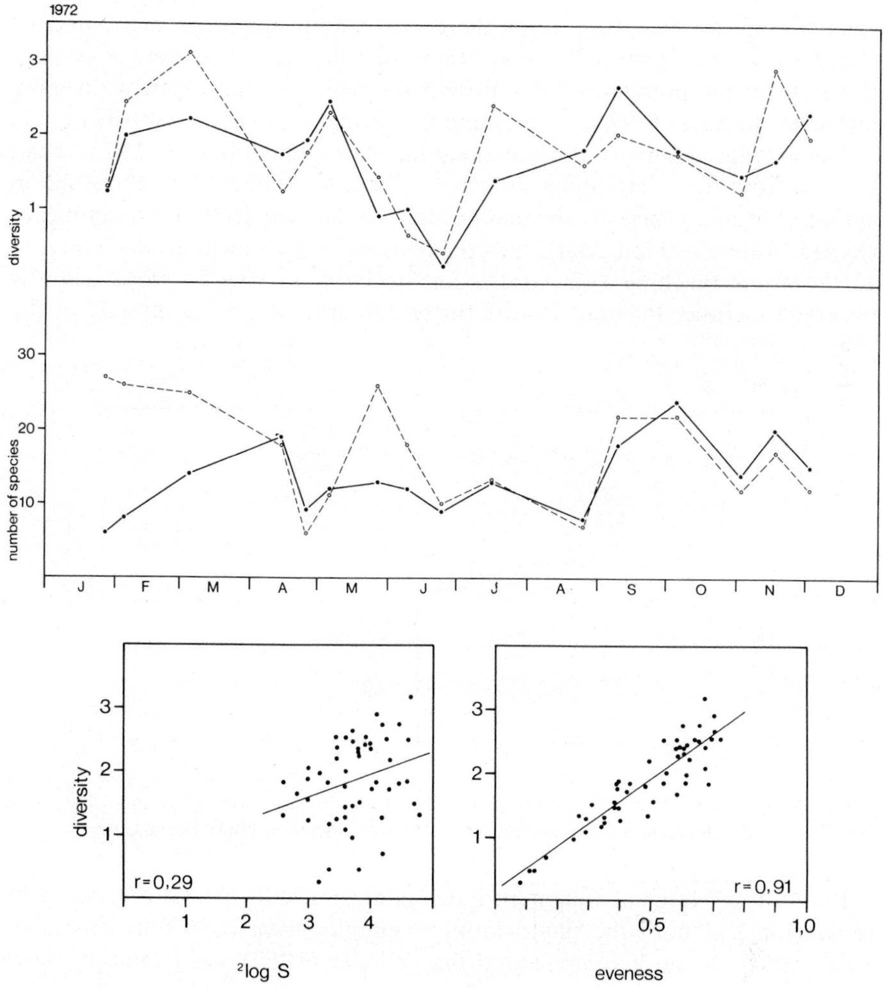

Fig. 17.7. Shannon's diversity index and number of species for the open lake and a bay situation (broken lines) for the year 1972. Correlations to species richness and evenness are for the open lake only.

correlation coefficients were 0.83 (diversity-richness, n = 22) and 0.94 (diversity-evenness, n = 22) whereas r = 0.29 and 0.91 respectively in 1972 (Fig. 17.5 and 17.6). It seems therefore that, species diversity in Neusiedlersee is more affected by evenness than by richness, as was also observed on Lake Hjälmaren (Willén 1976).

In 1975 and 1976 pigment diversity index (pigment ratio E_{430}/E_{665}) was estimated from spectrophotometer readings of acetone extracts according to suggestions by Margalef (1960). According to Margalef (1965) the ratio of the yellow pigments (E_{430}) to chlorophyll-a (E_{665}) is a measure of the structure (e.g. diversity) of a population. The pigment index also reflects the effect of nutrient depletion. Besides nitroge . deficiency, increased absorption at 430 nm may also mean increased chlorophyll degradation (Moss 1967).

Changes in the E_{430}/E_{665} ratio therefore cannot solely be attributed to changes in species composition since they are also reflections of the physiological state of the population. For further discussion of the pigment diversity index ref. to Moss (1973, p. 350) and Goltermann (1975, p. 240–42).

The average pigment ratio remained more or less constant in 1975 as can be seen from the overlapping standard deviations. Values were about 2.5 in spring, shifting to three by the end of May. In the year 1976 some significant changes were observed. Margalef's pigment ratio, 2.2 in winter, decreased to 1.7 in March reaching 3 in April. After declining in May a ratio of 1.8 was observed in June. Pigment indices fluctuated around 2.5 for the rest of the year.

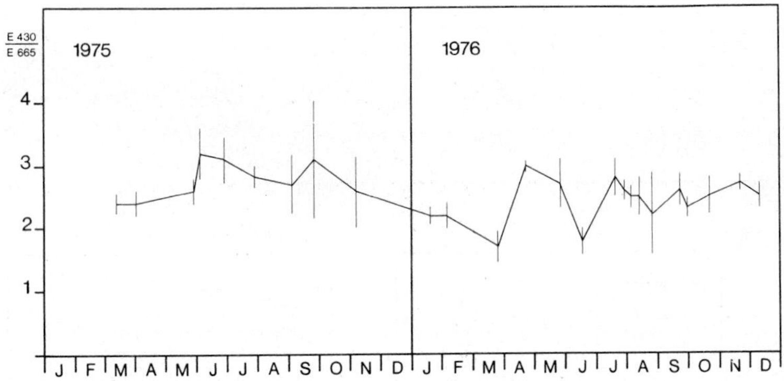

Fig. 17.8. Margalef's pigment diversity index (E_{430}/E_{665}) for the years 1975 and 1976. Each point represents the mean of 20 samples. Vertical lines indicate standard deviation.

If and how Margalef's pigment index conforms with species diversity will remain unclear until the phytoplankton samples have been fully analysed. However, it should be emphasized that Winner (1969) was unable to show any correspondence between diversity and the pigment ratio.

References

Ahlgren, G. 1970. Limnological studies of Lake Norrviken, a eutrophicated Swedish Lake. II. Phytoplankton and its production.– Schweiz. Z. Hydrol. 32: 353–396.

Bailey-Watts, A. E. 1974. Loch Leven research group. Annual Rep. 1974.

Bellinger, E. G. 1974. A note on the use of algal sizes in estimates of population standing crops.– Br. phycol. J. 9: 157–161.

Berman, T. & U. Pollingher. 1974. Annual and seasonal variations of phytoplankton, and photosynthesis in Lake Kinneret.– Limnol. & Oceanogr. 19: 31–54.

DeJong, T. M. 1975. A comparison of three diversity indices based on their components of richness and eveness.– Oikos 26: 222–227.

Devaux, J. 1973. Contribution a l'etude des populations phytoplanctoniques du Lac de Tazenat (Puy-deDôme).– Annal. Stat. Biol. de Besse-en-Chandesse 7: 1–101.

Dokulil, M. 1973. Planktonic primary production within the *Phragmites* community of Neusiedlersee (Austria).– Pol. Arch. Hydrobiol. 20: 175–180.

Dokulil, M. 1974a. Der Neusiedlersee (Österreich).– Ber. Naturhist. Ges. Hannover 118: 205–211.

Dokulil, M. 1974b. Die Seetrübe and ihre Bedeutung. In: H. Löffler Der Neusiedlersee. S. 52–54, Verlag Molden, Wien, München, Zürich.
Dokulil, M. 1975a. Bacteria in the water and mud of Neusiedlersee.– Symp. Biol. Hung. 15: 135–140.
Dokulil, M. 1975b. Planktonic primary and bacterial productivity in shallow waters within a large *Phragmites* community (Neusiedlersee, Austria).–Verh. Int. Verein. Limnol. 18: 1295–1304.
Dokulil, M. 1975c. Horizontal- und Vertikalgradienten in einem Flachsee (Neusiedlersee, Österreich).– Verh. Ges. Ökologie Wien 1975, 177–187.
Findenegg, I. 1971. The productivity of some planktonic species of algae in their natural environment (Ger.).– Arch. Hydrobiol. 69: 273–293.
Fondén, R., B. Grönberg & A. Tolstoy. 1968. Bakteriehalt, växplankton och klorofyll a i Vättern 1966 och 1967.– Mimeographed (Lake Mälaren research, Inst. of Limnology, Uppsala, Rep. No. 23).
Fott, B. 1967. *Chodatella* stages in *Scenedesmus*.–Acta Ur iv. Carolina-Biological 1967, 189–196.
Fott, B. 1971. Algenkunde. 581 S., 2nd Ed., VEB G. Fischer, Jena.
Geipel, E. & K. Bauer. 1976. In: Attersee. Mimeographed Report. pp. 179.
Gibson, C. E., R. B. Wood, E. L. Dickson & D. H. Jewson. 1971. The succession of phytoplankton in Loch Neagh 1968–70.– Mitt. Int. Verein. Limnol. 19: 146–160.
Goldman, C. R., M. Gerletti, P. Javornicky, U. Melchiorri & E. De Amazaga. 1968. Primary productivity, bacteria, phyto- and zooplankton in Lago Maggiore.–Me. Ist. Ital. Idrobiol. 23: 49–127.
Golterman, H. L. 1975. Physiological Limnology. pp. 489, Elsevier Sc. Publ., Amsterdam, Oxford, New York.
Granberg, K. 1973. The eutrophication and pollution of Lake Päijänne, Central Finnland.– Ann. Bot. Fennici 10: 267–308.
Grunow, A. 1860. Über neue oder ungenügend bekannte Algen.– Verh. k.k. Zool.-Bot. Ges. Wien 10: 503–582.
Grunow, A. 1862. Die österreichischen Diatomaceen nebst Anschluß einiger neuer Arten von anderen Lokalitäten und einer kritischen Übersicht der bisher bekannten Arten. – Ibidem 12: 315–473 and 545–588.
Grunow, A. 1863. Über einige neue und ungenügend bekannte Arten und Gattungen von Diatomaceen.– Ibidem 13: 137–162.
Hagmeier, E. 1961. Plankton – Äquivalente.– Kieler Meeresf. 17: 32–47.
Happey, C. M. 1968. Physico-chemical and phytoplankton investigations in Abbot's Pond, Somerset. Ph.D. thesis, Univ. Bristol.
Hickman, M. 1974. The seasonal succession and vertical distribution of the phytoplankton in Abbot's Pond, North Somerset, U.K.– Hydrobiologia 44: 127–147.
Hobro, R. & E. Willén. 1975. Phytoplankton countings and volume calculations from the baltic – a method comparison.– Vatten 4: 317–326.
Holmes, R. W. & T. M. Widrig. 1956. The enumeration and collection of marine phytoplankton.– J. cons. int. explor. Mer. 22: 21–32.
Hustedt, F. 1959a. Die Diatomeenflora des Neusiedlersees im österreichischen Burgenland.– Österr. Bot. Z. 106: 390–430.
Hustedt, F. 1959b. Bemerkungen über die Diatomeenflora des Neusiedlersees und des Salzlackengebietes.–Wiss. Arb. Burgenland 23 (Landschaft Neusiedlersee), 129–133.
Javornický, P. 1958. Revise nekterych metod pro zijstovani kvantity fytoplanktonu (Revised method for quantitative enumeration of phytoplankton).– Sc. Pap. Inst. Chem. Technol. Faculty of technol. of fuel and water 2: 283–367 (Prague).
Javornický, P. 1974. The relationship between productivity and biomass of phytoplankton in some oligotropic water-bodies in the German Democratic Republic.– Limnologica (Berl.) 9: 181–195.
Javornický, P. & J. Komárková. 1973. The changes in several parameters of plankton primary productivity in Slapy Reservoir 1960–1967, their mutual correlations and correlations with the main ecological factors.– Hydrobiol. Studies 2: 155–211.
Jørgensen, E. G. 1964. Chlorophyll content and rate of photosynthesis in relation to cell size of the diatom *Cyclotella meneghiniana*.–Physiol. Plant. 17: 407–413.

Jørgensen, E. G. 1970. The adaptation of plankton algae. V. Variation in the photosynthetic characteristics of *Skeletonema costatum* cells grown at low light intensity.–Physiol. Plant. 23: 11–17.

Kalff, J., H. E. Welch & S. K. Holmgren. 1972. Pigment cycles in two high-arctic Canadian lakes.– Verh. Int. Verein. Limnol. 18: 250–256.

Krey, J. 1957. Chemical methods of estimating standing crop of Phytoplankton.– Cons perm. int. explor. Mer. Rapp. proc.– verb. réunions 144: 20–27.

Kusel-Fetzmann, E. 1973. *Gyrosigma macrum* – neu für den Neusiedler See.–Österr. Bot. Z. 122: 115–120.

Kusel-Fetzmann, E. 1974. Beiträge zur Kenntnis der Algenflora des Neusiedler Sees I. – Sitz. Ber. Österr. Akad. Wiss., math.– nat. Kl., Abt. I, 183: 5–28.

Lorenzen, C. J. 1967. Determination of chlorophyll and pheopigments: spectrophotometric equations.– Limnol. & Oceanogr. 12: 343–346.

Loub, W. 1955. Algenbiozönosen des Neusiedler Sees.– Sitz. Ber. Österr. Akad. Wiss., math.– nat. Kl., Abt. I, 164: 81–107.

Lund, J. W. G. 1964. Primary production and periodicity of phytoplankton.– Verh. Int. Verein. Limnol. 15: 37–56.

Lund, J. W. G., C. Kipling, & E. D. LeCren. 1958. The inverted microscope method of estimating algal numbers and the statistical basis of estimations by counting.– Hydrobiologia 11: 143–170.

Margalef, R. 1960. Valeur indicatrice de la composition des pigments du phytoplankton sur la productivité, composition taxonomique et propriétés dynamiques des populations.– Rapp. Process-Verbaux C.I.E.S.M. 15: 274–281.

Margalef, R. 1965. Ecological correlations and the relationship between primary productivity and community structure.– p. 355–364. In: C. R. Goldman (ed.): Primary productivity in aquatic environments.–Mem. Ist. Ital. Idrobiol., Suppl. 18, 355–364 Univ. Calif. Press. Berkeley.

Mori, S. & G. Yamamoto (Ed.). 1975. Productivity of communities in Japanese inland waters.– JIBP Synthesis Vol. 10, pp. 436, Univ. Tokyo Press.

Moss, B. 1967. A spectrophotometric method for the estimation of percentage degradation of chlorophylls to pheo-pigments in extracts of algae.– Limnol. & Oceanogr. 12: 335–340.

Moss, B. 1973. Diversity of fresh-water phytoplankton.– Amer. Mid. Nat. 90: 341–355.

Nalewajko, C. 1966. Dry weight, ash and volume data for some fresh water planktonic algae.– J. Fish. Res. Bd. Canada 23: 1285–1287.

Neuhuber, F. 1971. Ein Beitrag zum Chemismus des Neusiedler Sees.– Sitz. Ber. Österr. Akad. Wiss., Math.– nat. Kl., Abt. I, 179: 225–231.

Overbeck, J. & E. M. Stange-Bursche. 1965. Experimentelle Untersuchungen zum Coenobienformwechsel von *Scenedesmus quadricauda* (Turp.) Bréb.–Ber. Dt. Bot. Ges. 78: 357–372.

Paasche, E. 1960. On the relationship between primary production and standing stock of phytoplankton.– J. cons. int. explor. Mer. 26: 33–48.

Paerl, H. W., M. M. Tilzer & Ch.R. Goldman. 1976. Chlorophyll-a versus adenosine triphosphate as algal biomass indicators in lakes.– J. Phycol. 12: 242–246.

Pantocsek, J. 1912. A fertö tó kovamoszat viránya (Bacillariae lacus peisonis) Preßburg, 1–43.

Parsons, T. R. 1961. On the pigment compositions of eleven species of marine phytoplankters.– J. Fish. Res. Bd. Canada 18: 1017–1025.

Rott, E. 1975. Phytoplankton und kurzwellige Strahlung im Piburgersee.–Ph.D. thesis, Univ. Innsbruck.

Ruttner-Kolisko, A. & F. Ruttner. 1959. Der Neusiedlersee.– Wiss. Arb. Burgenland (Landschaft Neusiedlersee) 23: 195–202.

Sager, P. E. 1967. Species diversity and community structure in lacustrine phytoplankton.– Ph.D. Thesis, Univ. Wisconsin.

Schiller, J. 1955a. Untersuchungen an den planktischen Protophyten des Neusiedlersees 1950–1954. I. Teil.– Wiss. Arb. Burgenland 9: 1–66.

Schiller, J. 1955b. Untersuchungen an den planktischen Protophyten des Neusiedlersees 1950–1954. II. Teil.– Wiss. Arb. Burgenland 18: 1–44.

Schiller, J. 1956. Untersuchungen an den planktischen Prototypen des Neusiedlersees 1950–1954. III. Teil: Euglenen.– Sitz. Ber. Österr. Akad. Wiss., math.– nat. Kl., Abt. I, 165:

547–582.

Schindler, D. W. & S. K. Holmgren. 1971. Primary production and phytoplankton in the Experimental Lakes Area, Northwestern Ontario, and other low-carbonate waters, and a liquid scintillation method for determining ^{14}C activity in photosynthesis.– J. Fish. Res. Bd. Canada 28: 189–201.

Scor/Unesco. 1966. Determination of photosynthetic pigments in seawater.– Unesco-Monographs on oceanographic methodology.

Smith, I. R. & I. J. Sinclair. 1972. Deep water waves in lakes.– Freshwat. Biol. 2: 387–399.

Straškraba, M. & P. Javornický. 1973. Limnology of two re-regulation reservoirs in Czechoslovakia.– Hydrobiol. Studies 2: 249–316.

Strickland, J. D. H. 1960. Measuring the production of marine phytoplankton.– Bull. Fish. Res. Bd. Canada No. 122, pp. 172.

Strickland, J. D. H. 1971. Microbiol activity in aquatic environments.– Symp. Soc. Gen. Microbiol. 21: 231–253. Univ. Press, Cambridge.

Stull, E. A., E. de Amazaga & Ch.R. Goldman. 1973. The contribution of individual species of algae to primary production of Castle Lake, California.– Verh. Int. Ver. Limnol. 18: 1776–1783.

Swale, E. M. F. 1965. Observations on a clone of *Lagerheima Subsalsa* Lemmerm. (*Chodatella subsalsa* Lemmerm.) in culture.–Nova Hedwigia 10: 1–10.

Talling, J. F. 1965. The photosynthetic activity of phytoplankton in East African Lakes.– Int. Rev. ges. Hydrobiol. 50: 1–32.

Thomas, H. K. & A. N. Dodson. 1972. On nitrogen deficiency in tropical Pacific oceanic phytoplankton. 2. Photosynthetic and cellular characteristics of a chemostat grown diatom.– Limnol. & Oceanogr. 17: 515–523.

Tilzer, M. M. & K. Schwarz. 1976. Seasonal and vertical patterns of phytoplankton light adaptation in a high mountain lake.– Arch. Hydrobiol. 77: 488–504.

Tinnberg, L. 1973. Fytoplanktons diversitet i Norrviken 1961–1972.– Stencilled report, Inst. of Limnology, Uppsala.

Tolstoy, A. 1974. Klorofyll som mått på växtplankton i Mälaren.– Nordforsk. pp. 45–57, Helsinki.

Trainor, Fr. R. & H. L. Rowland. 1968. Control of colony and unicell formation in a synchronized *Scenedesmus*.–J. Phycol. 4: 310–317.

Utermöhl, H. 1931. Neue Wege in der quantitativen Erfassung des Planktons.– Verh. Int. Verein. Limnol. 5: 567–596.

Utermöhl, H. 1958. Zur Vervollkommnung der quantitativen Phytoplanktonmethodik.–Mit. Int. Verein. Limnol. 9: 1–38.

Veszprémi, B. 1976. Vikémai, plankton és fenékfauna viszgálatok a Fertö tavon. (Investigations on water chemistry, the plankton and the benthic fauna of Lake Fertö = Neusiedlersee). Halaszat 22: 138–141.

White, A. W. 1976. Growth inhibition caused by turbulence in the toxic marine dinoflagellate *Gonyaulax Excavata*.–J. Fish. Res. Bd. Canada 33: 2598–2602.

Willén, E. 1976. Phytoplankton and environmental factors in Lake Hjälmaren, 1966–1973.– NLU Rapp. 87: 1–89, Uppsala.

Winner, R. W. 1969. Seasonal changes in biotic diversity and in Margalef's pigment ratio in a small pond.– Verh. Int. Verein. Limnol. 17: 503–510.

Wright, J. C. 1959. Limnology of Canyon Ferry Reservoir. II. Phytoplankton standing crop and primary production.– Limnol. & Oceanogr. 4: 235–250.

Zakovsek, G. 1961. Jahreszyklische Untersuchungen am Zooplankton des Neusiedlersees.– Wiss. Arb. Burgenland 27: 1–85.

8 Phytoplankton primary production

M. Dokulil

Primary productivity of the phytoplankton was evaluated both in the open lake and within the *Phragmites* stand using in situ carbon-14 uptake measurements.

In the shallow waters (0–0.7m) within the reed, plankton production increases rapidly after ice-break, coming to a peak in early spring. Later in the year production declines, in spite of increasing algal biomass, due to the shading effect of developing *Phragmites* stems and leaves. A second peak usually appears in autumn as a result of relatively better light conditions resulting from foliage loss of *Phragmites* (Dokulil 1975).

Absolute rates of carbon uptake are in the range of 1–25 mg C m^{-3} h^{-1}. Areal production is even lower (max. 7 mg C m^{-2} h^{-1}) because water depth is much less than one meter. Activity coefficients based on chlorophyll-a concentration (0.8–6 mg Chl a m^{-3}) range from 0.60 to 2.78.

Specific daily energy utilization is highest in spring and autumn whereas summer values are low due to strong light attenuation by *Phragmites*. On an average 0.16% per cent of the photosynthetically available radiation (PhAR) reaching the water surface is utilized (Dokulil 1973, 1975).

Planktonic primary production is governed by incoming radiation, development of the *Phragmites*, varying contributions of epiphytic and epipelic algae and horizontal transport of nutrients (Dokulil 1973).

Primary productivity of the open water is strongly affected by the underwater light conditions and the extent of the euphotic zone, both of which are altered by the amount of inorganic turbidity in the water column (Chapter 15). Thus day-to-day variation of carbon uptake can be strong even at constant biomass. In addition wind driven water circulation and seiches result in pronounced horizontal differences of biomass and production (Dokulil 1974). The pattern of annual change is therefore very irregular and strongly influenced by the conditions on the sampling days and their frequency.

Theoretical considerations suggest an enhancement of photosynthesis if algal cells are circulated through a light gradient, as is the case in shallow turbid waters of windy climatic areas (Baumert 1976a, b). In fact Jewson and Wood (1975) were able to verify a 10 per cent increase of integral production by artificial circulation experiments in Loch Neagh, Northern Ireland. Similar experiments by Dokulil, Hammer and Jewson (1977) in Neusiedlersee revealed no difference between stationary bottles and circulation.

Absolute figures of carbon uptake range from 2–100 mg C $m^{-3}h^{-1}$, with an energy utilization of about 0.1 per cent PhAR (Dokulil 1974).

References

Baumert, H. 1976a. Mathematische Modelle zur Deutung der durch intermittierende Belichtung von Phytoplankton hervorgerufenen Mehrleistung der Photosynthese.– Int. Rev. ges. Hydrobiol. 61: 517–527.

Baumert, H. 1976b. Abschätzung von Turbulenzkorrekturen für die planktische O_2-Produktion bei schwacher Turbulenz.– Int. Rev. ges. Hydrobiol. 61: 627–637.

Dokulil, M. 1973. Planktonic primary production within the *Phragmites* community of Neusiedlersee (Austria).– Pol. Arch. Hydrobiol. 20: 175–180.

Dokulil, M. 1974. Der Neusiedlersee (Österreich).– Ber. Naturhist. Ges. Hannover 118: 205–211.

Dokulil, M. 1975. Planktonic primary and bacterial productivity in shallow waters within a large *Phragmites* Community (Neusiedlersee, Austria).–Verh. Int. Verein. Limnol., 19: 1294–1304.

Dokulil, M., Hammer, L. and Jewson, D. H. 1977. Vergleichende Untersuchungen zur Primärproduktion des Phytoplanktons im Neusiedlersee: O_2, ^{14}C und Experimente mit künstlicher Zirkulation.– Ber. Forschungsinst. Burgenland, 29: 60–73.

Jewson, D. H. and Wood, R. B. 1975. Some effects on integral photosynthesis of artificial circulation of phytoplankton through light gradients.– Verh. Int. Verein. Limnol. 19: 1037–1044.

9 Submerged macrophytes in the open lake. Distribution pattern, production and long term changes

F. Schiemer

19.1 Introduction

Quantitative studies on distribution pattern and biomass of the submerged vegetation in the open lake outside the reed belt have been carried out within the framework of the 'International Biological Programme' between 1970 and 1972 (Schiemer & Weisser, 1972; Schiemer & Prosser, 1976). During this period a steady decrease of the macrophyte cover was observed, which resulted in an almost complete absence of vegetation in the open lake in 1976. Irrespective of the present situation (1976) a summary of the papers cited above is given together with additional information on biology and long term changes in the vegetation. The subject is discussed with respect to the ecology, the limits of growth, and the possible role of submerged macrophytes in the ecosystem of Neusiedlersee.

19.2 Composition of the submerged macrophyte vegetation

The literature on the aquatic vegetation of the lake was surveyed by Weisser (1970). The number of species of submerged macrophytes in the open lake is low, and only *Myriophyllum spicatum* L. and *Potamogeton pectinatus* L. have been found widely distributed outside the reed belt. *Ceratophyllum demersum* L., *Characeratophylla* Wallr. and *Najas marina* L. are restricted to a very few sheltered localities within reed bays (indicated in Fig. 19.1) and within the *Phragmites* belt. Details on the occurrence of these species and on the macrophyte distribution within the reed belt are given in the plant sociological survey of Weisser (1970).

The extreme situation of the open lake with respect to macrophyte growth is indicated by the low species number and the absence of a zone of floating leaved plants and charophytes. The two major species, *P. pectinatus* and *M. spicatum*, have a wide geographical and ecological range; the former is a true cosmopolitan species occurring in a wide variety of habitat types including brackish water and the latter is likewise eurytopic and of circumpolar distribution (see Sculthorpe, 1971 and Hutchinson, 1975).

A similar situation with respect to low species number and species composition is encountered in two large 'Pannonian' lakes (Maucha, 1931). In the open Lake Balaton (surface area: 585 km^2, z_m: 3.4 m) *Potamogeton perfoliatus* L. is the most important species in terms of biomass, followed by *M. spicatum* and *P. pectinatus*. Macrophyte growth in Lake Velence (surface area: 26 km^2, z_m: 1.2 m) is dominated by *M. spicatum*, followed by *Ceratophyllum submersum* L., *Chara* spp., *Najas marina* L. and *P. pectinatus* (Kiss, 1972).

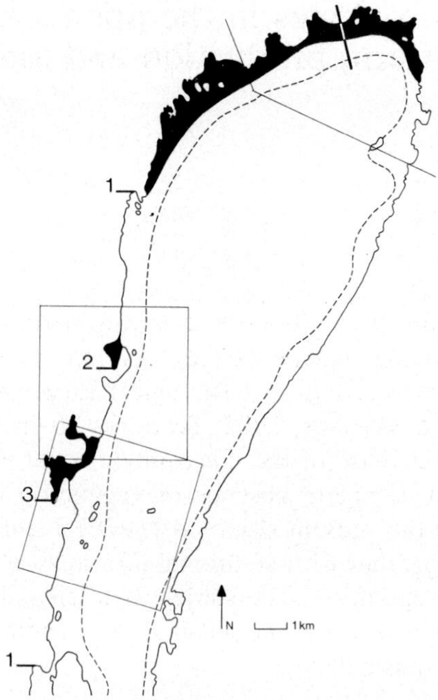

Fig. 19.1. Overall distribution of macrophytes in the northern part of the lake according to the phytosociological survey in 1970 (Schiemer and Weisser 1972), amended. The lake area depicted represents 85 per cent of the total lake area outside the reed belt. Black = submerged macrophytes, cover 20 per cent; broken line = outer limit of the macrophyte zone. For the delineated area at the northern end of the lake biomass values are given in Table 19.1. The transect refers to Fig. 19.3. The two areas delineated at the western shore are shown in Figs. 19.5 and 19.6. Locations of species with limited occurrence in the open lake: 1 = *Ceratophyllum demersum* L., 2 = *Characeratophylla* Wallr., 3 = *Najas marina* L.

19.3 Biological characteristics of *M. spicatum* and *P. pectinatus*

Only sporadic observations on the phenology and life cycle of the two species of Neusiedlersee can be presented. Growth starts with the warming up of the lake after ice break, usually in February. The shoots reach the water surface during May. From mid-June to mid-August phytobiomass is at its highest. The macrophyte belt largely disintegrates during September and October and disappears completely in November. However, even during the winter months and under the ice small but photosynthetically active plants of both species were occasionally observed.

Reproduction is both sexual and asexual. The formation of adventitious roots by broken off or uprooted shoots may be a means of intralacustrine dispersal in *M. spicatum* but was not observed in *P. pectinatus* (Weisser, 1970).

Most of the reproduction seems to occur asexually. An example is the enlargement of the ring formations of *P. pectinatus* which result from the centrifugal growth of rhizomes and tubers (see below). In *M. spicatum*

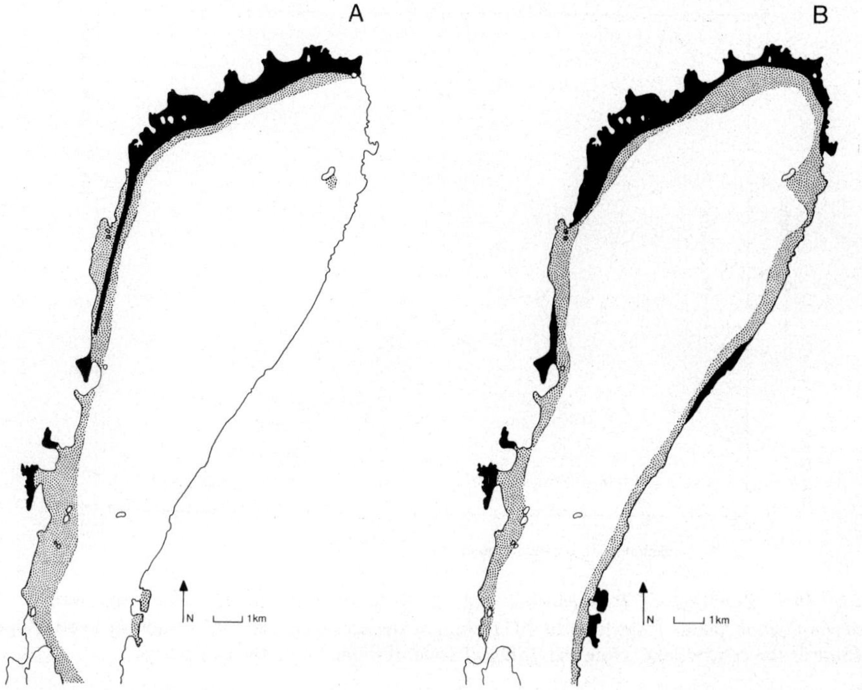

Fig. 19.2A. Distribution pattern of *M. spicatum*, according to Schiemer & Weisser, 1972. Black = plant cover >5 per cent. dotted <5 per cent. *B.* Distribution pattern of *P. pectinatus*, according to Schiemer & Weisser, 1972. Black = plant cover >1 per cent, dotted <1 per cent (especially ring formations).

flowering may occur twice annually, once in June and again in July (Weisser, 1970). In 1970 a strong initiation of young shoots was observed 4 times during the growth period (Dokulil, pers. comm.). *P. pectinatus* flowers in June and the seeds mature during August and September (Varga, 1931).

The growth form of the two species exhibits marked differences. In the case of *M. spicatum* individual plants form a mat of assimilating vegetation on the water surface (Fig. 19.7 and 19.8). An interesting morphological adaptation towards a life in such turbid water is the development of large aerenchymatic internodes of the near-surface shoots to increase the buoyancy of the assimilating organs. At the period of maximal biomass, in July, the underground biomass of the plants has been found to be only 6–12 per cent of the total (Schiemer & Prosser, 1976). The average plant weight and the number of shoots of *M. spicatum* increase with distance from the shore (Fig. 19.3).

P. pectinatus grows in patches and again the configuration of the stands depends on the locality: Adjacent to the reed belt the plants may form a closed cover with a high proportion of submerged growth. More frequently isolated patches several meters in diameter occur. With increasing distance from the reed belt, growth within such patches tends to decrease, leading to peculiar ring formations, which have been described in detail by Varga (1931,

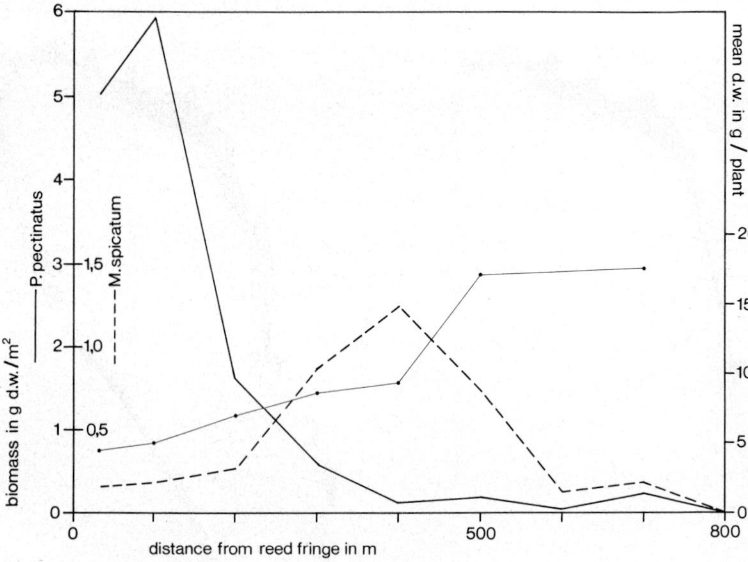

Fig. 19.3. Zonation of *P. pectinatus* and *M. spicatum* biomass and mean dry weight of *Myriophyllum* plants (thin line) in 1971 along a transect (see Fig. 19.1) from the reed fringe towards the central lake. Note the different scale of biomass for the two plants.

1933). At the time of Varga's observations (1928–1930?) the average diameter of such rings was 10–20 m (range: 2–50 m),[1] consisting of 20–30 individual plants. Further interesting details on the structure of these formations can be derived from aerial photographs from 1964. The peripheral zone of strongly developed plants is followed by a ring (3–6 m width in 1964) without vegetation. Towards the ring centre this zone is followed by weaker plant cover, which again is more marked at its periphery. This pattern is in some cases so well developed that formations of up to three concentric rings are distinctly recognizable on the 1964 pictures. Field observations carried out in 1970, showed that not only plant cover but also the size of individual plants is distinctly smaller in the centre of the rings when compared with plants from the periphery.

Aerial photographs from 1957 and 1964 (see Fig. 19.5) revealed that the maximum numbers of such rings occurred at a distance of 400–1,000 m from the reed. However, occasional rings were observed even in the central lake area.

Rings of *P. pectinatus* are perennial formations. A comparison of the above-mentioned aerial photographs, which have been made for cartographic purposes, allows the identification of individual rings in 1957 and 1964. Between these two dates there was practically no formation of new rings but a strong increase of ring diameter. This is demonstrated in Fig. 19.4 for an arbitrarily chosen area at the western shore. The average ring diameter was 14.0 m in 1957 (range 3–35 m, n = 26) and 26.8 (range 13–50) in 1964,

1. The largest ring found by us (1970) had a diameter of 70 m.

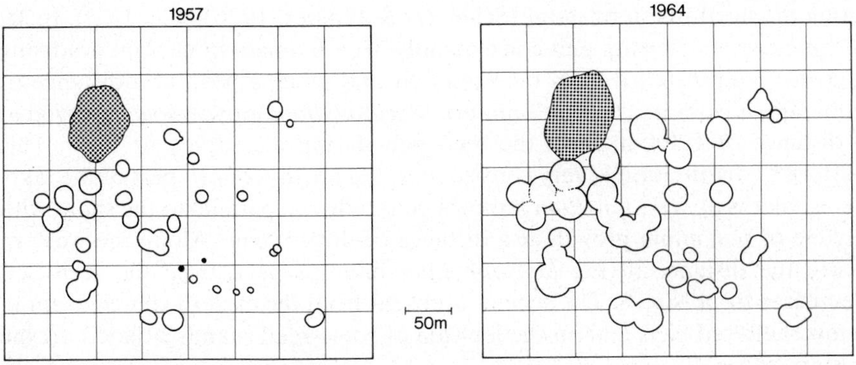

Fig. 19.4 Structure of *Potamogeton* rings in two different years at an arbitrarily chosen site situated 600–900 m from the reed belt at the western shore.

which represents an average increase of 1.8 m per year. Varga (op. cit.) found a ring growth of 2–3 m per year during the period of his study (1928–1930?). This shows conclusively that large *Potamogeton* rings are the product of decades of plant growth.

Such configurations are not limited to *P. pectinatus.* Karpati (pers. comm.) has found *M. spicatum* rings of a few meters' width in the Hungarian part of Neusiedlersee. However, records on ring formations in other lakes are few. In Balaton rings of *Potamogeton perfoliatus* have been described and illustrated by Karpati et al. (1971). On the other hand, such formations have not been reported from the more eutrophic but equally shallow lake Velence.

19.4 Distribution pattern of *M. spicatum* and *P. pectinatus*

The overall pattern of macrophyte distribution in the open lake, as found during a plant sociological survey in 1970 (Schiemer & Weisser, 1972), is shown in Fig. 19.1.

Although the depth of the water column is fairly uniform over the whole area of the open lake, macrophyte growth is restricted to a zone of 1.5 km maximum width, following the outer contour of the reed fringe. The central lake zone is free of macrophytes with the exception of few scattered *Potamogeton* rings. Such isolated stands can be found over the entire lake area.

In 1970 the highest overall plant density was encountered at the north-western end of the lake. Along the western shore line the plant cover decreased from N to S. Along the eastern shore the macrophyte growth is generally scarce and restricted to *P. pectinatus.* In the southern part of the lake the density and structure appeared to be similar to those in NW, but was not studied in detail.

The overall pattern of occurrence of the 2 species in the lake is similar (Fig. 19.2). However, considering the plant distribution on a more detailed scale, a distinctive inshore – offshore zonation in the two species was observed (Schiemer & Weisser, 1972), and best studied in the broad macrophyte belt along the north-western shore (Schiemer & Prosser, 1976) (Fig. 19.3). In *M. spicatum*, plant density and consequently area biomass in general gradually increase with distance from the reed belt and reach a peak at a distance of 300–500 m from it. The outer limit of *Myriophyllum* growth was observed at a distance of 1,300 m from the reed belt during the survey of 1970. This pattern of distribution is very consistent in the north-western part of the lake, i.e. isopleths of *Myriophyllum* biomass generally run parallel to the shore with a zone of maximum growth at a distance of 400–500 m. Along the western shore the distance of the *Myriophyllum* biomass peak from the reed belt decreases from N to S. The species is absent from the eastern shore except in a few sheltered bays and on the lee-side of some reed islands situated off the eastern shore.

In *P. pectinatus* biomass is generally highest close to the reed fringe and declines with distance from the shore. In contrast to *M. spicatum*, *P. pectinatus* also inhabits the eastern shore where, in addition to local denser stands in the immediate vicinity of the shore line, numerous rings were found in 1970.

Thus it appears that *P. pectinatus* has the wider habitat range in the open lake, as well as within the reed belt, where it is especially important in areas with poor *Phragmites* growth. In such zones, which mainly result from human activities (formation of canals, reed cutting), the species dominates over *Utricularia vulgaris* L., which is the main submerged macrophyte within denser *Phragmites* stands (see Chapter 21).

19.5 Biomass

Biomass estimates were carried out in 1971 and 1972 along transects from the reed fringe towards the lake centre. A detailed account of the methods and results of these studies has been given in Schiemer & Prosser (1976). Additionally, a tentative estimate of *Myriophyllum* biomass in the lake in 1970 was calculated, using the ratio of estimated plant cover/actual biomass and applying these values to the plant sociological estimates of 1970 (Schiemer & Weisser 1972). Further estimates of macrophyte biomass, especially of stands within the reed zone, were carried out for the Hungarian part of the lake by Karpati et al. 1972.

In Table 19.1 biomass estimations are given for the north-western end of

Table 19.1. Biomass and area of macrophyte cover in the years 1970 to 1976. Biomass (B) is given in tons dry weight, the area (A) in km^2. The data refer to the northwestern end of the lake (delineated in Fig. 19.1), with a total area of 14.75 km^2. Data for 1970–1972 from Schiemer & Prosser 1976.

	1970		1971		1972		1974		1976	
	A	B	A	B	A	B	A	B	A	B
Zone I										
M. spicatum	3.1	7.1	2.0	6.5	0.3	4.3	0	0	0	0
P. pectinatus	—	—	43.0	3.9	6.1	3.0	—	—	0	0

Neusiedlersee which is the main contributor to the total macrophyte crop. The data indicate that *P. pectinatus* was clearly dominant in terms of biomass, with values approximately twenty times higher than those of *M. spicatum*.

The absolute values of biomass per area of plant stands is low when compared with data from other lakes (Table 19.2). These comparatively low values are due to the generally loose nature of the stands.

Table 19.2. Comparison of submerged aquatic macrophyte biomass (kg dry weight/ha) for Neusiedlersee with a selection of other lakes.

Lake	Year	Area (ha)	Principal species	Biomass (kg/ha)	References
Neusiedlersee (Zone I)	1971	1,475	*Myriophyllum spicatum* L. *Potamogeton pectinatus* L.	30.5	Schiemer & Prosser (1976)
Neusiedlersee (Zone I)	1972	1,475	*Myriophyllum spicatum* L. *Potamogeton pectinatus* L.	4.3	Schiemer & Prosser (1976)
Balaton (Keszthely Bay)	1969	3,286	*Potamogeton perfoliatus* L.	8.5	Karpati and Varga (1970)
Balaton (Sziliget Bay)	1970	1,327	*Myriophyllum spicatum* L. *Potamogeton perfoliatus* L.	47.7	Karpati et al. (1971)
Marion Lake	1966	13	*Potamogeton natans* L. *Potamogeton epihydrus* Raf. *Nuphar polysepala* Engelm.	1,034	Davies (1970)
Sniardwy	1966	3,037*	*Potamogeton pectinatus* L. *Potamogeton perfoliatus* L. *Fontinalis antipyretica* L.	750	Bernatowicz et al. (1968)
Warniak	1966	38	*Potamogeton pectinatus* L. *Potamogeton compressus* L.	1,000	Bernatowicz (1969)

* Area of lake surface 0–5 m depth.

Values for cover of *Myriophyllum*, estimated in the 1970 survey, never exceeded 20 per cent. Cover values of 100 per cent were reached only for small patches of *P. pectinatus*. Within 10 × 10 m squares cover values for this species never exceeded 10 per cent.

However, it can be seen from aerial photographs (Fig. 19.5 and 19.6) that macrophyte growth in Neusiedlersee was at times much denser than during the period covered by our observations.

The comparison of biomass data from different lakes (Table 19.2) shows a similarity to Balaton while biomass per total lake area may be 1 to 2 orders of magnitude higher in other shallow lakes with high macrophyte production (e.g. L. Warniak and L. Sniardwy in Poland, L. Marion in Canada; for a more detailed discussion see Schiemer & Prosser, 1976).

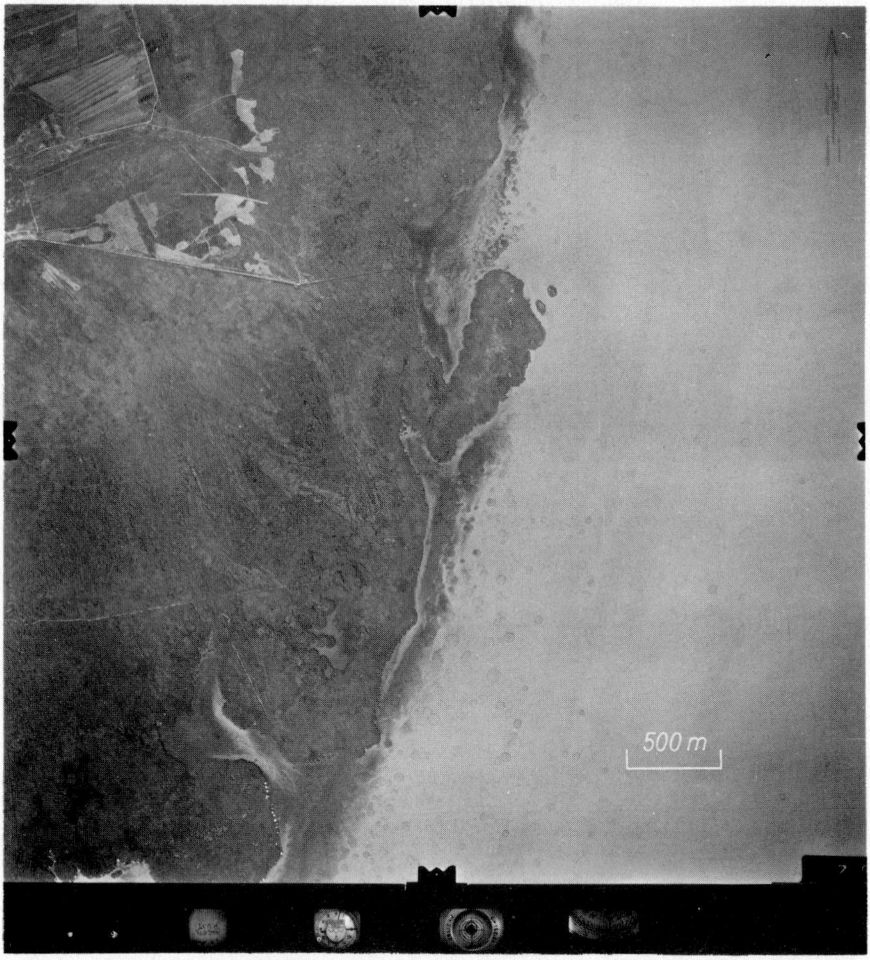

Fig. 19.5. Aerial photograph from the west shore, showing the reed belt and the adjoining macrophyte belt, especially the numerous *Potamogeton* rings (summer 1964).

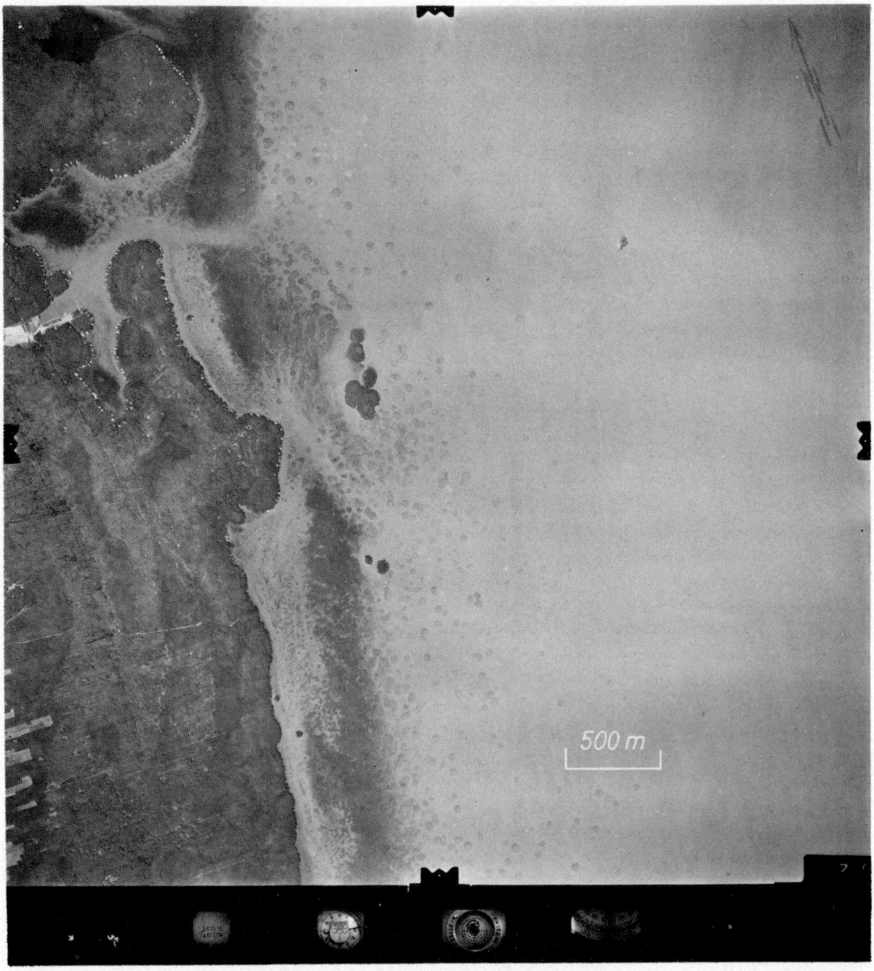

Fig. 19.6. As Fig. 19.5.

Fig. 19.7. Myriophyllum growth at the northern end of the lake (summer 1970).

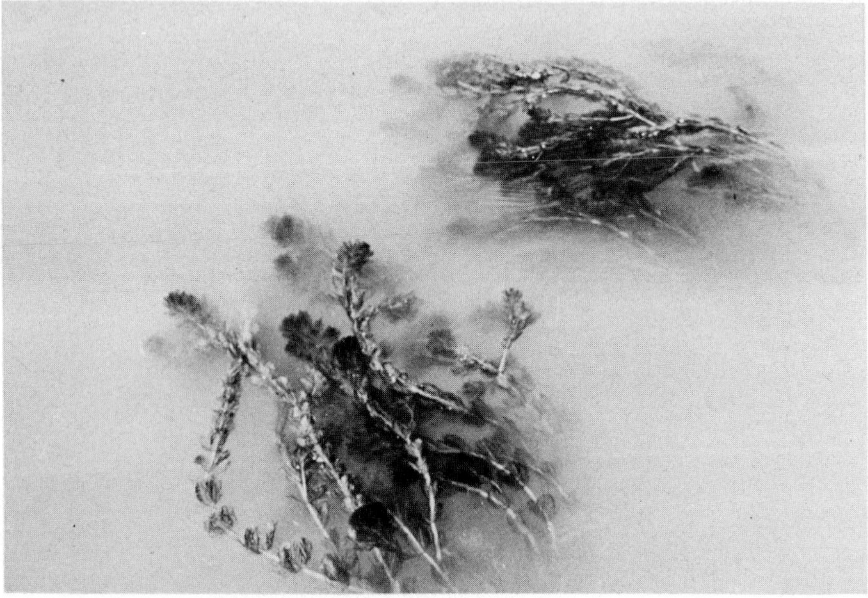

Fig. 19.8. Myriophyllum spicatum.

19.6 Long-term changes in the macrophyte belt

The definite decline in density and extent of the macrophyte belt observed during recent years (Table 19.1) leads to the question of the consistency of macrophyte growth under the specific conditions of Neusiedlersee.

Aerial photographs taken in summer 1957 and 1964 clearly show the extent and density of the macrophyte belt of a limited area on the western shore (between Mörbisch and Rust) and provide further information on long-term changes. The extent and structure of the macrophyte belt are similar in these 2 years, with *P. pectinatus* as the dominant species. However, a marked increase in density occurred over the seven-year period, besides the increase in size of *Potamogeton* rings, as discussed above. When in 1971 the same area was first surveyed, distinctly lower densities were observed and the phenomenon of long-term changes was discussed (Schiemer & Weisser, 1972); Extensive development of *P. pectinatus* in the lake obviously coincides with prolonged periods of low water level. The mean water depth, which was constantly low during the period from 1953–1964, increased in 1965 – due to a sluice regulation of the artificial affluent of the lake (Kopf, 1967) and due to a high precipitation in this particular year. The water level increased by approximately 40 cm, and remained high during the period of IBP. This increase was followed by a reduction of the macrophyte belt.

Varga's (1931) data on ring formation and ring growth show that growth conditions for *P. pectinatus* were favourable at the time of his studies (1928–1930, see above), when the water level was low.

19.7 Ecological factors governing the growth of macrophytes in Neusiedlersee

The macrophyte distribution in the lake observed during 1970–1972, as well as long-term vegetational changes, suggest 3 main factors:

1. Mechanical force of wave action
2. Sedimentation of silt on macrophytes
3. Light conditions.

These 3 factors are highly interdependent and provide in Neusiedlersee a situation typical for large, shallow lakes:

ad 1. The wind-driven currents cause the overall high turbidity of the lake water. The pattern of turbidity distribution in the lake depends on the wind direction and also strongly influences sedimentation rates and light conditions (see chapter 15).

Although the direct mechanical effect of water turbulence on the establishment and growth of the two species cannot be clearly defined, the distribution pattern of *M. spicatum* indicates a direct effect.

Water turbulence, on the other hand, exerts a positive effect by washing off the silt coating. (Varga, op. cit. e.g., explains the ring growth of *P. pectinatus* by a greater washing effect at the periphery of plant stands).

ad 2. A silt coating forms on the plant surface, especially in zones of lesser water turbulence, and such a coating is enhanced by periphyton. The screening effect of such a coating has been demonstrated by Weisser (1970) in some preliminary field experiments. In general, the coating is more marked on finely divided, feathery *M. spicatum* fronds than on the grass-like *P.*

pectinatus. The degree of coating in both species is highest in sheltered bays and decreases with distance offshore. The effect of the coating appears to be twofold:

(a) shading of assimilating parts
(b) increase of weight on buoyant fronds, pressing the assimilating parts deeper into the water column and thus decreasing assimilation.

ad 3. Light conditions are largely affected by the 2 factors discussed above.

The prevailing NW wind causes a specific distribution pattern of suspended soils and thus of light conditions within the lake (Chapter 15 – Dokulil 1979). In sum, light transmission is less at the eastern than at the northwestern shore. Also, light conditions can be especially critical during springtime, when the plants have to shoot up from the lake bottom.

The distribution pattern of both species in the open lake is governed by a combination of these 3 factors. Mechanical effects of wave action are considered to have a stronger effect on *M. spicatum* than on *P. pectinatus*. In contrast to the latter, *M. spicatum* is absent along the eastern shore except in a few sheltered bays and sometimes on the lee-side (south-east) of islands in the open lake, which act as a wind break. The distribution of *M. spicatum* along transects from the reed fringe towards the open lake is characterized by an offshore peak in plant density and biomass and an increase in average plant weight (Fig. 19.3, Table 19.2). This pattern suggests the effect of wave action, which increases, and the sedimentation rate of silt, which decreases with distance from the shore. The combined effects of these opposing factors permit abundant growth only in a narrow transitional zone. The mechanical effect of water turbulence is considered to cause the sharp reduction of *Myriophyllum* density towards the lake centre. Since the size of individual plants established at the outer limit of the species' occurrence is large (Fig. 19.3), it can be concluded that wave action negatively affects colonisation in zones with high water turbulence.

An interpretation of the causes of the distribution pattern of *P. pectinatus* on the basis of our field observations is not yet possible. Biomass maxima in sheltered situations indicate a greater indifference to the effects of silt-coating. Sedimentation of the fine silt has less effect on the grass-like *P. pectinatus* than on *M. spicatum* which has a feathery surface. On the other hand the occurrence of the former species along the wave-beaten eastern shore indicates that the species is also less dependent on the mechanical effects of water turbulence. In general, the sediment type seems to be of lesser importance since both species occur over a wide range of sediment depth, pH, Eh and nutrient content.

The cause of long-term dynamics must first be discussed in connection with aperiodic changes in the water level, which severely affect several limnological parameters. The most important factors influenced by water level fluctuations are light penetration and hydromechanical effects of wave action on the lake floor. Changes in the chemical condition in relation to water level are discussed by Neuhuber (1971).

Considering that for macrophyte growth the ecological conditions of

Neusiedlersee seem to be severe, such changes may be critical. Possibly a deterioration of underwater light conditions, which are most critical at the beginning of the growing season (Anderson, 1978), is the main cause of a macrophyte regression in connection with increased water levels.

It was pointed out earlier (Schiemer & Prosser, 1976), that the strong reduction in macrophyte biomass observed during recent years may find an additional explanation in the eutrophication, which is indicated by the increasing phytoplankton and epipelic algae biomass since 1972 (see Chapters 17 and 23). The increase in phytoplankton biomass causes a reduction in photosynthetically active radiation available for macrophytes. Furthermore, an enhanced growth of epiphytic algae exerts a screening effect as discussed above.

A reduction of macrophyte growth during phases of lake eutrophication, possibly caused by the attenuation of underwater light due to phytoplankton densities, has been recently described for several lakes (e.g. Jupp & Spence 1977). In Balaton a similar process was observed during recent years (Karpati, pers. comm.).

A further cause for the radical decrease of macrophytes from 1970 onwards is the grasscarp, probably introduced into the lake for the first time around 1970 (Schiemer 1979, Hacker 1979 – Chapter 28).

19.8 Role of submerged macrophytes within the ecosystem of the open lake

A general consideration of the importance of macrophytes with respect to the energetics of lakes and to the regulation of nutrient recycling was given by Wetzel & Allen, (1971) and Wetzel & Hough, (1973). In a lake of the Neusiedlersee type, which exclusively represents a littoral zone, the macrophytic community is of special importance in several respects.

The first of these concerns the quantitative importance of macrophytes as primary producers in the open lake beyond the reed belt. Reasonable estimates of mean biomass only exist for phytoplankton, macrophytes and epipelic algae. Epiphytic cover of macrophytes was low at the beginning of the IBP study (1967) but has increased since 1972. However, no quantitative information on biomass and production is available.

Within the ecosystem of the open lake the importance of submerged macrophytes as primary producers was small even in years with a higher standing crop (Table 19.3). This is surprising, considering the low overall productivity which results from the high turbidity of the lake. (See Chapter 28, Dokulil 1979).

The comparison shows that less than 10 per cent of the annual production is due to macrophytes. However, in terms of standing crop, the relatively small macrophyte zone nevertheless yielded appreciably higher values in 1971 than did the phytoplankton.

The pathway of energy fixed by macrophytes has not been analysed in detail. The importance of dissolved organic carbon released from photosynthetically active plants is discussed by Wetzel et al. (1972), Wetzel and Allan (op. cit), Wetzel and Hough (op. cit), (a.o.).

Table 19.3. Mean annual biomass (respectively biomass at period of maximal development of submerged macrophytes) and annual production of different primary producers of the open lake – in units of g dry weight/m^2. Detailed explanation in Schiemer 1979, Chapter 23, Table 23.14.

The values for epipelic algae and submerged macrophytes represent weighted means for the northernmost part of the lake as delineated in Fig. 19.1. Phytoplankton production according to Dokulil, 1974 (mean value for the years 1968–1970, Dokulil, pers. comm.). No data on epiphytic algae available.

	1970	1971	1972	1970	1971	1972
phytoplankton	0.24	0.32	0.72	142[1]	—	—
epipelic algae	—	0.45		—	—	—
submerged macrophytes	—	3.05	0.43	—	15.3	2.2

Particulate detritus, especially that produced at the end of the vegetational period from September to November, is considered to be of especial importance as an energy carrier for the benthic food chain in the sedimentation zones of the open lake (see Chapter 23). It forms the basis for a higher content of organic material and nutrients in the sediment and thus provides conditions for higher biomass of zoobenthos.

A second aspect is the importance of macrophytes for the nutrient budget of the open lake. The amount of phosphorus fixed by the macrophytes during the vegetation period is small in relation to the total available P in the water body (1–5 per cent based on values from 1970–72 and for the period of maximum phytobiomass). Similarly, the size of the total N pool accounted for by macrophytes during the period of investigation is negligible (Schiemer & Prosser, 1976). However, nutrient recycling of macrophytes should be a subject for further research both with regard to direct uptake of nutrients for the sediments or indirect recycling by oxygenation of the sediments. A further regulatory mechanism might be seen in the production of calcite which forms the main component of the sediments and inorganic turbidity. Adsorption processes of orthophosphate onto calcite are of prime importance in regulating the P-budget of the lake (Gunatilaka, 1978; see also Otsuki & Wetzel, 1972).

Besides their considerable quantitative importance, the macrophyte communities play a role in the lacustrine ecosystem in reducing the water turbulence in certain areas. This in turn means better light penetration within the macrophyte belt and a greater stability of the substrate, and in consequence a richer benthic fauna and epipelic flora (see Chapter 23). Additionally the macrophyte zone provides the structural basis for a rich epiphytic community.

Finally, the existence of a macrophyte belt is of importance for the fish fauna in several respects. Several species (all the Percidae, *Alburnus alburnus* L.) use the submerged macrophytes as a spawning substrate. For the pike, macrophytes are of importance as a habitat structure in the open lake.

Epiphytic food organisms are used by cyprinids and to some extent by the first year class of carnivorous fish like pike and pike perch. Higher macrobenthos densities in the soft sediment areas within the former macrophytic zone constitute the major food of several fish species (e.g. *Acerina cernua*).

[1]Mean value for the years 1968–70.

It is obvious that macrophytes affect fish in a variety of ways and their disappearance may represent one of the major causes for the extensive changes of the fish fauna over the past 10 years (see Chapter 28).

References

Anderson, M. G. 1978. Distribution and production of Sago pondweed (*Potamogeton pectinatus* L.) on a northern prairie marsh.– Ecology 59: 154–160.

Bernatowicz, S. 1969. Macrophytes in the lake Warniak and their chemical composition.– Ekol. Pol. A. 17: 447–467.

Bernatowicz, S., Pieczynska, E. and Radziej, J. 1968. The biomass of macrophytes in Lake Sniardwy.– Bull. Acad. Pol. Sci. Cl. II, 16: 625–629.

Davies, G. S. 1970. Productivity of macrophytes in Marion Lake, British Columbia.– J. Fis. Res. Board, Canada 27: 71–81.

Dokulil, M. 1974. Der Neusiedler See (Österreich).– Ber. Naturhist. Ges. Hannover 118: 205–211.

Dokulil, M. 1979. Optical properties, colour and turbidity.– In: Neusiedlersee, ed. H. Löffler. Junk, The Hague.

Dokulil, M. 1979. Seasonal pattern of phytoplankton.– In: Neusiedlersee, ed. H. Löffler. Junk, The Hague.

Dokulil, M. 1979. Phytoplankton primary produktion.– In: Neusiedlersee, ed. H. Löffler. Junk, The Hague.

Gunatilaka, A. 1978. Role of seston in the phosphate removal in Neusiedler See.– Verh. Int. Verein. Limnol. 20: 986–991.

Hacker, R. 1979. Fishes and fishery in Neusiedlersee.– In: Neusiedlersee, ed. H. Löffler. Junk, The Hague.

Hacker, R. & Meisriemler, P. (in press). The diet of two mainly benthos-feeding fish species, pope (*Acerina cernua* (L.) and white bream (*Blicca björkna* (L.)) in Neusiedlersee (Austria).– Env. Biol. Fish.

Hutchinson, G. E. 1975. A Treatise on Limnology.– Vol. III, Limnological Botany. John Wiley & Sons, 660 pp.

Jupp, B. P. & Spence, D. H. N. 1977. Limitations of macrophytes in a eutrophic lake, Loch Leven. I. Effects of phytoplankton.– J. Ecol. 65: 175–186.

Karpati, I. & Varga, G. 1970. Forschungsergebnisse der Laichkrautvegetation in der Balatonbucht von Keszthely, 1970.–Mitt. Hochsch. Landwirtsch., Keszthely, 12: 5, 1–67.

Karpati, I., Varga, G. & Novotny, I. 1971. Phytobiomassenproduktion der Laichkrautvegetation in der Brucht von Szigliget im Jahre 1970. (Produktionbiologische Erforschung der Laichkraut-vegetation des Balaton II).–Mitt. Hochsch. Landwirtsch. Keszthely 13: 1–42.

Karpati, I., Karpati, N. & Varga, G. 1972. Die methodischen Fragen der Auswertung der Phytomassen-Produktion und der Vegetations-kartierung von Potametea-Gesellschaften.– In: Grundfragen und Methoden in der Pflanzensoziologie, ed. R. Tüxen, Dr. W. Junk, Den Haag, 443–449.

Kiss, E. Cs. 1972. Hair-weed map of lake Velence.– Halászat 65: 20–21 (Hung.).

Kopf, F. 1967. Die Rettung des Neusiedler Sees.– Österr. Wasserwirtschaft 19: 139–151.

Maier, R. Production of *Utricularia vulgaris* L.–In: Neusiedlersee, ed. H. Löffler. Junk, The Hague.

Maucha, R. 1931. Sauerstoffschichtung und Seetypenlehre.– Verh. Intern. Ver. f. theor. u. angew. Limnol. 5: 75–102.

Neuhuber, F. 1971. Ein Beitrag zum Chemismus des Neusiedler Sees.– Sitz. Ber. Österr. Akad. Wiss., Math.– nat. Kl. I, 179: 225–231.

Otsuki, A. & Wetzel, R. G. 1972. Coprecipitation of phosphate with carbonates in a marl lake.– Limnol. Oceanogr. 17: 763–767.

Schiemer, F. 1979. The benthic community of the open lake.– In: Neusiedlersee, ed. H. Löffler, Junk, The Hague.
Schiemer, F. 1978. Vegetationsveränderungen im Neusiedlersee.–Österr. Wasserwirtschaft, 30: 252–253.
Schiemer, F. & Weisser, P. 1972. Zur Verteilung der submersen Makrophyten in der schilffreien Zone des Neusiedler Sees.– Sitz. Ber. Österr. Akad. Wiss., Math.– nat. Kl. I, 180: 87–97.
Schiemer, F. & Prosser, M. 1976. Distribution and biomass of submerged macrophytes in Neusiedlersee.– Aquatic Botany 2: 289–307.
Sculthorpe, C. D. 1976. The Biology of aquatic vascular plants.–London, Edward Arnold, 610 pp.
Straskraba, M. 1968. Der Anteil der höheren Pflanzen an der Produktion der stehenden Gewässer.– Mitt. Int. Ver. Limnol. 14: 212–230.
Varga, L. 1931. Interessante Formation von *Potamogeton pectinatus* L. im Fertö (Neusiedlersee).– Arb. Ung. Biol. Forsch. Inst. (Tihany), 4: 349–355.
Varga, L. 1933. Sonderbare Ringbildungen von *Potamogeton pectinatus* L. im Fertö (Neusiedlersee).– Int. Rev. ges. Hydrobiol. 28: 285–294.
Weisser, P. 1970. Die Vegetationsverhältnisse des Neusiedlersees. Pflanzensoziologische und ökologische Studien.– Wiss. Arb. Bgld. 45: 1–83.
Wetzel, R. G. & Allan, H. L. 1971. Functions and interactions of dissolved organic matter and the littoral zone in lake metabolism and eutrophication.– In: Productivity problems of freshwaters, ed. Kajak, I., Hillbricht-Ilkowska, A., Pwn, Warszawa-Krakow, 333–347.
Wetzel, R. G., Rich, P. H. Miller, M. C. & Allan, H. L. 1972. Metabolism of dissolved and particulate detrital carbon in a temperate hard-water lake.– In: Detritus and its ecological role in aquatic ecosystems, ed. Melchiorri–Santolini, U. & Hopton, J. W., Mem. Ist. Ital. Idrobiol. 29: 185–243.
Wetzel, R. G. & Hough, R. A. 1973. Productivity and role of aquatic macrophytes in lakes. An assessment.–Pol. Arch. Hydrobiol. 20: 9–19.

) The primary producers of the *Phragmites* belt, their energy utilization and water balance

K. Burian and H. Sieghardt

20.1 Vegetation conditions and production zones of the reed belt

The value of ecological research depends partly on the knowledge of the phytocenology of the region studied. The prevalence of *Phragmites communis* in the reed belt is so marked that at first sight the whole stand seems to be monospecific. The great competitive powers of *Ph. communis* in shallow waters (see Chapter 21) allow only a temporary occurrence of other emergent macrophytes (chiefly *Typha*, *Scirpus*, *Bulboschoenus*) in deficiency zones (stubble pools, pools, lacunae), but never a stronger of permanent populations except in the shore region that periodically dries out. This evident paucity of species in extensive zones of the reed belt is attributable to three external factors:

(a) the strongly anaerobic substrate of the root region (cellulose mud, Farahat & Nopp 1966, 1967);
(b) the possibility of freezing down to the bottom and of pressure from drifting ice;
(c) the strong fluctuations of the water level (Kopf 1964, 1966).

Ph. communis itself reduces the chances of other emergent macrophytes by additional internal factors:

(a) the deep-lying main horizon of the rhizomes (about 50 cm), which constitutes a mechanical obstacle for competitors;
(b) the high degree of light absorption in the upper third of the reed, the exponential vertical course of which (see p. 258) is rather that of a short-leaved dicotyledon than of a graminae association (cf. Larcher 1976);
(c) the growth rhythm with a complete generation of shoots ready to spout in autumn;
(d) the high productivity and the insensitivity to extreme temperatures, to edaphic peculiarities (optional halophyte) and to the stress of drought (optional xerophyte and xerohalophyte);
(e) the low degree of utilization for feeding purposes (Imhof & Burian 1972), which leads to a large surplus of biomass and to augmentation of the anaerobic cellulose mud (and thus to increased stress in the root area);
(f) the total cessation of generative propagation in shallow waters plus the advantage of a strong vegetative clone propagation by rhizomes and fragments of rhizomes; the delicate seedling stage can thus be avoided entirely.

Although this catalogue is undoubtedly incomplete, the aim is to provide a general understanding of the monospecific associations inhabiting shallow waters. At the same time it illustrates the fact that phenomena connected with competition among plants, whether inter- or intraspecific, cannot be explained by studying a single property or ecological factor alone (Went 1973).

Despite the high prevalence of reed, phytocenologists have been able to distinguish a number of different associations within the reed belt. Wendelberger (1941) in his study of the Neusiedlersee vegetation briefly considered the reed stands and in a later study (1964) he was particularly concerned with the reed of Pannonian lakes (no study of the Austrian part of the reed belt existed before this). Systematic classifications of the Hungarian reeds are to be found in papers by Tóth (1960) and Tóth & Szábo (1961), as well in a short report by Csapody (1965).

A detailed study of the Austrian portion of the reed belt and its surroundings was published by Weisser (1970). He describes the seemingly monospecific, monotonous reed belt as being highly differentiated in character:

In his catalogue of the flora of the reed belt Weisser itemizes 233 flowering plants, 15 mosses and 13 macroscopic species of algae. Within the association of *Scirpo-Phragmitetum* described by Koch (1926) Weisser again distinguishes 10 sub-associations specific for Neusiedlersee, whose characterizing species are mainly *Utricularia vulgaris*, *Typha angustifolia* and *Potamogeton pectinatus*. Two of these sub-assocations are of extreme ecological significance:

(a) *Scirpo-Phragmitetum utricularietosum*, occupies the main part of the reed belt between the outermost, lakeward monospecific *Phragmites* community and that situated furthest inland, which is typified by *Caricetae*.

(b) *Scirpo-Phragmitetum phragmitosum* (sn.: '*Phragmitetum nudum*'), the lakeward progression zone of the reed, which is greatly influenced by waves; a truly monospecific zone showing the highest primary productivity.

An essential part of this study (Maier, 1976) was devoted to an association that constitutes the landward progression zone of the reed. This riparian zone, subject to periodical drying-out, has a higher soil activity, less anaeorobic properties in the root region, smaller individual reed stalks, and permanent sedge populations. According to Weisser's classification it is a

(c) *Caricetum acutiformis-ripariae.*

For purposes of production ecology a clear order of priorities can be established for the most important primary producers involved:

(1) *Phragmites communis* Trin. (= Phragmites australis (Cav.) Trin. ex Steud.)

(2) *Utricularia vulgaris* L.

(3) *Carex*, ssp.: *acutiformis* Ehrh., *distans* L., *disticha* Huds., *elata* All. (unimportant for the Austrian part), *flacca* Schreb., *gracilis* Curt., *nutans* Host., *panices* L., *paniculata* L., *pseudocyperus* L., *riparia* Curt., *tomentosa* L., *otrubae* Podb.

Understandably, for production purposes *Carex* will be treated as a genus without further division into species.

The reed belt can be divided into three production zones (Burian 1973, Fig. 20.1), the activity of which is completely determined by *Phragmites*.

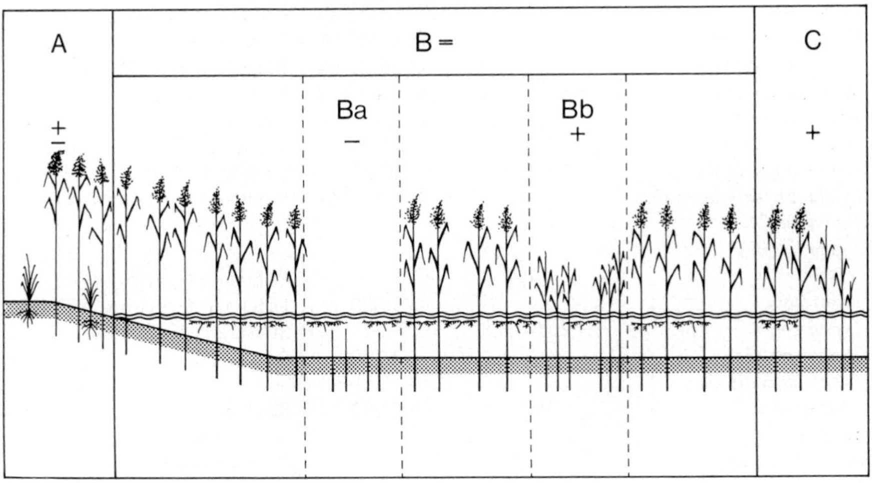

Fig. 20.1. Transect of the *Scirpo-Phragmitetum* from the shore to the open water
A±: Land ward progression zone
B: *Scirpo-Phragmitetum utricularietosum*
Ba−: Zone of negative production balance (Scirpo-Phragmitetum utriculariosum)
Bb+: Zone of positive production balance
C+: Zone of reed progression into the lake

(A) The zone that is gradually changing into firm land (*Scirpo-Phragmitetum utricularietosum* in transition to *Caricetum acutiformis-ripariae*). The term 'changing into firm land' is fully justified unless in the future a dramatic rise in water level occurs. The deposition stress due to biomass decomposition increases from year to year. Recent studies (Maier, pers. commun.) showed that in spite of strong competition due to *Caricetae*, which flourish in soils rich in oxygen, the reed would advance landwards into the neighbouring meadows if this were not prevented by deep ploughing and by cutting.

The unstable equilibrium between reed and agriculture depends solely upon the ground water level. In other words, the reed belt advances landward according to the quantity of rain per decade.

No doubt this part of the reed belt is extremely important for the entire ecosystem for it supplies the nourishment for the whole Neusiedlersee area. In this sense it is a classical eulittoral (see also Chapter 25).

(B) The reed-bladderwort zone (*Scripo-Phragmitetum utricularietosum*), Weisser 1970.

Due to the fact that the zone is permanently covered by water the competition from the emergent macrophytes is limited to the general *Phragmites* and *Typha*. *Utricularia* cannot be regarded as a competitor since, as a

Table 20.1. Calorific values of macrophytes

Species	cal/g	cal/g ash free (organic)	authors
Lemna spp.		4,810–5,140	Lautner & Müller 1954
Myriophyllum spp.		4,588–5,201	Gortner 1934
Phragmites communis		4,396–4,766	Birge & Juday 1922
Potamogeton spp.		4,235–4,881	Seidel 1956
Scirpus lacustris		4,372–4,740	Straskraba 1968
Scirpus lacustris juv.		4,790	Straskraba 1968
Typha angustifolia	3,836	4107	Grabowski 1973
Typha angustifolia stalks		4,490/5,070	Pribil and Dykyjova 1973
Typha latifolia	4,262		Boyd 1970
Typha latifolia			
stalks	4,490		Pribil and Dykyjovy 1973
rhizomes		4,020	Pribil and Dykyjova 1973
Typha spp.	4,207–4,523		Straskraba 1968
Typha spp.		4,340	Bray 1962
Typha angustifolia			
rhizomes	3,864	3,983	Sieghardt
roots			(unpublished)
Utricularia vulgaris	1,416		Grabowski 1973
Utricularia inflata	4,023		Boyd 1970
Utricularia vulgaris			
open lake area	2,905[1]		
on the reed bank	2,617		Sieghardt (quoted in Imhof and Burian 1972)

1. Sieghardt unpublished

submergent, rootless macrophyte it literally makes use of the 'interstices' thus occurring in the Neusiedlersee area as a synoecic plant of *Phragmites.*

Typha also settles in the larger gaps in the *Phragmites* stands, especially in those caused by reckless harvesting (Weisser 1970) below the water surface, or as a result of excessive rhizome development and consequent oxygen deficiency in the *Phragmites* stands. *Typha* has a considerably flatter rhizome horizon (Rudescu et al. 1965), which is one important reason why it cannot permanently hold its own against the reed.

The reed tends to breakdown in a patchwise manner similar to the situation observed in the boreal coniferous forests. The reeds are preclimactic and stable at this water level and are liable to loss of area. The rejuvenating parts of the ecosystem, however, quickly become overgrown once more. This kind of self-regulation of the preclimactic, stable stages of an ecosystem is otherwise not too frequent. Other ecosystems characterized by graminae, such as dry meadows, prairies, and even savannahs or semi-desert associations are not capable of re-attaining the preclimactic stage within such a short period of time. In the emergent graminae-macrophyte populations, however, the speed of recovery is in positive correlation to the degree of disturbance.

Phragmites, which is only of economic interest when dried, demonstrates

Table 20.2.

	Straskraba 1968	Grabowski 1973
	cal/g ash free	
submerged macrophytes	4,165–5,201 4,580	2,664
emersed macrophytes	4,207–4,987 4,480	4,006

that loss of area or expert harvesting only disturb possible competitors but not the dominant plant.

(C) The monospecific reed stand.

It is uncertain why *Utricularia*, the 'domestic plant' of the Neusiedlersee reed does not accompany it as far as the lakeward zone of progression (see Chapter 21). Indeed, the '*Phragmitetum nudum*', the reed zone advancing lakewards is devoid of plants other than *Phragmites* and its periphyton, mainly diatoms (see Chapter 16). It has not yet been ascertained how far this is due to vacation housing within the reed belt. But it is quite certain that if this activity is allowed to go on it will do irreparable harm to the lake. Settlements in the part of the reed near the open water bring problems which no sewage system can cope with. It should be noted that the region of the Bay of Rust, the site of a large part of our studies between 1965 and 1973, despite being a nature reserve (Biosphere Reserve, declared by UNESCO 1977), is now a biotope characterized by residues of lubricants and rows of tarred wooden huts.

The ecological characteristics of this zone can still be observed in the areas between the settlements, where the destructive processes are not yet in progress. *Ph. communis* is advancing lakewards which means that as a crop it shows a genuine net increase of biomass unlike zones (a) and (b).

As shown by the investigations of Kopf (1964, 1966, 1967, 1968), Riedmüller (1965), Weisser (1973) and others the increasing extent to which Neusiedlersee is becoming covered by reed has aroused interest for economic reasons. On the basis of the progression since 1926 (and even 1868/72, when the lake last completely dried up) Kopf calculated that if unmanaged the lake would be entirely overgrown with reed by 2120 (Fig. 20.2). One possible measure would be deep-ploughing along the lakeward reed margin. The asymmetrical distribution of the reed, with centres of density in the south, west and north-west, but weak development in the east illustrates that antagonistic factors characterize the dynamics of reed expansion.

(a) promotion: the development of soft anaerobic root areas where *Phragmites* flourishes free from competition is chiefly due to the drifting and accumulation of mud in the south-east and north-west (see Chapter 24). In these muddy zones Riedmüller found a reed progression of 20 m/year.

(b) inhibition: ice drift with its highly destructive mechanical power particularly affects the eastern shore. Stronger waves prevent sedimentation and thus the formation of a subsoil ideal for the reed.

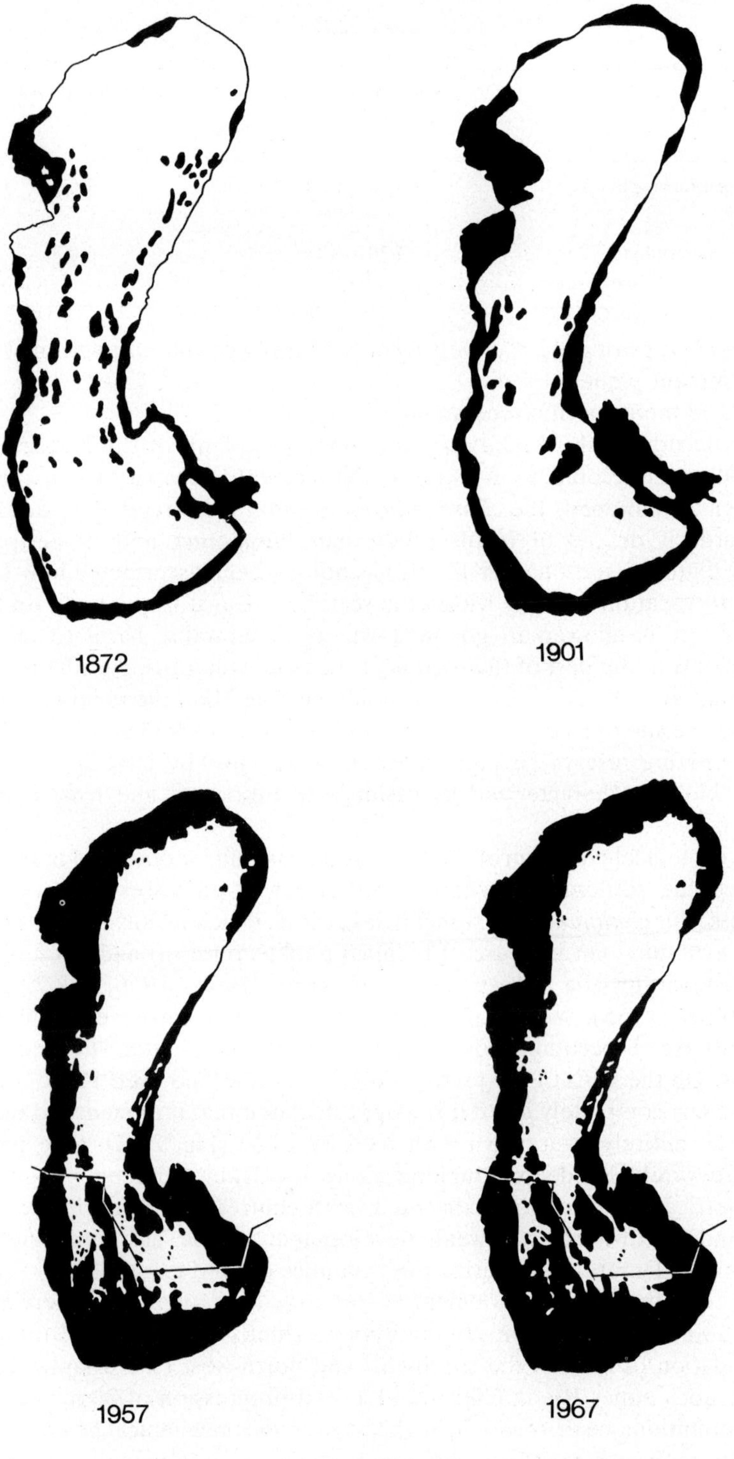

Fig. 20.2. Development of the reed belt from 1872–1967

Moreover a high water level may also inhibit the progress of the reed belt, though the critical depth of water is still under discussion. Only about 30 per cent of the reed on Neusiedlersee are suitable for harvesting, but even this is never totally exploited.

Phragmites is unable to tolerate regular cutting during its production period and thus it constitutes no danger if landside pastures are cut regularly. The effect of cutting in winter seems to differ depending upon the technique employed and upon the clone. Whereas Hübl (1966) detected favourable effects on production in the succeeding year, Weisser (1970) considers that the effect depends upon the swath length. If cut 20 cm above the ice surface almost no effect is noticed in the following year, whereas cutting on or below the ice is detrimental to development in the following year since the rhizomes die off as soon as water penetrates the stalks. This explains the formation of 'stubble pools' which quickly become overgrown once more from the margins. From a biological point of view the existence of such anthropogenic pools is by no means unfavourable since both primary and secondary production are promoted within them. The same is true for the natural reed gaps. However, a large increase in stubble gaps should not be permitted.

20.2 Production

20.2.1 CO_2 exchange analysis of *Phragmites communis*

In various plant associations that are inaccessible to complex harvesting measures and cannot be covered by allometric methods the only alternative is to measure gaseous exchange, which in the case of the reed crop is especially high. Strong development and/or inaccessibility of the subterranean biomass (both are true of *Phragmites*), make regular harvesting of the whole plant impossible and render the measurement of the gaseous exchange a necessity.

Field measurements were made for three years in the area of Rust. An Uras 1 (Hartmann & Braun) was employed in the open system. The measurements with the 2 m^3 crop cuvette (Burian 1969) were unsatisfactory whereas the values from non air-conditioned leaf cuvettes fully agreed with the results of model experiments performed in a phytotron.

The following data for the photosynthesis of *Phragmites* were obtained:

20.2.1.1 The highest spring values are around 25, the highest summer values around 30 mg CO_2/ leaf -dw.h^{-1}, which are high for an exposed C_3 grass.

2.2.1.2 The light saturation of photosynthesis fluctuates according to temperature and season:

June:	10–15°C	17 cal cm^{-2} h^{-1} (global)
	15–20°C	19 ,,
	20–25°C	18 ,,
	25–30°C	32 ,,

July:	10–15°C	16 cal cm^{-2} h^{-1} (global)
	15–20°C	15 ,,

20.2.1.3 The highest photosynthetic activity is reached in the middle and not in the best illuminated upper levels (cf. Koch et al. Solling project, 1973). (Fig. 20.3).

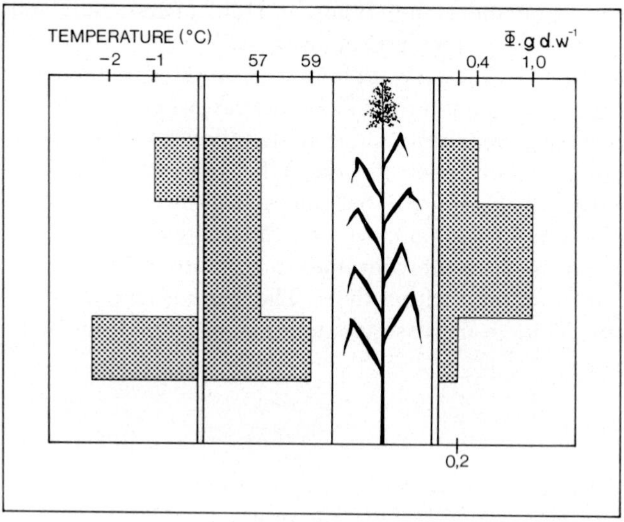

Fig. 20.3. Φ g dry weight^{-1}: left: vertical distribution of cold and heat resistance of a single reed stalk. right: vertical distribution of photosynthetic intensity of a single reed stalk

20.2.1.4 The daily course of photosynthesis shows no depression at noon but a rapid rise in the morning followed by a slightly declining plateau which comes to an abrupt end in the evening.

20.2.1.5 The photosynthesis-temperature behaviour of the reed shows a distinct bend at the beginning of summer (Fig. 20.3).

The spring temperature optimum (May) of 10°–15°C is replaced by a high temperature tolerance of 10°–30°C in June (Burian 1969, 1973).

20.2.2 *Production values calculated from the harvesting method*

It is almost impossible to estimate the gross primary production of a natural crop since, in addition to the biomass, respiration rates, feeding exploitation and detritus, root respiration etc. have to be considered.

The net production (in the sense of Boysen-Jensen as P_B-R) can only be approximately calculated because of the almost inaccessible subterranean biomass. An attempt was made to estimate it on the basis of regular epigeous harvests and occasional digging. The 'net production' in the sense of

Boysen-Jensen (1932) as a net increment of biomass within the ecosystem does not exist for *Scirpo-Phragmitetum utricularietosum*: as a preclimactic ecosystem it shows surpluses of substance only in an increasing cellulose-mud horizon, but does not gain real biomass over larger areas.

20.2.2.1 The net assimilation rate (NAR) in g of dry matter per 0.1 m^2 leaf surface and day (Fig. 20.4) of *Phragmites* starts above ground with the very high value of about 0.3 in April, a value that is not reached by cultivated graminae. But the epigeous NAR cannot be regarded as an expression of photosynthetic performance alone. Such values are only possible because of the sugar supply from the starch reservoirs within the rhizomes.

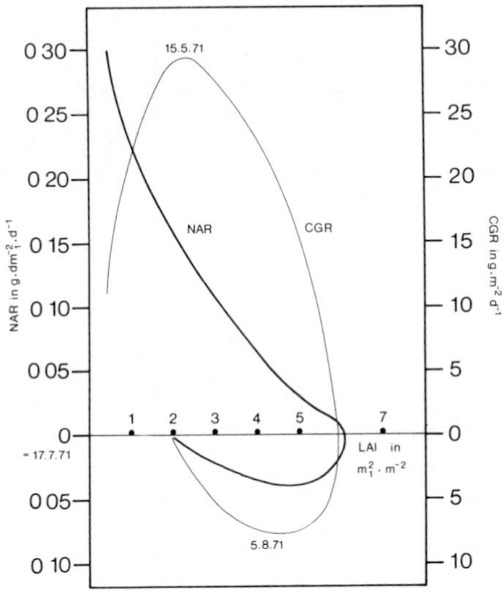

Fig. 20.4. Relation between LAI, NAR and CGR, April–September

The crop growth rate (CGR) in g of dry matter per m^2 ground area and day reaches its maximum of 30 in the middle of May (again in surface harvests). In Fig. 20.4 both values are plotted against the cumulative leaf area index (LAI)[1] in m^2_2 m^{-2} ground area. The fact that both quantities become negative does not mean a negative production balance of the crop, but, as Section 2.3.3 shows, a transport of the assimilates into the subterranean parts of the plant.

20.2.2.2 Epigeous harvesting results from many years and diggings give the following picture of biomass and production.

The total biomass (standing crop) of the reed from several (3–5) production periods amounts to 120 t/hectare dry matter. The turnover (= net production as P_B-R) includes parts that had died off or been eaten seems to be about 30 t/hectares. Cultivation experiments (Szepanski 1969, Sieghardt

1. LAI: $\frac{\text{double leaf area } (m^2)}{\text{ground area } (m^2)}$

1973) and diggings show that only one-fifth of the biomass consists of photosynthetic reproductive parts at the time when the epigeous crop reaches its maximum, whereas four-fifths are rhizome and root mass.

The maximum epigeous dry weight (Fig. 20.5) is reached in July and seems to correspond to an underground biomass minimum (death of the oldest rhizome following exhaustion of the starch reserves due to intensive growth in spring).

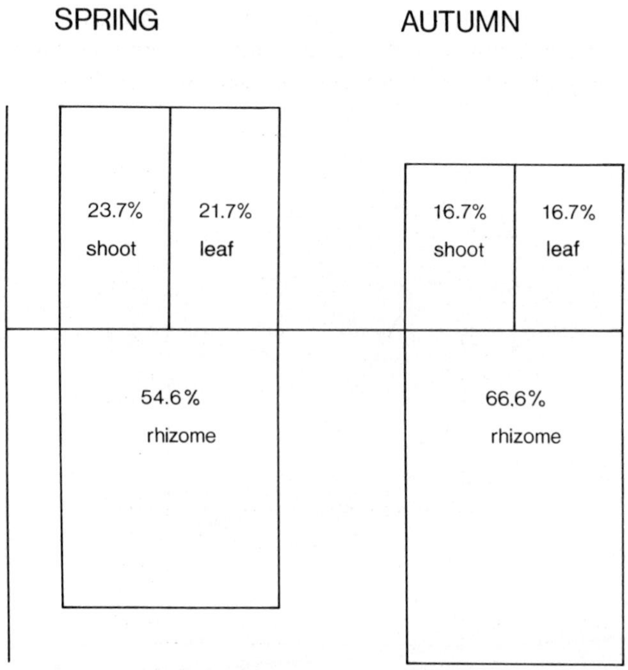

Fig. 20.5. Distribution of dry weight as percentage of total biomass

The maximum subterranean biomass, however, appears to be in autumn, when the entire photosynthetic capacity is used for filling the underground starch reservoirs.

20.2.2.3 The rhythm of assimilation: A comparison of the reed productivity based upon regular above-ground harvests and photosynthesis values corrected for respiration values and converted into dry substance production, Fig. 20.6, gives the following picture of the pattern of assimilates:

(a) In April, the period of growth (reed length up to 130 cm above the ground), great quantities of material are transferred from the rhizome reservoirs into the growing shoots. This is proved by the high NAR of the epigeous biomass. Probably the drying-off of the oldest generation of rhizomes is connected with such transport.

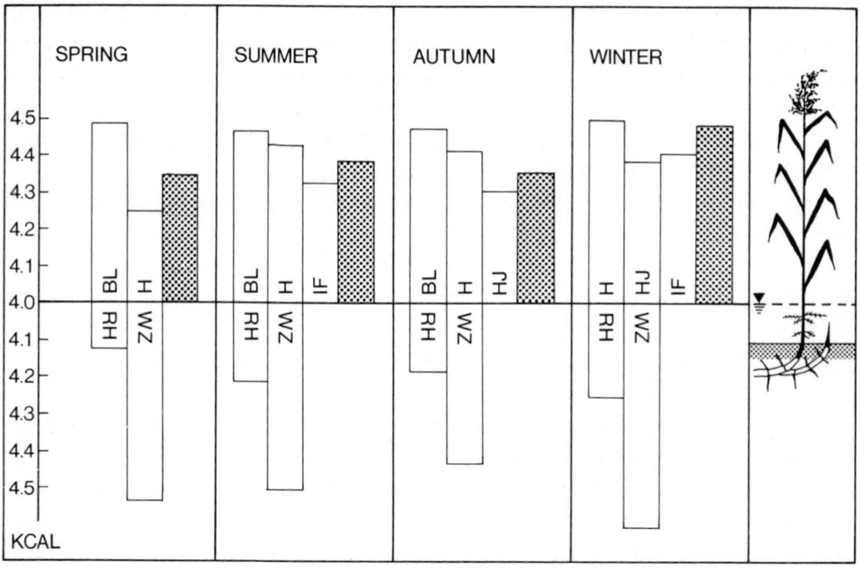

Fig. 20.6. Energy contents of different organs of *Phragmites communis* (kcal) (Kcal/g ash free). BL–leaf, H–stalk, RH–rhizome, WZ–root, HJ–young (autumn) shoot, IF–reproductive organs. ▒ Energy contents of the total phytomass (Kcal g^{-1}).

(b) In May and June photosynthesis and above-ground crop values correspond exactly. During these two months of extreme photosynthetic activity (average output in May 16 mg CO_2/g dw. h^{-1}; in June 13 mg CO_2/g dw. h^{-1} the assimilate gain is equally distributed between the parts of the plant below and above the ground.

(c) At the beginning of July (final length of stalks about 2.5 m) evidently a stronger growth of the rhizome sets in (Fiala 1973); the stream of assimilates is chiefly transported into the submersed portions.

(d) From August onwards harvesting of the epigeous parts of the crop alone implies loss of substance (leaf-fall), but measurement of gaseous exchange still indicates a photosynthetic activity of about 11 mg/g dw. h^{-1}, exclusively to the benefit of the growing elements of the rhizome and the lower parts of the stalks (Zax 1973). Production but no storage occurs above ground.

(e) Biochemical analysis of the assimilates (Krejci 1974) clearly shows that before metabolic activity ceases to be positive in October a large amount of starch from the rhizomes is invested in the newly developing generation of shoots.

Krejci confirmed this concept by a large number of biochemical analyses and worked out the most characteristic annual course for the essential monosaccharides, starch and total protein.

According to Krejci the physiological periodicity of *Phragmites* is most distinctly reflected in the starch content of the rhizomes. He specifies two phases in upward transportation of substance: the period of sprouting in spring and the development of inflorescences in summer. This second influx of sugar from the submersed parts cannot be detected in the gaseous exchange pattern. Krejci's studies (especially on the changes in cellulose and protein content of the roots) also prove that full growth of the rhizomes sets in August. He assumes that the decrease in the starch content of the rhizomes in late summer is due to the new batch of rhizomes. However, the new generation of shoots begins to grow at this time.

The temporal pattern of raffinose content is exactly correlated with frost resistance (Zax 1973).

20.3 Energy Utilization

20.3.1 *Utilization over the course of the year (1966—1971) (Fig. 20.7)*

The question as to the energy bound in the dry substance of *Phragmites communis* and the ecological utilization of radiation made it necessary to find reed biotypes of comparable productivity (Geisslhofer and Burian 1970, Burian 1973, Rodewald-Rudescu 1974). Samples were only taken from reed crops that had not been cut during the winter (Hübl 1966).

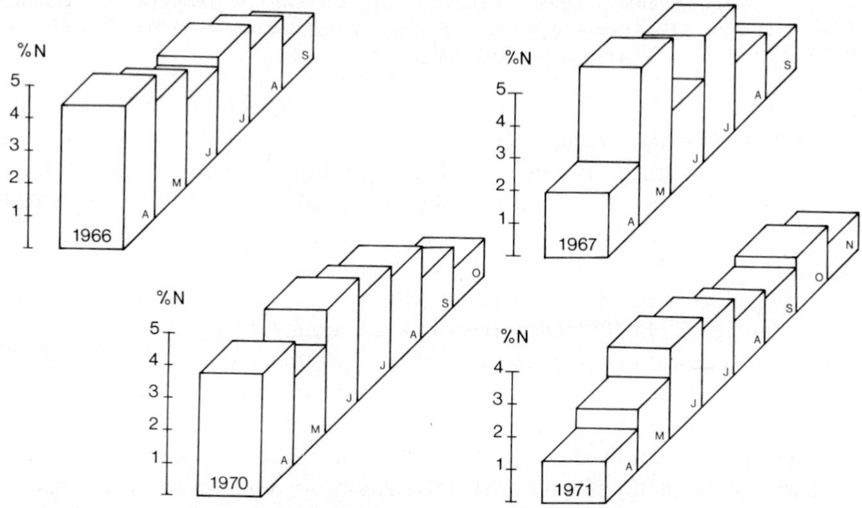

Fig. 20.7. Percentage light utilization (per cent N = per cent PhAR, monthly means) during the vegetation period

Radiation utilization was calculated for the standing crop (Westlake 1965, 1968) since the main part of the bound energy is stored in the shoots during the productive phase. According to Borutzki 1950 (quoted in Rodewald-Rudescu 1974) the loss of reed biomass during the flowering season amounts to only 2–3 per cent of the entire annual production (see Section 1). The energy content of the dry substance was measured in g.cal, the irradiation was calculated as PhAR (photosynthetic active radiation available for absorption, between 400 and 700 nm wavelength) taking 45 per cent global radiation as a long-term average (Monteith 1965, Ondok 1973).

The course of utilization during one period of vegetation gives an idea of the radiation energy utilized in photosynthesis and existing in the form of chemically combined energy in the dry substance. The different weather conditions in the years 1966–1971 are distinctly reflected in utilization values. Light climate and temperature conditions in the reed crop favour the utilization of radiation in the first half of the production period between April and June. The peak values in each are reached in May and June, when physiological activity in the reed crop is very high. Only in 1971 were the corresponding monthly values for utilization a little lower. In 1966 the highest utilization was registered in June (5 per cent PhAR). In October, when the last harvest of this production period is complete, the utilization sank to 1.3 per cent PhAR. It is remarkable that the curves for utilization and irradiation follow a similar course. The maximum irradiation and utilization are in June (250 cal cm^{-2} d^{-1} 5 per cent PhAR), a month of extremely high photosynthetic activity in the reed crop. The maximum energy content of the epigeous dry substance is also found in June (cp. Burian 1969). Most striking is the steep decline of the utilization rate in the second half of the production period. Crop dynamics also undoubtedly influence utilization: whereas during the period of stalk growth light conditions within the crop are still favourable, density and leaf area being low, this situation is completely changed from July onwards. With a maximum of between 90 and 110 stalks per m^2 soil and LAI values above 6 (Burian 1973), and a maximum average crop height of more than 250 cm from the ground to the tip of the inflorescence the light quality also alters. Although the PhAR in the crop is still relatively high (65–70 per cent) (Ondok 1973) at the beginning of the phase of stalk elongation it sinks to 15–26 per cent when LAI values are high in June and July. Towards late autumn it rises once more due to increasing leaf-fall. The 'thinning out' of the crop is only partly compensated by new leaf growth (Fig. 20.8).

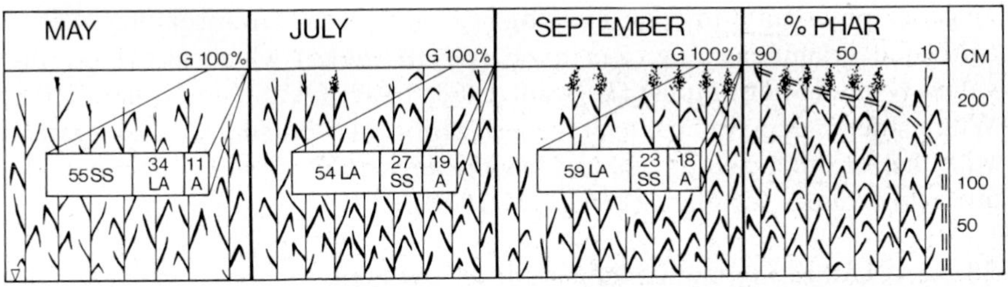

Fig. 20.8. Share of albedo (A), total absorption (LA), and mean light intensity (SS) at the zone of optimal production expressed as percentage of the total incoming radiation (G).

In April 1967 the initial value of utilization was very low, but it rose steadily until June when it reached 5 per cent PhAR as in 1966. The greatest intensity of irradiation was reached in mid July. In August the utilization, 1 per cent PhAR, remained well below that of the previous year, when the same value was not attained until 6 weeks later.

In 1970 utilization was at first relatively high in April about 4 per cent

PhAR), but then sank steadily until the beginning of June; the maximum of 4 per cent PhAR reached in mid June was below that of the years 1966 and 1967. In October 1 per cent recorded, which was approximately the same final value for radiation utilization as in the previous periods of production.

The situation in 1971 was similar: the monthly maximum was already reached in May with 3 per cent PhAR. The utilization rates then sank until the beginning of September, when a fresh increase (2 per cent) set in, the value in October being about 1 per cent PhAR. Despite stable average daily outputs of fixed organic substances this might as in 1970 have been caused by more intensive irradiation (about 32 per cent) especially in May and June, combined with low average daily temperature (see Section 3). This result, among others, seems to confirm Burian's finding that the temperature optimum for CO_2 absorption in the closed reed stand is distinctly lowered by unfavourable weather conditions (Chapter 2).

20.3.2 *Utilization over the course of 1972*

During the production period of *Phragmites*, utilization runs parallel to the amount of energy fixed in the dry substance, and to the increasing irradiation and development of the leaf area (Geisslhofer 1970, Sieghardt 1973). But the utilization value, which is always calculated at the beginning of the harvesting period in April, cannot directly reflect the actual photosynthetic activity of the reed. This 'initial value' includes about 1–2 per cent of additional energy reserves that partly belong to the previous production period. The maximum utilization coincides with the maximum of dry weight development (stalks and leaves) and more or less with that of global irradiation but it is reached before the maxima of the leaf area indices are attained. The final value for utilization amounts to about 1 per cent PhAR (October value).

In all cases 3 per cent PhAR was calculated as an average utilization for the epigeous phytomass for one growing season. This value resembles that for cultivated gramineae (e.g. *Zea mays*, *Oryza sativa*) which also have high values (Annual net output) (Hayashi 1968, 1969; Okubo and others 1968). Within limnic eco-systems average utilization of *Phragmites* is comparable with that of other helophytes (e.g. *Typha angustifolia, Scirpus lacustris*), but are below values for algae (Dykyjová 1971).

20.3.2.1 Utilization and development of leaf area

In April and May light conditions in the crop are much more favourable than in the following months and only improve when leaf fall sets in earnest. A rapidly increasing crop density during the growth phase of the reed stalks leads to a considerable increase in total leaf area and thus to a change in the light climate in the crop. As the stalks elongate the leaf area zone receiving an optimum light supply shifts vertically the middle and the lowest zones of leaf receive less intense light of a changed quality. In July and August the PhAR measures between 60 and 120 cm in a more or less dense crop (water reed) amounts to only 15–30 per cent of the PhAR measured above the crop

(Ondok 1973); higher values are reached again only after flowering when heavier leaf fall sets in. The productivity of this plant is influenced far more by light intensity and edaphic conditions than by general climatic conditions (Rudescu 1968).

The maximum LAI values are measured in the first half of July and lie between 5 and 9 according to the reed biotype (Geisslhofer and Burian 1970). The highest utilization values are reached some 4 or 5 weeks earlier, at a time when photosynthetic activity is generally high in the reed crop. Radiation utilization, however, seems to be optimum at leaf area indices between 3 and 4 (and in May–mid June) when the NAR and the LAR have already decreased by 66 per cent and 30 per cent respectively (Burian 1973).

20.3.2.2 Correlation between utilization and local factors

It has already been pointed out in Section 2.1 that the efficiency of primary production is strongly dependent on incoming radiation i.e. on the supply of radiation energy fit for photosynthesis.

Pechlaner (1971) found certain correlations between the utilization of the incoming radiation and the depth of a body of water; a higher efficiency of primary production is found at greater depths as compared to that of organisms in the uppermost strata. McColl and Cooper (1967) find that higher plants (young gramineae plants grown under controlled environmental conditions in a greenhouse) act in a similar way. Low light intensities (below 100 cal cm^{-2} d^{-1}) are better utilized whereas primary production is less efficient at higher light intensities (Fig. 20.9).

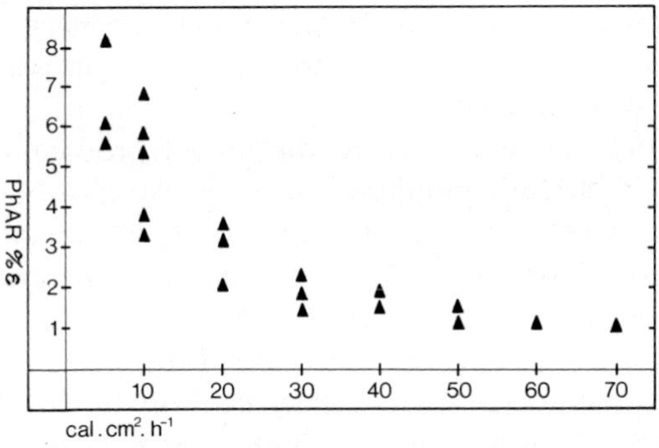

Fig. 20.9. Energy utilization (%ε) in relation to the incoming photosynthetic available radiation.

In the open, however, a good many uncontrollable ecological factors play a decisive role in the utilization of radiation. Besides the factor 'light', temperature, supply of mineral substances and their uptake (Hürlimann 1951, Haslam 1971, Dykyjová 1973) and the chemistry of the water and substrate (Rudescu and others 1965) are important for development and utilization. Fig. 20.9 shows a possible correlation between the utilization and intensity of irradiation. The better utilization of weaker light intensities found by McColl and Cooper seems also to hold for *Phragmites* in the open.

20.3.2.3 Energy storage in the individual organs

The change in energy content of the rhizomes indicates among other things their functions as carbohydrate reservoirs (see Section 2). In spring, when most of the mobilized energy reserves are used for building up the young shoot, the calorific value of the dry substance decreases (Fig. 20.6). In May, which is the most important month for productivity, the subterranean depots are temporarily refilled with translocated products of photosynthesis. In autumn the energy contents decreases again because energy is invested in the formation of the new autumn sprouts. Only relatively late in the year (November/December) are the energy reservoirs, the rhizomes, refilled. The energy reserves deposited in the subterranean organs permit a quick sprouting of the spring shoots when ecological conditions are favourable.

The energy content of the roots indicates the role that these organs play in absorbing mineral nutrients, a process which also demands energy, thus at least partly explaining the high calorific values (Zax 1973, Krejci 1974). The highest energy contents are measured in the first half of the year, when physiological activity in the aerial organs is high. According to Fiala (1973) these high calorific values are probably due to an intense root growth (Davidson 1969). Decreased energy content in the roots in summer might be due to the death of numerous roots, to the disintegration of the cortex parenchyma (Zax 1973), or simply to the cessation of stalk growth. The development of adventitious roots at a time when the stalks are elongating is directly connected with the absorption of mineral substances from the water (Rodewald-Rudescu 1974). Thus the percentage of energy rich lipo-protein complexes decreases; the energy content now sinks but rises again in winter (see also Albert & Kinzel 1973).

It is remarkable that even before the actual production period begins, energy reserves are already mobilized for intensive transportation into these sprouts, which have often not yet even reached the water surface. The young shoots seem to 'wait' for the onset of favourable local conditions in order to commence photosynthetic activity as quickly as possible.

The seasonal course of the calorific value of the stalks is similar to that of the dry weight. The increase of energy content in fall seems to have a functional significance: it may be due to the changed age composition of the crop – at this time young stalks occur in great numbers – but it also reflects the temporary storage function of submerged parts of the stalks. On an average the energy content of these organs lies only 1–2 per cent below that of emersed stalk portions at their summer maximum, but is 230–250 cal higher than the energy content of the autumn leaves.

The development of energy contents in the photosynthetic organs is the reverse of that in the stalks. Maximum values are registered at the onset of the production period. Evidently the young leaves contain organic compounds richer in energy than those in the other organs; the deposits of anorganic substances in the leaf blades are still negligible. Only later, as more and more anorganic substances are deposited, does the calorific value show a distinct decrease, reaching a minimum at the end of the production period (October). The coming of autumn is accompanied by a distinct rise in the energy content of the reproductive organs. This rise seems to parallel the ripening of seeds (Geisslhofer 1970).

The absolute energy content permits conclusions concerning the quality of the organic matter in the dry substance. The relatively low energy content of the rhizomes suggests that it is predominantly carbohydrate (cp. Kharasch 1929). The high values found in the leaves at the beginning of the production period indicate substances (lipo-protein complexes) containing more energy than simple carbohydrates. Almost the same is true of the roots (Tóth and Szabó 1958, Björk 1966, Simionescu and Rozmarin 1966 quoted in Rodewald-Rudescu 1974, Grabowski 1973). A vertical gradient in energy content is also observable: the upper zone with the greatest metabolic activity contains the most energy, whereas the middle part of the stalk contains distinctly less (cp. Tuschl 1970).

If the energy contents of the entire epigeous phytomass and the entire subterranean biomass are calculated at the time of maximum stalk dry weights (June–July) the ratio is approximately 1:1. This means that the reed plant seems to invest about the same quantity of chemically combined energy in the epigeous and the subterranean phytomass over a longer period of time (cp. Príbil and Dykyjová 1973).

20.3.4 *The utilization and binding of energy during one vegetation period*

About 3 per cent of the energy radiation entering during one single period of production (April to October, 356,000 kcal m^{-2} PhAR) is utilized by the reed crop. In other words: 5,000–6,000 kcal m^{-2} and year are deposited in aerial organs, i.e. the same or twice as much as in the underground parts of the shoots. Therefore one hectare of water reed app. contains 150.10^6–180.10^6 kcal of chemically combined energy. A substantial portion of the dry substance existing in the crop enters the cycle of decomposition as crop detritus. According to careful estimates *Phragmites* alone annually supplies organic detritus equivalent to about 15–25 tons of dry substance per hectare (Burian quoted in Weisser 1970). The energy content of the organic detritus consisting chiefly of leaves, parts of stalks and rhizome fragments of Phragmites, amounts on an average to 3,737 cal g^{-1}, which is about 1,000 cal lower than that of living epigeous organs. For comparison the calorific contents of strongly decomposed material of anaerobic lake mud lies between 5,000 and 5,200 cal g^{-1} (Gorham and Sanger 1967).

20.3.5 *Caloric content of Utricularia vulgaris L.*

The quantity of energy chemically combined in the dry substance of *Utricularia* depends not only on the season but also on habitat conditions. Plants growing in reedless lacunae contain approximately 300 cal g^{-1} more energy than those growing in the reed belt between old and new reed stalks (Sieghardt 1973).

The absolute energy content lies distinctly below that of *Phragmites*. Evidently *Utricularia* does not build up those energy-rich organic compounds that are responsible for higher calorific values e.g. as in *Phragmites communis* or in *Typha angustifolia* (cp. Grabowski 1973), (see Chapter 21).

A comparison of the energy bound in the dry substance distinctly shows the differences in production of *Utricularia*: plants in more or less dense, uncut reeds contain 36.8 kcal m^{-2}, whereas plants in cleared areas contain 113.0 kcal m^{-2}; both values were measured in the second half of June when biomass is at its maximum. Evidently a much larger amount of energy is invested in plants growing in reedless, unshaded areas of in smaller pools or channels (the ratio is about 3:1).

The calorific values of *Typha angustifolia* given in Table 20.1 are winter values for the rhizomes and roots (February 1971). A comparison with *Phragmites communis* shows a significant difference in calorific values: the energy content of winter rhizomes of *Phragmites* is 2–7 per cent higher than that of *Typhia angustifolia* (cp. 3.3).

20.3.5.1 Caloric content of the turions of *Utricularia vulgaris* L.

A special characteristic of the genus *Utricularia* is its ability to develop turions under certain ecological circumstances. These are vegetative stages, whose formation is not solely dependent upon the drop in water temperatures in autumn as Maier proved in 1973. Turions have the capacity of storing substantial energy reserves in the form of carbohydrates (such as sugar/starch) in their leaves and axes. In addition they possess a number of remarkable eco-physiological qualities: large quantities of mucilage, a low relative water content and a lower water emission compared to that of the fully developed plant (water loss in per cent of saturation):

	October	February
Average d.w. (in gram)	0.1869 ± 0.03	0.1234 ± 0.02
Energy content (Kcal g^{-1})	4.698 ± 15	4.719 ± 5
Ash content (per cent)	2.2	2.7

(Table according to Maier 1973)

From this, it appears that the energy content of October turions is about 300 cal higher (reference value is an average weight of turions in grams) than that of February turions, although the calorific value per gram of dry substance does not show any significant difference. Apparently no important qualitative changes occur in the phytomass of newly developed turions or those just about to sprout. The energy loss caused by respiration might well amount to about 300 cal as Maier points out, which would explain the relative loss of weight (approximately 34 per cent) of the February turions.

20.4 Production conditions of *Carex riparia* Curt.

The primary production of *C. riparia* was studied by Maier (1976) along a transect[1] (total length 250 m) from the landward marginal zone of the continuous reed belt as far as reedless pools near the lake. Six zones can be discerned, in which production conditions become increasingly unfavourable for *Carex* due to natural competition.

Zone I: a relatively dry meadow belt extending along the landward margin of the reed belt, at the same time limiting landward expansion of *C. riparia.*

Zone II: about 16 m west of zone I, towards the lake. A stagnant transitional area, which was not flooded in 1972. *C. riparia* grows densely, but the shoots and the individual plants are small compared to these in zone III–IV. A rich spectrum of species is characteristic of this area.

Zone III: is situated 90 m from zone I; an occasionally periodically flooded location with a dense population of *C. riparia* and *Phragmites* accompanied by submerged *Utricularia vulgaris.*

Zone IV: approximately 150 m lakeward of zone I: like zone III a periodically flooded location with *C. riparia, Ph. communis* and *U. vulgaris* as submergents. However, the number of individual plants of *C. riparia* per surface unit lies below that of zone II.

Zone V: about 220 m from zone I in the middle of dense of reed accompanied by crop of *C. riparia* and *U. vulgaris.*

Zone VI: About 250 m of zone I; a permanently flooded location, characterized by a thin reed growth, giving way to a reedless pool. A relatively dense growth of *U. vulgaris* is characteristic of this zone.

20.4.1 *The epigeous phytomass*

The maximum epigeous biomass (= standing crop) is attained in zone IV, the minimum in zone VI. The largest dry weight is measured in zone III at the end of June (29 June): crop value, calculated for a standard quadrat. Development of new *Carex* shoots in autumn seem to account for the large dry weight/m^2 in October in zone III–V (average).

The highest crop-growth rate (CGR) is found between 2 May and 29 June 1972, i.e. clearly in the first half of the production period of *C. riparia.* Only in zone VI, which can be regarded as a transition towards an open pool with sparse reed, does the CGR-maximum decrease after 29 June.

20.4.2 *Subterranean phytomass*

The ratio (quotient) – epigeous: subterranean biomass (as dry weight) is 0.14 (spring value) in a stagnant location (corresponds to zone II – wet transitional

1. In the deposition area of Neusiedlersee near Rust/See.

area); similar values can be found in a flooded sedge stand (comparable to zone IV) where the epigeous biomass amounts to about 11 per cent and the subterranean biomass to about 89 per cent (quotient = 0.12).

A quotient of 2.31 (epigeous: subterranean biomass) indicates a reversal of the situation in the flooded area in summer favouring the epigeous biomass (29 June 1972): 70 per cent of the dry weight are due to the epigeous phytomass, 30 per cent to the subterranean phytomass of *Carex riparia.*

References

Albert, R. & Kinzel, H. 1973. Unterscheidung von Physiotypen bei Halophyten des Neusiedlerseegebietes (Österreich).– Z. Pflanzenphysiol. 70: 138–157.

Birge, E. A. & Juday, C. 1922. The inland lakes of Wisconsin. The plankton I. Its quality and chemical composition.– Bull. Wisconsin geol. mat. hist. Survey 64: 1–222.

Björk, S. 1966. Report on ecologic investigations of *Phragmites communis* within the region around the lower parts of the Euphrates and Tigris.– Memograph. Limnol., Inst. Lund 130–140.

Boyd, E. C. 1970. Amino acid, protein and caloric content of vascular aquatic macrophytes.– Ecology 51: 902–906.

Boysen-Jensen, P. 1932. Die Stoffproduktion der Pflanzen.– G. Fischer Verlag, Jena.

Bray, J. R. 1962. Estimates of energy budgets for a *Typha* (cattail) marsh.–Science 136: 1119–1120.

Burian, K. 1969. Die photosynthetische Aktivität eines *Phragmites communis*-Bestandes am Neusiedler See (URAS-Messungen während der Vegetationsperiode 1967).– Sitz. Ber. Österr. Akad. Wiss., math.– nat. Kl. I, 178: 43–62.

Burian, K. 1973. *Phragmites communis* Trin. im Röhrict des Neusiedler Sees. Wachstum, Produktion und Wasserverbrauch.– In: Ellenberg (ed.) Ökosystemforschung, Springer Verlag, Berlin, Heidelberg, New York 61–78.

Csapody, I. 1965. Die Vegetation des Neusiedlersees und seiner Umgebung.– Wiss. Arb. Bgld. 32: 42–57.

Davidson, R. L. 1969. Effects of edaphic factors on the soluble carbohydrate content of *Lolium perenne* L. *Trifolium repens* L.–Ann. Bot. 33: 579–589.

Dykyjová, D. 1971. Productivity and solar energy conversion in reed-swamp stands in comparison with outdoor mass culture of algae in the temperate climate of Central Europe.– Photosynthetica 5, 4, 329–340.

Dykyjová, D. 1973. Content of mineral macronutrients in emergent macrophytes during their seasonal growth and decomposition.–In: Hejný (ed.). Ecosystem study on wetland biome in Czechoslovakia. IBT-PT-PP Report 3: 163–172.

Farahat, A. Z. & Nopp, H. 1966. Über die Bodenatmung im Schilfgürtel des Neusiedler Sees.– Sitz. Ber. Österr. Akad. Wiss., math.– nat. Kl. I, 75: 237–255.

Farahat, A. Z. & Nopp, H. 1967. Jahreszeitliche Änderungen der Bodenatmung im Schilfgürtel des Neusiedler Sees.– Anz. math.– nat. Kl. Österr. Akad. Wiss. 9: 220–223.

Fiala, K. 1973. Underground biomass and estimation of annual rhizome increments in two polycormones. Littoral of the Nesyt fishpond-Ecological Studies.– Ed. by Kvet, J., Studie CSAV 15, Academia 83–88.

Geisslhofer, M. 1970. Biometrische Untersuchungen an *Phragmites communis* im Verlauf der Produktionsperiode. (Laborversuche an Keimpflanzen und Untersuchungen im Schilfbestand des Neusiedlersees).– Diss. Univ. Wien.

Geisslhofer, M. & Burian, K. 1970. Biometrische Untersuchungen im geschlossenen Schilfbestand des Neusiedler Sees.–OIKOS 21: 248–254.

Gorham, E. & Sanger, J. 1967. Caloric values of organic matter in woodland, swamp and lake soils.– Ecology 48: 492–494.

Gortner, R. A. 1934. Lake vegetation as a possible source of forage.– Science 80: 531–533.

Grabowski, A. 1973. The biomass, organic matter contents and calorific values of macrophytes in the lakes of the Szeszupa drainage area.– Pol. Arch. Hydrobiol. 20: 269–282.

Haslam, S. 1971. The development and establishment of young plants of *Phragmites communis* Trin.– Ann. Bot. 35: 1059–1072.

Hayashi, K. 1968. Response of net assimilation rate to differing intensity of sunlight in rice varieties.– Proc. Crop. Sci. Soc. Japan 37: 528–533.

Hayashi, K. 1969. Efficiencies of solar energy conversion and relating characteristics in rice varieties.– Proc. Crop. Sci. Soc. Japan 38: 495–500.

Hübl, E. 1966. Stoffproduktion von *Phragmites communis* Trin. im. Schilfgürtel des Neusiedlersee im Jhre 1966 (Ergebnisse nach der Erntemethode).– Sitz. Ber. Österr. Akad. Wiss., math.– nat. Kl. I, 14: 271–278.

Hürlimann, H. 1951. Zur Lebensgeschichte des Schilfs an den Ufern der Schweizer Seen.– Geobot. Landesaufnahme d. Schweiz 30: 1–232. Verlag H. Huber Bern (84–120).

Imhof, G. & Burian, K. 1972. Energy-flow studies in a wetland ecosystem. (Reed belt of the lake Neusiedler See).– Österr. Akad. Wiss. in com. Springer Verlag Wien/New York, 1–15.

Kharasch, M. S. 1929. Heats of combustion of organic compounds.– Bur. Stands. J. Res. 2: 359–430.

Koch, W. 1964. Die Vegetationseinheiten der Linthebene.– Jahrb. St. Gallischen Naturwiss. Ges. 61: 1–146.

Kopf, F. 1964. Die wahren Ausmaße des Neusiedlersees.– Österr. Wasserwirtschaft 16: 11/12, 255–262.

Kopf, F. 1966. Der Neusiedlersee vor hundert Jahren.– Bgld. Heimatblätter 28, 2: 65–70.

Kopf, F. 1967. Die Rettung des Neusiedlersees.– Österr. Wasserwirtschaft 19, 7–8: 139–151.

Kopf, F. 1968. Der Schilffortschritt im Neusiedlersee.– Techn. Ber. Wien.

Krejci, G. 1974. Jahresperiodische Stoffwechselschwankungen in *Phragmites communis*.–Diss. Univ. Wien.

Larcher, W. 1976. Ökologie der Pflanzen.– 2. verb. Aufl. Eugen Ulmer, Stuttgart.

Lautner, V. & Müller, Z. 1954. Die Futterwerte einiger unserer Wasserpflanzen I.– Sbor. Ceskoslov. Akad. Zemedel. Ved. 27: 333–354.

Maier, R. 1973. Produktions- und Pigmentanalysen an *Utricularia vulgaris* L. In: Ellenberg (ed.) Ökosystemforschung, Springer Verlag Berlin, Heidelberg, New York, 87–101.

Maier, R. 1973. Das Austreiben der Turionen von *Utricularia vulgaris* L. Nach verschieden langen Perioden der Austrocknung.– Flora 162: 269–283.

Maier, R. 1976. Untersuchungen zur Primärproduktion im Grüngürtel des Neusiedlersees. Teil 1: *Carex riparia* Curt.–Pol. Arch. Hydrobiol. 23: 377–390.

McColl, D. & Cooper, J. P. 1967. Climatic variation in forage grasses III. Seasonal changes in growth and assimilation in climatic races of *Lolium*, *Dactylis* and *Festuca*.–J. appl. Ecol. 4: 113–127.

Monteith, J. L. 1965. Light distribution and photosynthesis in field crops.– Ann. Bot. N. S. 29: 17–37.

Okubo, T., Oizumi, K. & Hoshino, M. 1968. An observation on solar energy conversion in primary canopies of forage crops.– In: Photosynthesis and utilization of solar energy. Level III Experiments. JIBP/PP-Photosynthesis level III Group, Tokyo.

Ondok, J. P. 1973. Photosynthetically active radiation in a stand of *Phragmites communis* Trin., II. Model of light extinction in the stand. – Photosynthetica 7: 50–57.

Pechlaner, R. 1971. Factors that control production rates and biomass of phytoplankton in high mountain lakes.– Mitt. Intern. Ver. Limnol. 19: 125–145.

Pribil, S. & Dykyjová, D. 1973. Seasonal differences in caloric contents of some emergent macrophytes (A preliminary report). Ecosystem studies on wetland Biome in Czechoslovakia.– Czechosl. IBP/PT–PP Report 3: 97–99.

Riedmüller, G. 1965. Der Schilfgürtel des öster. Anteils des Neusiedler Sees 1938–1958.– Wiss. Arb. Bgld. 32: 58–59.

Rodewald-Rudescu, L. 1974. Das Schilfrohr, *Phragmites communis* Trinius.–Schweizerbartsche Verlagsbuchhandlung Stuttgart, 302 pp.

Rudescu, L. 1968. Der Einfluß der Umweltbedingungen auf die Produktivität (Photosyntheseintensität) des Schilfrohres (*Phragmites communis* Trin.) aus dem Donau-, Dnjeper- und

Wolgadelta, aus Mesopotamien, Ost- und Westpakistan und dem Nildelta.–Hidrobiologia 10: 77–87.
Rudescu, L., Niculescu, C. & Chivu, P. I. 1965. Monographie des Schilfrohres.– Edit. Acad. R.S.R. (Bukarest) 1–542.
Seidel, K. 1956. *Scirpus lacustris* im eutrophen See.–Z. Fisch. 5: 553–567.
Sieghardt, H. 1973. Strahlungsnutzung von *Phragmites communis*.–In: Ellenberg (ed.) Ökosystemforschung, Springer Verlag, Berlin, Heidelberg, New York 79–86.
Sieghardt, H. 1973. Utilization of solar energy and energy content of different organs of *Phragmites communis*. Trin.–Pol. Arch. Hydrobiol 20: 151–156.
Straškraba, M. 1968. Der Anteil der höheren Pflanzen an der Produktion der stehenden Gewässer.– Mitt. Internat. Verein. Limnol. 14: 212–230.
Szczepanski, A. 1969. Biomass of underground parts of the reed *Phragmites communis* Trin.– Bull. Akad. Polon. Sci. 17: 245–247.
Tòth, L. 1960. Phytozönologische Untersuchungen über die Röhrichte des Balaton-Sees.– Annal. Biol. Tihany 27: 209–242.
Tòth, L. & Szabò, E. 1958. Über die chemische Zusammensetzung verschiedener Schilfproben vom Balatonsee.– Ann. Biol. Tihany 25: 363–374.
Tòth, L. & Szabò, E. 1961. Zönologische und ökologische Untersuchungen in den Röhrichten des Neusiedlersees (Fertö-to).– Annal. Biol. Tihany 28: 151–168.
Tuschl, P. 1970. Transpiration von *Phragmites communis* Trin. im geschlossenen Bestand des Neusiedler Sees.– Wiss. Arb. Bgld 44: 126–186.
Weisser, P. 1970. Die Vegetationsverhältnisse des Neusiedler Sees. Pflanzenphysiologische und ökologische Studien.– Wiss. Arb. Bgld. 44: 126–186.
Weisser, P. 1973. Die Verschilfung des Neusiedlersees.– Umschau 73: 440–441.
Wendelberger, G. 1941. Die Vegetation der Salzlacken des Neusiedlersees.–Diss. Univ. Wien 216 pp.
Wendelberger, G. 1964. Vom Schilfröhricht pannonischer Steppenseen.–Natur u. Land 50, 3: 53–55.
Went, F. 1973. The competition among plants.– Proc. Nat. Acad. Sci. USA 70: 585–590.
Westlake, D. F. 1965. Some basic data for investigations of the productivity of aquatic macrophytes.– Mem. Ist. Ital. Idrobiol. 18: 229–248.
Westlake, D. F. 1968. Methods used to determine the annual production of reed-swamp plants with extensive rhizomes.– Int. Symp. USSR: Methods of productivity studies in root systems and rhizosphere organism Leningrad, Nauka.
Zax, M. 1973. Die Temperaturresistenz von *Phragmites communis* Trin.–Pol. Arch. Hydrobiol. 20: 159–164.

1 Production of *Utricularia vulgaris* L.

R. Maier

As a rootless plant the submerged carnivorous macrophyte *Utricularia vulgaris* is confined to calm water where it is safe from wave action. As a consequence it is invariably found in conjunction with reed, the only exception being the inshore and lakeward limits of the reed belt.

Utricularia thus inhabits an environment of striking uniformity, known as *Scirpo-Phragmitetum utricularietosum* in the plant sociological system (Weisser 1970). Neverthless a division of this apparently uniform plant society is possible on the basis of production. In view of the intimate connections existing between *Utricularia* and *Phragmites*, changes in the reed population can be considered as being equivalent to alterations in *Utricularia* density.

The following deserve mention as representing distinct sub-units within the uniform reed belt:

(1) More or less natural reed stands where human influence can be considered to have played a minimum role.
(2) Stands with sparse reed growth or small reedless areas within the reed belt itself.
(3) Reed stands near the shore with an undergrowth of large sedges.
(4) Reed stands in highly eutrophic water.
(5) Reed stands of widely varying form, usually harvested in winter for commercial purposes.

At the same time as providing *Utricularia* with a suitable environment *Phragmites* plays a decisive role in its development cycle. The appearance and subsequent development of the leaves of the *Phragmites* plants brings about a drastic alteration in the light climate within the stand. Two biotopes colonized by *Utricularia* provide us with extreme examples with respect to light climate. These are 1. the lacunae within the reed belt. Reed growth in these spots is either sparse or entirely lacking. In some cases the total global irradiation reaches the water surface (*Scirpo-Phragmitetum utriculariosum*, Weisser 1970). 2. The reed close to the shore, where shade is provided by *Phragmites* and *Carex riparia* or *Carex acutiformis* (*Caricetum acutiformis-ripariae,* Weisser 1970). At the height of leaf development of reed and sedges the global irradiation is qualitatively and quantitatively altered by *Phragmites* and only 3 per cent reach the water's surface. Reed harvesting in winter, which of course also involves *Carex*, ensures more favourable light conditions, as is shown in Table 21.1, by removing the shoots of the previous year.

Table 21.1. Relative light availability in *Utricularia* habitats of the various types of reed society

Plant society	rel. LA
Scirpo-Phragmitetum utriculariosum	100 per cent
Scirpo-Phragmitetum utricularietosum commercially exploited	11.1 per cent
Scirpo-Phragmitetum utricularietosum natural (not cut)	8.3 per cent
Caricetum acutiformis-ripariae commercially exploited	6.7 per cent
Caricetum acutiformis-ripariae natural	2.3–2.9 per cent

21.1 Production of *Utricularia* in the reedless areas

The lacunae which may be either of natural or anthropogenic origin (see Grünig 1975) support a plant society characterized as *Scirpo-Phragmitetum utriculariosum*, and are the only places in Neusiedlersee where *Utricularia* is found in reedless water. Lacunae habitats illustrate the fact that it is not temperature or light factors that bind *Utricularia* to the reed stands but mainly the necessity to avoid the drifting action of the water or the wave action in the open lake. Small lacunae, only a few square metres in area, are usually very densely populated by *Utricularia*, whereas in the larger reedless areas the population density is as a rule low and the individual plants are pushed towards the edges by the wind.

A comparison of the production of plants growing in habitats receiving all the global irradiation (lacunae) with that of plants in the natural (uncut) reed stands (Table 21.1) reveals that the former is considerably larger (Fig. 21.1). The energy fixed per gram dry weight in the plants of habitats receiving ample light is about 300 cal higher than that of plants growing in the reed itself (Imhof u. Burian 1972). Since the relatively low population density in both lacunae and natural reed stands is to some extent comparable (Fig. 21.1), it can be concluded that the conditions in the lacunae are more favourable to *Utricularia* production than those prevailing in the reed.

Where the reed is cut in winter the production of *Utricularia* rises and the phytomass curve resembles that of the reedless habitats (Fig. 21.1). If the dry weights of all three habitats are converted into relative values it becomes clear that the plants in the lacunae are the most productive up to the middle of July, whereas later in the year those growing in reed stands that are regularly cut are the most productive (Maier 1973b). This is due to the fact that the basal shoots of the plants in the lacunae die off much sooner than those in the reed (Maier 1973a). The peculiarity of *Utricularia* is that the basal end of the shoot dies off and growth is continued at the shoot tips.

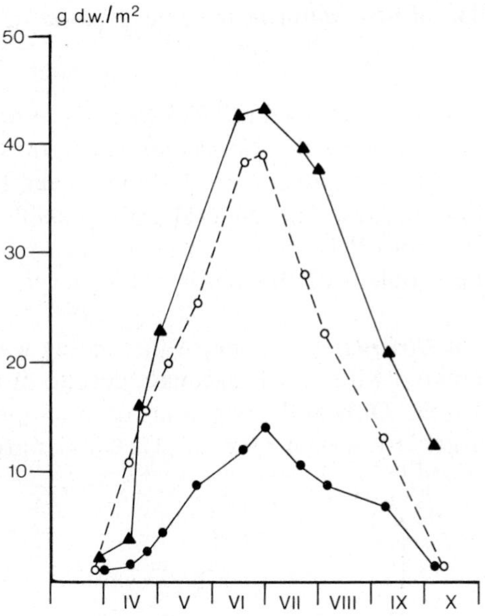

Fig. 21.1. Phytomass of *Utricularia* in 1971 in reedless areas (○ - - - - ○) and in uncut (● —— ●) and cut (▲ —— ▲) reed stands.

A plant that is equally capable of growth in poorly illuminated reed habitats and in the completely exposed lacunae would be expected to exhibit striking differences in pigment content. In fact lacunae plants do contain conspicuous amounts of anthocyan. Their chlorophyll content decreases over the course of the production period, as the incoming radiation increases, but that of the shaded plants rises (Fig. 21.2, see also Maier 1973a).

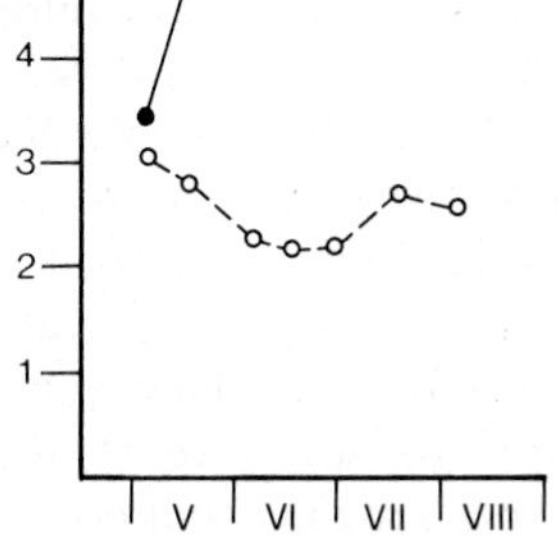

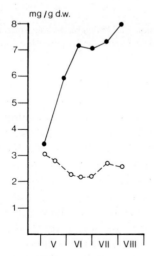

Fig. 21.2. Chlorophyll content of *Utricularia* in reedless lacunae (○ - - - - ○) and in the undergrowth of uncut reed stands (●——●).

21.2 The production of *Utricularia* in the zone exposed to fluctuating water conditions

As a result of the rise in water level in 1975 *Utricularia* was able to establish itself in a new plant society, the *Caricetum acutiformis-ripariae* (Weisser 1970). This is a society comprising reed stands, near the shore, with an undergrowth of large sedges. The undergrowth of reeds and sedges offers *Utricularia* extremely poor light conditions (Table 21.1), in addition to which it is faced with the problem of fluctuating water levels (see Maier 1973a, 1976).

The phytomass of *Utricularia* in zone of fluctuating water conditions was lower in 1971 compared with that in lacunae, despite of the relatively poor density in reedless areas. Only at the beginning of development do the former habitats yield a larger dry weight (Fig. 21.3). Good starting conditions for

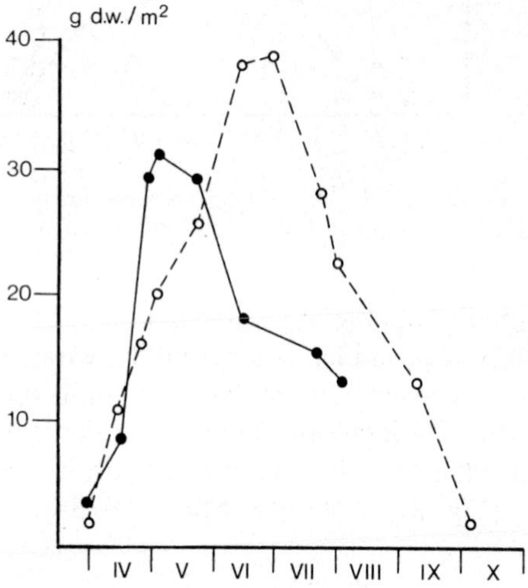

Fig. 21.3. Dry weight production of *Utricularia* in the *Caricetum acutiformis-ripariae* (uncut stands) (●——●) and in the *Scirpo-Phragmitetum utriculariosum* (lacunae) (○ - - - - ○), in 1971.

shoot growth are apparently provided by the higher population densities, the more rapid warming up of water near the shore and the still relatively good light conditions before the leaves begin to develop on the reed. Later in the year when both old and new shoots of reed and sedge overshadow *Utricularia*, the phytomass of the latter rapidly declines (Fig. 21.3). The lack of light expresses itself in a sharp rise in chlorophyll content (Fig. 21.2); anthocyanisation is either very weak or completely absent.

As mentioned above the production of *Utricularia* in the reed stands near the shore is determined not only by the lack of light but also by the periodic fluctuations in water level. In some years the habitat even dries out completely. The unfavourable growth conditions in the face of competition from reed and sedges for the available light lead to an almost complete reduction of the trapping bladders and to premature development of resting stages. In these habitats turions form as early as July, whereas in the lacunae their production is delayed until October (Maier 1973a).

Premature turion formation is one of the strategies by means of which *Utricularia* is able to colonize regions which are sometimes wet and sometimes dry. Whereas turions can survive in habitats without water (Maier 1973c) mature plants would die off. Part of the annual increase or losses of *Utricularia* are probably connected with inward or outward drift of turions. It can be assumed, however, that only as a result of the rise in water level in 1965, perhaps due to large-scale reed cutting, could a massive influx of turions into the shore zone have taken place. But under the conditions prevailing since 1971 this factor is probably of minor significance. In the region investigated the transport of turions is hindered by the shallowness of the water and the local reed cuttings. Again, although the spherical turions are easily dispersed on the water they are in some cases still attached to the dead shoot. They can only be wrenched away from an anchorage (e.g. from the shoots of Phragmites) by rougher water movements. Since only part of the region flooded at spring high water is colonized by *Utricularia* it appears that no large-scale transport of turions occurs. Water drift could carry turions up to the shore line in spring, but whatever the explanation, it is clear that *Utricularia* is gradually encroaching upon the shore, although on the whole the numbers of *Utricularia* in the zone with fluctuating water levels have been receding in recent years.

Fig. 21.4 shows the phytomass of *Utricularia* at various distances from the shore near Rust in a reed stand interspersed with sedges and cut in winter.

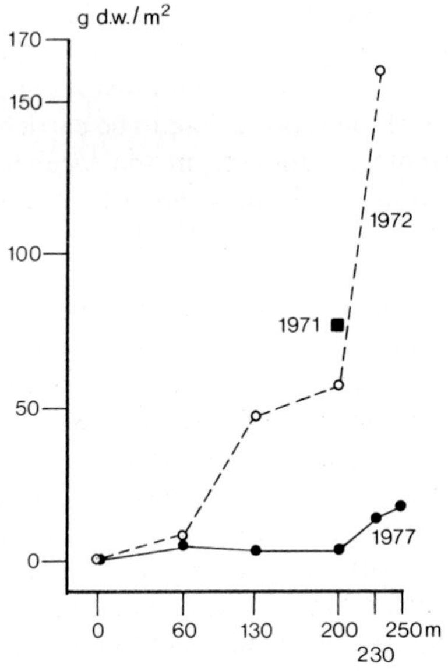

Fig. 21.4. Dry weight development of *Utricularia* at various distances from the shore at the time of maximum dry weight production (end of June/beginning of July) in 1971, 1972 and 1977 (means of four quadrats).

The production for 1972 can be described as follows: Where fluctuations in the water condition still exert an effect, the density of *Utricularia* is relatively low and the phytomass correspondingly small (about 60 m from the shore line at high water level). Beyond this, the phytomass rises and a fairly uniform population of *Utricularia* develops. At about 230 m from the shore the phytomass has increased by about threefold. This might be connected with the greater water depth although the light climate is probably the decisive factor. Only isolated specimens of *Carex riparia* are found at this point, and the reed stand has thinned out considerably since the habitat is in a state of transition to a lacunae.

A comparison of the phytomass in 1977 with that of 1972 for one and the same habitat (Fig. 21.4) reveals a considerable decline in the *Utricularia* populations. The decline cannot solely be attributed to the unfavourable conditions prevailing in alternately wet and dry habitats, since the habitats most affected are those that are constantly flooded. The habitat furthest inshore where conditions can be said to be extreme, because desiccation can exert its maximum effect, shows the least relative decline in numbers. In fact a comparison of the situations in 1972 and 1977 reveals a progression towards the shore in this habitat.

As already mentioned, the high production of the plants in 1971 finds its most likely explanation in a unique large-scale influx of turions. The decline in *Utricularia* density from 1971 onwards can be regarded as the consequence of the less favourable conditions in a *Caricetum* with reed as compared with a pure *Phragmitetum*. This assumption, however, would only apply to stands near the shore, in which the density of *Carex* is high (Fig. 21.4, 70–200 m) and on no account to the almost, and by 1977 completely, sedgeless habitats (constant flooding) with little reed near the lacunae (Fig. 21.4, 230 m). Loss of turions by outward drift might be the explanation for the population decline at such sites. The transect chosen for investigation is in a region where the reed is regularly cut. If the turions were to be carried out towards the open lake in spring they would accumulate in the lacunae bordering the areas where reed is cut. Beyond the lacunae no reed is cut. However, the 1977 production values for the lacunae (Fig. 21.4) indicate that this explanation is improbable.

The decline of *Utricularia* in the zone affected by the fluctuating water levels cannot be explained satisfactorily on the basis of light conditions either. The considerable reduction in light due to sedges and reeds and the consequently poor development of *Utricularia* (Fig. 21.3) over a number of years would be sufficient explanation for the collapse of the *Utricularia* population. In 1973 it was already apparent that *Utricularia* was suffering a decline, but this was thought to be the result of lack of reed cutting in 1972/73. However, the dry weights for 1977 show that the *Utricularia* population has not recovered in spite of reed cutting in the intervening years. Cutting the reed improves the light conditions and in this way provides *Utricularia* with a better chance to develop, as is evidenced by the progression of the plants towards the shore. Oddly enough, the decline in *Utricularia* is most conspicuous in habitats with the most light (Fig. 21.4, 230 m).

References

Grünig, A. 1975. Lochbildungen im Röhricht (dargestellt anhand des Rückganges des Schilfbestandes in Altenrhein (Bodensee) von 1926–1974).–Diplomarbeit Eidg. Techn. Hochsch., Geobot. Inst. Stftg. Rübel, Zürich, 66 pp.

Imhof, G. und Burian, K. 1972. Energy flow studies in a wetland ecosystem.– Sonderdr. Österr. Akad. Wiss, Wien, 15 pp.

Maier, R. 1973a. Produktions- und Pigmentanalysen an *Utricularia vulgaris* L.–In: Ökosystemforschung, ed. H. Ellenberg 87–101, Springer, Heidelberg.

Maier, R. 1973b. Aspects of production of *Utricularia vulgaris* L. in some vegetation types in the reed belt of lake Neusiedlersee.– Pol. Arch. Hydrobiol. 20/1: 169–174.

Maier, R. 1973c. Das Austreiben der Turionen von *Utricularia vulgaris* L. nach verschieden langen Perioden der Austrocknung.– Flora 162: 269–283.

Maier, R. 1976. Primärproduktionsuntersuchungen im ufernahen Schilfgürtel des Neusiedlersees.– Biol. Forschungsinst. Bgld. 13: 43–54.

Weisser, P. 1970. Die Vegetationsverhältnisse des Neusiedler Sees, pflanzensoziologische und ökologische Studien.– Wiss. Arb. Bgld. 45: 1–83.

2 The zooplankton of the open lake

A. Herzig

22.1 Introduction

About 80 years have passed since the first investigations on the zooplankton of Neusiedlersee were published. Daday (1890, 1897) was one of the first scientists to describe its rotifers and crustaceans. In a paper published in 1926 Varga described the rotifer fauna of Neusiedlersee and mentions that when he started his investigations (16 August 1918) very little was known about the 'Niedere Fauna des Wassers'. In 1929 Varga described a new, endemic rotifer species – *Rhinops fertöensis* (= synonym of *Rhinoglena fertöensis*). Later he presented some additional information about the rotifer plankton (Varga 1934), and in addition he discussed limnological aspects of Neusiedlersee and considered the influence of freezing to the bottom and of extreme wind situations on the different communities of the lake (Varga 1932, Varga & Mika 1937).

Some information on the zooplankton can be found in Geyer & Mann (1939) and Benda (1950), whilst the work of Pesta (1954) established the importance of planktonic crustaceans in Neusiedlersee.

The first paper dealing with quantitative aspects of the rotifer and crustacean plankton did not appear until 1961, and reported the investigations which the author (Zakovsek) had carried out in the years 1950/51/52. Although of interest as the first study of its kind on this particular subject, the inadequacies of sampling strategy and technique available at that time detract from the value of the quantitative data. But nevertheless it was the first description of the phenology of the plankton organisms and covered more than one year; the quantitative results can only be used by taking the seasonal means.

Samples from Neusiedlersee were analysed by Ruttner in 1956/57, 1958 (unpublished results); a species list of rotifers and crustaceans occurring in the plankton is given in Ruttner-Kolisko & Ruttner (1959).

Between 1968 and 1974 the zooplankton of the open water zone was the object of an IBP-study (IBP-PF – International Biological Program-Production of Freshwater Ecosystems) (Herzig 1973, 1974, 1975). Since 1974 the qualitative and quantitative changes in the zooplankton community occurring in recent years have been studied in a Man and Biosphere (MaB) project (Herzig 1977). Important information concerning the rotifer and crustacean communities of the reed belt can be found in papers by Donner (1979, Chapter 27), Löffler (1979, Chapter 26) and Ponyi & Dévai (1979).

Before embarking upon a detailed description of the zooplankton community a few words have to be said about the problems connected with attempts to quantify the plankton of such a large lake. Cassie (in Edmondson & Winberg 1971) has discussed logistic problems arising out of the need for representative sampling aimed at a population estimate. But very often limitations in man-power, time and money influence sampling strategy. The choice of the routine collecting tool is influenced by various practical considerations apart from cost and efficiency (Bottrell et al. 1976).

Frequently numerous horizontal and vertical stations have to be sampled to obtain a reasonable estimate of the abundance of the zooplankton. In the first years of the IBP project on Neusiedlersee (1968, 1969) samples were taken at 8–16 points and at three different depths (collecting gear: pump and bottle). But in shallow, large lakes with poor vertical variation, as in Neusiedlersee, integrated samples towed horizontally (using an Isaacs-Kidd high speed sampler) provide the most reliable population estimate (Herzig 1974, Bottrell et al. 1976). This sampling strategy was used from 1970 to 1974; 4–8 profiles with a tow length of an average of 1,000 metres were sampled at each date. Since 1975 point samples have again been taken, but the samples are vertically integrated (net tow), the aim being to show the horizontal distribution pattern of the various plankton animals and their developmental stages.

To determine the errors involved in the calculated mean population estimate the 95 per cent confidence limits were fitted to this mean for the whole water body. All calculations were performed with transformed values following the recommendations given in statistical handbooks (e.g. Sokal & Rohlf 1969, Elliott 1971) and in handbooks or reviews of zooplankton studies (e.g. Edmondson & Winberg 1971, Bottrell et al. 1976).

Table 22.1 shows some examples of the mean density obtained by different methods and the corresponding 95 per cent confidence limits. As can be seen the confidence limits are often less than 10 per cent of the transformed mean, which is in close agreement with the results obtained for other water bodies (Bottrell et al. 1976). Fig. 22.1 shows the difference in confidence limits between sampling at points and profile samples.

Two facts become obvious: firstly the confidence interval of profile samples is narrower than that of points; secondly it becomes apparent that 6–8 points or profiles provide a reliable estimate and the difference between sampling at 10 and 20 localities is almost negligible.

From the very beginning of this study (1968) interest was centred on the crustacean component of the zooplankton whereas the rotifer community was not studied until 1974. Thus results presented for rotifers are of a preliminary nature and no detailed analysis has been made.

22.2 Species composition and abundance of rotifers

Rotifers are the most numerous plankton species of the open lake. From 1950 to 1974, 19 species were recorded, 15 of which were present at all times, and 5 occurred in reasonable numbers in a few years only.

Table 22.1. Mean density and confidence limits of zooplankton in Neusiedlersee. Mean and 95 per cent confidence limits as percentages of mean. All numbers per m^3, transformed to natural logarithms.

collection	date	n	Crustacea	Rotifera
volume+	21.5.69	16	11.739 ± 2.4	10.613 ± 4.7
sampler	16.7.69	11	11.802 ± 3.4	10.055 ± 5.0
	27.9.69	8	11.494 ± 4.5	10.214 ± 7.6
Volume+	7.2.72	10	10.969 ± 2.0	6.215 ± 10.7
sampler	31.1.73	9	9.872 ± 13.4	9.701 ± 7.8
under ice	4.2.73	9	10.732 ± 5.6	9.788 ± 10.2
horizont.++	8.7.70	8	12.348 ± 3.6	10.361 ± 6.4
integrated	11.7.71	5	12.785 ± 2.4	11.999 ± 8.0
	21.8.71	4	12.988 ± 2.0	10.356 ± 9.8
vertically integrated+++	3.6.75	17	11.876 ± 2.5	8.377 ± 5.3
	20.1.76	11	10.737 ± 3.7	8.733 ± 5.8
	23.3.76	17	11.2947 ± 1.7	13.398 ± 1.7
	9.12.76	20	11.1227 ± 1.3	11.291 ± 1.7

+ pump samples taken at three depths, filtered through 50 μm

++ integrating by using an 'Isaacs-Kidd high speed plankton sampler', tow length 1,000 m (mean), mesh size 50 μm net

+++ integrating by using a plankton net, mesh size 30 μm

List of species and season of maximum occurrence

Brachionidae	
Rhinoglena fertöensis (Varga, 1928)	winter, spring
Brachionus angularis (Goose, 1851)	summer, autumn
Brachionus calyciflorus (Pallas, 1766)	spring
Brachionus quadridentatus (Hermann, 1783)	summer
Keratella quadrata (Müller, 1786)	winter, spring, autumn
Keratella quadrata f. valga	summer
Keratella cochlearis (Goose, 1851)	winter, spring
Kellicottia longispina (Kellicott, 1879)	spring, summer, autumn
Notholca acuminata (Ehrenberg, 1832)	winter, spring
Notholca striata (Müller, 1786)	spring, autumn
Anuraeopsis fissa (Gosse, 1851)	summer
Trichocercidae	
Trichocerca ruttneri (Donner, 1953)	summer
Asplanchnidae	
Asplanchna priodonta (Gosse, 1950)	summer, autumn
Synchaetidae	
Synchaeta tremula-oblonga group	winter, spring

Synchaeta pectinata (Ehrenberg, 1832)	summer
Polyarthra vulgaris (Carlin, 1943)	summer, autumn
Polyarthra dolichoptera (Idelson, 1925)	winter

Testudinellidae

Filinia longiseta (Ehrenberg, 1834)	summer, autumn
Filinia terminalis (Plate, 1886)	spring

Hexarthridae

Hexarthra fennica (Levander, 1892)	summer

Most of the species belong to the Brachionidae (10), others to the Synchaetidae (4), Testudinellidae (2), Asplanchnidae (1), Hexarthridae (1) and Trichocercidae (1). Almost all the species are tolerant towards alkaline, saline or brackish waters. A clear seasonal distribution is evident, 2–4 species dominating in each season (Table 22.2, 22.3).

Seasonal qualitative and quantitative aspects of the rotifer plankton.
Winter: Members of the *Synchaeta tremula-oblonga* group which normally show their maxima at low temperatures are dominant during the winter. From

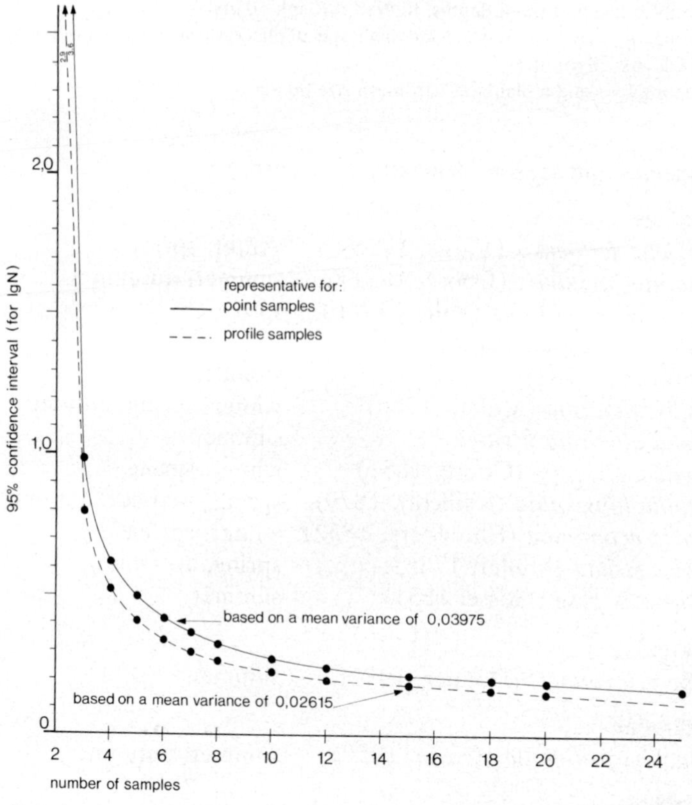

Fig. 22.1. Relation between number of samples and the 95 per cent confidence interval; all numbers (per m^3) transformed to logarithms.

1950 to 1972 these animals represented 28–66.2 per cent of the total number of rotifers (Table 22.3). Daday (1897) did not find any *Synchaeta* species, which in the case of *Synchaeta tremula oblonga* is due to a lack of winter and early spring samples. According to Varga (1926), animals of this group are very abundant during the winter, the maximum numbers amounting to 6 individuals l^{-1} (Feb. 1952) to 163 ind l^{-1} (Feb. 1951). Since 1973 their

Table 22.2. Relative abundance of rotifer species in spring and summer

Spring	1950–52	1956–58	1968–69	1970–72	1972–74
Rhinoglena fertöensis	0.2	—	—	1.8	29.0
Brachionus angularis	—	3.4	0.9	0.1	—
Brachionus calyciflorus	11.8	3.8	4.4	5.0	0.1
Brachionus quadridentatus	0.2	—	0.1	—	—
Keratella quadrata	33.2	26.0	7.7	44.6	58.6
Keratella quadrata f. valga	0.5	—	0.3	—	—
Keratella cochlearis	—	—	45.7	0.9	—
Kellicottia longispina	—	—	0.4	—	0.4
Anuraeopsis fissa	—	—	—	—	—
Notholca striata	1.8	—	4.2	—	—
Notholca acuminata	7.0	6.4	1.7	1.6	0.9
Asplanchna priodonta	—	—	0.4	—	—
Synchaeta tremula-oblonga	40.6	22.1	21.7	17.6	0.6
Synchaeta pectinata	—	—	—	—	—
Polyarthra vulgaris	0.9	3.3	2.2	—	—
Polyarthra dolichoptera	—	—	—	2.4	0.3
Filinia longiseta	—	—	0.2	12.4	—
Filinia terminalis	2.0	35.0	4.5	10.8	10.1
Hexarthra fennica	0.1	—	0.3	2.7	—

Summer	1950–52	1956–58	1968–69	1970–72	1972–74
Rhinoglena fertöensis	—	—	—	—	—
Brachionus angularis	2.0	9.3	9.4	24.6	1.3
Brachionus calyciflorus	1.0	—	0.1	—	—
Brachionus quadridentatus	0.2	15.6	0.2	—	—
Keratella quadrata	1.1	0.5	1.2	5.0	5.0
Keratella quadrata f. valga	0.4	—	22.2	27.4	8.4
Keratella cochlearis	—	—	6.1	0.6	0.3
Kellicottia longispina	—	—	0.6	0.4	—
Anuraeopsis fissa	50.7	—	0.1	—	—
Notholca striata	—	—	16.9	—	—
Notholca acuminata	—	—	—	—	—
Asplanchna priodonta	0.1	0.5	1.6	1.0	0.7
Synchaeta tremula-oblonga	—	—	—	—	—
Synchaeta pectinata	—	4.1	1.7	1.0	—
Polyarthra vulgaris	6.5	13.5	13.8	2.0	1.6
Polyarthra dolichoptera	—	—	—	—	—
Filinia longiseta	—	—	14.4	26.5	75.3
Filinia terminalis	24.9	35.9	4.6	0.5	—
Hexarthra fennica	10.3	20.6	6.1	11.0	6.6

Table 22.3. Relative abundance of rotifer species in autumn and winter

Autumn	1950–52	1956–58	1968–69	1970–72	1972–74
Rhinoglena fertöensis	—	—	—	—	0.2
Brachionus angularis	—	—	10.0	22.5	0.8
Brachionus calyciflorus	16.7	—	1.8	—	—
Brachionus quadridentatus	—	2.9	—	—	—
Keratella quadrata	32.0	41.7	3.3	5.4	73.6
Keratella quadrata f. valga	15.6	—	14.5	24.1	5.8
Keratella cochlearis	—	—	5.0	1.7	1.6
Kellicottia longispina	—	—	0.1	3.2	0.6
Anuraeopsis fissa	5.8	—	—	—	—
Notholca striata	—	—	19.0	—	—
Notholca acuminata	—	—	—	—	0.6
Asplanchna priodonta	0.5	—	1.8	3.0	1.4
Synchaeta tremula-oblonga	—	17.1	10.0	—	—
Synchaeta pectinata	—	—	—	1.0	—
Polyarthra vulgaris	6.3	10.3	13.0	1.8	1.8
Polyarthra dolichoptera	—	—	—	—	—
Filinia longiseta	—	—	15.5	20.8	11.2
Filinia terminalis	14.1	18.6	4.0	4.5	0.6
Hexarthra fennica	2.1	9.4	2.0	12.0	1.4
Winter	**1950–52**	**1956–58**	**1968–69**	**1970–72**	**1972–74**
Rhinoglena fertöensis	1.0	—	0.8	3.0	24.0
Brachionus angularis	—	1.0	—	—	—
Brachionus calciflorus	14.0	4.8	3.0	9.5	—
Brachionus quadridentatus	—	—	—	—	—
Keratella quadrata	31.5	16.6	1.1	18.0	69.8
Keratella quadrata f. valga	1.0	—	0.1	—	—
Keratella cochlearis	0.4	—	16.0	1.3	0.1
Kellicottia longispina	—	—	—	—	—
Anuraeopsis fissa	—	—	—	—	—
Notholca striata	6.1	—	1.0	0.8	—
Notholca acuminata	17.0	1.1	8.0	4.0	1.2
Asplanchna priodonta	—	—	—	—	—
Synchaeta tremula-oblonga	28.0	66.2	65.0	49.4	2.0
Synchaeta pectinata	—	—	—	—	—
Polyarthra vulgaris	—	—	—	—	—
Polyarthra dolichoptera	1.0	8.5	4.0	9.3	2.5
Filinia longiseta	—	—	—	—	—
Filinia terminalis	—	1.8	1.0	4.7	0.4
Hexarthra fennica	—	—	—	—	—

numbers have been decreasing and densities of 1 animal per litre and less have been recorded.

Since 1972/73 *Rhinoglena fertöensis* has become more and more abundant (−24 per cent) and nowadays dominates in the winter plankton (−90 per cent) whereas formerly (1950–1972) percentages of 0–3 per cent were common. *Rhinoglena* is a cold stenothermic species and occurs in the pelagic zone of shallow waters (Ruttner-Kolisko, 1974). In Neusiedlersee its occurrence is correlated with temperatures below 10°C (Varga, 1929). The following diagram shows the increasing importance of this species (1950–1976).

winter 1950/51/52	0.10 – 1.20 ind l^{-1}
winter 1967/68	0.05 – 3.50 ind l^{-1}
winter 1968/69	0.02 – 0.10 ind l^{-1}
winter 1970/71	0.02 – 0.06 ind l^{-1}
winter 1971/72	0.05 – 0.30 ind l^{-1}
winter 1972/73	0.50 – 7.00 ind l^{-1}
winter 1975/76	6.00 – 600.00 ind l^{-1}

During the cold season another frequent species is *Keratella quadrata*, accounting for 1.1–69.8 per cent (Table 22.3). The maximum densities range from 1 (Feb. 1969, 1971, 1972) to 59 ind l^{-1} (Jan. 1952).

In addition, *Brachionus calyciflorus*, *Notholca acuminata*, *Polyarthra dolichoptera* and *Filinia terminalis* should be mentioned (Table 22.3). In 1968 and 1969 *Keratella cochlearis* shows up and constitutes 16 per cent of the standing crop (Table 22.3).

The mean total numbers are between 5 and 89 individuals per litre, reaching maximum densities of 13–199 ind l^{-1} (Table 22.4).

Spring: From 1950–1972 the spring situation was characterized by a large quantity of *Synchaeta tremula-oblonga* (17.6–40.6 per cent) and *Keratella quadrata* (7.7–44.6 per cent). After 1972 the *Synchaeta* numbers decreased, *Keratella quadrata* became more important (58.6 per cent) and *Rhinoglena fertöensis* now formed the major part of the rest of the rotifer plankton (29 per cent) (Table 22.2). Maximum densities for *Synchaeta* varied between 25 (March 1971) and 500 (March 1968) ind l^{-1}. In recent years (1973–1976) no more than 10 ind l^{-1} were present. *Keratella quadrata* attained values of 4 (May 1969)–1,005 (March 1974) ind l^{-1} and *Rhinoglena fertöensis* occurred in maximum numbers ranging from 1 (May 1972) to 78 (March 1973) ind l^{-1} and even more than 600 ind l^{-1} in March 1976.

The highest numbers of *Filinia terminalis*, a cold stenothermic animal, were encountered in spring (1 – April 1972, 85 – May 1968, 222 – March 1958) and formed 2–35 per cent of the standing crop (Table 22.2).

Brachionus calyciflorus and *Notholca acuminata* also occurred. At the middle or end of May *Brachionus angularis* and *Hexarthra fennica* started to develop out of the resting eggs.

In 1968/69 *Keratella cochlearis* became dominant (−45.7 per cent) and a maximum density of 862 ind l^{-1} (May 1968) could be recorded.

Maximum total rotifer numbers observed were between 32.7 and 1945 ind l^{-1} and the mean values of the spring season showed a range of 21–784 ind l^{-1} (Table 22.4).

Summer: Filinia longiseta (14.4–75.3 per cent), *Keratella quadrata f. valga* (0.4–27.4 per cent), *Brachionus angularis* (1.3–24.6 per cent), *Polyarthra vulgaris* (2–13.6 per cent) and *Hexarthra fennica* (6.1–20.6 per cent) formed the major part of the summer rotifer plankton. All of them are found to be

Table 22.4. Rotifer numbers . $10^3 . m^{-3}$

	spring			summer			autumn			winter		
	mean	max	min	mean	max	min	mean	max	min	mean	max	min
1950/51/52	110.9	245.6	48.9	229.6	549.0	32.5	101.1	450.0	4.6	47.4	199.0	1.5
1956/57/58	784.0	1,945.0	62.0	199.2	2,400.0	25.5	239.8	2,300.0	18.0	69.3		
1968	401.4	1,015.0	103.8	167.3	288.0	66.2	56.2	73.0	34.6	88.9	148.8	8.2
1969	36.8	130.5	9.0	108.3	310.8	21.9	41.2	232.3	6.3			
1970				38.6	179.2	10.0	20.7	44.4	14.0	16.8	38.9	1.5
1971	21.3	32.7	10.5	63.0	162.6	20.0	19.8	26.4	13.1	5.0	12.0	0.5
1972/73/74	115.7	149.0	17.0	78.8	297.2	18.4	20.4	54.2	7.7	14.3	20.5	10.2

1950/51/52 – Zakovsek 1961
1956/57/59 – Ruttner unpublished data

warm stenothermic or at least eurythermic. Less important were *Brachionus calyciflorus* and *quadridentatus, Keratella cochlearis, Kellicottia longispina*, *Asplanchna priodonta*, *Synchaeta pectinata* and *Filinia terminalis* (Table 22.2).

Filinia longiseta showed maximum densities of 10 (Aug. 1969) – 170 (June 1951) ind l^{-1}, *Keratella quadrata* f. *valga* 5 (Aug. 1973) – 105 (Aug. 1970) ind l^{-1}, *Brachionus angularis* 17 (July 1971) – 168 (June 1956) ind l^{-1}, *Polyarthra vulgaris* 1 (Aug. 1970/71) – 69 (June 1968) and *Hexarthra fennica* 2 (Aug. 1973) –300 (Aug. 1958) ind l^{-1}.

In 1951 a 'bloom' of *Anuraeopsis fissa* showed up (50.7 per cent of the standing crop) and reached densities of 420 ind l^{-1} in July and 384 ind l^{-1} at the end of August, but with no numerous occurrence in the subsequent years. A similar event was observed in *Notholca striata* and accounted for 16.9 per cent of the total numbers in 1968/69.

Total numbers reached values between 162 and 2,400 ind l^{-1}; the summer means were in the range of 20–230 ind l^{-1} (Table 22.4).

Autumn: The situation in autumn was very similar to that in the summer months. The main constituents were *Keratella quadrata* (3.3–73.6 per cent), *Keratella quadrata f. valga* (0.4–24.1 per cent) and *Filinia longiseta* (11.2–20.8 per cent) (Table 22.3). Maximum densities of *Keratella quadrata* (including *K. qu. f. valga*) were 10 (Sept. 1971) to 141 (Nov. 1951) ind 1^{-1} and for *Filinia longiseta* (including *F. terminalis*) 1 (Sept. 1971/72) – 200 (Sept. 1958).

Polyarthra vulgaris was not unimportant for the autumnal plankton and occurred in numbers ranging from 1 (Sept. 1973, Oct. 1972, Sept. 1971) to 129 (Sept. 1969) or 1,500 (Sept. 1958) ind l^{-1}. *Brachionus angularis* was still present with densities of 1 (Oct. 1972, Sept. 1973) to 50 (Sept. 1951) ind l^{-1}.

Total rotifer numbers reached a maximum of 21 to 2,300 ind l^{-1}, and the mean values ranged from 20 to 240 ind per litre (Table 22.2).

Summary and Discussion

In general, 3 to 5 maxima can be recorded per year. The highest numbers per unit volume are present in spring (mainly in March or beginning of April) or in mid- and late summer (August, beginning of September). By comparing the seasonal means from 1950 to 1973 a tremendous decrease becomes evident. The highest values are from 1956, 1957 and 1958 (Ruttner unpublished data), possibly the result of too few sampling dates. Apart from this the years 1951 and 1968 yielded the maximum rotifer densities, with more than 1,000 individuals per litre. Estimates from 1974, 1975 and 1976 (not yet finally worked out) reflect a tendency to increasing numbers.

Comparing numbers of rotifers with those of crustaceans (see below) it becomes obvious that with increasing numbers of the latter a decrease in rotifer density occurrs and vice versa. This may be due to competition, but other possibilities such as long term fluctuations should not be excluded. More information on this problem could be obtained by elucidating the life history of each species and considering it in relation to the main abiotic and biotic factors.

As will be mentioned later there is strong evidence that Neusiedlersee is in a state of eutrophication. Parallel to this it is quite obvious that the diversity of the rotifer species in Neusiedlersee has changed and at the moment (1976) a more or less two-species community exists.

Table 22.5 gives a comparison of total rotifer numbers from different lakes. The examples chosen are from latitudes of about 60° North to 30° South, from high mountain lakes to tropical bodies of water, from freshwater lakes to saline ecosystems.

Table 22.5. Comparison of rotifer abundance from subarctic to subtropic regions

Lake	date of survey	spring	summer	autumn ind · l^{-1}	winter	author
Heimdalsvatn	1969		50			Larsson 1972
Peipsi-Pihkva	1965/66	17	60	2	6	Habermann 1974
Vortsjärv	1965/66	114	85	12	8	Habermann 1974
Gossenköllesee	1962/64/65	26	14	22	41	Eppacher 1968
V. Finstertalersee	1969/70/71	14	13	108	240	Pechlaner et al. 1972
Mikolajskie	1963–1970	1,800		1,000		Spodniewska et al. 1973
Mikolajskie	1964		430			Bottrell at al. 1976
Warniak	1967/68/69		1,650	440		Hillbricht-Ilkowska, Weglenska 1973
Neusiedlersee	1950/51/52	111	230	101	47	Zakovsek 1961
Neusiedlersee	1956/57/58	784	199	240	69	Ruttner pers. comm.
Neusiedlersee	1968/69	219	138	49	89	Herzig, this paper
Neusiedlersee	1970/71	21	51	20	11	Herzig, this paper
Neusiedlersee	1972/73/74	116	79	20	14	Herzig, this paper
Attersee	1974/75	5	9	6	2	Müller 1976
Titisee	1968/69/70	3	46	88	35	Szymanski-Bucarey 1974
Piburger See	1971–1975	238	297	73	126	Schaber 1974, 1976
Balaton	1965/66/67		140			Zankai-Ponyi 1972
Lake Suwa	1948	105	156	95	37	Kurasawa et al. 1952
Lake George	1967/68	455	804	458	302	Burgis 1969
Lake Werowrap	1968/69/70	650	2,100	1,500	200	Walker 1973

The figures from northern European lakes (like Øvre Heimdalsvatn, Peipsi-Pihkva or Vortsjärv) agree very well with those from high mountain lakes in central Europe. The season of maximum occurrence varies (and of course the species composition), but in general the total individual numbers are in the same range, with maxima of 60–240 ind l^{-1}.

Comparing lakes from the temperate region it is obvious that the eutrophic lakes show the highest rotifer densities (Mikolajskie, Warniak, Piburger See and sometimes Neusiedlersee). Numbers as high as those found in eutrophic lakes can occur in Neusiedlersee, but some very low individual densities have also been recorded. The results from Balaton are similar to those from Neusiedlersee. The fact that Attersee is an oligotrophic lake is reflected in its very low rotifer numbers. Piburger See, a more or less eutrophic lake, has relatively high individual densities, despite its location at an altitude of 915 metres above sea level (Schaber, 1974). Titisee values lie between those of the eutrophic and the oligotrophic lakes, the lake itself being in a state of progressive eutrophication (Szymanski-Bucarey, 1974). Lake Suwa is a eutrophic Japanese lake at a relatively high altitude (759 m above sea level) (Kurasawa et al. 1952) showing values like Neusiedlersee in 1950/51/52.

Lake George, a tropical lake, shows less variation in numbers throughout the year, the highest densities occurring in the summer months (= the rainy season, Burgis, 1969).

Lake Werowrap is a saline ecosystem with a rotifer plankton consisting of two species, one of them accounting for about 90 per cent of the total numbers (Walker, 1973). The densities are the highest mentioned in the course of this comparison. In this special biotope we find only a few species but these occur in a tremendously high density. The highest individual numbers occur in the summer months which form the rainy season (Walker, 1973).

Thus, geographically seen, rotifer numbers seem to be primarily correlated with temperature, but at identical temperatures the numbers are a function of food availability. Extreme biotopes show a low species diversity, but achieve extremely high animal densities ('biozönotisches Grundprinzip' – Thienemann, 1920; Sanders, 1968; Ruttner-Kollisko, 1974).

22.3 Species composition and abundance of crustaceans

The crustacean component of the zooplankton community comprises 7 cladoceran and 4 copepod species. One of the two dominant species – *Arctodiaptomus spinosus* – occurs all year round, whereas the second – *Diaphanosoma brachyurum* – appears in the summer months.

List of species:

Leptodora kindti (Focke, 1844)[1]
Diaphanosoma brachyurum (Liévin, 1848)
Daphnia pulex (Scourfield, 1942)
Daphnia longispina s.l.*
Ceriodaphnia reticulata (Jurine, 1820)
Ceriodaphnia quadrangula (O. F. Mueller, 1785)
Bosmina longirostris (O. F. Mueller, 1785)

Arctodiaptomus spinosus (Daday, 1891)

Acanthocyclops robustus (G. O. Sars, 1863)
Mesocyclops leuckarti (Claus, 1857)
Thermocyclops crassus (Fischer 1853)

1. Since June 1977 Leptodora has been found in the plankton and has reached maximum densities of 57 individuals (September 1977) and 72 individuals (end of July 1978) per cubic metre. It was found as resting eggs in February 1978 (up to 940 eggs m^{-2} mud surface) and adult animals in the stomach of pike perch (length 13.2 cm and 17.2 cm) in June 1978.

*Daphnia cucullata and Daphnia galeata, according to Hrbaček pers. comm.

All of these planktonic species tolerate higher alkalinities (Herzig, 1975); *Arctodiaptomus spinosus* is a typical crustacean of sodium lakes (Löffler, 1959, 1961; Kiefer, 1972). Its upper tolerance limit is 700 meq l^{-1}, but reproduction stops at 500 meq l^{-1}. At these very high alkalinities the death rate of *Arctodiaptomus* rises with increasing temperatures (Herzig, 1973).

The second important species, *Diaphanosoma brachyurum*, can tolerate alkaline waters up to 50 meq l^{-1}, under which conditions the embryos can complete their development, whereas the free-swimming instars die off immediately. Normal development occurs up to 25 meq l^{-1} (Herzig, 1975).

THE RELATIVE ABUNDANCE OF SPECIES (1968–1973; Fig. 22.2; Table 22.6)

The dominant species, *Arctodiaptomus spinosus*, accounts for maximum percentages of 51.2–99.1 per cent of the total individual numbers, with annual means between 57.2 and 74.3 per cent. By comparing the percentages over the six-year-period it becomes obvious that *A. spinosus* is increasing in numbers. This species is dominant throughout the year, except for the summer months when *Diaphanosoma brachyurum* forms the major component of the crustacean plankton; the latter shows maximum percentages of 34.2–61.9 per cent, and summer means of 8.4–40 per cent.

Until 1971 the cyclopoids accounted for an average of 24.3–28.5 per cent. They have been steadily decreasing since 1972 and the figures sank from 16.5 per cent in 1973 to between 1 and 8 per cent in 1974/75/76.

Daphnia, *Ceriodaphnia* and *Bosmina* are negligible and show annual means of 0.9–5.2 per cent; the maximum occurrence was recorded in autumn 1970 (12.6 per cent) when *Bosmina* became relatively numerous.

The winter crustacean plankton consists of a pure copepod population with *Arctodiaptomus* attaining 45.8–95.6 per cent and the cyclopoids 4.4–54.2 per cent.

INDIVIDUAL NUMBERS AND BIOMASS 1968–1973

Total standing crop (Tables 22.7, 22.8, Fig. 22.3a, 22.4a)

In general the highest densities can be recorded in the summer months and exhibit one (July or August) or two peaks (June and August). The lowest numbers were observed in winter when copepods only are present. Beside the maxima in summer a smaller peak showed up regularly in spring and rather rarely another in autumn.

The mean estimates for the winter months vary between 2.1 and 50×10^3 ind m^{-3}, reaching a maximum density of 94.3×10^3 m^{-3}. In spring the mean densities are between 16.6 and 296.6×10^3 ind m^{-3}, with maximum numbers of $31.8–435.8 \times 10^3$ ind m^{-3}. At this time the Cladocera – namely *Diaphanosoma brachyurum* – become increasingly important.

For the summer season mean values range from 66.6 to 484.6×10^3 ind m^{-3}, with maximum densities between 129.3 (1968) and 693.4×10^3 ind m^{-3} (1971). At this time the Cladocera are prevalent.

The range of mean densities for the autumn months is $33.2–168.6 \times 10^3$

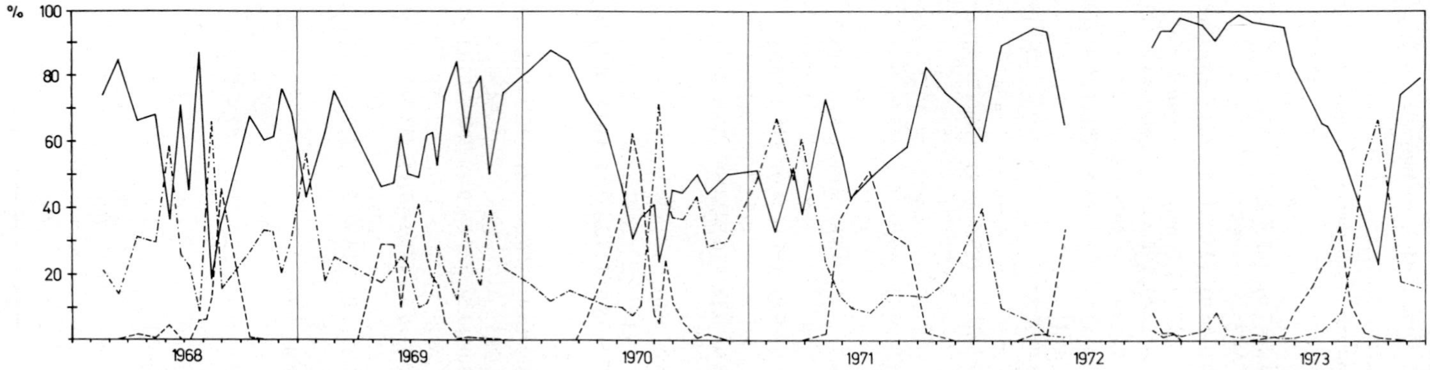

Fig. 22.2. Relative abundance of the components of the crustacean plankton. —— *Arctodiaptomus spinosus*, ---- *Diaphanosoma brachyurum*, -·-·-·- Cyclopidae.

Table 22.6. Relative abundance of crustacean species (seasonal mean percentage, maximum and minimum in parenthesis)

		spring	summer	autumn	winter	year
Arctodiaptomus spinosus	1968	70.2 (84.4–47.0)	52.6 (85.8–18.3)	57.1 (67.2–36.0)	61.5 (76.4–43.8)	60.9 (85.8–18.3)
	1969	58.5 (74.0–46.0)	56.3 (75.2–47.6)	70.4 (85.4–50.6)	82.1 (88.0–76.5)	62.4 (85.4–43.8)
	1970	72.2 (86.0–52.5)	38.2 (52.8–24.1)	46.9 (50.2–44.3)	45.8 (51.2–32.8)	57.6 (88.0–24.1)
	1971	54.5 (73.2–37.8)	50.0 (56.2–43.5)	71.3 (81.8–56.4)	73.5 (91.0–59.9)	57.2 (81.8–32.8)
	1972	87.1 (94.0–62.3)			95.6 (98.0–92.2)	
	1973	95.1 (99.1–82.9)	67.3 (82.5–48.4)	45.0 (76.0–22.5)		74.3 (99.1–22.5)
Diaphanosoma brachyurum	1968	0.9 (3.8– 0.1)	8.4 (46.9– 0.2)	12.2 (45.0– 0.1)		7.2* (46.9– 0.1)
	1969	11.3 (29.3– 0.1)	23.7 (40.8– 4.0)	0.7 (5.0– 0.0)		11.9* (40.8– 0.0)
	1970	10.8 (33.0– 0.0)	31.7 (61.9– 4.0)	2.7 (11.0– 0.1)		15.1* (61.9– 0.0)
	1971	7.0 (38.0– 0.1)	40.0 (50.8–30.6)	11.2 (30.0– 0.1)		19.9* (50.8– 0.1)
	1972	8.3 (38.0– 0.0)				
	1973	1.3 (7.3–0.1)	20.4 (34.2– 7.4)	2.2 (10.1– 0.1)		8.0* (34.2– 0.1)
Cyclops spp.	1968	27.0 (48.0–13.4)	33.0 (66.2–18.2)	25.7 (32.3–15.0)	33.5 (56.2–18.1)	28.5 (66.2– 8.2)
	1969	23.1 (25.0–17.5)	18.8 (29.0– 9.9)	24.4 (39.6–13.1)	17.3 (22.0–12.0)	24.3 (56.2– 9.9)
	1970	16.8 (17.9–10.5)	28.7 (71.1– 7.1)	37.8 (43.9–27.9)	49.6 (67.2–31.0)	26.5 (71.1– 7.1)
	1971	38.0 (61.5–11.0)	10.0 (13.7– 8.0)	15.3 (22.0–10.8)	26.5 (40.1– 9.0)	27.7 (67.2– 8.0)
	1972	4.5 (9.0– 1.8)			4.4 (7.8– 0.9)	
	1973	2.5 (3.0– 0.4)	7.5 (22.5– 0.8)	46.4 (67.9–17.8)		16.5 (67.9– 0.4)
Rest	1968	1.9	6.0	5.0	5.0	4.1
Ceriodaphnia+	1969	7.1	1.2	3.5	0.6	4.4
Daphnia+Bosmina	1970	0.2	1.4	12.6	4.6	5.2
	1971	0.5	0.0	2.2	0.0	0.9
	1972	0.1			0.0	
	1973	1.1	4.8	6.4	0.0	3.2

* mean percentage for the 'growing season'

Table 22.7. Crustacean plankton, numbers $10^3\ m^{-3}$

	Jan	Feb	March	April	May	June	July	Aug	Sept	Oct	Nov	Dec
1968			8.9	17.4	23.6	34.1	72.9	93.0	55.9	29.4	14.3	3.2
1969	1.6	1.6	5.2	23.2	96.8	152.2	128.5	139.0	128.5	67.4	30.4	10.4
1970	9.7	10.5	21.8	57.0	140.2	253.5	185.1	232.5	157.6	76.4	35.7	27.0
1971	19.8	11.6	30.3	111.8	266.5	414.2	610.5	429.0	248.3	162.2	95.3	54.0
1972	34.7	61.3	205.1	384.9	299.8						122.8	39.6
1973	16.6	38.2	116.2	319.3	416.2	386.5	356.4	328.4	272.4	122.3	55.7	
1968–1973	16.5	24.6	65.0	152.2	207.2	248.1	270.7	244.4	172.6	91.5	59.0	26.9

	spring			summer			autumn			winter		
	mean	max	min	mean	max	min	mean	max	min	mean	max	min
1968	16.6	31.8	7.2	66.6	129.3	33.0	33.2	74.2	7.8	2.1	6.4	1.5
1969	41.7	165.2	2.3	139.9	199.3	97.5	75.4	150.2	12.2	10.2	12.1	9.6
1970	73.0	192.0	13.8	223.7	288.2	87.1	89.9	220.0	30.0	19.5	28.1	7.6
1971	136.2	313.8	26.5	484.6	693.4	290.0	168.6	321.0	74.2	50.0	94.3	31.1
1972	296.6	427.0	100.0							31.5	72.1	13.5
1973	283.9	435.8	41.4	357.1	400.0	320.2	150.1	330.9	47.9			
1968–1973	141.3	435.8	2.3	254.4	693.4	33.0	103.4	330.9	7.8	22.7	94.3	1.5

Table 22.8. Crustacean plankton, biomass mg dry weight m^{-3}

	Jan	Feb	March	April	May	June	July	Aug	Sept	Oct	Nov	Dec
1968			20	36	50	50	82	172	152	69	23	10
1969	2	3	23	62	171	282	316	238	160	126	75	24
1970	20	16	23	71	222	686	477	454	256	125	91	76
1971	70	88	93	280	815	1,312	1,855	1,628	881	846	499	228
1972	116	125	413	1,016	892					329	426	142
1973	47	92	341	1,046	1,260	986	854	700	654	316	238	
1968–1973	51	65	152	420	568	663	717	638	421	302	225	96

	spring			summer			autumn			winter		
	mean	max	min	mean	max	min	mean	max	min	mean	max	min
1968	36	54	17	101	215	36	81	210	12	5	17	2
1969	85	260	12	279	460	200	120	195	33	19	32	15
1970	105	355	17	539	970	146	157	391	84	78	101	44
1971	396	980	84	1,598	2,014	1,000	742	1,300	340	156	320	109
1972	774	1,200	204							94	280	43
1973	882	1,440	100	847	1,050	648	403	700	210			
1968–1973	380	1,440	12	673	2,014	36	301	1,300	12	70	320	2

(a)

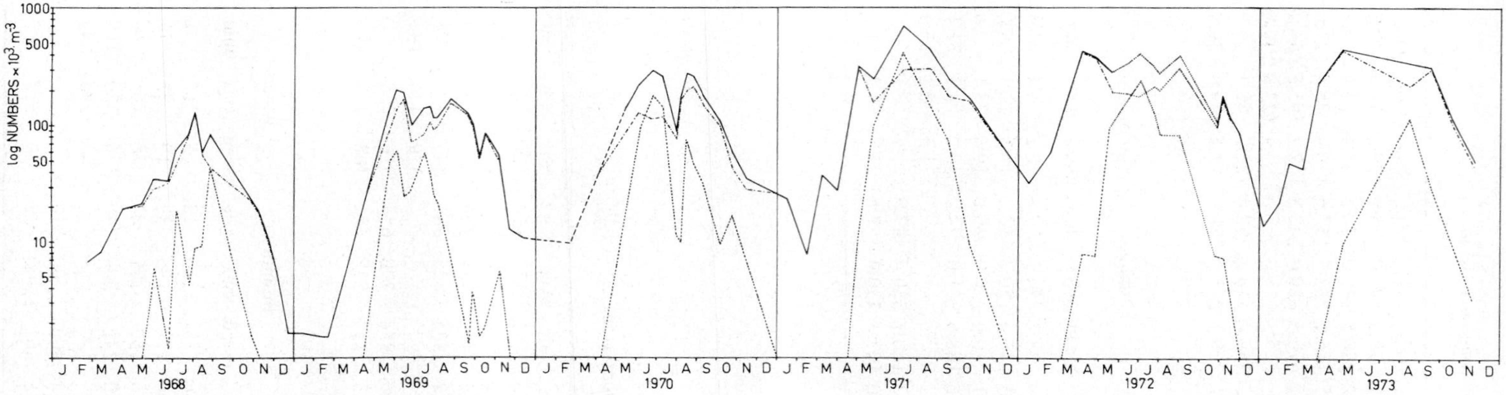

Fig. 22.3. Crustacea – abundance; numbers m^{-3} (a) —— total Crustacea, -·-·-·- Copepoda, –––– Cladocera; (b) ——*Arctodiaptomus spinosus*, ––––– *Diaphanosoma brachyurum*, -·-·-·- Cyclopidae.

(b)

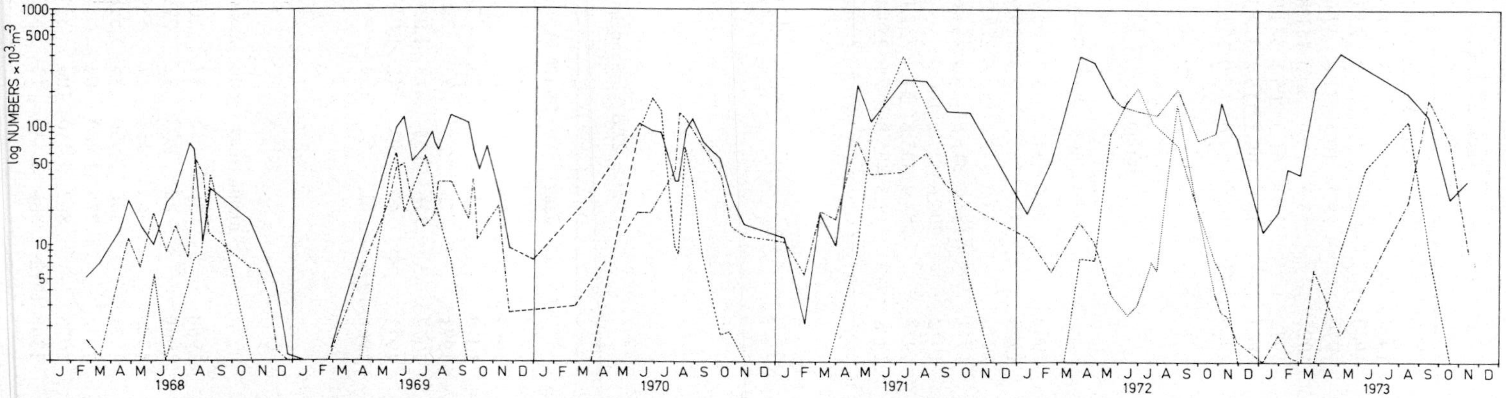

(a)

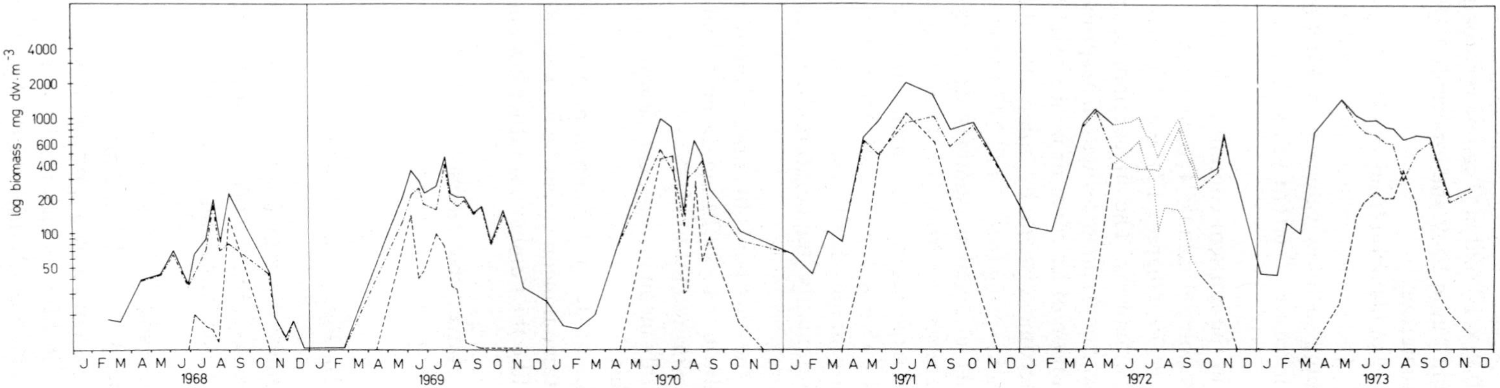

Fig. 22.4. Crustacea – biomass; mg dry weight m^{-3}: (a) —— total biomass, -·-·-·- Copepoda, ---- Cladocera; (b) —— *Arctodiaptomus spinosus*, ---- *Diaphanosoma brachyurum*, -·-·-·- Cyclopidae.

(b)

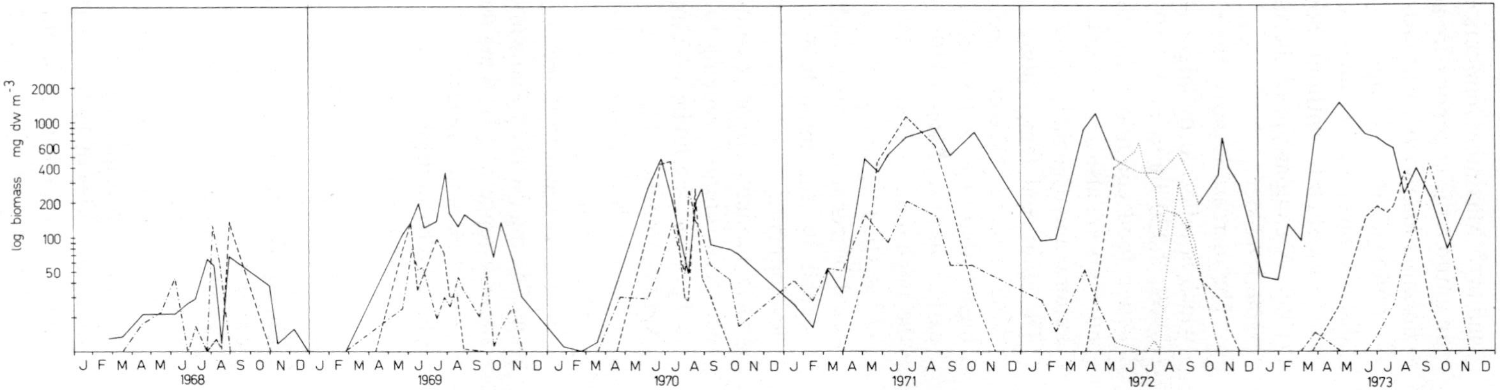

ind m^{-3}. Maximum numbers of 74.2 – 330.9 × 10^3 ind m^{-3} can be recorded. In autumn the copepods are again dominant but in 1969 and especially 1970 *Bosmina longirostris* was also relatively numerous.

Following the development of the crustacean plankton from 1968 to 1973 a distinct increase becomes obvious until 1971, after which the numbers decrease slightly, without, however, reaching such low values as in 1968 or 1969. A comparison of the monthly means of the different years shows the same tendency.

The increase is partly the result of rise in the nutrient content of the lake, which is reflected by larger algal biomass (annual means: 426 mg fw m^{-3} – 1969; 2,220 mg fw m^{-3} – 1972; Dokulil, 1979, Chapter 17), which creates better food conditions for the plankton organisms. The higher individual numbers are also a reaction to the better (warmer) climatic situation in spring, autumn and especially in winter (short duration of ice cover or no ice cover). The improved temperature conditions become evident by comparing the rapid increase of the crustacean population in spring time. Until 1970, values higher than 100 × 10^3 ind m^{-3} were only attained in summer or at the earliest in May. After 1970 such values were seen in April (1971) or even at the beginning of March (1972), at which time spring maximum values of 427 × 10^3 (beginning of April 1972) and 435.8 × 10^3 (end of April 1973) could be recorded. Such high numbers are mainly due to the shorter time needed by the animals for their predominantly temperature-dependent development.

The biomass curve resembles that of the numbers. Maximum biomass occurs in summer, ranging from 215 mg dw m^{-3} to 2,014 mg dw m^{-3}. Only in 1973 did the highest biomass occur in spring (1,440 mg dw m^{-3}; end of April/beginning of May). The highest monthly mean biomass can be recorded either in July/August or in June and August.

Mean spring values range from 36 to 882 mg dw m^{-3}, summer means vary between 101 and 1598 mg dw m^{-3} and autumnal mean values lie between 81 and 742 mg dw m^{-3}. The winter biomass, like the numbers, is rather low and shows a range of 5–156 mg dw m^{-3}.

For the same reasons as mentioned above, the biomass estimates – obtained by multiplying individual numbers by their corresponding weight – again show a tendency to increase.

Arctodiaptomus spinosus (Table 22.9, 22.10, Fig. 22.3b, 22.4b)

This species shows one maximum in abundance in spring (April or May) and one or two in summer. Additional peaks are rare, but can occur in March and October. In summer there is either one distinct maximum in July, or one in June and one in August.

Mean spring values vary between 10.1 and 267.6 × 10^3 ind m^{-3}, the maximum densities being between 24 and 424.7 × 10^3 ind m^{-3}. The high spring densities of 1972 and 1973 may be explained by a short, mild winter, and a rapid increase in water temperature in March. This results in a high survival rate of adults throughout the winter months and a short turnover

Table 22.9. *Arctodiaptomus spinosus*, numbers 10^3 m^{-3}

	Jan	Feb	March	April	May	June	July	Aug	Sept	Oct	Nov	Dec
1968			5.8	12.4	12.2	13.8	37.2	31.5	20.0	14.4	7.3	2.1
1969	0.7	0.8	2.4	11.7	45.2	90.5	68.9	92.1	102.2	51.1	19.2	8.2
1970	9.3	14.8	23.9	40.9	74.1	98.4	68.8	75.2	74.1	38.9	16.3	13.1
1971	9.6	4.1	14.2	63.8	148.0	165.6	242.9	226.0	146.8	126.5	75.6	38.2
1972	22.9	46.9	178.9	367.5	222.9						115.6	40.0
1973	15.6	37.2	112.3	312.4	378.3	305.3	241.8	185.9	116.6	33.8	30.5	
1968–1973	11.7	20.8	63.1	154.2	161.9	135.4	133.8	123.7	92.9	53.7	44.4	20.4

	spring			summer			autumn			winter		
	mean	max	min	mean	max	min	mean	max	min	mean	max	min
1968	10.1	24.0	5.6	34.3	72.6	10.0	17.4	28.0	5.6	1.4	5.0	0.7
1969	20.0	73.2	2.0	83.8	121.6	48.7	57.5	104.8	9.4	10.8	18.0	7.4
1970	46.3	96.0	18.5	80.8	117.5	34.3	43.1	90.0	15.0	8.9	13.5	2.1
1971	75.4	226.1	10.0	211.5	249.5	135.0	116.3	160.0	56.0	36.0	76.0	19.4
1972	256.4	403.0	80.0							31.0	65.0	12.9
1973	267.6	424.7	41.0	244.4	340.0	150.0	60.3	148.4	25.3			
1968–1973	125.9	424.7	2.0	131.0	340.0	10.0	58.9	160.0	5.6	17.6	76.0	0.7

Table 22.10. *Arctodiaptomus spinosus*, biomass mg dw m^{-3}

	Jan	Feb	March	April	May	June	July	Aug	Sept	Oct	Nov	Dec
1968			14	20	21	24	45	40	58	43	18	12
1969	1	2	18	41	90	145	200	164	132	97	53	21
1970	13	10	14	42	137	337	156	174	122	71	52	36
1971	29	41	42	148	395	499	707	805	574	702	449	199
1972	98	109	362	915	555					267	409	146
1973	46	89	317	1,018	1,164	780	598	329	255	102	179	
1968–1973	37	50	128	364	394	357	341	302	228	214	193	83

	spring			summer			autumn			winter		
	mean	max	min	mean	max	min	mean	max	min	mean	max	min
1968	19	21	13	36	64	11	40	63	12	5	15	1
1969	50	120	11	170	357	119	94	148	27	15	27	9
1970	64	225	12	222	476	49	82	250	45	35	50	26
1971	195	469	33	670	867	400	575	800	300	136	290	90
1972	611	1,140	168							94	250	41
1973	833	1,410	95	569	900	225	179	380	76			
1968–1973	295	1,410	11	333	900	11	194	800	12	57	290	1

time of the *Diaptomus* population in spring.

Mean summer densities range from 34.3 to 244.4 × 10^3 ind m^{-3}, the highest values being 72.6–340.1 × 10^{-3}. A dramatic breakdown of the population as was seen in July 1970 is mainly caused by an extreme wind situation (wind speed reaching 50 km/h) in which the animals are crushed by the suspended particles.

Autumn means are between 17.4 and 116.3 × 10^3 ind m^{-3}, with maximum densities of 28 – 160 × 10^3 ind m^{-3}.

During winter time nauplii and copepodides represent about 90 per cent of the total population. In recent years (1971/72/73) the change towards a higher proportion of adults is due to better conditions for their development in autumn and low mortality rates of adults in winter. Seasonal mean densities vary between 1.4 and 36.1 × 10^3 ind m^{-3}, reaching maximum numbers of 5–76 × 10^3 ind. $m.^{-3}$.

Taking into consideration the duration of the different instars at various temperatures (known from laboratory experiments, Herzig in prep.) and their phenology it becomes evident that in Neusiedlersee the *Arctodiaptomus spinosus* population produces 4–6 generations per year, one in spring and autumn, and 2–4 in summer. Reproduction takes place throughout the year, with the biggest clutch sizes – 10–14 eggs per egg sac – in spring, and 4–7 eggs per egg sac over the rest of the year. No resting eggs or resting stages have been found throughout the investigation period, although the Neusiedlersee population is apparently able to produce them (according to laboratory experiments, Herzig in prep.). *Arctodiaptomus spinosus* from the small, shallow, alkaline, temporary lakes east of Neusiedlersee ('Seewinkel') shows quite regular production of resting eggs (Newrkla, 1974).

The distribution of biomass is very similar to that of the numbers. Normally the highest biomass values occur in summer. Since 1972 a tendency towards higher biomass estimates has been seen in spring for the same reasons as mentioned earlier. Spring mean values are between 19 and 833 mg dw m^{-3}; the maxima range from 21–1,410 mg dw m^{-3}. Mean summer estimates vary between 36 and 670 mg dw m^{-3} (max.: 64–900 mg dw m^{-3}), autumn mean biomass values range from 40–575 mg dw m^{-3} (max.: 63–800 mg dw m^{-3}) and the winter values show means between 5 and 136 mg dw m^{-3}, (max.: 15–290 mg dw m^{-3}).

For the reasons already mentioned the increase in total numbers and biomass observed for crustaceans is also seen in the *Arctodiaptomus* population from 1968–1973. If the growth under identical temperature conditions (this is valid for the summer months) is considered to be purely food dependent and the weights of adult animals for the years 1968–1970 and 1971–1973 are compared a marked difference is found. The mean summer weight for an adult female was 7.7 μg dw (4.9–10.8 μg dw) in the years 1968–1970, and 11 μg (8–12.5 μg) in 1971–1973, i.e. an increase of 42.8 per cent. The difference in the males is smaller but still considerable; 6.8 μg dw for 1968–1970 and 7.2 μg dw for 1971–1973, which is an increase of 5.9 per cent. Even this rough comparison reflects the improvement in food conditions since 1972.

Diaphanosoma brachyurum (Tables 22.11, 22.12, Fig. 22.3b, 22.4b)

Table 22.11. *Diaphanosoma brachyurum*, numbers 10^3 m^{-3}

	Jan	Feb	March	April	May	June	July	Aug	Sept	Oct	Nov	Dec
1968			0.2	0.9	1.9	3.5	3.4	16.1	19.9	6.3	0.2	
1969				5.6	32.1	34.6	42.9	13.4	2.8	0.5	0.1	
1970				11.5	47.5	129.8	95.7	34.3	10.3	2.5	0.9	
1971				2.6	62.0	232.9	341.4	183.6	80.0	24.1	2.4	
1972			3.2	8.7	67.2						2.2	
1973				4.7	21.5	47.4	81.9	86.9	27.6	1.3	0.1	
1968–1973			1.4	5.5	38.7	89.6	113.1	66.8	28.1	6.9	1.1	

	spring			summer			autumn			winter		
	mean	max	min	mean	max	min	mean	max	min	mean	max	min
1968	0.9	2.2	0.1	7.7	39.3	0.3	9.7	35.0	0.05			
1969	21.5	50.0	0.06	30.3	58.7	7.3	1.3	5.2	0.03			
1970	33.1	76.2	0.1	86.6	175.5	8.3	4.6	24.0	0.5			
1971	38.2	138.2	0.07	252.6	391.4	120.0	35.5	110.0	0.1			
1972	29.2	89.5	0.03									
1973	13.1	34.0	0.2	72.1	113.6	36.0	9.6	63.0	0.08			
1968–1973	22.7	138.2	0.03	89.9	391.4	0.3	12.1	110.0	0.03			

Table 22.12. *Diaphanosoma brachyurum*, biomass, mg dw m^{-3}

	Jan	Feb	March	April	May	June	July	Aug	Sept	Oct	Nov	Dec
1968				0.3	1.1	6	4	37	39	8		
1969				9	56	63	64	19	4	0.5	0.1	
1970				16	60	220	273	90	25	8	2	
1971				13	253	633	858	598	270	79	4	
1972			1.7	19	252					36	14	
1973			3.0	10	59	143	228	270	95	8	0.5	
1968–1973			2.4	14	136	213	285	203	86	19	5	

	spring			summer			autumn			winter		
	mean	max	min	mean	max	min	mean	max	min	mean	max	min
1968	0.3	1	0.1	18	135	0.4	22	123	0.2			
1969	40	108	0.2	54	134	9	2	8	0.1			
1970	42	100	0.2	194	446	29	12	38	0.1			
1971	153	540	0.4	696	1,088	470	130	420	0.3			
1972	100	510	0.2									
1973	34	117	0.4	214	359	125	38	201	0.1			
1968–1973	75	540	0.1	234	1,088	0.4	40	420	0.1			

As a rule *Diaphanosoma brachyurum* is typical of the summer plankton. In deep, stratified, temperate lakes it first appears in spring (April or May); in Neusiedlersee, as in southern European lakes, this takes place in March at temperatures between 4 and 10°C. The first males show up in late August at the earliest, usually by mid September, and the production of resting eggs starts one to two weeks later. *Diaphanosoma* disappears rapidly in November or December, when temperatures drop beyond 5°C. This species overwinters in the form of resting eggs. Parthenogenetic females with continuous egg production are present over the entire period (Herzig, 1974). The highest numbers can be recorded in June or July, with sometimes a second maximum in August. In deep, stratified, temperate, central European lakes *Diaphanosoma* reaches its maximum population density at the end of August, at the earliest, but usually in the first half of September (Odermatt, 1970; Mittelholzer, 1970; De Bernardi, 1976; a.o.). The mean summer values recorded vary between 7.7 and 252.5×10^3 ind m^{-3}, with maxima of $39.3–391.4 \times 10^3$ ind m^{-3}. Sometimes it happens that the population breaks down for a short time, as in June 1968 or July 1970. This is due to wind-generated turbulences resulting in large amounts of suspended material.

Although the abundance of *Diaphanosoma* is influenced predominantly by temperature and food, the mechanical effect of the turbid material also plays a very important role in mortality, especially in shallow lakes like Neusiedlersee (Herzig, 1975). Fish predation – in early summer by all the newly-hatched fry – should not be ignored as an important elimination factor for this population.

Autumn is characterized by the rapid decrease of the whole population, the appearance of the males (in rather small numbers) and the production of resting eggs. The mean number of the latter found in the mud in winter varies between 26.6 and 71.7×10^3 m^{-2}. Mean individual numbers are between 1.3 and 35.5×10^3 ind m^{-3} (max.: $5.2–110 \times 10^3$ ind m^{-3}).

The outburst and the exponential growth phase of the population takes place in spring; maximum densities are reached in June ($2.2–138.2 \times 10^3$ m^{-3}) and the seasonal means range from $9–38.2 \times 10^3$ ind m^{-3}.

The distribution of biomass exactly parallels that of the numbers. The highest biomass estimates are found in summer, varying between 125 and 1,088 mg dw m^{-3}, the means ranging from 18 to 696 mg dw m^{-3}. The mean biomass values are 0.3 to 153 mg dw m^{-3} (max.: 8–420 mg dw m^{-3}) for the spring season and 2–130 mg dw m^{-3} (max.: 8–420 mg dw m^{-3}) for the autumn.

Daphnia longispina s.l. (Fig. 22.5b, 22.6b)

This species is a typical member of the plankton community of temperate lakes. In Neusiedlersee it is an unimportant species and since 1975 it has been increasingly replaced by *Daphnia pulex*. It shows either two maxima (1970) or one (1971), normally reaching maximum densities in autumn. The maximum density recorded is 5,280 animals per cubic metre (September 1970) with a corresponding biomass of 17 mg dw m^{-3}.

(a)

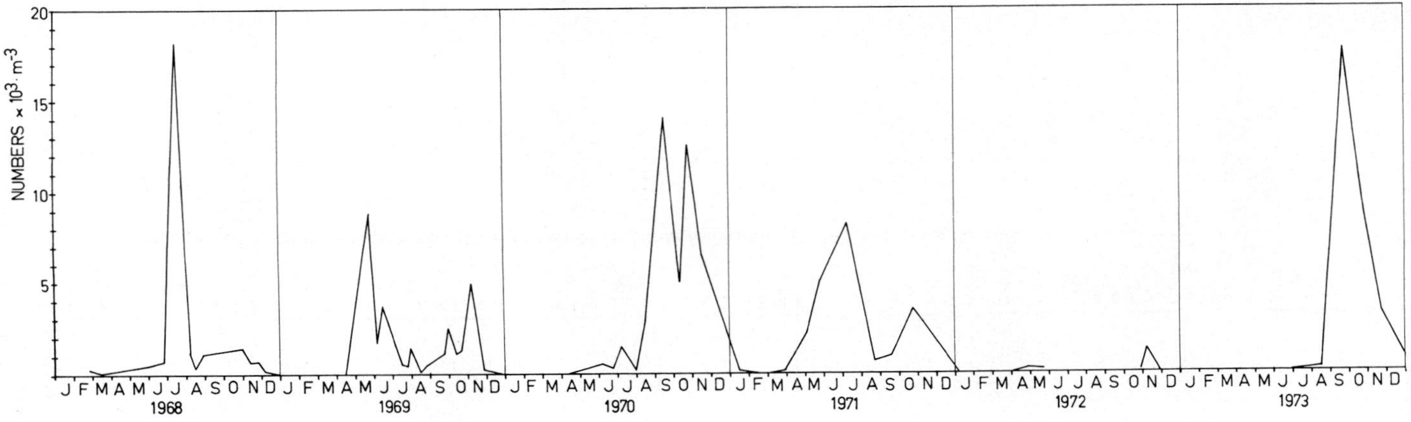

Fig. 22.5. (a) Abundance of *Bosmina longirostris*; (b) Abundance of *Daphnia longispina* (——) and *Ceriodaphnia quadrangula* (----).

(b)

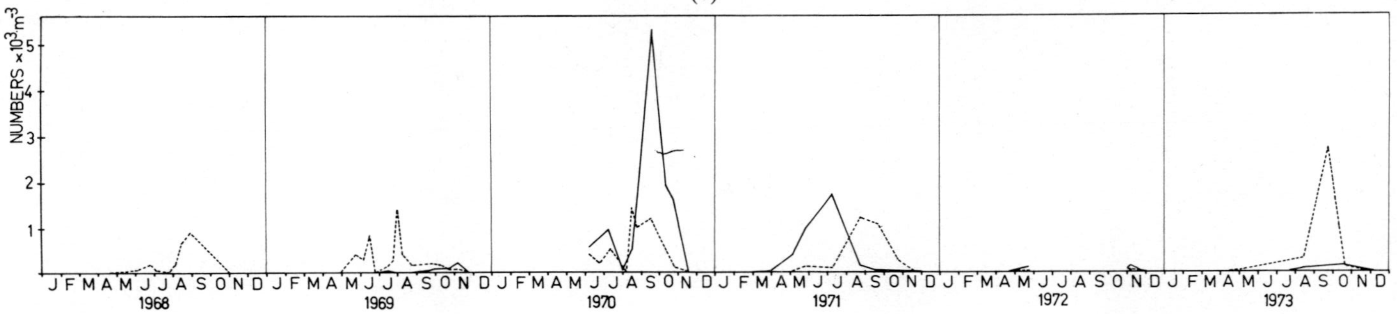

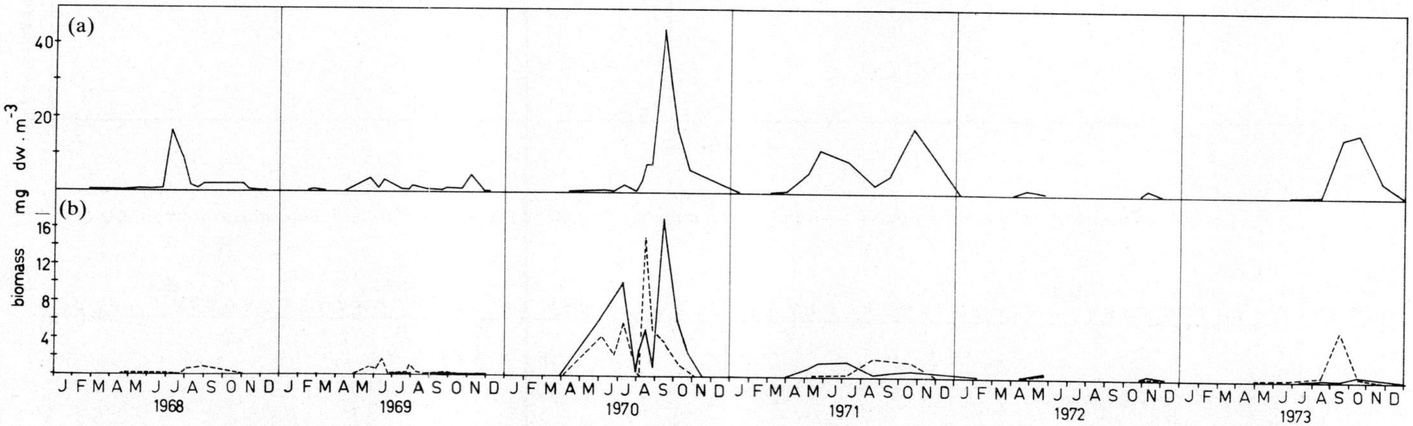

Fig. 22.6. (a) Biomass of *Bosmina longirostris*; (b) Biomass of *Daphnia longispina* (——) and *Ceriodaphnia quadrangula (————).*

Its occurrence is mostly restricted to summer and autumn; *Daphnia longispina* overwinters as resting eggs (ephippia) which start hatching in April. Some specimens can be found even in winter.

Daphnia pulex

Described as a typical member of the ponds in the *Phragmites* belt (Pesta, 1954) it is very rare in the plankton of the open lake. Since 1975 *Daphnia pulex* has been replacing *Daphnia longispina* in the open lake community.

Ceriodaphnia quadrangula (Fig. 22.5b, 22.6b)

Individuals of this species prefer smaller lakes or ponds and are rare in deep and stratified lakes, where they are replaced by *Ceriodaphnia pulchella*, especially under more eutrophic conditions (Flössner 1972).

The maximum numbers in Neusiedlersee vary between 900 and 2,700 animals per cubic metre; in terms of biomass 0.8–15 mg dw m^{-3} can be found.

During the winter months only resting eggs are present (ephippia). The hatching of the ephippia starts in March/April and lasts until the end of May. The highest animal densities are reached in late summer or autumn.

Ceriodaphnia reticulata

It is a member of the *Phragmites* – pond plankton and is found occasionally in the open water in extremely low numbers.

Bosmina longirostris (Fig. 22.5a, 22.6a)

These animals prefer small, eutrophic lakes or ponds, but are also found in the plankton of large but not too oligotrophic lakes as well (Flössner, 1972). The phenology shows great variation, most probably dependent upon the development of phytoplankton (Berg & Nygaard, 1929). In Neusiedlersee this species produces either one maximum in summer (July 1958), or one in spring plus another in autumn (1969 and 1971) or in autumn alone (1973). The densities vary widely over the six-year period. In 1968 the highest numbers occurred in the middle of July, amounting to 18,200 animals per cubic metre or, expressed as biomass, 16 mg dw m^{-3}. A similar density was recorded for September 1972 with 17,700 individuals and a biomass estimate of 18 mg dw m^{-3} in October. The lowest values recorded are 8,750 ind m^{-3} in May and 4,900 ind m^{-3} in November 1969; the biomass estimate from this time is 4 mg dw m^{-3}. In September 1970 the highest biomass found in the 6 years was recorded; it is 44 mg dw m^{-3}.

Bosmina overwinters as resting eggs (ephippia) and hatching takes place in April/May. Sometimes a few species can be found even in winter time.

Cyclops spp. (Fig. 22.3b, 22.4b)

For the following reasons the cyclopoids are treated as a group and not studied at the species level.

(1) Since the majority of the animals found are in the nauplius stage it is very difficult to identify different species (1968–1973: nauplii: 75.8 per cent; adults and copepodides: 24.2 per cent).
(2) Adults and copepodides very rarely reach densities higher than 20×10^3 ind m^{-3}.
(3) At the time of maximum occurrence *Mesocyclops leuckarti* is the predominant species (1970). In addition *Acanthocyclops robustus* and *Thermocyclops crassus* are present; in recent years *Acanthocyclops robustus* has become the dominant cyclopoid species.

The cyclopoids mainly emerge from the *Phragmites* belt or the macrophyte zones. This is shown by a comparison of samples taken from the open lake and from near the vegetation (Table 22.13). As can be seen, the mean

Table 22.13. *Cyclops* spp.: Abundance in summer; a comparison between open lake and near macrophytes. (Numbers $\times 10^3$ m^{-3})

Date	open lake	near macrophytes
26.6.1969	6.0	25.3
1.8.1969	10.5	40.8
7.8.1969	11.1	45.0
22.8.1969	17.3	49.6
8.7.1970	16.4	40.4
31.7.1970	24.1	61.7
6.8.1970	101.0	153.2
12.8.1970	89.6	138.1
21.8.1970	76.7	107.3
22.5.1971	11.7	31.3
11.7.1971	34.4	53.4
21.8.1971	41.8	93.5

numbers for the open lake are always lower than the others, although this might partly be the result of the wind situation. An extreme wind blowing from NW can result in higher numbers near or in the vegetation. But even in the absence of wind cyclopoid numbers are higher near the vegetation (1.8.1969, 7.8.1969, 8.7.1970, 6.8.1970, dates in table 22.13).

Over a longer period of time it becomes obvious (as was pointed out earlier) that their numbers are steadily decreasing, a fact which is associated with the rapid decrease of the macrophytes (Schiemer & Prosser, 1976) (Table 22.14). According to Schiemer (pers. commun.) the macrophyte density is still decreasing: the same can be observed for the cyclopoids. Throughout the investigation period nauplii predominated (75.8 per cent of

Table 22.14. *Cyclops* spp.: Mean summer percentages and numbers per cubic metre compared with the total biomass of *Myriophyllum spicatum* and *Potamogeton pectinatus* (according to Schiemer & Prosser 1976, Table II)

Year	mean summer standing crop % of total N	mean summer standing crop $N \times 10^3\ m^{-3}$	biomass tons dw in Zone I *M. spicatum*	biomass tons dw in Zone I *P. pectinatus*
1968	33.0	22.0		
1969	18.8	26.3		
1970	28.7	64.2	3.12	
1971	10.0	48.5	1.98	43.00
1972	8.7	37.1	0.32	6.05
1973	7.5	26.8		
1976	4.4	7.6	0.00[1]	0.00[1]

[1] with the methods used no estimate possible

the total cyclopoid number) in the plankton of the open lake. Possibly the older stages prefer areas within the vegetation because food conditions are better; this idea is supported by the numbers of adults and copepodides in the open lake and near the vegetation (Table 22.15). In general, older stages are found in larger numbers near the macrophytes or the *Phragmites*, whereas only half as many or even less (mean 30 per cent) are found in the open lake samples.

In general the maximum densities occur in spring (April/May) and late summer (August/September); the highest value ever recorded is $177.6 \times 10^3\ m^{-3}$ or, expressed as biomass, 427 mg dw. m^{-3}.

Table 22.15. *Cyclops* spp.: Adults + copepodides in the open lake and near macrophytes (during summer)

Date	numbers $\times 10^3\ m^{-3}$ open lake	numbers $\times 10^3\ m^{-3}$ near macrophytes
26.6.1969	5.3	12.3
1.8.1969	6.2	22.2
7.8.1969	0.7	21.1
22.8.1969	1.5	38.9
8.7.1970	30.7	65.8
31.7.1970	32.1	29.4
6.8.1970	21.2	50.2
12.8.1970	55.9	68.9
21.8.1970	40.1	90.1
22.5.1971	1.8	48.1
11.7.1971	17.2	52.7
21.8.1971	24.4	86.4

22.4 Horizontal and vertical distribution of crustaceans

The horizontal pattern of zooplankton largely reflects the water movements, which depend upon the wind regime and the morphology of the lake. But physical, chemical and biological factors can also affect the distribution of the animals. The direct effects of wind on plankton distribution have been demonstrated e.g. by Ragotzkie & Bryson 1953, Langford & Jermolajev 1966, Burgis et al. 1973 or George & Edwards 1976, whilst Colebrook (1960) has shown the influence of internal seiche movements on the horizontal distribution of zooplankton.

The development of local aggregations in regions of upwelling and downwelling water shows whether these animals are capable of orientation. If they are positively buoyant they concentrate in regions of downwelling water – as long as their ascending velocity is greater than that of the descending water – or if negatively buoyant they accumulate in regions of upwelling water (Stavn 1971, George & Edwards 1976). On the other hand it is unlikely that the zooplankton can offer much resistance to horizontal movements of water.

In Neusiedlersee a comparison of the actual wind situation and the corresponding water currents – obtained from the lake model simulating different situations – with the distribution pattern of the zooplankton shows close agreement. The prevailing winds on Neusiedlersee are from north-west and south-east (Steinhauser 1970), and the examples given apply mainly to such conditions.

The maps given in figures 22.7, 22.8, 22.9 show the results from integrated samples taken at 19–22 sites on the same day, from the bay of Neusiedl down to the region of Illmitz-Mörbisch and over the whole water column (net-tow, 30 μm mesh size). The data were put into a PDP-11 minicomputer and programmed according to Davis (1973) in a modification by Winkler (unpublished results). The output was used to draw the maps given in the figures.

In smaller lakes, like Eglwys Nynydd, it proved to be sufficient to take into account the wind situation for the previous 24 hours only (George & Edwards 1976), whereas in Neusiedlersee it seemed advisable to consider the last four days.

Major plankton density gradients can be recorded along the wind axis, but local concentrations due to circular movements (e.g. in the centre of the northern part of the lake) can also arise. In general the maximum occurrence of *Arctodiaptomus spinosus* and *Diaphanosoma brachyurum* is found upwind and the distribution pattern of both is very similar (Fig. 22.7, 22.8). However, differences in the distribution pattern of several stages of *Arctodiaptomus spinosus* (adults, copepodides, nauplii) become evident (Fig. 22.9), the adults showing a well established upwind concentration, that of the copepodides being less pronounced. It seems that the adults and, to a slighter degree, the copepodides, show a tendency to concentrate deeper in the water in accordance with their negative phototactic reaction; they try to resist upward displacement and are concentrated upwind due to the reverse water currents. The nauplii show a clear downwind concentration, reflecting the absence of

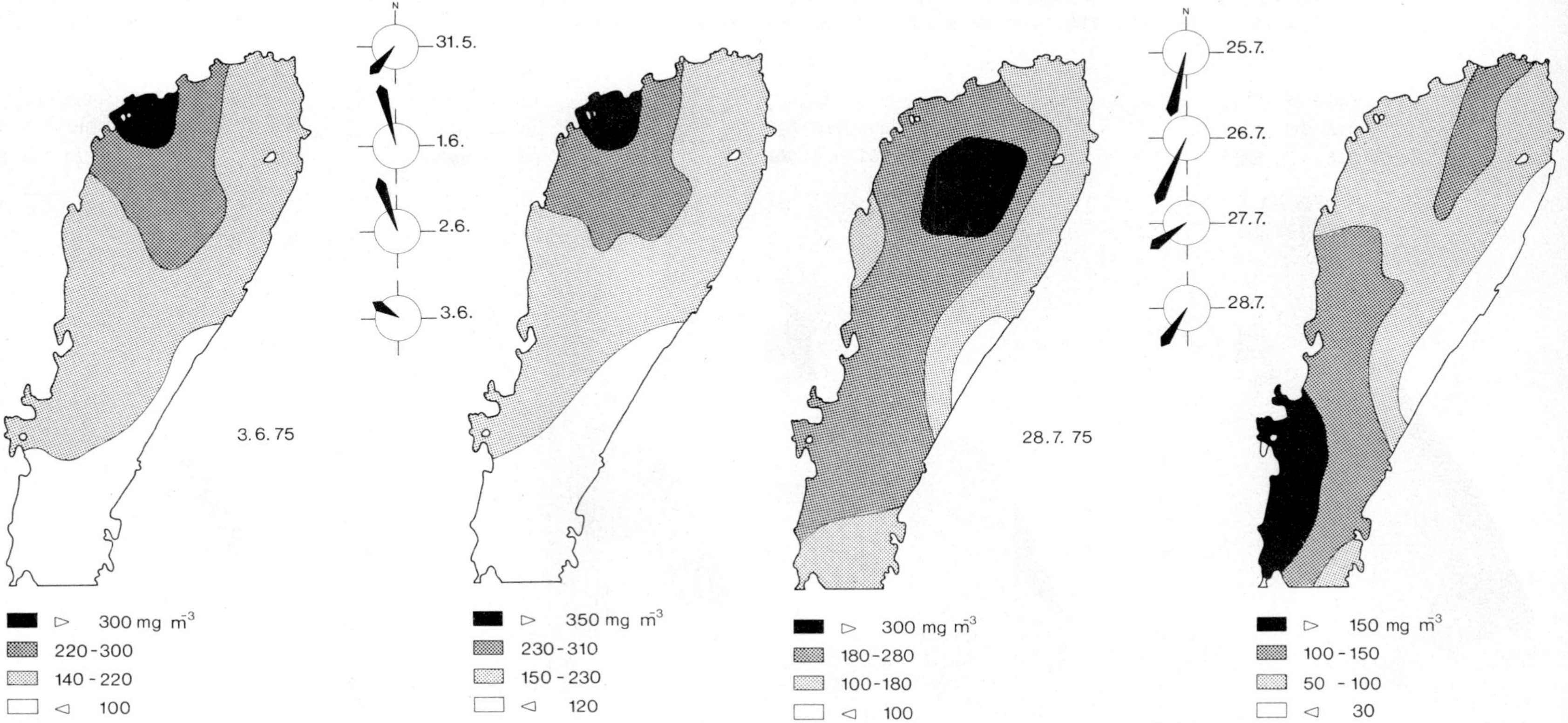

Fig. 22.7. Horizontal distribution of *Arctodiaptomus spinosus* and *Diaphanosoma brachyurum*, 3.6.1975: left–*A. spinosus*; right–*D. brachyurum*. 28.7.1975: left–*A. spinosus*; right–*D. brachyurum*, (mean wind direction and wind speed (per day) are indicated by a vector, circle indicates a wind speed of 2 m/sec.).

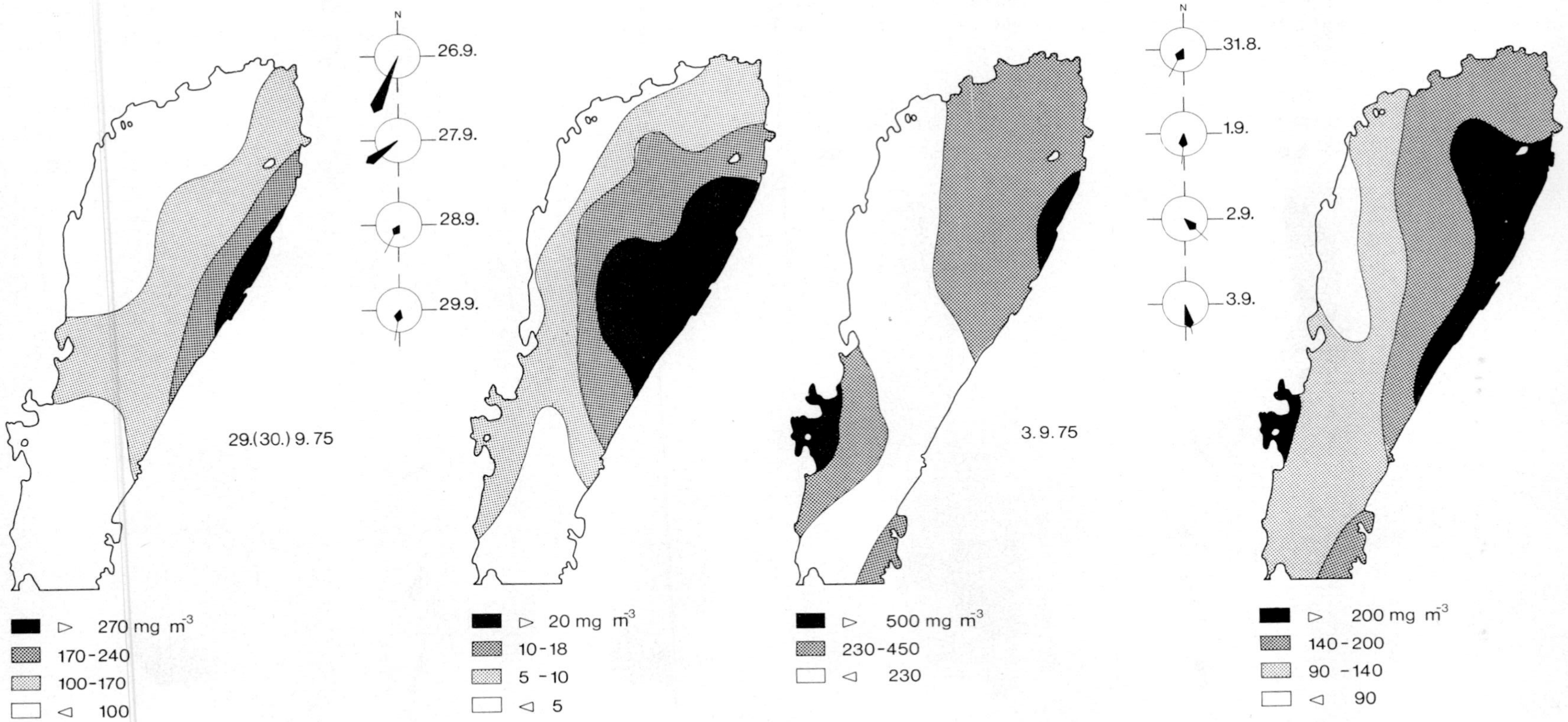

Fig. 22.8. Horizontal distribution of *Arctodiaptomus spinosus* and *Diaphanosoma brachyurum*, 29.9.1975: left-*A. spinosus*; right-*D. brachyurum*, 3.9.1975: left–*A. spinosus*; right–*D. brachyurum*, (mean wind direction and wind speed (per day) are indicated by a vector, circle indicates a wind speed of 2 m/sec).

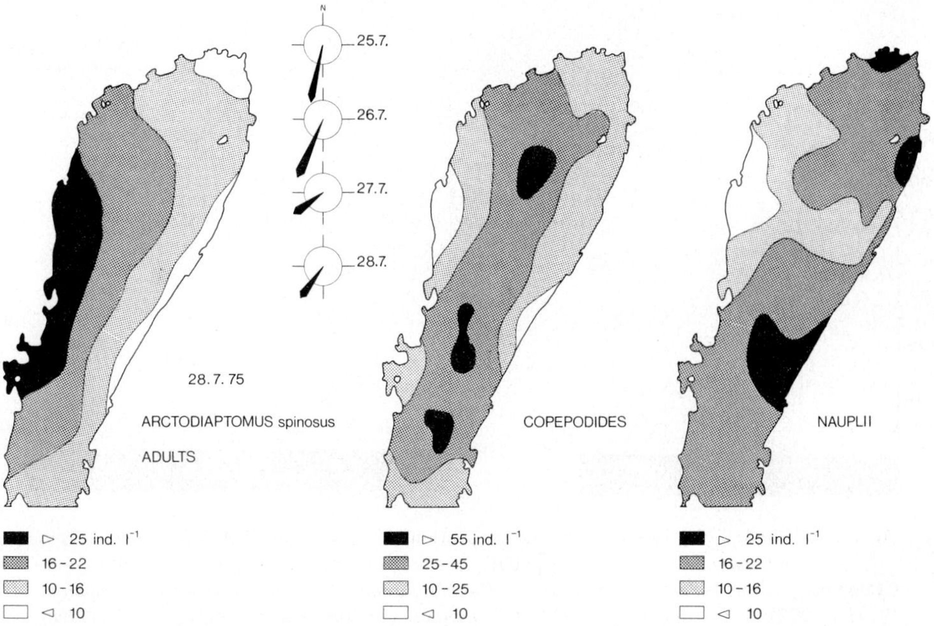

Fig. 22.9. Horizontal distribution of the stages of *Arctodiaptomus spinosus* (mean wind direction and wind speed (per day) are indicated by a vector, circle indicates a wind speed of 2 m/sec).

depth selection; they are transported passively downwind and are partly returned by the deeper reverse currents. The distribution of the animals in wind-sheltered situations is very much influenced by the previous wind situation, but despite this a tendency towards a less pronounced horizontal distribution can be seen.

Comparing the horizontal distribution of zooplankton with that of phytoplankton and inorganic particles (Dokulil 1979, Chapter 15, 17) a striking difference becomes evident. The highest densities of the latter are always found concentrated downwind (like the nauplii), whereas the zooplankton, except the less mobile and neutrally buoyant stages (nauplii), is concentrated upwind.

Summarizing the results it can be stated that the distribution patterns of the animals are the result of an interaction of wind-induced water movement and consequent transport of the organisms, and their power of vertical movement and depth selection.

Vertical stratification is very rare in Neusiedlersee, because the water is so well mixed. Even after a week of calm weather (Fig. 22.10a, first week of August 1970) no distinct stratification is seen. There is a tendency for adult diaptomids and cyclopoids to prefer the bottom layers, but the copepodides, nauplii and *Diaphanosoma* show a random distribution throughout the entire water column. In winter, on the other hand, when the whole lake is covered with ice, vertical stratification is established (Fig. 22.10b). All animals, except the nauplii, prefer the bottom layers, which is of two-fold advantage. In the first place, water temperatures are higher near the bottom than in the upper

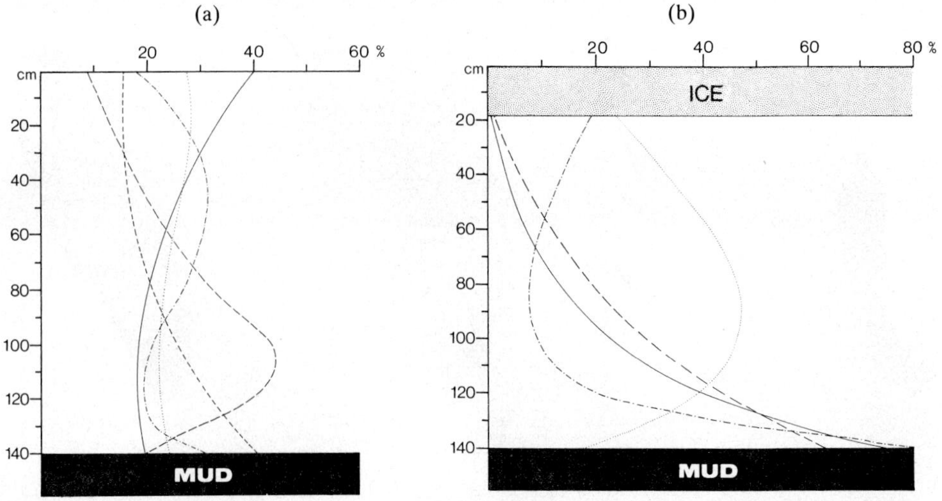

Fig. 22.10. Vertical distribution of crustaceans. (a) summer situation (6.8.1970): —— nauplii, ········ *Arctodiaptomus spinosus*, copepodides, ------- *Arctodiaptomus spinosus*, adults, -------- Cyclopidae, adults + copepodides, -·-·-·- *Diaphanosoma brachyurum*; (b) winter situation (9./31.1.1973): ······· nauplii, —— *Arctodiaptomus spinosus*, copepodides, —— *Arctodiaptomus spinosus*, adults, -·-·-·- Cyclopidae, adults + copepodides.

layers and secondly there is more food. Whereas the phytoplankton values are at their lowest in winter, growth of the epipelic algae on the mud surface is at a maximum at this time. Gut analysis of male and female *Diaptomus* confirmed this assumption regarding bottom feeding in winter.

22.5 A comparison for the last 25 years

As was mentioned earlier, all species show a higher tolerance limit for alkalinity and/or salinity. Comparing the species list from 1967 and 1973 with those from earlier years (Geyer & Mann 1939, Benda 1950, Pesta 1954, Ruttner-Kolisko & Ruttner 1959, Zakovsek 1961) it becomes obvious that *Moina rectirostris* (= synonym of *Moina micrura* according to Goulden 1968 and Smirnov 1976), which was described as being very abundant until 1950 (Pesta 1954), has not been found since (Zakovsek 1961, Herzig 1973, 1974, 1975). *Moina rectirostris* can tolerate high salinities (Löffler 1961) and alkalinities above 200 meq l^{-1} (Herzig 1975).

On the other hand, it is well known that in the last 40 years the salt content of the lake water has been steadily decreasing in parallel to the rising water level, as illustrated by a few examples given below.

Author	depth in cm	salt content, g l^{-1}
Varga & Mika 1937	50– 70	3.5–20.0
Hock 1957	80–100	1.8
Knie 1958	80–100	1.5
Neuhuber 1971	130	1.0

The disappearance of *Moina rectirostris* is closely connected with the decreasing salt content of the water. A similar result is recorded for *Moina hutchinsoni* (Edmondson 1969) in Lake Lenore (Washington, USA). But *Moina rectirostris* is still an important component of the plankton communities of the shallow, more alkaline lakes east of Neusiedlersee (Seewinkel) (Löffler 1959, Herzig unpublished data), where *Diaphanosoma brachyurum* is rather rare.

When the data for the years 1970–1973 are compared with those obtained by Zakovsek (1961) for 1950–1952 a striking difference in the *Diaphanosoma* density becomes apparent. It seems that since the disappearance of *Moina rectirostris* and since the salt content of the water has sunk to its present low level, *Diaphanosoma* has taken over the niche previously occupied by *Moina*. Nowadays *Diaphanosoma* is an important component of the lake plankton.

Figure 22.11 and Table 22.16 show the changes in total crustacean numbers and biomass from 1950–1973 (values are seasonal means). There is obviously a marked difference between the 50's and the 70's. Numbers and biomass have increased dramatically since 1970, reaching maximum values in 1971, subsequently decreasing somewhat, but not returning to the low values of the years 1968 or 1969. A strongly marked increase is evident, especially for the spring and summer months, and to a lesser extent for autumn and winter.

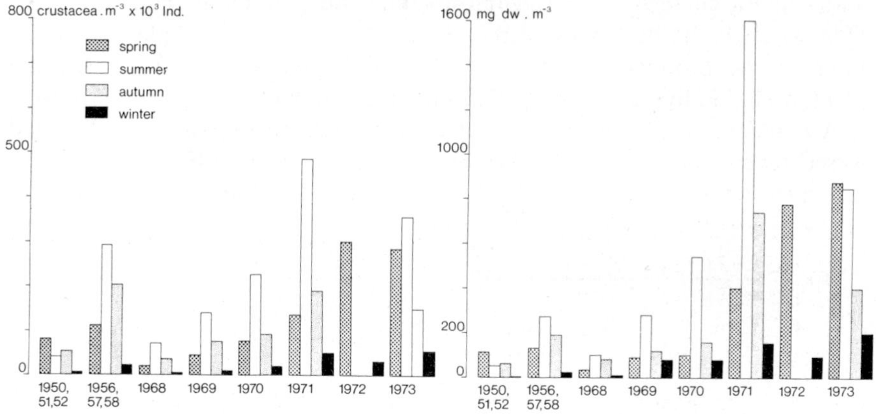

Fig. 22.11. Total crustacean numbers and biomass (values are means per season) 1950/51/52 – Zakovsek 1961, 1956/57/58 – Ruttner unpublished results.

Again it must be mentioned that the rather high values of 1956/57/58 are most likely due to the low number of sampling dates.

In general it seems that since 1970 conditions have been much more favourable to growth. Since the wind influence has not altered, these tremendous changes may to some extent be due to better climatic conditions (mild winter, short duration of ice cover) but in the main to a better food supply. This is reflected in the phytoplankton development in these years (Dokulil 1979, Chapter 17). The development of both phyto- and zooplankton is shown in Fig. 22.21 (annual means).

Table 22.16. Seasonal means of numbers and biomass

a) Crustacea mean numbers × 10^3 m^{-3}

	1950/51/52	1956/57/58	1968	1969	1970	1971	1972	1973
spring	81.6	107.5	16.6	41.7	73.0	136.2	296.6	283.9
summer	39.2	292.2	66.6	139.9	223.7	484.6		357.1
autumn	51.9	200.0	33.2	75.4	89.9	168.6		150.1
winter	2.5	19.5	2.1	10.2	19.5	50.0	31.5	55.0

b) Crustacea mean biomass mg dw (dry weight) m^{-3}

	1950/51/52	1956/57/58	1968	1969	1970	1971	1972	1973
spring	112	127	36	85	105	396	774	882
summer	52	268	101	279	539	1,598		847
autumn	57	190	81	120	157	742		403
winter	3	26	5	19	78	156	94	180

On a larger geographic scale, long-term studies from Europe and Asia (Lake Dalnee, Baikal, Rybinsk Reservoir, Lake Balaton) reveal similar fluctuations but at different times, which may support the idea of a connection with climatic changes. A period of low values can be found in the mid 50's and 60's, whereas the beginning of the 70's is characterized by higher values for biomass and production (Smirnov 1973, Sebestyén 1958, Zánkai & Ponyi 1971, 1972, Ponyi & Zánkai 1972, Ponyi 1975, Nauwerck et al. in press).

A comparison of the numbers of rotifers and crustaceans from Neusiedlersee reveals an inverse relationship, as can be seen in Figs. 22.12/22.13. Apart from the data of 1956/57/58, which result from too few sampling dates,

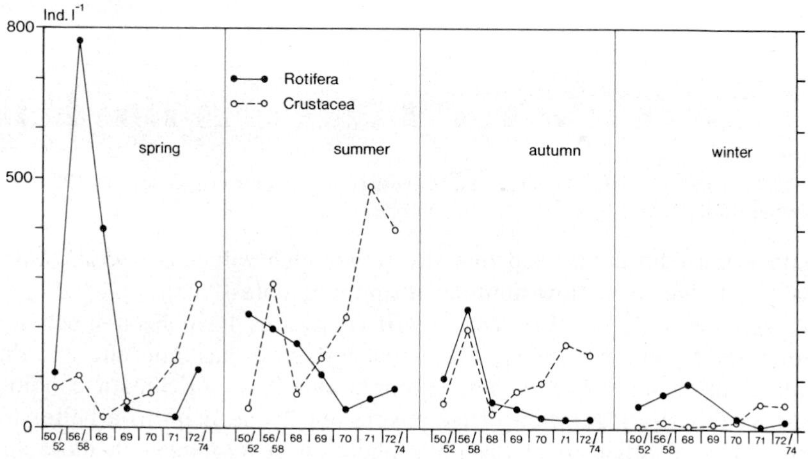

Fig. 22.12. Seasonal occurrence of rotifers (●——●) and crustaceans (○——○) from 1950–1974.

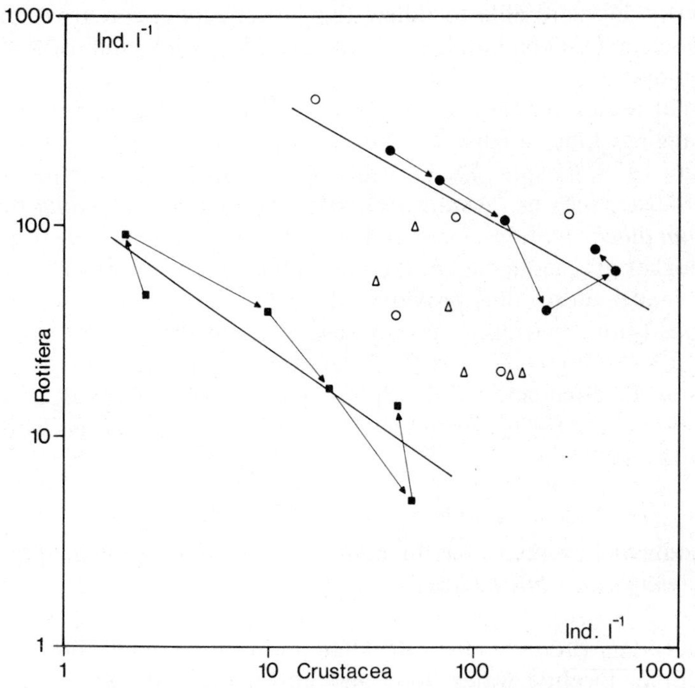

Fig. 22.13. Numerical relation between the abundance of rotifers and crustaceans (1950/51/52, 1968, 1969, 1970, 1971, 1972/73/74) ● summer, ■ winter, ○ spring, △ autumn.

the same tendency can be seen in all seasons, especially in summer and winter. If the same comparison is made for two other shallow lakes, Lake Balaton (Sebestyén 1958, Zánkai & Ponyi 1971, 1972, Ponyi & P. Zánkai 1972) and Lake d'Endine (Barbanti et al. 1974), a similar inverse relation, particularly well marked in the colder season (December–April), can be found. In deeper lakes, however, e.g. Piburger See (Schaber 1975, 1976) or Lago Mergozzo (De Bernardi & Soldavini 1976), an inverse relationship is not found, but the numbers of crustaceans and rotifers frequently parallel one another.

Although there is little direct evidence of competition between rotifers and crustaceans (Dumont 1977), this seems nevertheless to be the reason for the inverse relation between these two plankton components, as shown in Fig. 22.12/22.13. Especially in summer, high temperatures are responsible for the rapid turnover and high metabolic rates seen in both groups. It is possible that there is too little food and therefore the two populations cannot develop equally well.

In winter, on the other hand, bacteria and phytoplankton biomass are at their lowest (Dokulil 1975), but the highest biomass of epipelic algae is found (Prosser unpublished data). Animals like the copepodides and adults of copepods, which are able to utilize this bottom food, can grow quite well; animals feeding on planktonic material only, like nauplii and rotifers, are at a disadvantage.

Another reason for these alterations in zooplankton composition might be the changing food quality in Neusiedlersee over recent years (Kusel-Fetzmann 1979, Chapter 16, Dokulil 1979, Chapter 17). The percentage of colonial algae, such as *Pediastrum duplex*, *Lobocystis dichotoma* or *Dictyosphaerium pulchellum*, has increased remarkably. This may have been caused by the higher grazing impact of the more abundant herbivorous zooplankton on the smaller algae, thus providing the colonial algae with better growing conditions (more nutrients). In any case some of the colonial algae can be used as food by copepods and Cladocera but not by the rotifers. A detailed analysis of the abundance of the different size classes of the algae and more information about the life histories of the various rotifer species would throw light on this problem.

22.6 Production estimates for the main species: *Arctodiaptomus spinosus* and *Diaphanosoma brachyurum*

Since secondary production is calculated and not measured directly, time of investigation, locality, water body and information concerning life history including different larval stages are important data. Two essential facts particularly influence the reliability of production estimates: the number of samples per time unit (Hillbricht-Ilkowska & Weglenska 1970) and information on growth rates. Some reviews on this subject already exist (Hillbricht-Ilkowska & Patalas 1967, Edmondson & Winberg 1971, Winberg 1971, Edmondson 1974, Bottrell et al. 1976, The Plankton Ecology Group in prep.).

Turnover time or turnover rate is commonly used to give an estimate of production although only in the case of continuously reproducing animals (e.g. Wright 1965, Heinle 1966, Cummins et al. 1969, Burgis 1974, George & Edwards 1974, Herzig 1974). This is a fairly direct method requiring only information on population size and duration of embryonic development. It implies the basic assumption of a steady state, which in fact never exists in natural populations. The great advantage of this method is that only one parameter, the duration of embryonic development, is taken from laboratory experiments and this is purely temperature dependent. Other methods, e.g. Winberg's method (Winberg 1971), which employs the growth increments of the different stages, require more additional information about the duration of different postembryonic stages. These postembryonic development times are highly dependent upon food (Weglenska 1971). Up to the present, little information has been available on this subject and very often production is overestimated due to using laboratory data obtained under optimum food conditions.

For the reasons mentioned above, production estimates for the Neusiedlersee populations were calculated from turnover times. Since age distribution is not stable it is assumed that in the one day to which the calculation applies, no particular change in age distribution and weight of the different animals takes place. The values for a week, month or year are found by integrating the curve resulting from as many daily estimates as possible. In the course of this investigation it turned out that in addition to the problem of age distribution, calculations at the time of low egg (embryo) numbers result in low birth- and turnover rates and thus in an underestimate of production. As long as this can be excluded, the results obtained by this method are in close agreement with Winberg's method (The Plankton Ecology Group, in prep.).

Figure 22.14 shows the daily production for the two species throughout the investigation period and the total production per species and year. Two facts

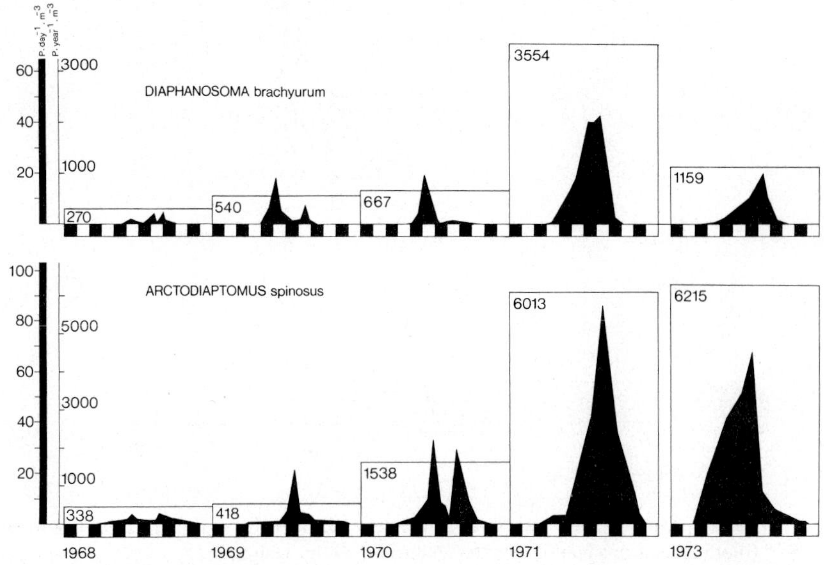

Fig. 22.14. Daily and annual production of *Arctodiaptomus spinosus* and *Diaphanosoma brachyurum* (production expressed as dry weight per m^3).

become obvious: first, the highest values occur in summer, when good food conditions and high temperatures result in a rapid turnover of the population. Secondly, an increase in both daily and annual production becomes apparent, for the same reasons as discussed earlier in connection with numbers and biomass changes. In addition the importance of *Arctodiaptomus spinosus* as a plankton organism of Neusiedlersee is revealed.

Figures 22.15/22.16 show the relation of production to biomass throughout different years (1969 and 1971). They illustrate – for both species – the increase in biomass as well as in production, and the dynamic aspect over longer periods of time becomes apparent.

In the case of *Arctodiaptomus spinosus* (Fig. 22.15) it can be seen that production increases in spring – much earlier in 1971 than in 1969 – and is followed by a marked increase in biomass. A rather stable biomass situation

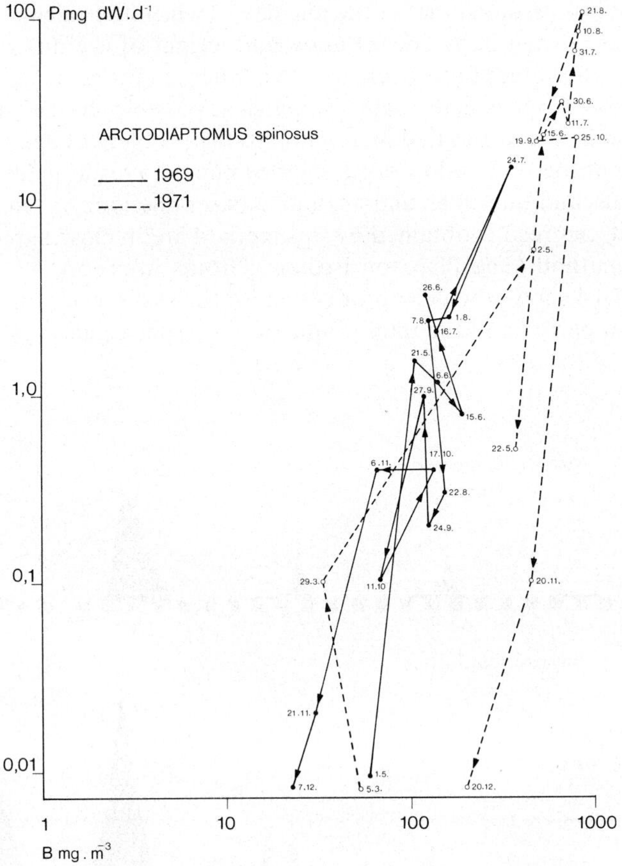

Fig. 22.15. Arctodiaptomus spinosus: Relation between biomass (B) and production (P) for 1969 and 1971 (biomass and production per day expressed as dry weight m^{-3}).

occurs in summer, whereas the production per day varies tremendously. The fact that there is no increase in biomass may indicate fish predation. In autumn, production decreases rapidly, whilst the biomass values drop more slowly.

Diaphanosoma brachyurum (Fig. 22.16) shows a similar pattern. An increase of production in spring (April/May) is followed by high biomass values. In summer again, production values are high, while biomass values increase only slightly. The reason for this is probably that fish predation checks the growth of the *Diaphanosoma* population. At the end of August and beginning of September production drops rapidly due to the decrease in temperature; the biomass shows a similar but less dramatic tendency.

In Figs. 22.17/22.18 again production of *Arctodiaptomus* and *Diaphanosoma* is plotted against biomass for the summer months only. The same trend towards steadily increasing production and biomass values can be recognized in both species. The relation between biomass and production

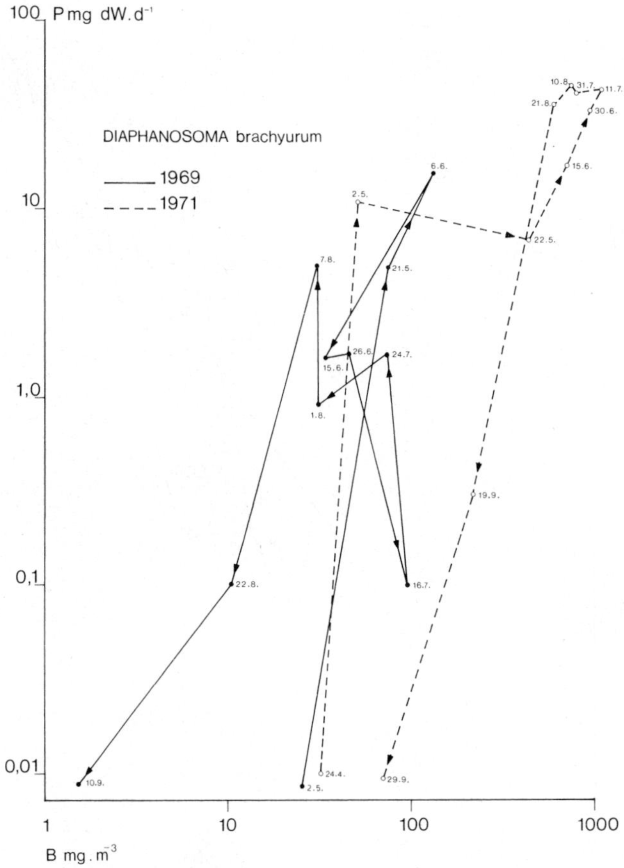

Fig. 22.16. Diaphanosoma brachyurum: Relation between biomass (B) and production (P) for 1969 and 1971 (biomass and production per day expressed as dry weight per m^3).

over a period of 6 years can be defined by calculating for each species the regression for all the values obtained. A logarithmic scale was used in the graph and for the calculation; the following equations are obtained by transformation to a linear scale:

$P = 0.0028 \times B^{1.427}$ ....*Arctodiaptomus spinosus*
$P = 0.029 \times B^{1.01}$ ....*Diaphanosoma brachyurum*

(P-production and B-biomass are expressed as mg dry weight m^{-3}) Hillbricht-Ilkowska (1977) finds a similar relation between production and biomass for *Daphnia cucullata* from Mazurian lakes. She gives examples for a mesotrophic lake (Taltowisko: $P = 0.0132 \times B^{1.251}$) and two eutrophic lakes (Mikolajskie: $P = 0.397 \times B^{1.099}$ and Śniardwy: $P = 0.0119 \times B^{1.302}$).

Table 22.17 summarizes some equations of the P to B relation, calculated for different lakes and different species using data from the literature. It becomes obvious that the slopes of all the regressions are very similar. On the other hand, it can be seen that for lakes with a similar range of biomass values

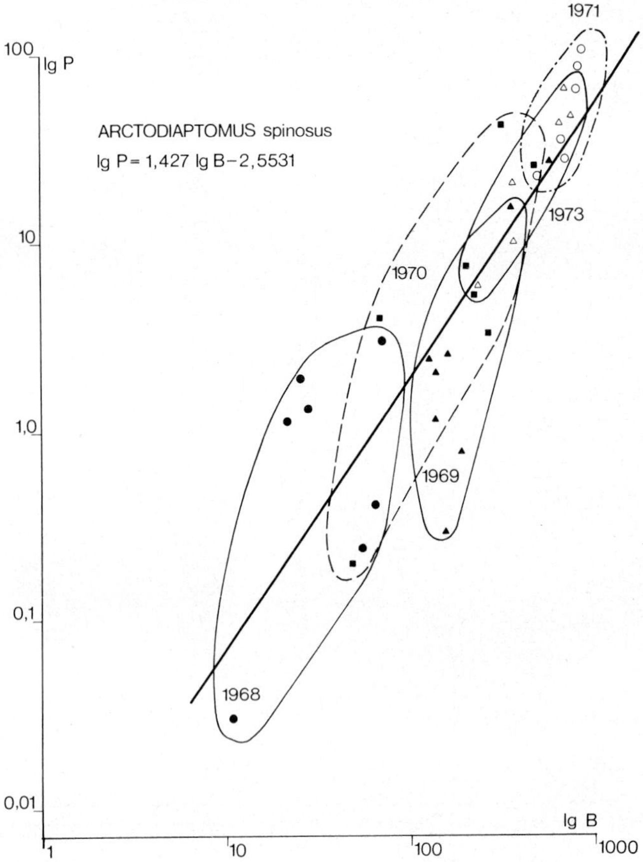

Fig. 22.17. Arctodiaptomus spinosus: Relation between biomass (B) and production (P) (June–September) (biomass and production as mg dry weight m^{-3}).

and similar slopes, the second character of the regression reflects to some extent the nutritional capacity of the water body, e.g. the *Eudiaptomus* population from Millstätter See (mesotrophic) and Ossiacher See (eutrophic), or the *Diaphanosoma* populations from Mikolajskie (eutrophic) and Neusiedlersee (mesotrophic). The relation of the *Eudiaptomus* population from an eutrophic, temperate lake (Mikolajskie) and the *Thermocyclops crassus* population from an eutrophic, tropical lake (Lake George, Uganda) are very similar. This shows that the summer values for a copepod population of an eutrophic, temperate lake are in the same order of magnitude as the values of the tropical lake whatever the species studied.

These relationships can be used to obtain probable production values of a species, as long as reasonable estimates of the mean biomass are available and the equations are based on long-term observations.

Another relation between production and biomass, the P/B (production/biomass) coefficient, is used as an index of productivity with respect to a given population under known conditions (Winberg 1971). It is fully realistic

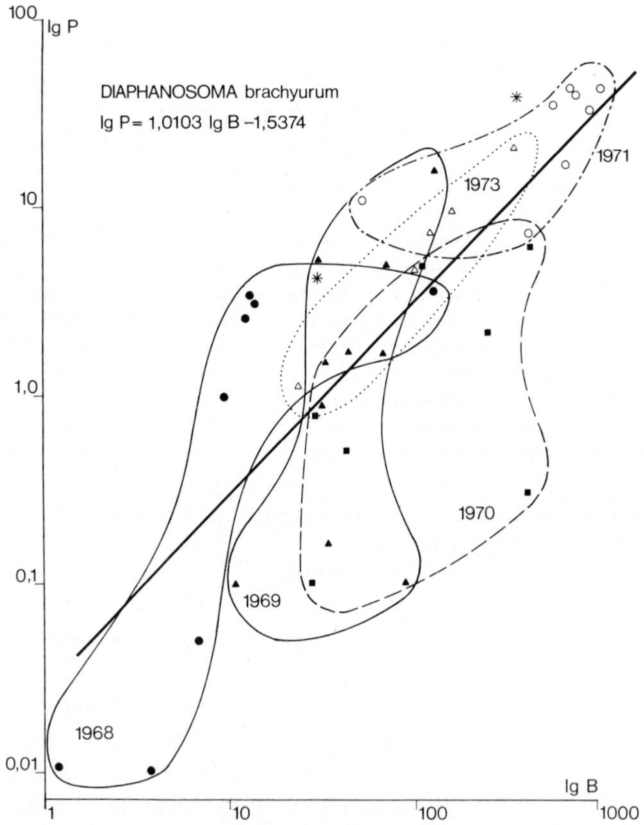

Fig. 22.18. Diaphanosoma brachyurum: Relation between biomass (B) and production (P) (May–September) (biomass and production as mg dry weight m^{-3}).

and significant for stationary populations, i.e. populations with a constant age structure and biomass, which applies only over a short time interval (e.g. one day, see above). It is computed per day, month or year, which gives different values, but for the reason mentioned above the most valid ratios are those calculated on a daily basis.

Comparing P/B values for several diaptomids over a large geographic area and a longer period of time (at least one year or season) a difference between populations from subarctic, temperate and tropical lakes becomes apparent (Table 22.18). The same can be shown for *Diaphanosoma brachyurum* (Table 22.19). Lakes within the same climatic region differ according to their trophic status. In temperate regions the *Diaptomus* populations from eutrophic lakes show P/B values between 0.1 and 0.218 (mean for May–October). This agrees very well with the idea of Patalas (1970) who found that the P/B ratio tends to increase in proportion to the productivity of a lake. Pečen (1965) and Shushkina (1966) note the same tendency of higher P/B ratios in lakes with better food conditions. On the other hand, Hillbricht-Ilkowska (1977) mentions that this coefficient is subject to relatively high variability and does not correlate either with temperature or with the amount of food available. This statement is based mainly on results from eutrophic lakes and from one

Table 22.17. Relation between biomass and production from different lakes for the summer months

species	lake	trophic state	equation
*Eudiaptomus graciloides**	Mikolajskie (Poland)	eu	$P = 0.0677 \times B^{1.082}$
Arctodiaptomus spinosus	Neusiedlersee (Austria)	meso	$P = 0.0028 \times B^{1.427}$
*Eudiaptomus gracilis***	Millstätter See (Austria)	meso	$P = 0.0117 \times B^{1.256}$
*Eudiaptomus gracilis***	Ossiacher See (Austria)	eu	$P = 0.0728 \times B^{1.136}$
*Eudiaptomus gracilis***	Wörthersee (Austria)	meso	$P = 0.0091 \times B^{1.57}$
*Thermocyclops crassus****	Lake George (Uganda)	eu	$P = 0.0924 \times B^{0.972}$
*Daphnia cucullata**	Mikolajskie (Poland)	eu	$P = 0.0326 \times B^{1.319}$
*Diaphanosoma brachyurum**	Mikolajskie (Poland)	eu	$P = 0.0918 \times B^{1.185}$
Diaphanosoma brachyurum	Neusiedlersee (Austria)	meso	$P = 0.0290 \times B^{1.01}$

* calculated from Hillbricht-Ilkowska & Weglenska 1970 and Weglenska 1971
** calculated from Herzig, unpublished data
*** calculated from Burgis 1974, for the whole year

area of Poland (Mazurian lakes) only, and this may account for the failure to find a clear-cut relationship.

Higher P/B ratios are found in tropical lakes which show no marked variability in P/B throughout the year. But even in oligotrophic, subarctic lakes (Krugloe) relatively high P/B values are found, e.g. for *Eudiaptomus graciloides*, and are probably caused by relatively high production values in combination with low biomass values. The high production might be the result of more favourable food conditions than those found in the oligotrophic subarctic lake Krivoe. Thermal stratification can occur in Lake Krivoe, which has a mean depth of 12 m, but not, or only to a slight extent, in Lake Krugloe, which has a mean depth of 1.5 m. The lakes differ considerably in the total content of organic matter, the sediments of Krugloe being much richer in organic matter than those of Krivoe. Lake Krugloe, being polymictic, offers the plantkon animals an additional benthic food supply and hence the possibility of higher productivity than those from Lake Krivoe. This is reflected in the relatively high P/B values.

The 6-year mean of the P/B ratio for the period May–October calculated for the *Arctodiaptomus spinosus* population of Neusiedlersee is found to lie between values calculated for populations – mainly of the genus *Eudiaptomus* – from oligotrophic and mesotrophic lakes (Table 22.18). The P/B of the Neusiedlersee population is nearly three times higher than that of the population of *Arctodiaptomus spinosus* and *Arctodiaptomus bacillifer* from Lake Sevan. The mean P/B (May–October) increases from less than 0.02 (1968/69) to more than 0.034 (1975/76). The highest ratios are found in summer, with high temperatures and the best food supply (Table 22.20); the mean summer P/B is similar to values found for eutrophic, temperate lakes, maximum P/B ratios even reaching figures calculated for populations from tropical lakes.

Table 22.18. Diaptomidae; mean daily P/B from different lakes (mean for the period May–October)

Lake	trophic state	species	P/B	author
Krivoe (USSR) 1968	oligo	*Eudiaptomus graciloides*	0.03	Winberg 1972
Krivoe (USSR) 1969	oligo	*Eudiaptomus graciloides*	0.039	Winberg 1972
Krugloe (USSR) 1968	oligo	*Eudiaptomus graciloides*	0.07	Winberg 1972
Krugloe (USSR) 1969	oligo	*Eudiaptomus graciloides*	0.08	Winberg 1972
Krasnoe (USSR)	meso	*Eudiaptomus gracilis*	0.068	Andronikova et al. 1972
Erken (Sweden)	meso	*Eudiaptomus graciloides*	0.01	Nauwerck 1963
Naroch (USSR)	meso	*Eudiaptomus graciloides*	0.05	Winberg 1972
Myastro (USSR)	eutr	*Eudiaptomus graciloides*	0.055	Winberg 1972
Batorin (USSR)	eutr	*Eudiaptomus graciloides*	0.064	Winberg 1972
Warniak (Pol.) 1967	eutr	*Eudiaptomus graciloides*	0.106	Hillbricht-Ilkowska & Weglenska 1973
Warniak (Pol.) 1968	eutr	*Eudiaptomus graciloides*	0.218	Hillbricht-Ilkowska & Weglenska 1973
Mikolajskie (Pol.)*	eutr	*Eudiaptomus graciloides*	0.12	Weglenska 1971
Neusiedlersee (Austr.)**	meso	*Arctodiaptomus spinosus*	0.03	Herzig this paper
Millstätter See (Austr.)***	meso	*Eudiaptomus gracilis*	0.026	Herzig unpublished data
Weissensee (Austr.)***	oligo	*Eudiaptomus gracilis*	0.026	Herzig unpublished data
Ossiacher See (Austr.)***	eutr	*Eudiaptomus gracilis*	0.117	Herzig unpublished data
Wörthersee (Austr.)***	meso	*Eudiaptomus gracilis*	0.059	Herzig unpublished data
Clear Lake (USA)	oligo	*Leptodiaptomus minutus*	0.009	Schindler 1972
Sevan (USSR)	oligo	*Acanthodiaptomus denticornis*	0.014	Meshkova 1952
Sevan (USSR)	oligo	*Arctodiaptomus bacillifer*	0.014	Meshkova 1952
Sevan (USSR)	oligo	*Arctodiaptomus spinosus*	0.011	Meshkova 1952
Tchad (Tchad)****	eutr	*Tropodiaptomus incognitus*	0.062	Leveque et al. 1972
Tchad (Tchad)****	eutr	*Tropodiaptomus incognitus*	0.066	Leveque et al. 1972

* August only
** mean for the period 1968–1976
*** summer mean for the years 1967–1971
**** July only

Table 22.19. *Diaphanosoma brachyurum*; mean daily P/B ratio from different lakes (means for May–October)

Lake	trophic state	P/B	Author
Krasnoe (USSR)	meso	0.128	Andronikova et al. 1972
Warniak (Pol.) 1967	eutr	0.165	Hillbricht-Ilkowska & Weglenska 1973
Warniak (Pol.) 1968	eutr	0.288	Hillbricht-Ilkowska & Weglenska 1973
Mikolajskie (Pol.)*	eutr	0.24	Weglenska 1971
K. P. St. Cool. Res. (USSR)**		0.136	Pidgaiko et al. 1972
K. P. St. Cool. Res. (USSR)***		0.211	Pidgaiko et al. 1972
Neusiedlersee (Austria)****	meso	0.143	Herzig this paper
Millstätter See (Austria)*****	meso	0.086	Herzig unpublished data
Weissensee (Austria)*****	oligo	0.086	Herzig unpublished data
Ossiacher See (Austria)*****	eutr	0.136	Herzig unpublished data
Wörthersee (Austria)*****	meso	0.155	Herzig unpublished data
Tsimliansk Res. (USSR)		0.132–0.769	Glamazda 1971

* August only
** Kurakhov's Power Station Cooling Reservoir, July, unheated part
*** K. P. St. C. R., July, heated part
**** mean for the period 1968–1976
***** summer mean for the years 1967–1971

Values for Cladocera are generally higher and, even in colder waters, ratios of 0.1 can be reached (Krasnoe, Table 22.19). This is probably because their postembryonic development is shorter than that of the copepods. The range of the P/B ratios for Cladocera is about 0.003–0.3 (Nauwerck in prep.) For *Diaphanosoma brachyurum* the mean P/B varies between 0.086–0.288 (Table 22.19). Summer means for the temperate regions again reflect the trophic status of the lakes. The highest values of any year are reached when growth of the population is exponential, as can be seen for *Diaphanosoma brachyurum* from Neusiedlersee (Table 22.20); at the same time the population shows a very high proportion of young animals which is in agreement with Zaika and Malatovtskaja (both cited according to Winberg 1971), who found that P/B values are higher in younger than in older populations. Bearing this and the value for the Neusiedlersee population in mind, the extremely high P/B ratio for the Tsimliansk Reservoir population (Glamazda 1971) – 0.769 – is probably due to its having been estimated at the time of exponential growth (Table 22.19), rather than to erroneous calculation.

The importance of temperature for *Diaphanosoma brachyurum* is shown by the data of the Kurakhov Power Station Cooling Reservoir, where the ratios calculated for the unheated part (23°C) are lower than those for the heated part (28°C). The populations from Neusiedlersee and Lake Krasnoe provide another example. Both lakes are shallow and of the same trophic status, but summer water temperatures are approximately 10°C higher in Neusiedlersee than in Lake Krasnoe, and this is reflected by higher P/B ratios (Table 22.19).

Table 22.20. *Arctodiaptomus spinosus* and *Diaphanosoma brachyurum*; mean daily P/B ratio for different seasons; means are for the years 1968–1976; in parenthesis 95 per cent confidence limits

Arctodiaptomus spinosus	spring	summer	autumn
	0.018 (0.026–0.010)	0.048 (0.059–0.038)	0.013 (0.019–0.008)
Diaphanosoma brachyurum	time of exponential growth	summer	autumn
	0.434 (0.530–0.340)	0.084 (0.108–0.059)	0.016 (0.029–0.002)

The mean summer P/B in Neusiedlersee is very similar to the values for Millstätter See and Weißensee; the lowest values of the year are recorded in autumn, when *Diaphanosoma* slows down parthenogenetic reproduction and starts the production of resting eggs.

Summarizing the results on production it can be said that in Neusiedlersee both species show the same tendency towards higher production values. The P/B ratios obtained for Neusiedlersee are in good agreement with those found in mesotrophic lakes for the same or related species.

22.7 A comparison of the phytoplankton – zooplankton relationship in deep stratified and shallow lakes

The morphometry of the water body, water depth and type of circulation, changes in radiation and temperatures all influence secondary production. Above all, the amount of food (bacteria, algae, detritus) undoubtedly sets an upper limit to the biomass and production of zooplankton.

The plankton of the pelagic zone of deep, stratified lakes can be characterized as a system independent of the littoral and bottom zone. Except for resting stages, which are stored in the bottom layers and represent valuable resources for the pelagic population (ephippia or resting eggs of Cladocera and resting eggs of Copepoda, resting stages of Cyclopoida, e.g. 4th or 5th copepodid stage) it can be regarded as a system with a considerable degree of independence. In such cases lakes with more phytoplankton show higher densities of zooplankton. It is also well known that changes in zooplankton, such as abundance, fecundity or size, are closely correlated with the amount of planktonic algae (Pennington 1941, Czeczuga 1959, Elster 1954, Edmondson 1962, 1964, 1965, Einsle 1967). In deep, stratified lakes phytoplankton is the basic, and frequently the only, food source of the zooplankton community. The two trophic levels – phytoplankton as the producer and zooplankton as the consumer – are fairly directly connected via the 'grazing food chain' (Odum 1971). This is clearly shown in Fig. 22.19; higher phytoplankton biomass results in higher zooplankton biomass.

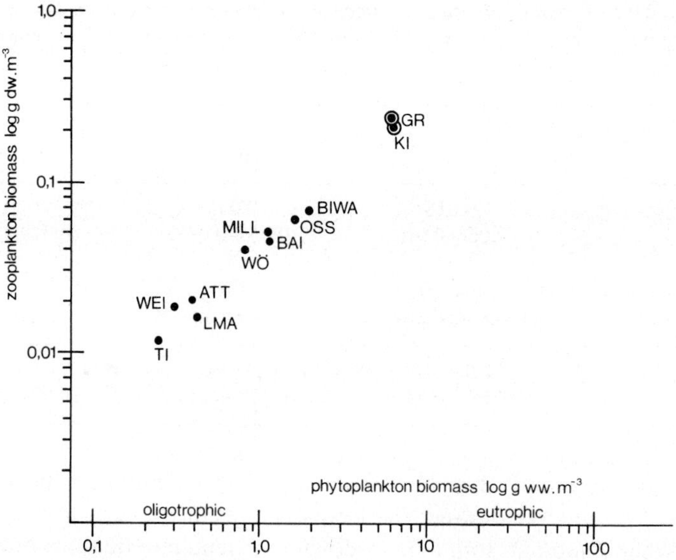

Fig. 22.19. Relation between phytoplankton biomass (g fresh weight m^{-3}) and zooplankton biomass (g dry weight m^{-3}) in deep, stratified lakes. ◉ indicates eutrophic lakes TI – L. Titicaca, Widmer et al. 1975; LMA – Lago Maggiore (ISPRA), Ravera 1969; ATT – Attersee, Müller 1976; BAI – L. Baikal, Moskalenko 1972; BIWA – L. Biwa, Mori & Yamamoto 1975; GR – Greifensee, Mittelholzer 1970; KI – L. Kinneret, Berman et al. 1972; WEI – Weissensee, WÖ – Wörthersee, MILL – Millstätter See, OSS – Ossiacher See, Herzig unpublished results, Sampl et al. 1976, 1977, Schultz unpublished results.

In the Figs. 22.19, 22.20, 22.21, the biomass values (phytoplankton expressed as fresh weight and zooplankton as dry weight; literature data were converted by using the factors of Winberg 1971) of the zooplankton are plotted against the phytoplankton biomass; a log-log scale is used and the values given are means of single years.

The role of the littoral zone has to be taken into account in shallow (mean depth less than 5 m, mostly polymitic, thermal stratification a rare event) or not very deep, stratified lakes (mean depth about 10 m, maximum depth 20–25 m). For example in Lake Erken (Nauwerck 1963) material found on the plants of the littoral zone, or the decomposed plants themselves, drift into the open water. Very often production is higher inshore than offshore (e.g. Lake George, Uganda, or Neusiedlersee), but a large proportion enters the open lake as detritus. This detritus is additional food for the zooplankton, and the interaction of dead organic material-microorganisms-detritivores, called the 'detritus food chain' (Odum 1971), is of very great importance in such lakes.

Many scientists have found that zooplankton feeds on detritus (e.g. Bogatova 1966, Erman 1962, Galkovskaja 1963, Gliwicz 1969a, b, Nadin-Hurley & Duncan 1976, Nauwerck, 1963, Saunders 1969, 1972, Smirnov 1969). The question of the nutritive value of the detritus has to be considered. Nauwerck (1962) showed that *Eudiaptomus gracilis* from Lake Erken could

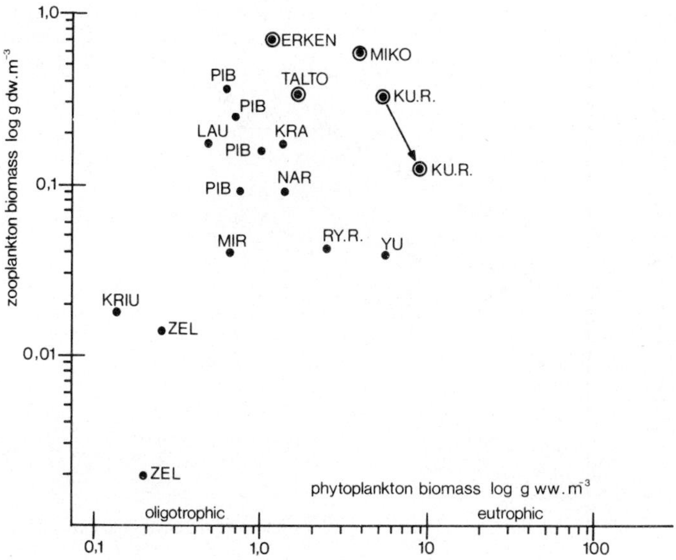

Fig. 22.20. Relation between phytoplankton biomass (g fresh weight m^{-3}) and zooplankton biomass (g dry weight m^{-3}) in lakes (reservoirs) with a mean depth of about 10 m. ⦿ indicates eutrophic lakes. Zel – Zelenetzkoye 1970, 1971, Winberg et al. 1973; Kriv – Krivoe, Alimov et al. 1972; Mir – Mirror Lake, Jordan & Likens 1975; Ry.R. – Rybinsk Reservoir, Sorokin 1972; Yu – L. Yunoko, Mori & Yamamoto 1975; Pib – Piburger See 1972, 1973, 1974, 1975, Rott 1975, 1976, Schaber 1975, 1976; Nar – L. Naroch, Winberg et al. 1972; Lau – Lauerzersee, Odermatt 1970; Kra – Krasnoe, Andronikowa et al. 1972; Talto – Taltowisko, Kajak et al. 1972; Ku.R. – Kurakhov Power Station Cooling Reservoir (heated and unheated), Pidgaiko et al. 1972; Miko – Mikolajskie, Kajak et al. 1972; Erken – L. Erken, Nauwerck 1963.

live on organic detritus alone, without algal food; the same has been found for *Arctodiaptomus spinosus* and *Diaphanosoma brachyurum* from Neusiedlersee (Herzig unpublished results). Although optimum growth is not achieved and mortality rates are higher, this form of nutrition nevertheless ensures survival and even reproduction. Saunders (1972) mentions bacteria and detritus as buffer food, consumed when algae are in short supply.

According to Nauwerck (1963) the zooplankton of Lake Erken had to rely on food sources other than phytoplankton to cover their needs for growth and reproduction. He suggests that bacteria and detritus filled the gap; as a result the mean zooplankton biomass is higher than in some eutrophic lakes of Poland (Taltowisko, Mikolajskie, e.g.).

Hillbricht-Ilkowska (1977) mentions that very often small, edible algae cover only 50 per cent of the food requirement of the animals. Gliwicz (1969a, b) found in 'in situ' experiments that the food particles filtered by a natural zooplankton population of an eutrophic lake consisted to 5–15 per cent of small algae, 10–20 per cent detritus and 70–85 per cent bacteria, whereas in oligo- and mesotrophic lakes the percentage of small algae may be as much as 50 per cent, whilst bacteria account only about 30 per cent. Gak (1972) estimated the part played by bacteria in very eutrophic reservoirs as 77 per cent.

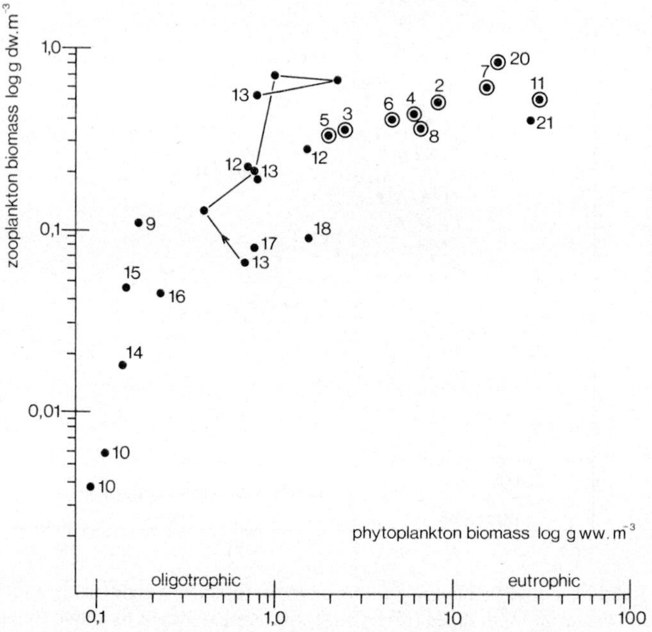

Fig. 22.21. Relation between phytoplankton biomass (g fresh weight m^{-3}) and zooplankton biomass (g dry weight m^{-3}) in shallow lakes (reservoirs) (mean depth <5 m) ◉ indicates eutrophic lakes. 1 – L. George, Burgis 1974, Burgis et al. 1973; 2 – Tjeukemeer, Beattie et al. 1972; 3 – Loch Leven, Morgan 1972; 4 – Flosek, Kajak et al. 1972; 5 – Śniardwy, Kajak et al. 1972; 6 – Kiew Reservoir, Gak et al. 1972; 7 – Batorin, Winberg et al. 1972; 8 – Myastro, Winberg et al. 1972; 9 – Krugloe, Alimov et al. 1972; 10 – Akulkino, Winberg et al. 1973; 11 – Suwa, Mori & Yamamoto 1975; 12 – Queen Elizabeth II Reservoir, Duncan 1975; 13 – Neusiedlersee 1968–1973, Herzig this paper and Dokulil 1979; 14 – Krivoe, Alimov et al. 1972; 15, 16, 17, 18 – Balaton, 1947, 1949, 1951, 1965, 1967, Tamás 1955, 1974, Sebestyén 1958 a b, Ponyi 1975; 20 – L. Endine, Barbanti et al. 1974; 21 – Tchad, Leveque et al. 1972.

Sorokin (1967; cit. acc. Hillbricht-Ilkowska 1977) states that the more trophic the water body and the greater the input of allochthonous material, the greater is the abundance and production of bacteria and thus their contribution to the total food of the zooplankton.

This is shown in Fig. 22.20. In the case of the subarctic Lake Zelenetzkoye low phytoplankton biomass is correlated with low zooplankton biomass; all the other lakes given as examples, mainly from temperate regions, reflect the more or less complex situation; the importance of the 'detritus food chain' as compared with the direct 'grazing food chain' seems to vary greatly according to the degree of trophy, the influence of the littoral zone and of allochthonous material. Low phytoplankton biomass does not imply low zooplankton biomass and vice versa.

In shallow lakes an additional input to the open water community is provided by material stirred up from the bottom. Material from the bottom may be rich in organic substances, possibly nitrogenous (Dowgiallo 1970), with a high calorific content (Rybak 1969), and is inhabited by large numbers of bacteria (Henrici-McCoy 1938, cit. acc. Nadin-Hurley & Duncan 1976). In shallow lakes, which are polymictic, this material is regularly brought into the pelagic zone and represents an additional food supply for the zooplankton. In addition, such shallow lakes offer better temperature conditions to the plankton animals than deep stratified lakes. The pelagic zone of shallow lakes might be compared with the epilimnion of stratified lakes.

As can be seen in Fig. 22.21 low phytoplankton biomass does not imply low zooplankton biomass (e.g. Lake Krugloe, subarctic, oligotrophic, shows a relatively high organic content of the sediment and high allochthonous input). Only in the subarctic, ultraoligotrophic Lake Alkulkino is a low phytoplankton biomass found in combination with low zooplankton biomass. In general, high zooplankton values are already reached at a level of 1 g fw m^{-3} algal biomass, a value still found in oligotrophic lakes (according to Vollenweider 1970: <1 g fw m^{-3} = ultra oligotrophic; >10 g fw m^{-3} = highly eutrophic lakes). With higher phytoplankton biomass the zooplankton biomass levels out and the relationship resembles a 'saturation curve', where even in highly eutrophic situations and tropical lakes (Tschad, L. George) higher zooplankton biomass values are not reached. The fact that in shallow lakes phytoplankton development may be of lesser importance for the plankton animals can be seen in Neusiedlersee; the development of both trophic levels is shown for the years 1968–1973. Excluding extremely low (1969: 0.4 g fw m^{-3} algal biomass) and high values (1972: 2.22 g fw m^{-3}) an increase from 0.67 to 0.987 g fw m^{-3} algal biomass is evident, whereas the zooplankton biomass increases from 0.067 to 0.716 g dw m^{-3}, which means that the value for 1971 is ten times higher than that for 1968.

Just the opposite can be shown for Balaton, where a tremendous increase in algal biomass (10 times) is accompanied by a small increase in zooplankton biomass (Fig. 22.21).

Summarizing the results concerning the phytoplankton-zooplankton relation from different lakes it can be said that, if lakes of similar morphometry, depth and type of circulation are compared, phytoplankton is seen to be an important producer for the filter-feeding zooplankton. Passing from deep, stratified lakes to shallow ones, with little or no stratification, the direct 'grazing food chain' seems to be increasingly replaced by the indirect 'detritus food chain', i.e. the role of phytoplankton as the main food source is taken over by bacteria and detritus, very often originating from the bottom.

For this comparison the planktonic algae were of course treated as the phytoplankton in toto and were not separated into size classes, or into edible and non-edible algae (even species, genus or higher categories were ignored). The impact of invertebrate and vertebrate predators on the zooplankton was also ignored. Despite the known shortcomings, the comparison of biomass data of different types of lakes revealed the complexity of the interaction between different trophic levels.

References

Andronikova, I. N., Drabkova, V. G., Kuzmenko, K. N., Michailova, N. F., Stravinskaja, E. A. 1972. Biological productivity of the main communities of the Red Lake. Proc. IBP/UNESCO Symposium on Productivity Problems of Freshwaters (ed. Kajak & Hillbricht-Ilkowska), Kazimierz Dolny, Poland, 57–71.

Alimov, A. F., Boullion, V. V., Finogenova, N. P., Ivanova, M. B., Kuzmitskaya, N. K., Nikulina, V. N. Ozeretskovskaya, N. G., Zharova, T. V. 1972. Biological productivity of Lakes Krivoe and Krugloe. Proc. IBP/UNESCO Symposium on Productivity Problems of Freshwaters (ed. Kajak & Hillbricht-Ilkowska), Kazimierz Dolny, Poland, 39–56.

Barbanti, L., Bonacina, C., Calderoni, A., Carollo, A., De Bernardi, R., Guilizzoni, P., Nocentini, A. M., Ruggiu, P., Saraceni, C., Tonolli, L. 1974. Indagini ecologiche sul lago d'Endine. Ed. Ist. Ital. Idrobiol. verb. Pallanza 1974, 304 pp.

Beattie, M., Bromley, H. J., Chambers, M. Goldspink, R., Vivjerberg, J., van Zalinge, N., Golterman, H. L. 1972. Limnological studies on Tjeukemeer – a typical Dutch 'polder reservoir'. Proc. IBP/UNESCO Symposium on Productivity Problems of Freshwaters (ed. Kajak & Hillbricht-Ilkowska) Kazimierz Dolny, Poland, 421–446.

Benda, H. 1950. Fischereibiologisches über den Neusiedlersee. Öst. Fischerei 3, 8/9: 189–194.

Berg, K., Nygaard, G. 1929. Studies on the plankton in the lake of Frederiksborg Castle. D. K. Danske Vidensk. Selsskab. Skr. Naturvidensk. math.– Afd. 9(1): 223–316.

Berman, T., Pollingher, V., Gophen, M. 1972. Lake Kinneret: planktonic populations during seasons of high and low phosphorus availability. Verh. Int. Ver. Limnol. 18, 2: 588–599.

Bogatova, J. V. 1966. Eksperimentalnoe issledovanie pitanija *Daphnia longispina* O. F. Müller i *Daphnia pulex* (de Geer). Trudy vses. naučissled. Inst. prud. ryb. Choz. 14: 83–93.

Bottrell, H. H., Duncan, A. Gliwicz, Z. M., Grygierek, E., Herzig, A., Hillbricht-Ilkowska, A., Kurasawa, H. Larsson, P., Weglenska, T. 1976. A review of some problems in zooplankton production studies. Norw. J. Zool. 24: 419–456.

Burgis, M. J. 1969. A preliminary study of the ecology of zooplankton in Lake George, Uganda. Verh. Int. Ver. Limnol. 17, 1: 297–302.

Burgis, M. J., Darlington, J. P. E. C., Dunn, I. G., Ganf, G. G., Gwahaba, J. J., McGowan, L. M. 1973. The biomass and distribution of organisms in Lake George, Uganda. Proc. Roy. Soc. Lond. B, 184: 271–298.

Burgis, M. J. 1974. Revised estimates for the biomass and production of zooplankton in Lake George, Uganda. Freshwat. Biol. 4: 535–541.

Colebrook, J. M. 1960. Plankton and water movements in Windermere. J. Anim. Ecol. 29: 217–240.

Cummins, K. W., Costa, R. R., Rowe, R. E., Moshiri, G. A., Scanlon, R. M., Zajdel, R. K. 1969. Ecological energetics of a natural population of the predaceous zooplankter *Leptodora kindti* (Focke) (Crustacea, Cladocera) Oikos 20: 189–223.

Czeczuga, B. 1959. Oviposition in *Eudiaptomus gracilis* G. O. Sars and *Eudiaptomus graciloides* Lill. (Diaptomidae, Crustacea) in relation to season and trophic level of lakes. Bull. Acad. pol. Sci. Cl. II, 7: 227–230.

Daday, J. 1890. Übersicht der Diaptomusarten Ungarns. Termrajzi Füzetek 8: 114–180.

Daday, J. 1897. A magyarországa tavak halainak természetes tápláléka (Die natürliche Nahrung der Fische in den ungarischen Teichen). Kgl. ung. Naturw. Ges., Budapest, 1897: 297–306.

Davis, J. C. 1973. Statistics and data analysis in Geology (with Fortran programs by R. J. Sampson). Kansas Geological Survey. John Wiley & Sons Inc., New York, London, Sydney, Toronto.

De Bernardi, R., Soldavini, E. 1976. Long term fluctuations of zooplankton in Lake Mergozzo, Northern Italy. Mem. Ist. Ital. Idrobiol. 33: 345–375.

Dokulil, M. 1975. Bacteria in the water and mud of Neusiedlersee (Austria). Symp. Biol. Hung. 15: 135–140.

Dokulil, M. 1979. Seasonal pattern of phytoplankton. In: Neusiedlersee, ed. H. Löffler, Junk, The Hague.

Donner, J. 1979. The Rotifers of Neusiedlersee. In: Neusiedlersee, ed. H. Löffler, Junk, The Hague.

Dowgiallo, A. 1970. Water, organic matter resources of high dispersion. Polskie Arch. Hydrobiol. 17: 121–131.

Dumont, H. J. 1977. Biotic factors in the population dynamics of rotifers. Arch. Hydrobiol. Beiheft (Ergebn. Limnol.) 8: 98–122.

Duncan, A. 1975. Production and biomass of three species of *Daphnia* co-existing in London reservoirs. Verh. Int. Ver. Limnol. 19: 2858–2867.

Edmondson, W. T. 1962. Food supply and reproduction of zooplankton in relation to phytoplankton population. Int. Cons. l'Explor. Mer. Rapp. Proc. Verb. 153: 137–141.

Edmondson, W. T. (1964). The rate of egg production by rotifers and copepods in natural populations as controlled by food and temperature. Verh. Int. Ver. Limnol. 15: 673–675.

Edmondson, W. T. 1965. Reproductive rate of planktonic rotifers as related to food and temperature in nature. Ecol. Monogr. 35: 61–111.

Edmondson, W. T. 1969. The present condition of the saline lakes in the Lower Grand Coulee, Washington. Verh. Int. Ver. Limnol. 17, 1: 447–448.

Edmondson, W. T. 1974. Secondary production. Mitt. Int. Ver. Limnol. 20: 229–272.

Edmondson, W. T., Winberg, G. G. 1971. A manual on methods for the assessment of secondary productivity in freshwaters. IBP-Handbook No. 17, Blackwell, Oxford, 358 pp.

Einsle, U. 1967. Über einige Auswirkungen der Eutrophierung des Bodensee-Obersees auf seine planktisch lebenden Copepodenpopulationen. Schweiz. Z. Hydrol. 29: 305–310.

Elster, H.-J. 1954. Über die Populationsdynamik von *Eudiaptomus gracilis* G. O. Sars und *Heterocope borealis* Fischer im Bodensee-Obersee. Arch. Hydrobiol. Suppl. 20: 487–523.

Elliott, J. M. 1971. Some methods for the statistical analysis of samples of benthic invertebrates. Scient. Publs. Freshwat. Ass. 25: 1–144.

Eppacher, T. 1968. Physiographie und Zooplankton des Gossenköllesees. Ber. Nat.– med. Ver. Innsbruck, Bd. 56, Festschr. Steinböck, 31–123.

Erman, L. A. 1962. Ob ispolzovaniju trofičeskich resursov vodoemov planktonnymi lolovrotkami. Bull. mosc. Obsc. Isp. Prir. Otd. Biol. 67, 4: 32–47.

Flössner, D. 1972. Krebstiere, Crustacea; Kiemen- und Blattfüßer, Branchiopoda; Fischläuse, Branchiura. In: Die Tierwelt Deutschlands. Gustaf Fischer Verlag, Jena, pp. 501.

Gak, D. Z., Gurvich, V. V., Korelyakova, I. L., Kostikova, L. W., Konstantinova, N. A., Olivari, G. A., Priimachenko, A. D., Tseeb, J. J., Vladimirova, K. S., Zimbalevskaya, L. N. 1972. Productivity of aquatic organisms communities of different trophic levels in Kiev Reservoir. Proc. IBP/UNESCO Symposium on Productivity Problems of Freshwaters (Ed. Kajak & Hillbricht-Ilkowska), Kazimierz Dolny, 447–455.

Galkovskaja, G. A. 1963. K voprosu o pitanii plynktonnych kolovratok. Dokl. Akad. Nauk. SSSR 7: 203–205.

George, D. G. 1976. Life cycle and production of *Cyclops vicinus* in a shallow eutrophic reservoir. Oikos 27: 101–110.

George, D. G. and Edwards, R. W. 1974. Population dynamics and production of *Daphnia hyalina* in a eutrophic reservoir. Freshwat. Biol. 4: 445–465.

George, D. G., Edwards, R. W. 1976. The effect of wind on the distribution of chlorophyll a and crustacean plankton in a shallow eutrophic reservoir. J. Appl. Ecol. 13: 667–690.

Geyer, F., Mann, H. 1939. Limnologische und fischereibiologische Untersuchungen am ungarischen Teil des Fertö (Neusiedlersee). Arb. Ung. Biol. Forschungsinst. Tihany 11: 64–193.

Glamazda, V. V. 1971. Zooplankton of the Tsimliansk Reservoir in 1966–1968. Trudy Volgograd. otd. Gos. Niorkh. 5 (in Russian).

Gliwicz, Z. M. 1969. Studies on the feeding of pelagic zooplankton in lakes with varying trophy. Ekol. pol. A 17: 663–708.

Gliwicz, Z. M. 1969. The share of algae, bacteria and trypton in the food of the pelagic zooplankton of lakes with various trophic characteristics. Bull. Acad. pol. Sci. Cl. II. 17 159–165.

Goulden, C. E. 1968. The systematics and evolution of the Moinidae. Trans. Am. Philosoph. Soc. 58(6): 101 pp.

Haberman, J. 1974. On the changeability of the role of different groups of zooplankton (Cladocera, Copepoda, Rotatoria, Mollusca) in Lake Peipsi-Pihkva and Lake Vortsjärv. Est. Contr. IBP VI, Tartu, 107–134.

Heinle, D. R. 1966. Production of a calanoid copepod, *Acartia tonsa*, in the Patuxent River Estuary. Chesapeake Sci. 7: 59–74.

Henrici, A., McCoy, E. 1938. The distribution of heterotrophic bacteria in the bottom deposits of some lakes. Trans. Wis. Acad. Sci. 31: 323–361.

Herzig, A. 1973. Phänologie, Populationsdynamik und Produktion des Crustaceenplanktons im Neusiedlersee. Ph.D. Univ. Vienna.

Herzig, A. 1974. Some population characteristics of planktonic crustaceans in Neusiedlersee. Oecologia (Berl.) 15: 127–141.

Herzig, A. 1975. Der Neusiedlersee – charakteristische Eigenschaften und deren Auswirkungen auf das Zooplankton. Verh. Ges. Ökol. Wien 1975, 189–96.

Herzig, A. 1977. Qualitative und quantitative Veränderungen im Zooplankton des Neusiedlersees 1968–1975. Biol. Forsch. Inst. Burgenld. Ber. 24: 28–34.

Herzig, A. 1977. Einfluß von Landwirtschaft und Tourismus auf die Eutrophierung des Neusiedlersees (MaB Projekt 7045). Man on Biosphere, project 5, Report 1977, 68–105.

Hillbricht-Ilkowska, A. 1977. Trophic relations and energy flow in pelagic plankton. Pol. ecol. Stud. 3, 1: 1–98.

Hillbricht-Ilkowska, A., Patalas, K. 1967. Methods of estimating production and biomass and some problems of quantitative calculation methods of zooplankton. Ekol. pol. B 13: 139–172.

Hillbricht-Ilkowska, A., Weglenska, T. 1970. Some relations between production and zooplankton structure of two lakes of varying trophy. Pol. Arch. Hydrobiol. 17: 233–240.

Hillbricht-Ilkowska, A. Weglenska, T. 1973. Experimentally increased fish-stock in the pond type Lake Warniak. VII: Numbers, biomass and production of zooplankton. Ekol. pol. 21: 533–552.

Hock, R. 1957. Ein Beitrag zur Chemie des Neusiedler Seewassers. Prakt. Chemie, Wien, 8/6: 163–166.

Hudson, C. T., Gosse, P. H. 1889. The Rotifera; or wheel-animalcules. 2 vols. Longmans, Green, London.

Kajak, Z., Hillbricht-Ilkowska, A., Pieczyńska, E. 1972. The production process in several Polish lakes. Proc. IBP/UNESCO Symposium in Productivity Problems of Freshwaters (ed. Kajak & Hillbricht-Ilkowska) Kazimierz Dolny, Poland, 129–147.

Kiefer, F. 1972. Revision der *bacillifer*-Gruppe der Gattung *Arctodiaptomus* Kiefer. Mem. Ist. Ital. Idrobiol. 27: 113–269.

Knie, K. 1959. Über den Chemismus der Wässer im Seewinkel und des Neusiedlersees. Wiss. Arb. Burgenland, Eisenstadt 23: 65–68.

Kurasawa, H., Kitazawa, Y., Shiraishi, Y. 1952. Studies on the biological production of Lake Suwa IV. The stratification, the seasonal succession and the standing crop of zooplankton (1). Misc. Rep. Res. Inst. Nat. Resourc. 27: 29–39.

Jordan, M., Likens, G. E. 1975. An organic carbon budget for an oligotrophic lake in New Hampshire, USA. Verh. Int. Ver. Limnol. 19: 994–1003.

Langford, R. R., Jermolajev, E. C. 1966. Direct effect of wind on plankton distribution. Verh. Int. Ver. Limnol. 16: 188–193.

Larsson, P. 1972. Distribution and estimation of standing crop of zooplankton in a mountain lake with fast renewal. Verh. Int. Ver. Limnol. 18, 1: 334–343.

Lévêque, Ch., Carmouze, J. P., Dejoux, C., Durand, J. R., Gras, R., Iltis, A., Lemoalle, J., Loubens, G., Lauzanne, L., Saint-Jean, L. 1972. Recherches sur les biomasses et al productivité du Lac Tchad. Proc. IBP/UNESCO Symposium on Productivity Problems of Freshwaters (ed. Kajak & & Hillbricht-Ilkowska) Kazmiierz Dolny, Poland, 165–181.

Löffler, H. 1959. Zur Limnologie, Entomostraken- und Rotatorienfauna des Seewinkelgebietes (Burgenland, Österreich). S.– B. öst. Akad. Wiss. math.– nat. Kl. Abt. I 168: 315–362.

Löffler, H. 1961. Beiträge zur Kenntis der iranischen Binnengewässer II. Int. Rev. Hydrobiol. 46: 309–406.

Löffler, H. 1974. Der Neusiedlersee, Naturgeschichte eines Steppensees. Verlag Fritz Molden, Wien-Müchen-Zürich. 175 pp.

Löffler, H. 1979. The crustacean fauna of the *Phragmites* belt. In: Neusiedlersee, ed. H. Löffler, Junk, The Hague.

Meshkova, T. M. 1952. Zooplankton ozera Sevan (biologiya i produktivnost'). The zooplankton

of Lake Sevan (biology and productivity). Trudy sevan. gidrobiol. Sta. 13: 5–170.
Mittelholzer, E. 1970. Populationsdynamik und Produktion des Zooplanktons im Greifen- und Vierwaldstättersee. Schweiz. Z. Hydrol. 32, 1: 90–149.
Morgan, N. C. 1972. Productivity studies at Loch Leven (a shallow, nutrient rich lowland lake). Proc. IBP/UNESCO Symposium on Productivity Problems of Freshwaters (ed. Kajak & Hillbricht-Ilkowska) Kazimierz, Dolny, Poland, 183–205.
Mori, S. and Yamaoto, G. 1975. Productivity of communities in Japanese Inland Waters. JIBP Synthesis Vol. 10, 9: 379–419.
Moskalenko, B. K. 1972. Biological productivity system of Lake Baikal. Verh. Int. Ver. Limnol. 18, 2: 568–573.
Müller, G. 1976. Zooplankton. In: Attersee, vorläufige Ergebnisse des OECD Seeneutrophierungsund des MaB-Programmes, 1976, 116–128.
Nadin-Hurley, C. M., Duncan, A. 1976. A comparison of daphnid gut particles with the sestonic particles present in two Thames Valley Reservoirs throughout 1970 and 1971. Freshwat. Biol. 6: 109–123.
Nauwerck, A. 1962. Nicht algische Ernährung bei *Eudiaptomus gracilis* (Sars). Arch. Hydrobiol. Suppl. 25: 393–400.
Nauwerck, A. 1963. Die Beziehung zwischen Zooplankton und Phytoplankton im See Erken. Symb. bot. upsal. 17, 5: 1–163.
Nauwerck, A. (in press). Zooplankton, in Le Cren (Ed.) The Structure and Functioning of Freshwater Ecosystems. Cambridge University Press, Cambridge.
Neuhuber, F. 1971. Ein Beitrag zum Chemismus des Neusiedlersees. S.– B. öst. Akad. Wiss., math.– nat. Kl. Abt. I. 179, 8–10: 225–231.
Newrkla, P. 1974. Populationsdynamik, Production und Respiration von *Arctodiaptomus spinosus* (Daday) in einem alkalischen Kleingewässer. Birnbaumlacke, Zicklackengebiet, Burgenland). Ph.D. Univ. Vienna.
Odermatt, D. M. 1970. Limnologische Charakterisierung des Lauerzersees mit besonderer Berücksichtigung des Planktons. Schweiz. Z. Hydrol. 32, 1: 1–75.
Odum, E. P. 1971. Fundamentals of Ecology (Third edition). W. B. Saunders, Comp., Philadelphia-London-Toronto, 574 pp.
Pechen, G. A. 1965. Produktsiya vetvistousykh rakoobraznykh ozernogo zooplanktona (The production of the cladoceran crustaceans in lake zooplankton). Gidrobiol. Zh. 1, 4: 19–26.
Pechlaner, R., Bretschko, G., Gollmann, P., Pfeifer, H., Tilzer, M., Weissenbach, H. P. 1972. The production processes in two high-mountain lakes (Vorderer and Hinterer Finstertaler See, Kühtai, Austria). Proc. Sympsium on Productivity Problems of Freshwater (ed. Kajak & Hillbricht-Ilkowska) Kazimierz Dolny, Poland, 239–269.
Pennington, W. 1941. The control of the numbers of freshwater phytoplankton by small invertebrate animals. J. Ecol. 29: 204–211.
Pesta, O. 1954. Studien über die Entomostrakenfauna des Neusiedlersees. Wiss. Arb. Burgld. 2: 1–82.
Pidgaiko, M. L., Grin, V. G., Kititsina, L. A., Lenchina, L. G., Polivannaya, M. F., Sergeeva, O. A., Vinogradskaya, T. A. 1972. Biological productivity of Kurakhov's Power Station Cooling Reservoir. Proc. IBP/UNESCO Symposium on Productivity Problems or Freshwaters (ed. Kajak & Hillbricht-Ilkowska) Kazimierz, Dolny, Poland, 477–491.
Ponyi, J. E. 1975. The biomass of zooplankton in Lake Balaton. Symp. Biol. Hung. 15: 214–224.
Ponyi, J. E., Dévai, I. 1979. The Crustacea of the Hungarian Area of Lake Fertö. Opŭsc. Zool. Budapest, XVI, 1–2, 107–127.
Ponyi-Zankai, N., Ponyi, J. E. 1972. Quantitative relationships of the Rotatoria plankton in Lake Balaton during 1965–1966. Annal. Biol. Tihany 39: 189–204.
Ragotzkie, R. A., Bryson, R. A. 1953. Correlation of currents with the distribution of adult *Daphnia* in Lake Mendota. J. mar. Res. 12: 157–172.
Ravera, O. 1969. Seasonal variation of the biomass and biocoenotic structure of plankton of the Bay of Ispra (Lago Maggiore). Verh. Int. Ver. Limnol. 17: 237–254.
Rott, E. 1975. Phytoplankton und kurzwellige Strahlung im Piburgersee. Jber. Abt. Limnol. Innsbruck 1: 51–62.

Rott, E. 1976. Phytoplankton und kurzwellige Strahlung im Piburgersee. Jber. Abt. Limnol. Innsbruck, 2: 63–69.

Ruttner-Kolisko, A. 1974. Plankton rotifers (biology and taxonomy). In: Die Binnengewässer Vol. XXVI/1 Suppl., Stuttgart, E. Schweizerbart'sche Verlagsbuchhandlung (Nägele & Obermiller), 146 pp.

Ruttner-Kolisko, A., Ruttner, F. 1959. Der Neusiedlersee. In: Landschaft Neusiedlersee (Grundriß der Naturgeschichte des Großraumes Neusiedlersee). Wiss. Arb. Burgld. 23: 195–201.

Rybak, J. I. 1969. Bottom sediments of the lakes of various tropic types. Ekol. pol. A 17: 611–662.

Saunders, G. W. 1969. Some feeding aspects of zooplankton. In: Eutrophication, causes, consequences, corrections. Wash. Nat. Acad. Sci., 553–573.

Saunders, G. W. 1972. The transformation of artificial detritus in lake water. Mem. Ist. Ital. Idrobiol. Suppl. 29: 261–288.

Saunders, H. L. 1968. Marine benthic diversity: a comparative study. Amer. Naturalist 102: 243–282.

Schaber, P. 1975. Rotatorien und Crustaceen des Piburger Sees (1971–1973). Jber. Abt. Limnol. Innsbruck, 1: 59–72.

Schaber, P. 1976. Rotatorien und Crustaceen des Piburger Sees (1973–1975). Jber. Abt. Limnol. Innsbruck 2: 78–94.

Schiemer, F., Prosser, M. 1976. Distribution and biomass of submerged macrophytes in Neusiedlersee. Aquatic Botany 2: 289–307.

Schindler, D. W. 1972. Production of phytoplankton and zooplankton in Canadian Shield Lakes. Proc. IBP/UNESCO Symposium on Productivity Problems of Freshwaters (ed. Kajak & Hillbricht-Ilkowska) Kazimierz, Dolny, Poland, 311–331.

Sebestyén, O. 1958. Mennyiségi planktonanulmányok a Balatonon. VIII: Biomassza számítások nyíltvízi Rotatóriákon. Annal. Biol. Tihany 25: 267–279.

Sebestyén, O. 1958. Quantitative plankton studies on Lake Balaton. IX: a summary of the biomass studies. Annal. Biol. Tihany 25: 281–292.

Shushkina, E. A. 1966. Sootnoshenie produktsii i biomassy zooplanktona ozer (The ratio of production to biomass in lake zooplankton). Gidrobiol. Zh. 2, 1: 27–35.

Smirnov, N. N. 1974. Chydoridae. In: Fauna of the U.S.S.R., Crustacea, Vol. I, No. 2, Israel Program for Scientific Translations, 644 pp.

Smirnov, N. N. 1976. Macrothricidae and Moinidae. In: Fauna of the U.S.S.R. (in Russian), T. I., 3: 237 pp.

Sokal, R. R., Rohlf, F. J. 1969. Biometry; the Principles and Practice of Statistics in Biological Research. Freeman, San Francisco 776 pp.

Sorokin, J. I. 1967. Nekotorye itogi izucenija troficeskoj roli bakterii v vodoemach. Gidrobiol. Zh. 3, 5: 32–42.

Sorokin, J. I. 1972. Biological productivity of the Rybinsk Reservoir. Proc. IBP/UNESCO Symposium on Productivity Problems of Freshwaters (ed. Kajak & Hillbricht-Ilkowska), Kazimierz, Dolny, Poland, 493–503.

Spodniewska, I., Hillbricht-Ilkowska, A., Weglenska, T. 1973. Long-term changes in the plankton of eutrophic Mikolajskie Lake as an effect of accelerated eutrophication. Bull. Acad. Pol. Ser. scien. biol. Cl. II, 21, 3: 215–221.

Stavn, R. H. 1971. The horizontal-vertical distribution hypothesis: Langmuir circulations and *Daphnia* distributions. Limnol. Oceanogr. 16, 2: 453–467.

Steinhauser, F. 1970. Kleinklimatische Untersuchung der Windverhältnisse am Neusiedlersee. 1. Teil: Die Windrichtungen. Idöjárás 74: 76–88.

Szymanski-Bucarey, E. 1974. Untersuchungen über die Eutrophierung des Titisees und ihre Auswirkung auf die Populationsdynamik des Zooplanktons. Teil 2. Arch. Hydrobiol. Suppl. 47: 167–238.

Tamás, G. 1955. Mennyiségi planktonanulmányok a Balatonon, VI. A negyvenes évek fitoplanktonjának biomasszája. Annal. Biol. Tihany 23: 95–109.

Tamás, G. 1974. The biomass changes of phytoplankton in Lake Balaton during the 1960ies. Annal. Biol. Tihany 41: 323–342.

Thienemann, A. 1920. Die Grundlagen der Biozönotik und Monards faunistische Prinzipien. Festschrift für Zschokke, Nr. 4, Basel.
Varga, L. 1926. Die Rotatorien des Fertö (Neusiedlersee). Arch. Balat. I, 181–225.
Varga, L. 1928. Allgemeine limnologische Charakterisierung des Fertö (Neusiedlersee). Int. Rev. Hydrobiol. 19: 289–294.
Varga, L. 1929. *Rhinops fertöensis*, ein neues Rädertier aus dem Fertö. Zool. Anz. 80: 236–253.
Varga, L. 1932. Katastrophen der Biozönosen des Fertö. Int. Rev. Hydrobiol. 27: 130–150.
Varga, L. 1934. Neue Beiträge zur Kenntnis der Rotatorienfauna des Neusiedlersees. Allatani Közlemènyek 31: 139–150.
Varga, L., Mika, F. 1937. Die jüngsten Katastrophen des Neusiedlersees und ihre Einwirkungen auf den Fischbestand des Sees. Arch. Hydrobiol. 31: 527–546.
Vollenweider, R. A. 1970. Scientific fundamentals of the eutrophication of lakes and flowing waters, with particular reference to nitrogen and phosphorus as factors in eutrophication. OECD-report, Paris 30th September 1970, 159 pp.
Walker, K. F. 1973. Studies on a saline lake ecosystem. Austr. J. Mar. Freshwat. Res. 24: 21–71.
Weglenska, T. 1971. The influence of various concentrations of natural food on the development, fecundity and production of planktonic crustacean filtrators. Ekol. pol. 19:427–473.
Widmer, C., Kittel, T., Richerson, P. J. 1975. A survey of the biological limnology of Lake Titicaca. Verh. Int. Ver. Limnol. 19: 1504–1510.
Winberg, G. G. 1971. Methods for the estimation of production of aquatic animals. Academic Press, London & New York; (translated from the Russian by A. Duncan). 175 pp.
Winberg, G. G. 1971. IBP-PF Productivity of Freshwaters: Symbols, units and conversion factors in studies of freshwater productivity. Cable printing Services Ltd. London, G, B. 23 pp.
Winberg, G. G. 1972. Etudes sur le bilan biologique énergetique et la productivité des lacs en Union Soviétique (Edgardo Baldi Memorial Lecture) Verh. Int. Ver. Limnol. 18: 39–64.
Winberg, G. G., Babitsky, V. A., Gavrilov, S. I., Gladky, G. V., Zakharenkov, I. S., Kovalevskaya, R. Z., Mikheeva, T. M., Nevyadomskaya, P. S., Ostapenya, A. P., Petrovich, P. G., Potaenkio, J. S., Yakushko, O. F. 1972. Biological productivity of different types of lakes. Proc. IBP/UNESCO Symposium on Productivity Problems of Freshwaters (ed. Kajak & Hillbricht-Ilkowska), Kazimierz, Dolny, Poland, 383–404.
Winberg, G. G., Alimov, A. F., Boullion, V. V., Ivanova, M. B., Korobetzova, E. V., Kuzmitzkaya, N. K., Nikulina, V. N., Fursenko, M. V. 1973. Biological productivity of two subarctic lakes. Freshwat. Biol. 3: 177–197.
Wright, J. C. 1965. The population and production of *Daphnia* in Canyon Ferry Reservoir, Montana. Limnol. Oceanogr. 10: 583–590.
Zakovsek, G. 1961. Jahreszyklische Untersuchungen am Zooplankton des Neusiedlersees. Wiss. Arb. Burgld. 27: 1–85.

23 The benthic community of the open lake

F. Schiemer

23.1 Introduction

This paper is an attempt to summarize the information on the benthic community of the open lake obtained between 1967 and 1973 as part of the International Biological Programme. Prior to IBP, knowledge of the benthos of the lake was very limited. Geyer and Mann (1939) had analysed the macrobenthos of a small series of Ekman-Birge samples taken during a survey on the fishery biology of the small Hungarian part of the lake. A few comments concerning benthos were made by Findenegg (1959). Even faunistic studies of benthic animals were very scarce (Daday, 1891, Pesta, 1954).

On account of this lack of knowledge the IBP study had to start with faunistics, an analysis of the overall distribution pattern of benthic parameters and fauna, and a study of the population dynamics and life cycles of predominant benthic animals of the lake. The final aim was the assessment of the role of benthos, especially with respect to trophic relationships in the ecosystem of the open lake beyond the reed belt.

In the course of the IBP study it was often impossible to study different aspects of the benthic community simultaneously as would have been desirable. Nevertheless the combined results obtained in different years and at different sampling grids, allow us to outline the general structure and dynamics of the benthic community in Neusiedlersee.

23.2 Faunistics

The fauna of the reedless areas differs distinctly from that of the reed belt. However, a number of reed forms are found on the edge of the open lake region, particularly in sheltered bays with abundant vegetation. Delineating the benthic fauna is additionally complicated by the difficulty of distinguishing between benthic, pelagic and periphytic species on account of the thorough mixing of the whole water mass by the wind. This applies particularly to rotifers and crustaceans.

The species list is still incomplete (see Table 23.1). Detailed studies on rotifers, nematodes and chironomids have been carried out, but other quantitatively important groups, particularly the protozoa, have been neither systematically nor ecologically investigated. Faunistic information available is summarized in Table 23.1.

Protozoa: Rhizopoda shells are frequently found in meiobenthos samples from the soft mud areas (*Difflugia* sp., *Cyphoderia* sp., *Actinophrys* sp., Donner (1972)). Benthic ciliates are also frequent in soft sediment habitats;

Table 23.1. Survey on the bottom fauna of the open lake

Systematic group	Species number	Quant. importance	Authors
Rhizopoda	not studied	loc. abundant	Donner 1972, named 3 spp.
Ciliophora	4	?	Dietz 1966, Donner 1972
Turbellaria	not studied	rare	this chapter
Nematoda	26	abundant	Schiemer et al. 1969, Schiemer 1970, Schiemer 1978
Gastrotricha	1	rare	Schiemer et al. 1969
Rotatoria	42	loc. abundant	Donner 1968, 1972; includes epiphytic and some plankton spp.
Mollusca	2	rare	Schubert 1968, Hacker & Herzig 1970
Oligochaeta	6	abundant	this chapter
Tardigrada	1	loc. abundant	Schiemer et al. 1969
Hydracarina	not studied	rare	this chapter
Cladocera	7	rare	Daday 1891, Pesta 1954, this chapter
Copepoda	2	rare	
Ostracoda	2	loc. abundant	
Chironomidae	13	abundant	includes epiphytic spp., Findenegg 1959, Schiemer in press, this chapter
Ceratopogonidae	not studied	rare	
other Insecta	11	rare	this chapter

however, except for the reference made to certain readily-recognizable forms (*Coleps hirtus*, *Stentor* sp., *Stylonychia* sp.) by Donner (1972) and by Dietz (1966) (*Cinetochilum margaritaceum*, the only species occurring in low abundance in the benthos of the open lake, off Illmitz), faunistic data are lacking.

Turbellaria: An unpublished list of the turbellarians of Neusiedlersee was placed at our disposal by Professor Reisinger (Graz). The majority of the species listed are probably confined to biotopes within the reed belt. It is not known how many of them may also occur in the benthos of the open lake. Of the 19 species listed 18 have been reported from wells, channels, gravel pits and salt pans in the Seewinkel region, on the eastern shore of Neusiedlersee (see Kraus, 1965). In the course of a seasonal study of the meiofauna in the soft mud zone near the reed fringe, Rhabdocoela were found consistently but in low numbers. The material was not examined systematically.

Gastrotricha: *Chaetonotus* cf. *maximus* was frequently found in the soft mud zone on the edges of the reed.

Rotatoria: A detailed systematic account on benthic rotifers of the open lake was given by Donner (1968, 1972). The material originated mainly from soft mud habitats of the northern end of the lake, but a smaller set of samples also included localities in the central and eastern parts (see Chapter 27).

The species list comprises 45 species, 17 of which are considered to be genuine benthic forms.

Dicranophorus forcipatus (O. F. Müller)
Dicranophorus uncinatus (Milne)
Encentrum putorius putorius Wulfert

Encentrum diglandula (Zawadowsky)
Encentrum sp.
Notommata cyrtopus Gosse
Paradicranophorus sordidus Donner
Paradicranophorus sudzukii Donner
Paracentrum saundersiae (Hudson)
Resticula nyssa (Harring et Myers)
Philodina megalotrocha Ehrenberg
Rotaria laticeps Wulfert
Rotaria neptunia (Ehrenberg)
Rotaria rotatoria Pallas
Rotaria tardigrada (Ehrenberg)
Rotaria tridens Montet
Rotaria cf. *curtipes* (Murray)

Two conspicuous benthic forms belonging to the genus *Paradicranophorus* were described from Neusiedlersee by Donner, 1968. So far they have been found in no other lake. Both of these species, as well as *Encentrum diglandula*, *Paracentrum saundersiae* and *Rotaria tardigrada*, are stenotopic benthic forms. 9 of the 45 species are planktonic, whilst the remainder occur on various substrates and in the periphyton.

Nematoda: This group was studied in some detail during the IBP project (Schiemer et al. 1969, Schiemer, 1978). A survey covering all major bottom types of the open lake revealed 26 species. Of these the following 13 are common and typical inhabitants of the bottom of the open lake. They are listed according to their quantitative importance. The number in brackets indicates the number of localities – of a total of 44 – at which they were found.

Paraplectonema pedunculatum (Hofmänner)	(41)
Tobrilus gracilis (Bastian)	(39)
Chromadorita leuckarti (De Man)	(18)
Monhystera macramphis Filipjev	(11)
Theristus flevensis Schuurmans-Stekhoven	(12)
Tripyla glomerans Bastian	(18)
Prodesmodora circulata (Micoletzky)	(11)
Nygolaimus sp.	(16)
Theristus setosus (Bütschli)	(15)
Monhystrella cf. *marina* (Timm)	(7)
Prismatolaimus intermedius (Bütschli)	(9)
Monhystera filiformis Bastian	(5)
Aphanolaimus cf. *aquaticus*.	(5)

Paraplectonema pedunculatum and *Tobrilus gracilis* are the two most widely distributed species within the lake and usually exhibit the highest population densities. Several of the other species show a restricted distribution pattern but may be locally dominant (see below).

Oligochaeta: Oligochaetes are represented by the 3 families Aeolosomatidae, Naididae and Tubificidae. The following 4 species of

Tubificidae were identified by R. O. Brinkhurst from material of 13 localities distributed over the whole area of the open lake.[1]

Potamothrix bavaricus (Oschmann)
Psammoryctides barbatus (Grube)
Limnodrilus profundicola (Verrill)
Limnodrilus udekemianus (Claparède)

Highest numbers of Naididae (*Chaetogaster langi* Bret., *Homochaeta naidina* Bret. det O. Pfannkuche) are attained in the soft mud inshore zones (Schiemer et al., 1969). Aeolosomatidae are restricted to such localities but have been only rarely encountered.

Mollusca: Though several species of gastropods are important components of the reed belt ecosystem (Imhof & Burian, 1972), genuine species of the open lake have only been recognized during the last decade. The occurrence of *Anodonta cygnea* (L.) was first mentioned by Schubert in 1968. During the IBP study the species was found to occur at low population densities in the soft mud areas near the reed fringe.

Both this species and *Dreissena polymorpha* (Pallas), first mentioned by Hacker & Herzig (1970), are undoubtedly newcomers to the lake. Possible reasons for these additions to the fauna are discussed below (see Graefe et al. 1972). Stanczykowska (1977) and Stanczykowska & Schiemer (in prep.) discussed the ecology of *D. polymorpha* in Neusiedlersee.

Tardigrada: This phylum is represented by only one species *Hypsibius* (*Isohypsibius*) *augusti* (J. Murray) – which exhibits a distinct spatial and seasonal pattern of occurrence (see below) (Dollfuss in manuscript).

Hydracarina: Köfler (thesis Graz 1975, unpubl.) named 17 species from different habitats within the reed belt. How many of these occur in the open lake is not yet known. Outside the reed belt the only place where they occur is in the region bordering the reed, and then only in low numbers. The material collected has not yet been systematically examined.

Crustacea: The benthic crustaceans collected during the IBP investigations included 6 Cladocera, 2 Harpacticoida and 2 Ostracoda (det. H. Löffler).

Macrothrix laticornis (Jurine)
Ilyocryptus sordidus (Liévin)
Ilyocryptus agilis Kurz
Leydigia acanthocercoides (Fischer)
Alona rectangula Sars
Chydorus sphaericus (O. F. Müller)

Nitocra hibernica (Brady)
Attheyella crassa (Sars)

Limnocythere inopinata (Baird)
Ilyocypris bradyi Sars

A. rectangula, *C. sphaericus*, (Pesta, 1954); *I. sordidus* (Donner, 1972) and

1. *Tubifex* sp. in Schiemer et al. 1969 stands for Tubificidae spp.

L. inopinata (Schiemer et al. 1969) were already known from the lake. The other species are reported for the first time and had probably been overlooked previously on account of their low population densities.

Pesta (1954) listed still more benthic crustaceans from Neusiedlersee, which are confined to the reed belt and to submerged macrophytes of the reedless areas (*Macrocyclops fuscus* (Jurine), *Eucyclops speratus* (Lilljeborg), *Acanthocyclops vernalis* (Fischer), *Pleuroxus aduncus* (Jurine)). *Macrothrix hirsuticornis* (Norman et Brady) recorded by Pesta (1954) from the open lake was not found in our material.

Chironomidae: The following species have been identified by F. Reiss and M. Hirvenoja from emergence trap material obtained in 1968 and 1969 (Schiemer in press). This faunal list is not complete but includes the more abundant species of the open lake, as well as the bottom-dwelling forms and the species living on submerged macrophytes (marked with asterisk).

Procladius sp.[1]
Tanypus punctipennis (Meig.)
**Cricotopus sylvestris* (Fabr.)
**Psectrocladius* sp.
**Paratanytarsus inopertus* (Walk.) Edw.
Chironomus plumosus L.
Cryptocladopelma virescens (MG.)
Cryptotendipes usmaensis (Pag.)
**Dicrotendipes nervosus* (Staeg.)
Harnischia pseudosimplex G.
Leptochironomus tener (K.)
**Parachironomus arcuatus* G.
Polypedilum nubeculosum (MG.)

Previous records on chironomid species of the open lake are restricted to those of Geyer & Mann (1939): *Phytochironomus* sp., *Paratanytarsus* ex. gr. *lauterborni* (det. Lenz) and to Findenegg (1959): *C. plumosus, T. punctipennis*, *Procladius* sp. (det. Strenzke).

Other insecta: Larvae of Ceratopogonidae occurred regularly but in very low numbers in the reed fringe zone. *Micronecta pussilla*, a common hemipteran on submerged weeds, was occasionally found in bottom samples from the same inshore localities.

The material from emergence traps, operated in 1968 and 1969 in areas with a partial cover of submerged macrophytes (see Fig. 23.11) contained, besides the predominant chironomids, small numbers of Trichoptera (det. H. Malicky), Ephemeroptera and Odonata (det. S. Andrikovits) – all common inhabitants of weeds.

Trichoptera:

Agrypina pagetana Curt.
Cyrnus crenaticornis Kol.

1. The systematics of this species will be treated by Aagaard in a revision of the genus *Procladius*.

Ecnomus tenellus Ramb.
Holocentropus picicornis Steph.
Oecetis furva Ramb.
Oecetis ochracea Curt.
Orthotrichia costalis Curt.

Ephemeroptera:

Cloeon dipterum L.
Caenis horaria L.

Odonata:

Erythromma najas (Hansem.)
Ischnura pumilio (Charp.)

The unique limnological character of Neusiedlersee is expressed in the composition of its benthic fauna which differs strikingly from that of other European waters.

Due to the shallowness of the lake the entire body of water achieves high average summer temperatures, a fact conducive to the occurrence of polythermic forms in both plankton and benthos. Of particular interest is the finding of nematode species hitherto known only from the tropics and subtropics (*Plectus andrassyi*, *Paraphanolaimus microstomus*, *Udonchus* cf. *tenuicaudatus*) (Schiemer 1978). Similar observations have been made for planktonic rotifers (Ruttner-Kolisko & Ruttner, 1959).

Although the ion concentration in the lake has sunk during the last 15 years due to the rising water level, the influence of a high salt content on the faunal composition is still clearly recognizable. A striking number of halophilic forms is found within certain animal groups, and can probably be regarded as relics from a time when the salt concentration was considerably higher (e.g. 1902, Emszt (1904)).

Especially among the well-investigated nematode and rotifer fauna, species typical of salt and brackish water, such as *Cephalodella catellina maior* and *Rotaria laticeps*, *Theristus flevensis* and *Monhystera hallensis* occur beside a series of euryhaline freshwater species.

Similarities to waters of the soda type, expressed for example in the diatom flora (Hustedt, 1959) and in the presence of *Arctodiaptomus spinosus*, are also seen in nematode fauna in the occurrence of the genus *Monhystrella.* This genus has so far mainly been reported as being characteristic for soda lakes, occurring sometimes in large population densities (Schiemer, 1965, 1978).

Characteristic halobionts are missing from the chironomid and tubificid fauna whilst halophilic chironomids occur in the small salty pans east of the Neusiedlersee (Schiemer unpubl.).

The occurrence of polythermic species is probably a result of the combination of exceptional conditions, involving chemical components, temperature, astatics and sediment movements, which assists such forms in competing with the more common and widespread freshwater species.

The sediments of Neusiedlersee exhibit a horizontal zonation with respect to trophic conditions. This is largely connected with sediment erosion and sedimentation, which are dependent upon the wind and upon deposition of organic material from the reed belt and macrophyte zone (see below).

In the highly organic sediments of the deposition zone, organisms indicative of eutrophication, such as *Chironomus plumosus* and *Tobrilus gracilis*, exhibit maximum development. Additionally *Limnodrilus udekemianus* and *Psammoryctides barbatus*, both tubificids typical of polluted habitats (Milbrink, 1973), are found. However, the tubificid species with the highest tolerance of pollution, *Tubifex tubifex* and *Limnodrilus hoffmeisteri*, are missing.

A comparison with fauna lists of the large Hungarian Lakes, Balaton and Lake Velence indicates to what extent benthic faunal associations characteristic of Pannonian lakes may be defined.

Comparable data on the meiobenthic fauna is limited to the nematodes and crustaceans of the Balaton. Biro (1973) lists 46 nematode species from Balaton, of which 31 were found in sediments from the open lake. If this fauna is compared with the 26 species found in Neusiedlersee and the 27 species found by Preys (1977) in the litori-profundal of Mikolajskie, species agreement amounts to 30 per cent in all cases.

Paraplectonema pedunculatum, *Paraphanolaimus behningi*, *Ironus tenuicaudatus*, *Theristus setosus* and *Monhystera paludicola* appear to be most frequent in the sediment of Balaton. It is interesting that whereas *Paraplectonema pedunculatum*, a form hitherto rarely encountered, is dominant in both lakes, *P. behningi* and *I. tenuicaudatus*, two forms occurring in large numbers in Balaton, are absent from Neusiedlersee.

Benthic crustaceans found in the open lake area of Neusiedlersee comprised 7 cladocera, 2 harpacticoids and 2 ostracods, of which the majority are limited to the region immediately bordering the reed belt. All of these species are widely distributed eurytopic forms.

Ponyi (1969) listed 10 Cladocera, 6 copepods and 4 ostracods as typical mud inhabitants of the open water of Balaton. Cladocera common to both lakes are the eurytopic species *Macrothrix laticornis*, *Iliocryptus sordidus* and *Leydigia acanthocercoides.* Oddly enough the copepods and ostracods of the two lakes show no agreement. Species playing an important quantitative role in Balaton, such as *Ectinosoma abrau* (Kritschagin), *Paracyclops fimbriatus* (Fischer), *Nannopus palustris* Brady and *Monospilus dispar* G. O. Sars, have not been found in Neusiedlersee. Conversely, *Limnocythere inopinata*, which is the dominant crustacean in Neusiedlersee and frequently encountered in the salt pans east of the lake is unknown from Balaton.

A certain amount of agreement is detectable between the chironomid fauna of the lakes: *Chironomus plumosus* and *Tanypus punctipennis* are characteristic species in the macro-benthos of both Balaton and Lake Velence. Lenz (1926) described from Balaton species of Chrionomini which are systematically and ecologically closely related to those known from the open lake of Neusiedlersee (*Cryptotendipes anomalus*, close to *C. usmaensis*; *Harnischia conjugens*, close to *Leptochironomus tener*).

In summing-up the results of this comparison it can be said that the benthic fauna of Neusiedlersee is poorer in species than Balaton, and that the faunistic similarities between the large shallow lakes of the Pannonian region are, with a few exceptions, (*Paraplectonema pedunculatum*) limited to some eurytopic forms.

The extent of long-term changes in the benthic fauna cannot be decided owing to lack of data from past years. In this context, two aspects are of interest: firstly, long-term changes in the water level and the major limnological characteristics of the lake associated with it, and secondly that the region is one of the main European transit regions for migrant aquatic birds. Both of these factors obviously contribute to changes in the fauna.

Over the past few years a strong tendency to eutrophication has become apparent. This is manifest in the composition and productivity of the phytoplankton and zooplankton (see Chapters 18 and 22). Changes in the zoobenthos have not so far been detected.

The only certain indication that changes are in progress is the appearance of the two molluscs *Anodonta cygnea* L. and *Dreissena polymorpha* Pallas. Especially in the case of the latter the question arises as to why it did not appear in the lake much sooner, and whether under the present limnological conditions a rapid development and distribution of the species is to be expected, comparable with that occurring in Balaton from 1932 onward (Sebestyen, 1937). Judging from the population development and body growth there seems to be no reason to expect mass development of the species under the conditions prevailing at the moment (Stanczykowska & Schiemer, in press).

23.3 Spatial pattern of benthic conditions

In spite of the uniform shallowness of Neusiedlersee the sediment conditions within the open lake are very diverse. The upper sediment layers consist predominantly of fine clay and silt material, with a highly variable water content. Along thc eastern shore line sediments are coarser due to a higher content of quartz sands. Present knowledge on the petrography of Neusiedlersee sediments has been summarized by Preisinger (see Chapter 12).

The water content of the fine-grained surface sediments exhibits a definite pattern (Schiemer et al. 1969, Schiemer 1978, see Chapter 13). This is essentially the result of wind induced water turbulence. Water turbulence can lead to erosion of the lake bottom in areas of the lake that are exposed to the wind (Tauber 1959, Tauber & Wieden 1959). The material eroded from the bottom, dying plankton and inorganic particles of biogenic origin (chiefly protodolomite, (see Chapter 12), sediments in places where the water is less agitated. Such places are found mainly in the northwestern area, as a consequence of the northwesterly winds, and in extensive areas at the southern end of the lake where the open water is split up by reed islands that provide shelter from the wind. Water turbulence on the lake bottom, and hence the tendency for particles to be stirred up from it and to drift off, increases from northwest to southeast (see also Dokulil 1975).

Although plant cover and biomass are relatively low (Schiemer & Prosser 1976, see Chapter 19), submerged macrophytes growing adjacent to the reed belt influence sediment conditions in several ways. Macrophyte growth, extending to the water surface, restricts water turbulence and has led to increased sedimentation within this zone. Furthermore, decomposing macrophytes contribute considerably to the detritus pool of the sediments. During the period of study the macrophyte zone, which consists exclusively of *Myriophyllum spicatum* L. and *Potamogeton pectinatus* L., was well developed at the northern end and in the southern part of the lake whereas along the eastern shore macrophyte growth is scarce and restricted to *P. pectinatus*.

The macro-zonation of bottom conditions for the purpose of eludicating the distribution of the fauna, as depicted in Fig. 23.1A) is based on the data of Jungwirth (see Chapter 13) and on the author's own fields observations. Three zones can be distinguished according to the thickness of the soft mud layer. The term 'soft mud', according to Preisinger (see Chapter 12), indicates a water content of 60–90 per cent.

zone 1: Sedimentation bottoms with soft mud layers of 5–40 cm thickness. Such accumulations of soft mud, with a high organic content, are typical for sheltered bays as well as zones which formerly had denser stands of submerged macrophytes especially in the NW and S parts of the lake.

zone 2: Bottoms with soft mud layers of 1–5 cm thickness, adjoining zone 1. During the period of the IBP study, this zone was to a large extent (see Fig. 23.1A) covered loosely with submerged macrophytes.

zone 3: Compact bottoms covered with less than 1 cm of soft mud. These are typical of areas with considerable windfetch, especially from northwesterly winds. Bottoms of this type prevail in the central and eastern parts of the lake. The sediments along the eastern shore are characterized by a higher content of sand.

This is, however, an over-simplification of the situation, particularly in the case of zone 3, and does not indicate the full extent of the heterogeneity of the lake bottom. Besides areas of soft mud in some offshore regions, apparently the result of long-term current systems[1] (see Fig. 8.1a), the bottom can change from soft mud to hard bottom within a very short distance. The broad spectrum of pH values in the sediments of zone 3 shown in Fig. 23.1B indicates the extent of this heterogeneity.

The bottoms also vary quite distinctly with respect to their content of organic substances. Water content and organic material present in the sediments of the open lake are strongly positively correlated (see Chapter 13). The following mean values of organic contents (measured as loss on ignition at 450°C for 24 h) were found for the different regions of the lake (see Chapter 13).:

1. Evidence for this assumption is provided by the large quantities of pollen and of resting stages of animals in these deposits.

zone 1,	soft mud in sheltered bays and in the reed fringe zone	8 per cent (range 4–10 per cent)
zone 2,		6 per cent (range 2–9 per cent)
zone 3,	central lake area, hard bottom	2–4 per cent
	eastern shore, sand	2 per cent

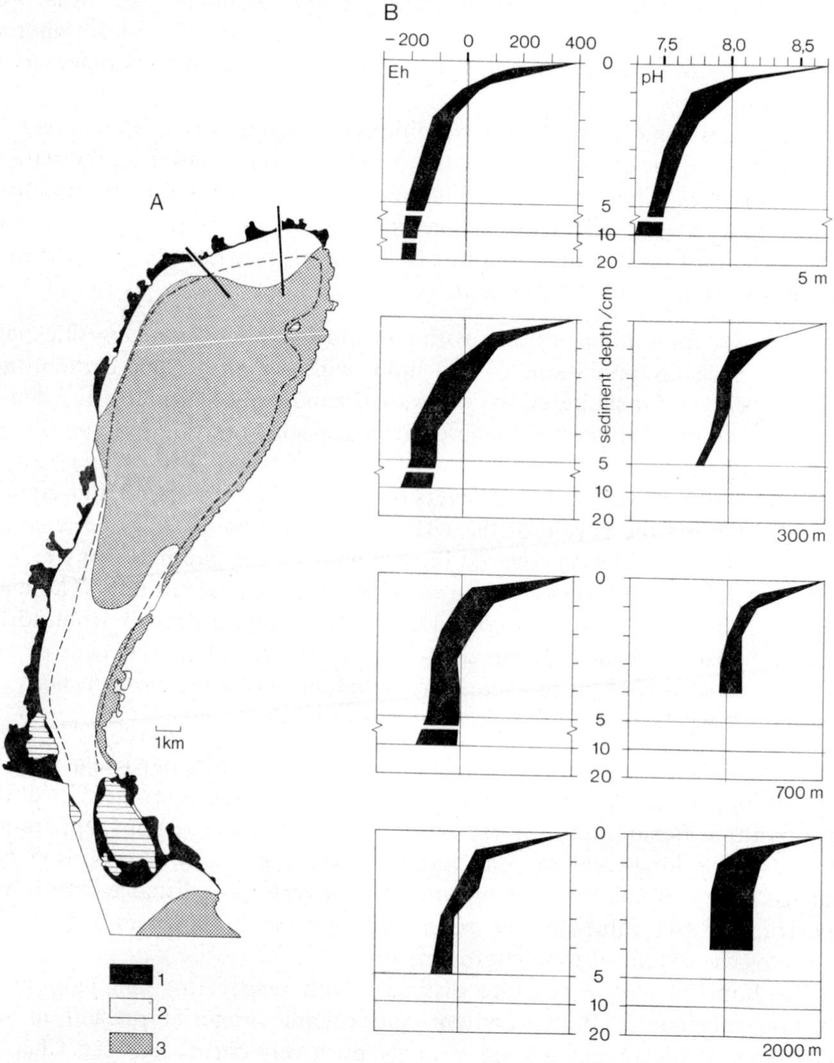

Fig. 23.1A. Spatial pattern of benthic conditions in the open lake outside the reed belt. 3 zones are distinguished (see text): zone 1 = black, zone 2 = white, zone 3 = dotted, striated area = reed islands. The outer border of the submerged macrophyte growth – as found in 1970 (Schiemer & Weisser 1972) – is indicated by broken line.
B. Redox potential (E_h in mv) and pH profiles in sediments from various distances from the reed belt along 2 trasects (indicated in Fig. 23.1A) in the north western part of the lake. The bands represent the range of measurements from 5–10 cores.

A study of several sediment characteristics was concentrated on the NW part of the lake and was combined with a detailed study of the zoobenthos. A very distinct inshore-offshore gradient could be observed (Fig. 23.1B).

Table 23.2 indicates the decrease in thickness of the soft mud layer and in the total phosphorus and Kjeldahl-N in the upper sediment layers along a transect from the reed fringe towards the open lake. Both the P and N contents show a high degree of correlation with the organic content of the sediments (see Chapter 10).

Table 23.2. Sediment characteristics and macrophyte biomass at different distances from the reed fringe along a transect in the NW part of the lake (transect 1 in Fig. 23.3). Kjeldahl-N and total phosphorus in per cent of sediment dry weight (from Schiemer and Prosser, 1976).

Distance from reed belt	Depth of soft sediment (cm)	N	P	Total macrophyte biomass (g dry weight/ 1,000 m^2)
bay	15–25	0.270	0.060	—
30 m	8.6	0.255	0.056	3,200
100 m	7.0	0.220		3,140
300 m	4.7	0.190	0.055	760
400 m	4.0	0.190		533
500 m	2.9	0.170	0.051	115
600 m	2.0	0.165		47

pH and redox potential profiles of sediments exhibit similar gradients (Fig. 23.1B). The pH (see Table 23.3) and E_h values found in deeper sediment strata were lowest in the reed fringe zone and increased towards the open lake as a result of reduced microbial activity.

The pH in sediments decreases with sediment depth due to the accumulation of metabolic endproducts. In the NS sediments the steepest gradients occur within the upper 2 centimeters (Fig. 23.1B), in parallel with the redox profiles. Diurnal fluctuations in pH may occur at the sediment surface in localities with a high biomass of epipelic algae, particularly during the winter months (see below).

Table 23.3. pH values measured in different sediment depths at 3 stations along transects from the reed fringe towards the lake centre at the northwestern end of the lake (see Fig. 23.1A). Measurements were carried out by pushing glass electrodes (tip diameter 10 mm) into sediment cores.

Sediment depth	Distance offshore			
	5	300	700	2,000 m
1 cm	7.7–8.0	8.0–8.2	8.1–8.2	7.9–8.3
4 cm	7.4–7.6	7.8–7.9	7.9–8.1	7.9–8.2
10 cm	7.3–7.5	—	—	—

Redox potential measurements[1] are the best means of determining the demarcation line between oxic and anoxic conditions in sediments. The position of the redox discontinuity layer and E_h values below +200 mv (Mortimer, 1942) indicate the onset of anoxic conditions. Different bottom types had in common the steep gradient in E_h values measured within the upper centimeter. The thickness of O_2-depleted sediment layers is between 3 and 7 mm and does not show distinct territorial differences. This is in contrast to the situation in Balaton where the oxidised stratum is thinner in sediments with a higher organic content (Olah, 1975). The situation found in Neusiedlersee is explainable by the negative correlation between organic content and density (i.e. exchange of interstitial water) of the sediments (see Chapter 13).

In summarizing it can be stated that in the sedimentation zones of the lake the bottom shows an increase in organic content and in bacterial activity towards the reed belt (see Table 23.15). This is clearly illustrated in the O_2 consumption processes registered beneath the ice cover (Neuhuber 1974, see Chapter 11). O_2 depletion occurs only near the reed fringe and during the night, when the benthic algae are not producing O_2 by photosynthesis. The oxygen conditions become especially critical for the bottom fauna in the vicinity of the reed in periods when the photosynthesis of the benthic algae is drastically reduced under a winter cover with low light transmission (see below).

23.4 The distribution pattern of the fauna

The distribution pattern of the benthic fauna corresponds to the zonation of benthic conditions as described above. Such a zonation generally exists with respect to population densities, but is also expressed in the limitation of the occurrence of some species to particular areas (Schiemer et al. 1969, Schiemer 1978 and in press).

High population densities of both meiobenthos and macrobenthos occur in zones 1 and 2, lower densities in zone 3. Nematodes are generally numerically dominant in the meiobenthos although in zone 1 – the inshore zone with deep layers of soft mud – other taxa may be of additional seasonal importance (i.e. *Hypsibius augusti* – Tardigrada, *Chaetogaster langi* – Oligochaeta, *Paradicranophorus sordidus* – Rotatoria).

A significant species zonation was found with respect to nematodes. On the basis of regularly occurring species, albeit sometimes of low abundance, five species associations could be defined, which show a topographical sequence from NW to SE, corresponding to the main wind direction (Schiemer, 1978).

Monhystera macramphis was regularly and exclusively encountered in zone 1. In well protected locations, e.g. in reed bays and in the marginal fringe of the reeds this species may be numerically dominant in the meiobenthos.

1. Measurements were carried out with a Methrom E 488 by pushing a Pt-electrode through a slit in the plexiglas corer. The slit is sealed with silicon rubber. A calomel electrode served as a reference.

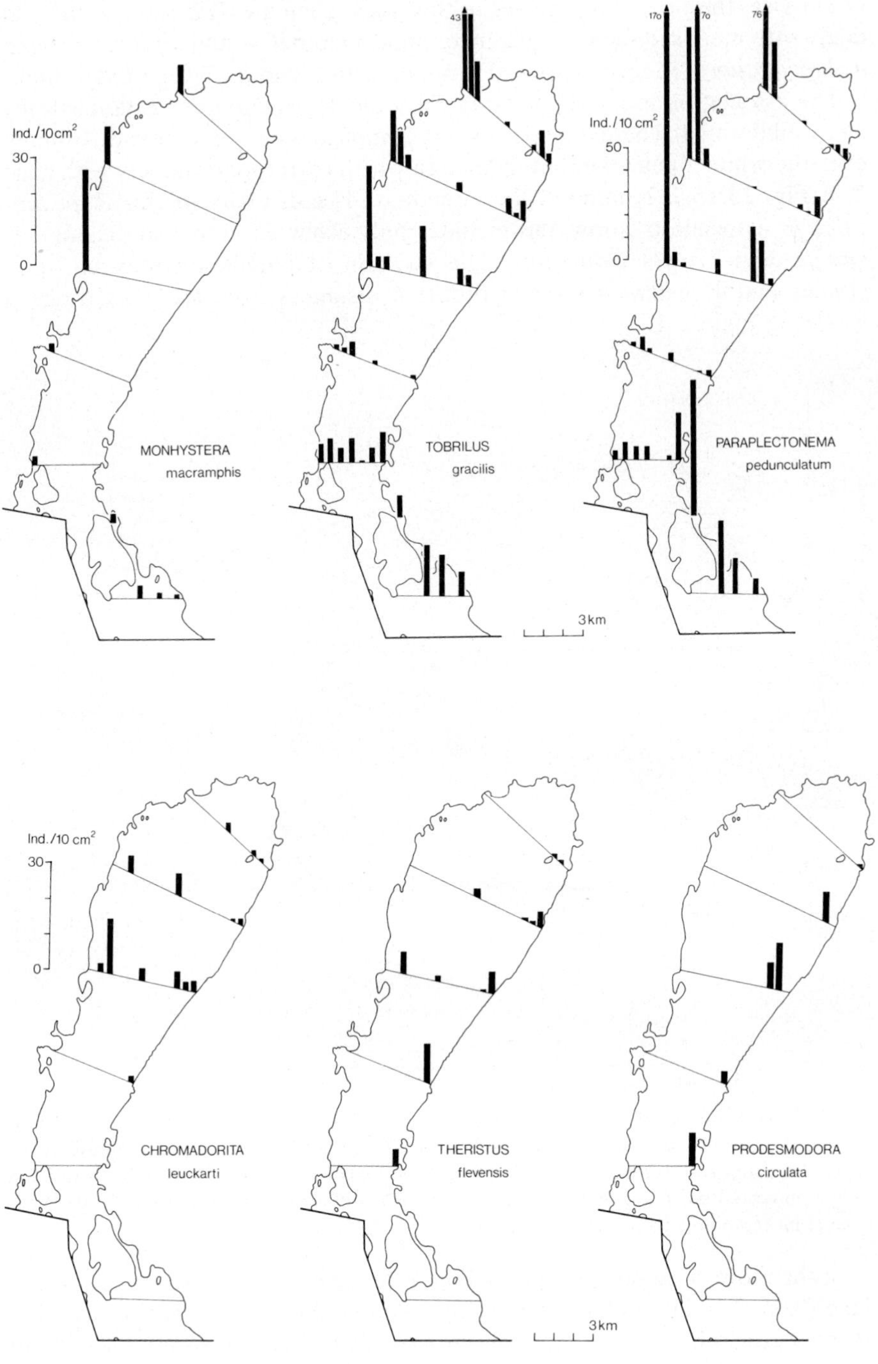

Fig. 23.2. Distribution pattern of 6 species of nematodes as found during a survey in October 1968. Each column represents the mean of 3 samples. Note the different scale in *Paraplectonema pedunculatum* as compared to the other 5 spp.

Otherwise, the prevailing species in this association are *Tobrilus gracilis* – in deep soft mud sediments, rich in organic material – and *Paraplectonema pedunculatum* – in more exposed grounds with a thinner layer of soft mud.

The association found adjacent to the zone *M. macramphis* is faunistically less well defined. The main species – *P. pedunculatum* – is widely distributed over the whole open lake, but in an area which corresponds largely with zone 2 in Fig. 23.15 it is numerically abundant. The diversity of this *P. pedunculatum* association is low and includes only a few eurytopic species like *T. gracilis* and *Tripyla glomerans.* The location of population maxima of *T. gracilis* and *P. pedunculatum* at different distances from the reed fringe is shown in Fig. 23.3

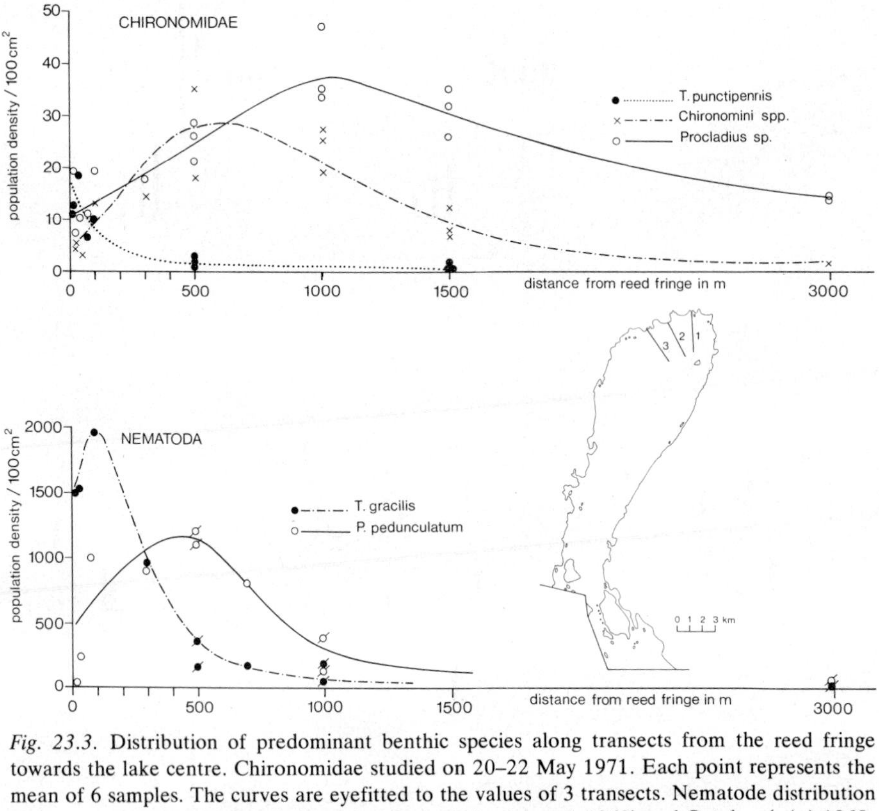

Fig. 23.3. Distribution of predominant benthic species along transects from the reed fringe towards the lake centre. Chironomidae studied on 20–22 May 1971. Each point represents the mean of 6 samples. The curves are eyefitted to the values of 3 transects. Nematode distribution along transect 3 studied in July (o •; data from Schiemer et al. 1969) and October (ȯ •̸) 1968. Each point represents the mean of 3 samples.

In the more solid bottoms prevailing in the central and eastern part of the lake ('zone 3'), in addition to *P. pedunculatum* and *T. gracilis*, various other forms occur which are completely missing from the nematode associations mentioned above. Of these species *Chromadorita leuckarti* is the most common. Other species, e.g. *Theristus setosus*, *Nygolaimus* sp. and *Prismatolaimus*, are similar distributed but less frequent. A further topographic differentiation exists in 'zone 3' in the appearance of additional nematode species in sequence from W to E. In this direction *Theristus flevensis* is the first to occur, whereas the occurrence of *Prodesmodora circulata* is confined to the more coarsely grained sediments along the eastern shore line. The quantitative occurrence of six nematode species, as found during a survey in October 1968, is shown in Fig. 23.2.

Of the other meiobenthic species only *Hypsibius augusti* (Tardigrada) and *Limnocythere inopinata* (Ostracoda) have been studied in greater detail (see Chapter 24). The topographic pattern of occurrence of *H. augusti* is restricted to 'zone 1' and thus exactly resembles that of *M. macramphis. L. inopinata*, the most abundant crustacean, is distributed over the whole area of the open lake. High population densities, however, are also restricted to 'zone 1' (see Chapter 24).

The distribution pattern of benthic rotifers has been studied in less detail. From two W–E transects analysed (Donner, 1972) it appears that species numbers decline from the soft mud areas along the western reed fringe to the central area and to sandy bottoms along the eastern shore line. However, in contrast to nematodes, the fauna on the eastern coastline is not specific but represents only an impoverished fauna of the soft mud zones.

The distribution pattern of macrobenthic species was studied in greater detail at the north-western end of the lake (see Fig. 23.3). Chironomidae is the most abundant macrobenthos taxon in all localities studied and representatives of the Tanypodinae form 79–86 per cent of the total chironomid fauna, based on annual means for different localities.

The two main species of Tanypodinae, *Tanypus punctipennis* and *Procladius* sp. show a clear difference in habitat choice.

The population maximum of *T. punctipennis* was found to be at the edge of the reed belt. The abundance declines sharply with distance offshore. A similar pattern, however, at an overall lower density is found in several other species, e.g. in *C. plumosus* and the benthic cladocera like *Leydigia acanthocercoides*, *Macrothrix laticornis* and *Iliocryptus sordidus*.

Tubificidae are quantitatively of less importance. The annual means of population density (Table 23.4) indicate a decrease from inshore to offshore,

Table 23.4. Annual mean (1971/72) of population densities (individuals/100 cm^2) at three localities along a profile from the reed fringe towards the lake centre in the northern part of the lake. The profile is shown in Fig. 23.3 (1). The localities are indicated in metres from the reed fringe.

	10 m	450 m	2,000 m
Chironomidae total	88.1	66.0	24.2
Tanypus punctipennis	65.7	4.7	0
Procladius sp.	17.8	47.5	20.9
Chironomus plumosus	4.1	0	0
'small Chironomini'	6.7	13.3	3.1
Tanytarsini spec.	3.8	0.5	0.2
Ceratopogonidae	1.0	0	0
Tubificidae	6.8	4.2	2.6
Benthic Cladocera	4.1	0	0

which is, however, less pronounced than in the case of the chironomids. The tubificid material has not yet been analysed with respect to species composition.

Procladius sp. and 'small Chrionomini ssp.' show similarities with the pattern found in *Paraplectonema pedunculatum*, namely a population maximum at some distance offshore (Fig. 23.3). Maximum population densities have been found in the outer zones of the loose macrophyte belt. *Procladius* sp. is by far the most common macrobenthic organism in the open lake area.

In summarizing the results of the distribution of meio- and macrobenthos, different types of distribution pattern and faunal assemblages can be distinguished.

(a) species restricted to 'zone 1': *M. macramphis*, *H. augusti*, *C. plumosus*
(b) species occurring over the entire area of the open lake but with maximum abundance in soft bottom sediments high in organic content and bacterial activity near the reed fringe (zone 1), and a steep decrease in abundance within 500 m offshore: *T. gracilis*, *L. inopinata*, *T. punctipennis*
(c) species occurring over the entire area of the open lake with, however, maximum densities at some distance between 300 and 1,500 m from the reed fringe along the north-western shore line (representative for zone 2): *P. pedunculatum*, *Procladius* sp. *Harnischia* complex ssp.
(d) species with a restricted occurrence in the central and eastern part of the lake (largely zone 3), missing in the soft mud – sedimentation zones. Such a pattern was so far encountered only with respect to nematodes (see above).

A full explanation of the pattern described above is not yet possible since the ecological requirements of the predominant species are not sufficiently known.

The steepest gradients in fauna and ecological conditions are found within 500 m lakeward of the reed fringe along the western shore. This gradient is expressed in a decrease of
- the thickness of the soft mud layer
- the bacterial activity
- the organic content
- the abundance and biomass of epipelic algae

For species with maximum population densities at or near the reed fringe (pattern a and b) strong correlations of population densities with high values of all the factors listed above can be found (see Chapter 24).

Maximum sedimentation of phytoplankton and highest densities of benthic algae in zone 1 provide algivorous species with optimum feeding conditions (see below).

Offshore peaks of species like *P. pedunculatum* and *Procladius* sp. are possibly a response to perodical deoxygenation phenomena in zone 1 under ice cover (see Chapter 11), or may be the result of specific nutritional requirements. However, the food of the two species differ drastically: *P. pedunculatum* is a bacterivorous species. *Procladius* is omnivorous but with a high proportion of animal prey. Sedimenting zooplankton appears to be the main food source of this species in Neusiedlersee. The distribution pattern of population densities of this species is probably correlated with the rate of zooplankton sedimentation in the different zones of the lake.

On the other hand, predation of *Procladius* could be an important factor for the zonation of benthos, causing low densities of other macrobenthic species in offshore areas.

The NW-SE sequence of nematode zones indicates the wind influence and its effect on particle size, organic content and stability of the substrate. However, compared with its erosion effect on the lake bottom, water turbulence appears to have very little direct influence on the distribution via drifting of eggs and animals. The occurrence of benthic animals in plankton samples is low even after periods of strong water movement (Dokulil, Herzig, pers. commun.). The distinct pattern of nematode distribution in the lake also points to a low mixing of fauna. Furthermore a study on the seasonal dynamics of macrobenthos at three stations at different distances from the reed did not indicate strong shifts in populations due to the wind effect.

The vertical distribution of the benthic fauna in the sediments of Neusiedlersee has already been described (Schiemer et al. 1969) and discussed with respect to ecophysiological adaptations (Ott & Schiemer, 1973; Schiemer & Duncan, 1974). Most species of meiobenthos in Neusiedlersee are confined to the surface layer of the sediments (e.g. *Hypsibius augusti*, *Monhystera macramphis*, *Chaetogaster langi*). A few species of nematodes, however, have their maximum in deeper sediment strata. A study of the vertical distribution of meiobenthos in soft sediments in the vicinity of the reed belt (Schiemer et al. 1969) revealed that the population maximum of *Tobrilus gracilis* is clearly below the redox discontinuity layer. The depth distribution of this species seems to be limited by the decreasing water content.

In *Paraplectonema pedunculatum* a vertical zonation of life stages could be observed. The population maximum of young development stages is closer to the surface than that of the adult, parthenogenetic females. The depth distribution gives econological evidence for anaerobiosis in *T. gracilis* and adults of *P. pedunculatum*, while the larvae of the latter species might be more sensitive to anoxic conditions.

A study of the respiration rate in *T. gracilis* revealed distinctly lower values as compared with the general metabolic level of free-living nematodes. This can be assumed to be an ecophysiological adaptation either in the form of a generally lower metabolic level of the species, or a partially fermentative metabolism even in the presence of oxygen (Schiemer & Duncan, 1974).

The macrobenthic species, *Chironomus plumosus* and the Tubificidae form deep passages within the soft sediments along the reed fringe. The two main macrobenthos components, *Tanypus puctipennis* and *Procladius* sp., however, are free-living on the sediment surface. This differences in vertical distribution appears to be of importance for prey selection in benthivorous fish (see p. 379).

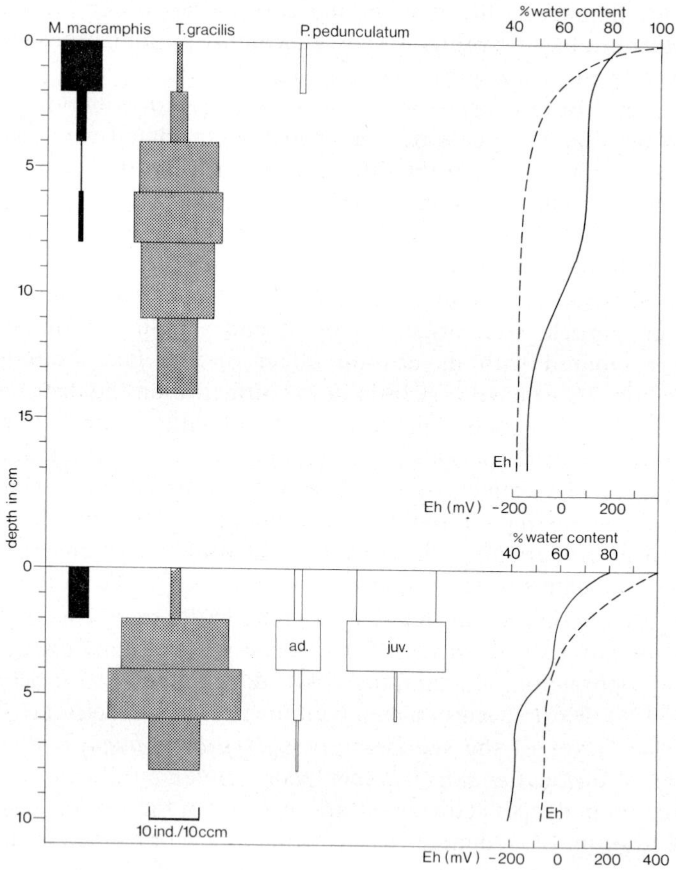

Fig. 23.4. Vertical distribution of 3 species of nematodes together with representative profiles in water content and redox potential of the sediments (data from Schiemer et al. 1969) 5 and 70 m distance from the reed fringe.

23.5 Seasonal dynamics of benthic fauna

23.5.1 *Localities studied and methods used*

After initial surveys on the distribution patterns of meio- and macrobenthos, detailed studies on annual cycles, biomass and trophic relationships have concentrated on the northernmost part of the lake (see Fig. 23.5).

Seasonal dynamics of meiobenthos was studied 1967 and 1968 at a locality situated within the soft mud area, 50 m from the reed fringe. From September 1967 until September 1968 29 sets of core samples (3 samples each, opening 5.3 cm^2) were taken. The soft mud layer of the sediment was processed with a sieving apparatus as described by Löffler (1961). The meiofauna was separated from the residue in the sieves by Baermann's method and investigated in vivo. Animals were fixed in formalin, the nematodes subsequently transferred to glycerine, mounted on slides and their body size calculated on the basis of camera lucida drawings.

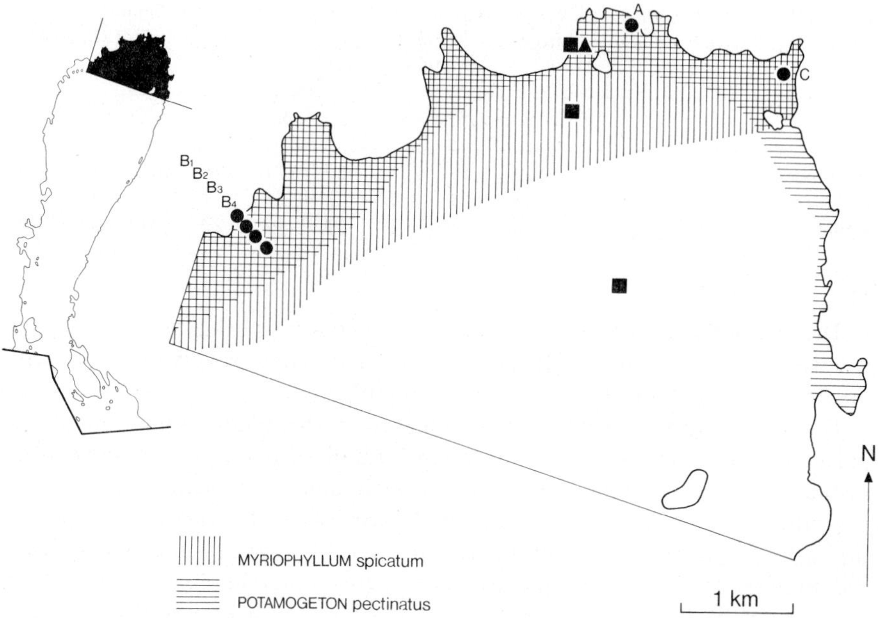

Fig. 23.5. Position of sampling sites of macrobenthos (■), meiobenthos (▲) and chironomid emergence traps (●; letters – A, B_1 – B_4, C – refer to Fig. 11). Areas of submerged macrophyte growth in 1970 after Schiemer & Weisser 1972.

Macrobenthos was studied in 1971 and 1972 at 3 localities situated along a transect 10, 450 and 2,000 metres from the reed fringe towards the centre of the lake. The 10 m station is situated in a sheltered bay with a deep layer of soft mud, the 450 m station within an area dominated by *Myriophyllum* growth (see Chapter 19). The 2,000 m station is representative for the generally denser bottom of the central lake, devoid of macrophyte growth. The 3 stations are representative for the 3 benthic zones of the open lake as discussed above. From May 1971 until May (July at the 450 m station) 1972, 21, 23 and 18 sets of 10 samples each were taken by means of a core sampler with a 16.6 cm^2 opening.

The samples were washed through a 100 μ sieve. Animals were floated from the sediment with a sugar solution of 1.25 density, after which they were sorted alive under the microscope and preserved in 5 per cent formalin.

The chironomids were separated into instar classes on the basis of the size of the head capsules. Dry weight determinations of individual larvae of *Procladius* sp. and *Tanypus punctipennis* were made after drying at 95–98°C using a Mettler ME 22 electro-magnetic balance. Body length – body dry weight regressions were established and used for biomass calculations.

23.52 *Seasonal changes in meiobenthos*

The composition and density of the fauna proved to be subject to great temporal changes (Table 23.5). Nematodes in total showed a maximum in May and June. *Paraplectonema pedunculatum* and *Tobrilus gracilis* constitute approximately 90 per cent of the nematode fauna throughout the year with *P. pedunculatum* being generally the dominant form.

Tables 23.5. Monthly means of the population density (per 10 cm^2) since September 1967. Values which have been obtained from less than 6–12 samples are set out in italics (Schiemer et al. 1969).

	S	O	N	D	J	F	M	A	M	J	J
nematodes	78	85	112	149	*235*	200	*171*	*148*	405	618	315
tardigrades	127	55	46	35	*11*	2	*1*	*1*	2	15	27
naidids	10	2	6	8	*4*	22	*94*	*29*	29	42	2

The average population density of *P. pedunculatum* at the inshore sampling site was 196 ind./10 cm^2. A distinct maximum occurred in May and June, with maximal average densities of 600 ind./10 cm^2.

The proportion of reproductive females in the total population of the parthenogenetic species varied between 20 and 70 per cent, but showed no distinct seasonal pattern. The population dynamics indicate a continuous reproduction throughout the year. The life history of *T. gracilis* is of specific interest considering the anaerobic or oligoaerobic way of life of the species. An analysis of the size frequency pattern (Schiemer in prep.) suggest one overwintering generation with a reproduction phase in May and June and 2 additional summer generations, indicated by a high proportion of adults in July and September respectively. However, the material available is insufficient to prove the proposed pattern. Pehofer (1977) found *T. gracilis* to be univoltine in the litori-profundal of a meromictic lake in Austria, with maximal reproductive acivity in autumn and winter. Such a prolonged life cycle of a small sized animal has to be discussed in respect to the inefficient energy gain of anaerobic species.

A pronounced seasonal cycle was found in *Hypsibius augusti* (Tardigrada) (Dollfuss in manuscript). In both years of study (67–69) the species was most abundant during autumn and very scarce in the period from January to May.

Naididae showed a complementary pattern, with high population densities from February to June and low numbers throughout the rest of the year.

23.5.3 *Seasonal changes in macrobenthos*

The macrobenthos data are presented in Table 23.6. Mean densities at the 10 m, 450 m and 2,000 m station are $10 \cdot 10^3$, $7 \cdot 10^3$ and $2.7 \cdot 10^3$ individuals/m^2 respectively, with seasonal fluctuations in the range of 25–250 per cent of these annual means. The fluctuations are mainly caused by the population dynamics of the predominant heterotopic chironomids, presented in Figs 23.6 and 23.10. A detailed discussion of population dynamics of *Tanypus punctipennis* and *Procladius* is given in connection with a description of their life histories.

The seasonal changes in abundance of *T. punctipennis* near the reed fringe as well as of *Procladius* sp. at the offshore stations (450 m, 2,000 m)

Table 23.6. Density fluctuations of macrobenthos in 1971/72 at 3 localities in the northern end of the lake. Each value (ind./100 cm^2) represents the mean of 10 samples. The population dynamics of the dominant chironomids *Tanypus punctipennis* and *Procladius* sp. is presented in Figs. 23.10 and 23.6 respectively.

	21.5.1971	15.6.1971	25.6.1971	7.7.1971	21.7.1971	1.8.1971	14.8.1971	26.8.1971	8.9.1971	20.9.1971	5.10.1971	22.10.1971	4.11.1971	19.11.1971	21.12.1971	26.1.1972	3.3.1972	31.3.1972	21.4.1972	4.5.1972	19.5.1972	13.6.1972	14.7.1972	mean
Station 1																								
10 m off reed fringe																								
T. punctipennis	16.2	70.5	82.5	49.4	54.2	50.6	65.1	63.2	56.0	27.1	53.6	161.4	94.6	119.3	117.5	103.0	51.2	42.2	37.3	50.6	13.9			65.7
Procladius	9.3	11.4	12.0	9.0	9.7	6.0	4.2	0	5.4	4.2	4.8	14.4	3.6	10.8	10.8	7.2	6.0	10.2	10.2	14.4	0.6			7.8
Other Chironomidae	4.5	27.1	11.4	11.4	22.3	10.2	6.0	5.4	3.6	6.0	13.9	26.5	27.7	32.5	41.0	16.3	11.4	21.0	3.6	3.0	1.2			14.6
Ceratopogonida	0.3	0.6	0.6	1.8	0.6	0.6	0	0.6	0.6	1.2	0.6	4.0	0.6	3.6	0.6	1.2	0	1.5	0.6	0.6	1.2			1.0
Tubificidae	10.7	11.4	8.4	0	6.6	5.3	3.6	1.8	3.0	1.8	6.6	10.8	3.0	10.8	16.3	10.2	3.6	12.0	6.0	1.8	8.4			6.8
Benthic Cladocera	–	4.2	12.6	7.8	13.8	2.4	1.2	3.6	7.2	7.8	4.2	0.6	4.2	1.2	0.6	0.6	0	10.5	0	0	0			4.1
	41.0	125.2	127.5	79.4	107.2	75.1	80.1	74.6	75.8	48.1	83.7	217.7	133.7	178.2	186.8	138.5	72.2	97.4	57.7	70.4	25.3			100.0
Station 2																								
450 m off reed fringe																								
T. punctipennis	1.8	6.6	3.0	5.4	3.6	0.6	4.2	6.0	3.6	0	4.2	6.6	20.5	17.5	3.6	6.0	1.8	1.8	3.6	0	1.2	2.4	3.0	4.7
Procladius	24.7	26.5	37.3	56.6	62.0	49.4	41.6	33.1	42.8	20.4	24.7	93.4	96.4	97.0	39.8	78.9	75.3	50.6	43.4	36.7	15.7	21.0	25.3	47.5
Other Chironomidae	31.9	17.5	13.2	6.6	7.8	1.8	15.1	8.4	10.8	3.6	3.0	23.5	45.2	36.1	8.4	18.7	25.3	7.8	18.7	1.8	3.6	5.4	4.2	13.8
Tubificidae	1.5	13.2	1.2	2.4	0.6	0.6	22.9	1.2	3.0	1.8	4.2	7.8	12.0	1.2	1.2	0.6	7.8	2.4	2.4	0.6	1.2	0	6.0	4.2
	59.9	63.8	54.7	71.0	74.0	52.4	83.8	48.7	60.2	25.8	36.1	131.3	174.1	151.8	53.0	104.2	110.2	62.6	68.1	39.1	21.7	28.8	38.5	70.2
Station 3																								
2000 m off reed fringe																								
Procladius	25.5	9.0	27.1	27.1			25.3	19.9	27.1	7.8	9.6	31.9	41.0	41.6	18.1	19.9		9.0	11.4	10.8	14.4			20.9
Other Chironomidae	7.5	1.8	10.2	0			5.4	1.8	1.2	2.4	0	3.0	6.0	7.8	2.4	4.5		0.6	1.8	0.6	3.0			3.3
Tubificidae	0	6.0	6.0	1.2			10.2	0.6	1.2	1.8	0	0.6	1.8	7.8	3.0	2.4		0.6	1.8	0	1.8			2.6
	33.0	16.8	43.3	28.3			40.9	22.3	29.5	12.0	9.6	35.5	48.8	57.2	23.5	26.8		10.2	15.0	11.4	19.2			

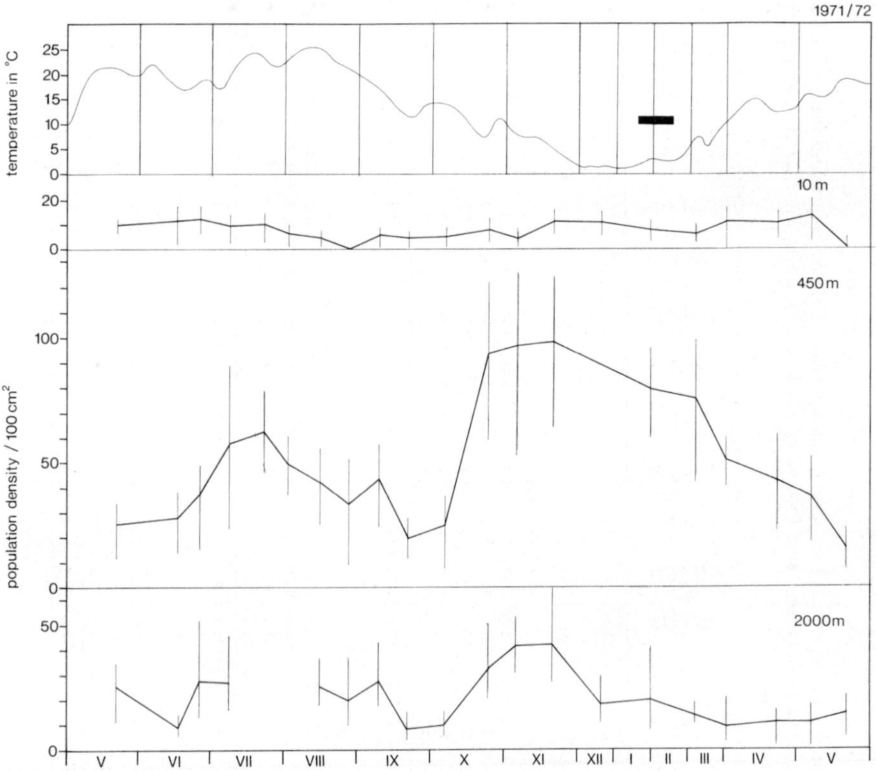

Fig. 23.6. Population dynamics (+95 per cent confidence intervals) of *Procladius* sp. at 3 sampling sites (see Fig. 23.5). Seasonal variation in temperature (1 m depth, weekly means) and extent of ice cover (black bar).

are characterized by generally lower values during the summer, as compared with the period from October to March. Both species achieved their highest population densities at the end of the emergence period, during October, due to the impact of young stages.

Fluctuations in abundance of *Procladius* larvae during the summer months coincide with the sequence of generations, emergence and reproductive phases. The patterns of seasonal changes are similar at the 450 m and the 2,000 m locality.

In *T. punctipennis*, however, the population density remains at a relatively constant level during the summer months (see Fig. 23.10). Population losses due to elimination and emergence are balanced during this period by continuous reproduction (see below). Low numbers in May and September are caused by emergence peaks.

From November onward the whole chironomid population is in the larval stage and, since no emergence occurs from November to March, the decrease in abundance during this period is due to mortality and predation by fish. Elimination rates are relatively constant over the whole winter period in

Procladius sp. and range from 4–8 per cent losses during one month. A large decrease in abundance in March marks the start of the flying period. The main emergence period of the overwintering population at the beginning of May leads to a further reduction of larval numbers.

For the population of *T. punctipennis* at the inshore station, monthly losses from the end of October until the end of January average 11 per cent. However, in the period from 27 January–3 March 1972 the population losses increase to 40 per cent per month (see below).

No distinct seasonal changes in frequency can be observed in tubificids. This may be due to low population density and the strongly aggregated type of distribution, which considerably detracts from the accuracy of the numerical data.

Benthic Cladocera, as would be expected, are less often encountered in the winter months than at other times of year.

The seasonal periodicity of the macrobenthos gives us some indication as to which ecological factors exert a major influence on the population dynamics of fauna in this particular type of lake. The general picture of a lower abundance during late spring and summer and higher densities in autumn and winter reflects the life history of the two main species, *T. punctipennis* and *Procladius* sp. However, the low summer values are not only the result of continuous emergence but also of higher elimination rates due to increased predation by benthos-feeding fish (especially the ruffe, *Acerina cernua* L.). A quantitative consideration of the elimination of chironomids by fish will be given in the context of a discussion of the whole benthic food chain (see below).

Elimination during the winter months is generally low. A remarkable increase of elimination, however, was observed at the inshore station during the period from 27 January–3 March 1972. A glance at the meteorological data for this particular winter shows that the lake was ice covered from 14 January – 14 February and was covered by deep snow from 22 January onwards. Deoxygenation processes occur under such conditions in the inshore zone (see Chapter 11) and are responsible for the high larval mortality during this period. However, even in such periods with deoxygenation no marked migrations from the marginal zone of the reed towards the offshore regions, better supplied with O_2, have been observed.

The parallel course of the population dynamics at the 450 m and 2,000 m sites throughout the year suggests that there is no strong horizontal shift in the populations due to migrations or passive drift.

23.6 Life cycles of predominant species

Life histories of *Procladius* sp. and *Tanypus punctipennis* have been studied on the basis of the larval material sampled during 1971–72. Additional information concerning the flying period of the adults can be derived from emergence trap catches made during 1968 and 1969 at different positions in the northern part of the lake (Fig. 23.5). The results concerning the phenology of the chironomids, including subdominant benthic species, e.g. *Chironomus plumosus*, have been discussed elsewhere (Schiemer, in press).

23.6.1 *Procladius sp.*

The most extensive material originated from the habitat in the macrophyte zone at a distance of 450 m from the reed fringe. This material (1,828 individuals) was not only investigated as to the component stages, but, in addition, the L4, ranging in size from 2.4–7.8 mm were differentiated into length classes differing by 0.3 mm. An interpretation of the population dynamics of this habitat is contributory to a better assessment of the dynamics of the less densely populated localities (10 m, 2,000 m).

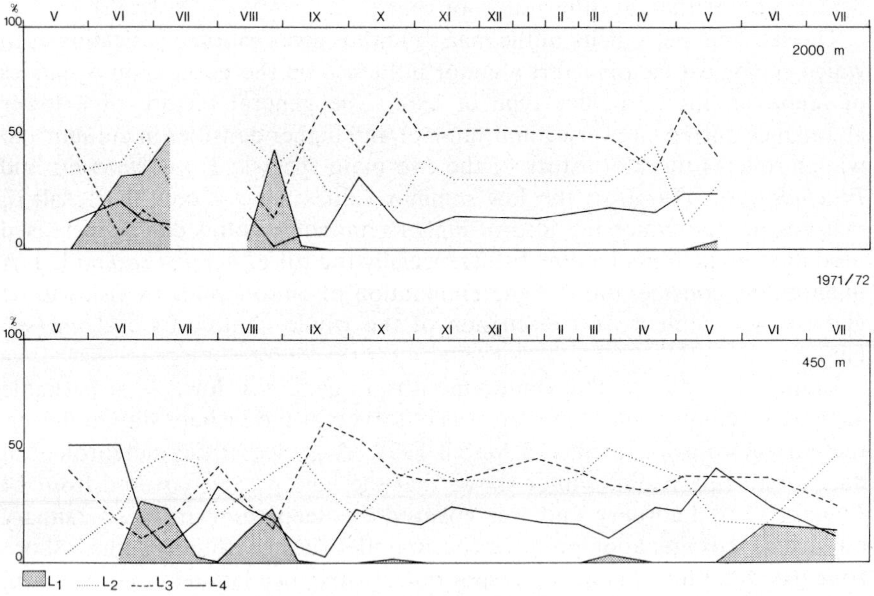

Fig. 23.7. Seasonal change in larval composition of *Procladius* sp. at 2 sampling sites.

Sampling was started in the second half of May 1971. A small population was in an advanced stage with a higher proportion of L4 and no L1. The population density rose during June due to a large increase of young larvae. Thereafter a distinct succession of the larval stages was observed leading to a further L4 maximum in August. The period with a high proportion of L4 is somewhat extended as a consequence of the heterogeneous state of development of the population, where some animals of one generation begin to hatch while others are still at a very young stage.

The data on the size distribution of larvae (Fig. 23.8) suggest that the first summer generation extended until the middle of September with a main emergence period throughout August. A decrease of larval abundance in August is the result of this flying period.

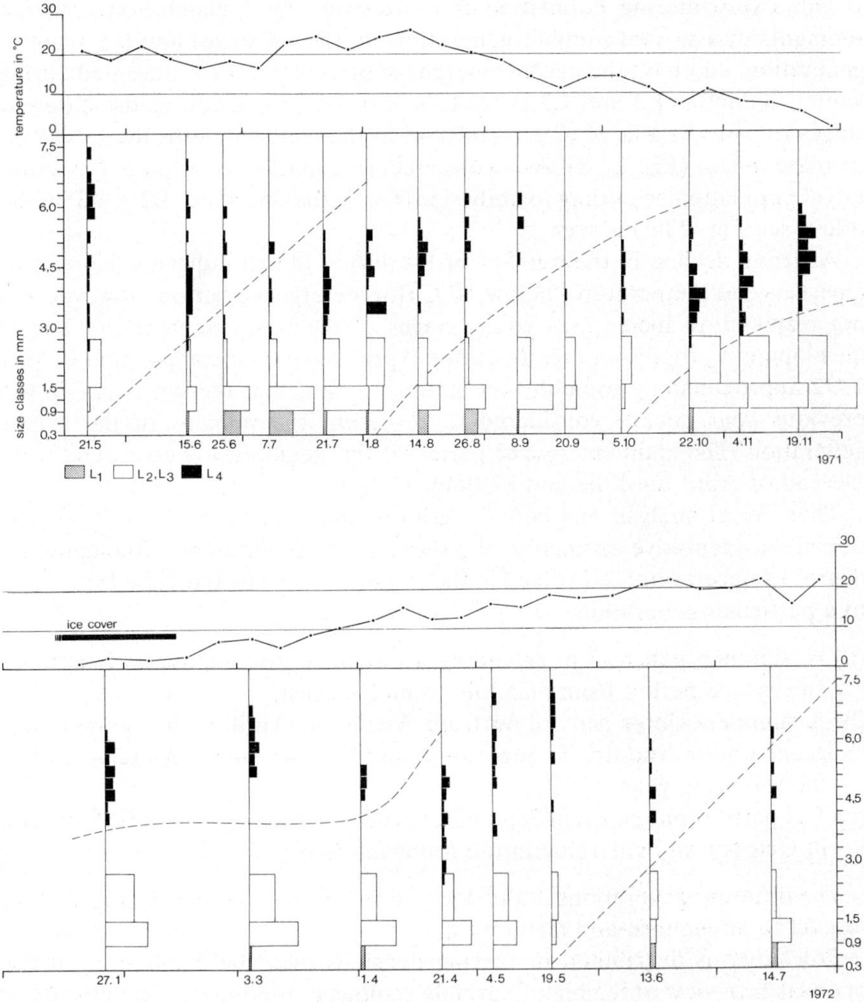

Fig. 23.8. Seasonal change in size distribution (body length in mm) of larvae of *Procladius* at the 450 m station. Absolute numbers per 10 samples. Broken lines indicate the delineation of different generations. Weekly mean temperatures at 1 m water depth.

The next distinct maximum of L1 larvae occurred at the end of August. It represents a generation which only partially attains maturity during autumn. A fraction of it overwinters as larvae (in the L4 stage exclusively). Adults of this generation emerge in September and at the beginning of October and their offspring are responsible for a large increase in the number of larvae during October.

From the end of October onward the changes in larval composition are very slow. The average composition from 20 October 1971–3 March 1972, including 6 sampling periods (800 individuals) is: L1 = 0.5 per cent; L2 = 41 per cent; L3 = 39.5 per cent; L4 = 18 per cent.

This overwintering population is represented by 2 generations, (a) the remnants of a second summer generation (in the L4 stage) and (b) a winter generation, based on the partial emergence of the former (represented during winter mainly as L2 and L3 larvae). Whilst the proportion of the different stages remains to a large extent constant in the winter months the L4 larvae increase in size (Fig. 23.8). As a consequence a marked division of the entire larval population according to body size (L4 on the one hand, L2 + L3 on the other, see Fig. 23.6) is seen.

A strong decline in the number of L4 during March indicates the start of emergence at temperatures below 10°C (this emergence period, however, did not result in an increase in young stages of the larval population). Larval development progresses rapidly during April. The larval composition in May 1972 approximately resembles the pattern obtained at the same time in the previous year, and is considered to represent the remains of the winter generation. The main emergence period of this generation must extend from the end of April until the end of June.

Thus, instar analysis and body length distribution of the 1971–72 population allow a tentative distinction of 3 overlapping generations, although some degree of uncertainty is involved in the assignment of any individual specimen to a particular generation.

(a) A summer generation extending from June to August with a main emergence period from mid July to mid August.
(b) A summer–winter generation from August to April, with 2 main emergence phases, the first in September and the final one in March–April of the following year.
(c) The partial emergence in September produces a winter generation starting in October and emerging in the following May.

The different generations in the population are distinguished in Fig. 23.9 in respect to abundance and mean weight.

Procladius is distributed in varying densities over the whole area of the open lake. In view of the highly variable ecological conditions within the open lake the question arises as to the effect of local differences on larval development. The question cannot be satisfactorily resolved on the basis of field analysis since growth differences induced by habitat factors may be masked by exchange between populations of different localities.

Data on larval composition at varying distances from the reed belt, obtained in the second half of May, showed a very distinct trend to increasing percentage of young stages with distance from the shore. This trend is most pronounced within the first 1000 metres (Table 23.7).

A comparison of the population dynamics at the 450 m and the 2,000 m station for 1971–72 demonstrates the earlier onset of the summer generation at the latter. During the summer, however, development is fairly well synchronized at the two stations. Although larval composition at the beginning of September is the same at both stations, it becomes obvious that thereafter development proceeds more rapidly at the 2,000 m station. This results in a different structure of the overwintering population (Table 23.8).

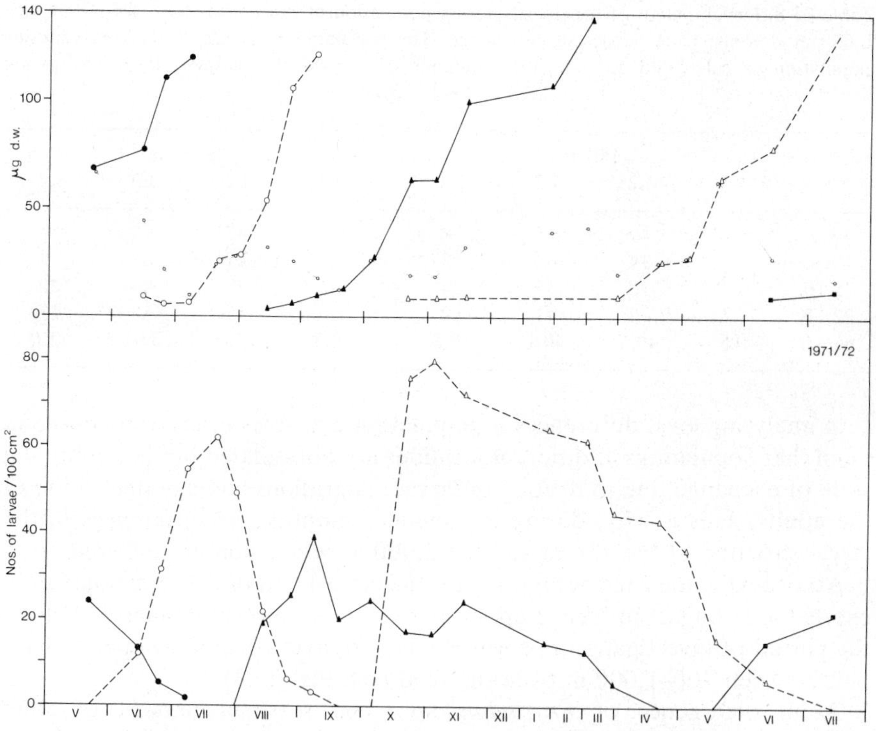

Fig. 23.9. Abundance and mean larval weight of successive and overlapping generations of *Procladius* sp. at the 450 m station. The thin, full line indicates the mean larval weight of the total population.

Table 23.7. Population structure of *Procladius* sp. (larval stages in percentage) at different distances from the reed fringe on 20–21 May 1971.

distance from reed fringe (m)	n	L1	L2	L3	L4
20	63	0	0	24	76
500	49	0	4	43	53
1,000	153	0	25	50	25
1,500	101	0	30	51	19
3,000	49	0	35	53	12

During the winter the larval population at both sites is composed of animals representing 2 different generations (a 'summer–winter' generation and a resulting 'winter' generation, see above), although the 'winter' generation is present in a higher proportion at the 2,000 m station and is in a more advanced state of development. The consequence should be an earlier beginning of the ensuing summer generation at 2,000 m that at 450 m, which is in full agreement with the data given in Table 23.7.

Table 23.8. Development of the overwintering population of *Procladius* sp. at the 450 m and 2,000 m stations. Instar stages in percentage. The population structure of the overwintering population is calculated for the total number of individuals analysed from the period 22.10.1971–3.3.1972

	450 m				2,000 m			
	L1	L2	L3	L4	L1	L2	L3	L4
8.9.	1	49	41	8	2	47	45	6
20.9.		34	63	3		27	67	7
5.10.		20	56	24		22	44	33
22.10.	2	38	41	19		19	68	13
winter	0.5	40.1	40.1	19.3	0.3	31.6	53.1	15.0

In analysing local differences in population dynamics it has to be borne in mind that populations at different localities are not isolated but in a continual state of exchange due to drifting of larvae, migrations and egg deposition of the adults. This is why, during the summer months, the differences in the larval structure of the 450 m and the 2,000 m population are indistinct.

Accordingly, the zone accounting for the largest part of the total population exerts the strongest influence upon the overall population dynamics. During the period of investigation this was the macrophyte zone and in particular a belt between 300–1,000 m from the reed (see Fig. 23.3).

Adult emergence of *Procladius* sp.: The seasonal pattern of adult emergence was studied in 1968 at 6 different sites from mid April until mid October. All traps except one were situated in the inshore zone, not more than 150 m off the reed. A description of the localities studied (see Fig. 23.5), methods used and results obtained have been given elsewhere (Schiemer, in press). In 1969 trapping was repeated at one locality for limited periods of time (stronger underlined in Fig. 23.11).

Emergence extended over the whole period of trap operation. Emergence before mid April may occur at certain localities (see above). Although at practically no time was flying activity absent, periods of higher and lower emergence can be distinguished. At the inshore station a first maximum was encountered at the beginning of May at temperatures of 13–14°C. Further peaks occurred during May and June. Throughout July flying activity was low and increased again during August. In September emergence gradually decreased and stopped in October at water temperatures below 13°C.

Marked differences have been observed at a station 300 m offshore where trapping was started in July. These differences confirm that the development of the populations is not synchronous over the entire of the open lake. Very high emergence values have been encountered here from mid August until the end of September, which is the period of flying activity to be expected on the basis of larval development at the 450 m station in 1971.

Fig. 23.11 compares the emergence at an inshore station in 1968 with data from 1969. Distinct differences between the 2 seasons correspond with differences in the water temperature and indicate that synchronous development from year to year is not to be expected.

23.6.2 *Tanypus punctipennis*

The seasonal periodicity in adult emergence and larval development show some similarities with that of *Procladius* sp. The entire larval population at the beginning of the sampling period (21 May 1971) is considered to represent the remnants of the overwintering generation (see Fig. 23.10). The abundance of larvae was low; L3 and L4 stages were clearly dominant.

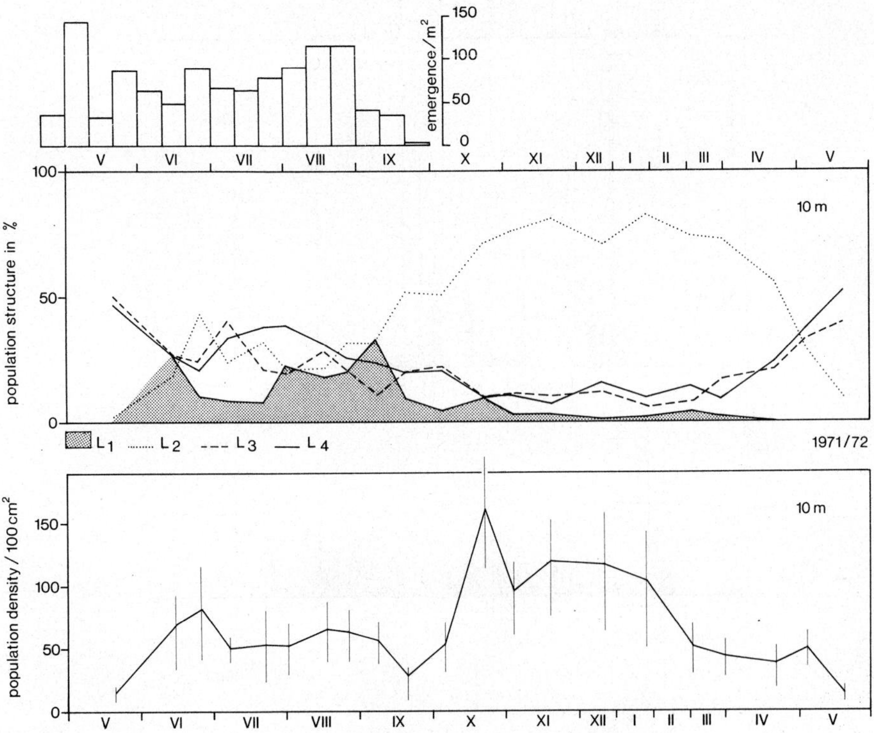

Fig. 23.10. Population dynamics of *T. punctipennis* at the inshore station in 1971–72. The emergence pattern (numbers of adults hatched per 10 days) obtained in 1968 is shown for comparison.

The emergence period of the overwintering larval population probably extends from April until the end of June. On account of this long emergence and reproduction phase it is difficult to distinguish the course of development of the successive generations, although a succession of poorly developed maxima of consecutive stages can be recognized (L1: 15.6.; L2: 25.6.; L3: 7.7.; L4: 21.7.–17.8.). The emergence period of such a first summer generation starts during June and thus overlaps that of a foregoing winter generation.

No delineation of cohorts or generations is possible during August and September. Continuous reproduction balances the losses in the larval popula-

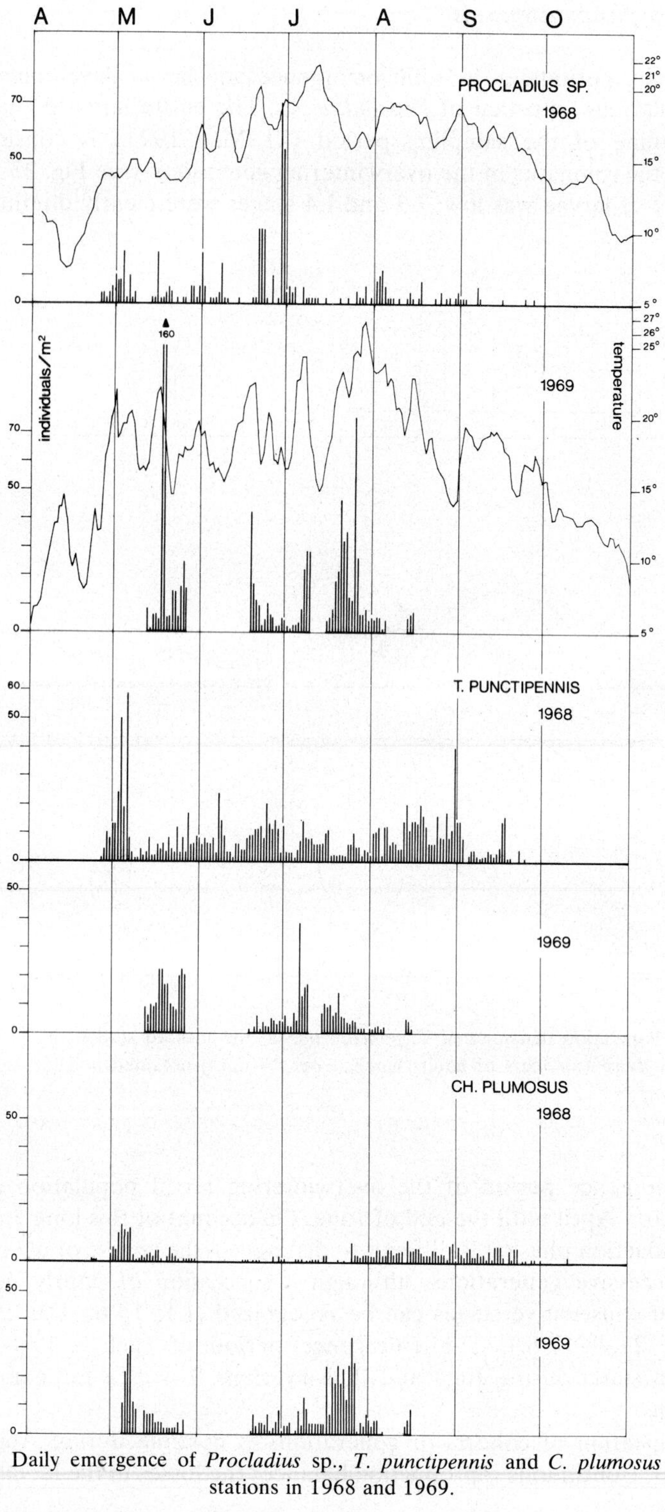

Fig. 23.11. Daily emergence of *Procladius* sp., *T. punctipennis* and *C. plumosus* in inshor stations in 1968 and 1969.

tion. The proportions of all four larval stages remain fairly constant during this period, which indicates that they must be of similar duration. This is in contrast to the findings for *Procladius* and to the overall developmental pattern in chironomids in which the younger larval stages are of shorter duration than the older.

The larval abundance increases in autumn after cessation of adult emergence. During the winter months the structure of the larval population remains constant until the end of March (see Fig. 23.10). L2 larvae are clearly dominant (see Table 23.9) in the overwintering population.

Table 23.9. Structure of the overwintering population of *T. punctipennis* (6 sampling periods from 20.10.71–3.3.72, n = 1,040 individuals). For purposes of comparison, the population structure of *Procladius* at the 450 m station is given.

	L1	L2	L3	L4
T. punctipennis	4.0	75.7	9.6	10.7
Procladius sp.	0.5	40.1	40.1	19.3

Adult emergence (studied in 1968 and 1969) started in the last week of April and continued without interruption until the end of September. The first distinct peak occurred during the first days of May. The main flying period of the species thus starts a few days later than that of *Procladius* sp. These differences in the time of onset of emergence in the 2 main chironomid species of the open lake benthos correspond well with the differences in the composition of larval stages during the winter months. In general, emergence of *T. punctipennis* is more evenly spread over the whole flying period than that of *Procladius*. This indicates a steady supply of pupating larvae and is in very good agreement with the findings regarding larval development.

A delineation of generations in this species is not possible, although the course of development during spring and early summer suggests that the speed of development is similar to, or possibly even somewhat slower than, that of *Procladius*. Probably individuals run through 2 to 4 parental life cycles.

In a recent study of the population of *T. punctipennis* in Balaton, Olah (1976) found only L3 and L4 stages present in the sediments. Highest population densities occurred during the winter, lowest during summer months. The author assumes – on the basis of Kajak and Rybak's (1966) field study of the life history of the species in Polish lakes – a sequence of at least 5 generations per year. Our discrepant interpretation of development rates must be clarified by experimental studies.

23.6.3 *Chironomus plumosus*

The emergence trap data suggest that the species had 2 distinct generations in 1968. The overwintering generation emerged in May with a few individuals flying in the last days of April. The flying period of the summer generation extended from mid June to the end of September, with a few occasional catches in October. The incomplete trapping series in 1969 showed a second distinct emergence peak in the second half of July indicating the possibility of the occurrence of a partial or complete third generation in years with higher summer temperatures (Fig. 23.11).

Records on emergence of the species are available for the large 'Pannonian' lakes Balaton and Velence, both in Hungary. Berczik (1967) estimated 3 generations in Lake Velence, where *C. plumosus* is by far the most important benthic species (see Table 23.12). Flying periods have been found at the end of April, at the beginning of July and during September.

In Balaton, Entz (1965) distinguished a summer generation with a flying period in the second half of September and an overwintering period with emergence starting in May.

A detailed account of the life cycle of the species in different waters has been given by Borutzky et al. 1971.

23.7 Biomass

A comparison of macro- and meiobenthic biomass shows that the latter is only of minor importance (Table 23.10). Meiobenthos accounts for roughly 10 per cent of the total biomass in the different zones of the lake.

Table 23.10. Biomass of macrobenthos and meiobenthos at different areas in the open lake of Neusiedlersee. Biomass in units of mg dry weight m^{-2}. The values for macrobenthos are annual means from the seasonal study 1971–72. Values for meiobenthos include only nematodes. One set of data (BA 68) is from a survey in October 1968. A seasonal study of biomass was established for a site (NS_2, see Fig. 23.5) within the macrophyte zone, 50 m from the reed fringe.

	zone 1	zone 2	zone 3
macrobenthos	550 mg	160 mg	50 mg
meiobenthos (BA 68)		7 mg	0.5 mg
meiobenthos NS_2, annual cycle		9–36 mg	

The average biomass declines from zone 1 to zone 3. The annual mean biomass value of the central lake bottoms is only about 10 per cent of that of the peripheral soft mud areas.

Tanypus punctipennis and *Procladius* sp., the two dominant macrobenthic species, are by far the most important contributors to total zoobenthic biomass (Table 23.11). A higher biomass in 'zone 1' than in 'zone 2' is due to differences in body weight of fully grown larvae of the dominating species (*Tanypus punctipennis* ~ 500 μg dry weight). Additionally, even though its population densities are low, *C. plumosus* may contribute significantly to the benthic biomass (20 per cent) in the inshore zone on account of its size. The percentages, of the open lake area accounted for by zones 1–3 were estimated for the northernmost part of the lake (Fig. 23.5) as 5 per cent, 40 per cent and 55 per cent respectively.

Table 23.11. Biomass of *Tanypus punctipennis* and *Procladius* sp. at different stations in the open lake during the seasonal cycle 1971–72. Biomass values in mg dwt m^{-2}. $\bar{B}$ = annual means, B_{min}, B_{max} = minimal and maximal biomass values observed; t_{min}, t_{max} = months at which minima and maxima occurred.

	$\bar{B}$	B_{min}	B_{max}	t_{min}	t_{max}
T. punctipennis					
10 m (zone 1)	456	158	681	March	August
450 (zone 2)	22	0	71	Sept., May	Nov.
Procladius sp.					
10 m (zone 1)	30	0	106	Aug.	May
450 m (zone 2)	131	27	318	Sept.	Nov.
2,000 m (zone 3)	47	13	105	Aug.	Sept.

Weighting of the biomasses according to these figures yields 24 per cent, 52 per cent and 24 per cent respectively, as the contributions of the individual zones to the total biomass of this whole lake area.

This means that zone 2 with 52 per cent is the main contributor to the total benthic biomass of the open lake. The reed fringe (zone 1) and the central lake area contribute the same percentage to the total benthic biomass despite the large differences in the area covered by each. The distribution pattern of

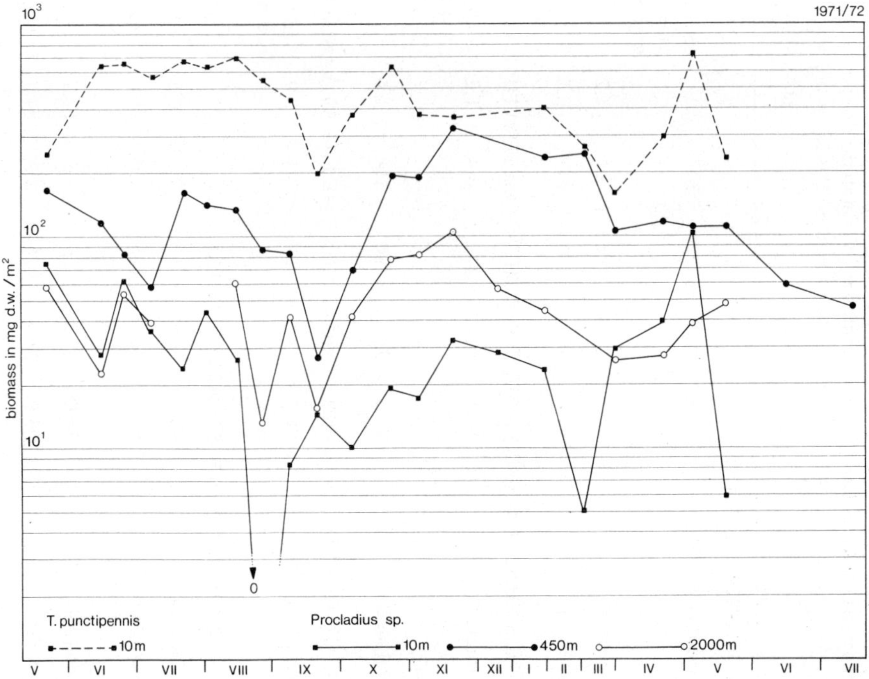

Fig. 23.12. The change of biomass of *Procladius* sp. (10,450 and 2,000 m station) and *T. punctipennis* (10 m station) during the sampling period in 1971–72.

Table 23.12. Comparison of population densities, mean annual biomass and annual production of macrobenthos in different types of limnic habitats within the temperate zone. Fresh weight has been converted into dry weight by a factor of 0.15; 1 g dry weight was assumed to be 19.7 kJ. I– latitude, II–longitude, III–altitude (m), IV–area (km^2), V–mean depth (m), VI–stratum, VII–population densities (ind. $10^3/m^2$), VIII–biomass (mg dry weight/m^2), IX–production (kJ/m^2; values based on P/B-ratios are set in parentheses), X–source of information.

	I	II	III	IV	V	VI	VII	VIII	IX	X
'PANNONIAN' LAKES										
1. Neusiedlersee, A	47.8	16.8	113	300	1	zone I	8.8	500	(148)	this paper
						zone II	6.6	150	(44)	
						zone III	2.4	50	(15)	
2. Velence, H	47.1	18.4	106	24	1.2		1.8	5,341	(1,600)	Berczik, 1967
3. Balaton, H	47	17.3	105	600	3.3		8.5			Franko, in litt.
OTHER SHALLOW LAKES WITHOUT THERMAL STRATIFICATION										
4. Warniak, P.	54.1	21.8	120	0.4	1.2	0.5–0.7 m		800	154	Kajak in litt.
						1.2–1.5 m		2,200	423	
						2.5 m		1,400	277	
5. Karakul, Su	43.8	78.1	480	1.4	1.4		8.6	390	110	Khusainova et al. in litt.
LITTORAL AND LITORI-PROFUNDAL ZONE OF LAKES; SANDS OR SOFT SEDIMENTS										
6. Loch Leven, UK	56.2	3.5	107	13.3	3.9	sand zone Ø 2 m	24.0	9,500	915	Maitland & Hudspith 1974
						mud zone Ø 4.5 m	11.0	7,500	682	Charles et al. 1974
PROFUNDAL ZONE OF STRATIFIED LAKES										
7. Krasnoe, Su	60.3	29.1	40	9.1	6.6			1,012	85	Andronikova et al. 1972
8. Mikolajskie, P	53.5	21.3	116	4.6	11.0	16 m		930	63	Kajak in litt.
						24 m		400	32	
9. Esrom, DK	56.0	12.2	9	17.3	12.3		20.4	16,500	420	Jonasson, 1972
10. Taltowisko, P	53.9	21.6	116	3.3	14.0	12 m		970	64	Kajak in litt.
						16 m		580	40	
						36 m		10	0.6	

Table 23.12. (contd.)

11. L. Ontario, C	Big Bay 5 m	1,030	312	Johnson & Brinkhurst, 1971
	Glenora 25 m	5,480	977	
	Conway 30 m	4,310	276	
	L. Ontario 30 m	5,000	212	
PONDS				
12. Sewage lagoon, OregonUS		19,200	3,400	Kimerle & Anderson 1971
13. Fish ponds, P	Pond no 7	4,590	1,693	Zieba 1971
	Pond no 8	5,030	1,661	
	Pond no 9	6,610	1,937	

Comments:

ad 1. the values for P are a rough estimate using a P/B ratio of 15
ad 2. offshore stations; 99.8 per cent *C. plumosus*; P calculated assuming a P/B ratio of 15
ad 3. Franko in litt, IBP data, collected by Kajak et al. in press; *C. plumosus* predominant (see Entz, 1966)
ad 4. IBP data collected by Kajak et al. in press
ad 5. IBP data, collected by Kajak et al. in press.; *C. plumosus* most important contributor in benthic biomass
ad 6. sand zone: *Glyptotendipes paripes*, mud zones: *C. anthracinus* (87 per cent of B) predominant
ad 7. *C. plumosus* and *C. anthracinus* predominant
ad 8. IBP data collected by Kajak et al. in press, *C. plumosus* and *C. anthracinus*
ad 9. 62 per cent of biomass = *C. anthracinus*
ad 10. IBP data collected by Kajak et al. in press; at 36 m *Sergentia longiventris* and *Prodiamesa bathyphila* (see Kajak et al. 1972)
ad 11. the 4 stations (Big Bay to L. Ontario) represent a gradient from gross pollution to oligotrophy
ad 12. population of *Glyptotendipes barbipes* (Chironomidae)
ad 13. carp fingerling ponds, no 7 unfertilized, no 8 and no 9 mineral fertilizers added. B and P values for a 120 day period from June to Oct. 1967. Most important species *C.* cf. *thummi* and *C.* cf. *plumosus*

the macrobenthos clearly shows that *Procladius* sp. is the most important species in the open lake, in terms of numbers and in biomass.

The weighted mean biomass for the open lake (calculated for the northern part) (area in Fig. 23.5) is 0.115 g dry weight m^{-2}.

Seasonal changes in biomass are of significance especially with respect to benthos feeding fish (see below). Strong fluctuations have been found for *Procladius* sp. especially during the warm water period, reflecting the population dynamics and the sequence of generation (Fig. 23.12). Biomass is high during winter due to large numbers of L4 larvae. Seasonal changes are parallel at the 450 and 2,000 m stations. An increase of biomass at the 10 m station during spring in contrast to the two offshore stations is due to the retarded development of the overwintering generation in this zone (see above).

The seasonal changes in biomass of the *T. punctipennis* population are less well expressed (Fig. 23.12). During the period June–August biomass was at a constant high level, which is of importance for benthos grazing fish.

23.8 Comparison of benthic biomass and production of various bodies of water within the temperate zone

How do these biomass data compare with values from other water bodies? Table 23.12 gives a comparison of benthic data from lakes and ponds within the temperate zone. Mean annual biomass ranges from 0.01–19.2 g dwt/m^2, annual secondary production from 0.6–3,400 kJ/m^2, e.g. the values vary by more than 3 orders of magnitude.

The lowest values originate from the deep profundal zone (36 m depth) of mesotrophic Taltowisko lake (Poland) i.e. 0.01 g d wt/m^2 biomass, 0.6 $kJm^{-2}\,yr^{-1}$ production. The main species occurring in this zone are *Sergentia longiventris* Kieff. and *Prodiamesa bathyphila* (and oligochaetes).

The highest value was found for a population of *Glyptotendipes paripes* (Chironomidae) in a sewage lagoon with a mean biomass of 19.2 g d wt m^{-2} and a seasonal production of 3,400 kJ $m^{-2}\,yr^{-1}$. In general, high biomass values have been found for benthic biotopes dominated by large chironomid species such as *C. plumosus* (Lake Velence, Hungary; fish ponds), *C. anthracinus* (Esrom, Loch Leven mud zone), *Glyptotendipes* (Loch Leven sand zone, waste water lagoon).

Such conditions are found not only in littoral zones of lakes and in eutrophic ponds but also in the seasonal deoxygenated profundal of lakes (e.g. Lake Esrom).

Biomass and secondary production are usually higher in the littoral zone than in the profundal. Biomass in the littoral zone, apart from dense macrophyte stands is generally above 1 g dw/m^{-2} independent of the trophic state of the lake (Kajak et al. in press).

In shallow, wind-exposed lakes like Neusiedlersee and Tjeukemer (Beattie et al. 1972) biomass values are very low, except in sheltered inshore zones. But even for such localities and despite high population densities, biomass values in Neusiedlersee are at the lower end of the biomass range (0.5 g dwt/m^2) due to the small size of the predominant chironomid species.

Within the group of large, shallow 'Pannonian' lakes Neusiedlersee has the lowest values of benthic biomass. This corresponds to its lower primary production and also to the higher degree of sediment instability of the lake (smaller wind fetch in Lake Velence, greater average depth in Balaton). In both, Lake Velence and Balaton, *C. plumosus* forms a major part of the benthic population. A rough estimate of production, assuming 3 generations per year and applying a P/B coefficient of 15, yields an annual production of 16,000 kJ m^{-2} yr^{-1} in Lake Velence. This would be the highest value recorded for a natural lake and is within the range typical for fish ponds.

23.9 Trophic relationships

In Neusiedlersee, the primary energy source within the benthic food chain consists of epipelic algae and 'detritus' of various origin.

A quantitative study of epipelic algae was carried out by M. Prosser (in prep.) in 1971–72 parallel to the sampling programme for macrozoobenthos. Diatoms proved to be predominant throughout the year. As in the benthic fauna a distinct inshore–offshore pattern was found in regard to species distribution and to total biomass: biomass is in general highest near the reed and decreases strongly with distance from the shore (Table 23.13). Seasonal fluctuations in biomass are very marked, with distinct peaks during the winter months under ice, and with lowest numbers in early summer, indicating that light and sedimentation are the governing ecological conditions for benthic algal growth.

The main sources of the detritus of the sediments are:

- phyto- and zooplankton
- material produced within the belt of submerged macrophytes
- outdrift of the reed belt
- allochthonous, airborne material

The quantitative make-up of these different sources and their specific role as energy carriers for the benthic food chain have not been analysed in detail; their quantitative role as energy carriers and their pathways in the benthic food chain are poorly understood. Comparisons of the mean annual biomass and – as far as available – the annual production of different autochthonous energy sources are compiled in Table 23.14, to facilitate a tentative evaluation of their importance.

It is of interest to note that during the first years of the study, before a regression of the macrophyte growth in the open lake took place (Chapter 19), the standing crop of macrophytes by far exceeded that of phytoplankton and epipelic algae. In terms of production, however, the phytoplankton is dominant and sedimenting algae therefore must be considered to be the main energy carrier for the benthic ecosystem in Neusiedlersee.[1]

Because of the overall shallowness of the lake and the lack of a stable

1. Considering that the strong quantitative decline of macrophytes since 1970 might actually have started in 1965 with the rise of the water level it may well be that the annual production of macrophytes in the years before 1970 was of the same order of magnitude as that of the phytoplankton (see Chapter 19).

Table 23.13. Biomass of benthic algae at 3 distances (10, 450 and 2,000 m) from the reed fringe in the northwestern part of the lake (see Fig. 23.5). The values are given as algal volumes in units of $\mu^3\ 10^8/cm^2$. Values in brackets represent a station at 1,000 m distance from the reed fringe (unpublished data of M. Prosser).

Date	10 m	450 m	2,000 m
21. 5.71	2.84	1.03	(0.55)
9. 6.71	3.25	0.87	(0.90)
25. 6.71	11.20	0.34	0.25
7. 7.71	8.74	0.31	(0.22)
22. 7.71	3.37	1.76	(1.2)
26. 8.71	5.81	1.16	0.93
8. 9.71	6.49	2.06	0.39
20. 9.71	4.24	1.41	1.05
5.10.71	8.10	1.43	0.14
22.10.71	10.09	1.61	0.36
4.11.71	14.54	2.74	0.25
19.11.71	10.42	3.54	0.47
26. 1.72	31.74	2.33	3.15
10. 2.72	14.76	6.92	(2.82)
3. 3.72	8.9	6.10	—
21. 3.72	7.27	—	2.66
5. 4.72	3.89	0.65	0.20
4. 5.72	16.86	1.96	0.43
7. 6.72	3.90	—	0.11
14. 6.72	—	0.48	—
30. 6.72	1.61	0.24	0.17
14. 7.72	1.97	0.44	0.13
28. 7.72	4.27	0.35	0.16
13. 8.72	4.61	0.35	0.08
19. 9.72	7.59	0.43	0.18
3.10.72	50.69	0.90	0.30
17.10.72	45.44	0.69	0.47
30.10.72	26.35	0.87	0.50
15.11.72	21.53	0.67	0.26
29.11.72	12.61	0.69	0.65
9. 1.73	13.66	0.86	0.29
31. 1.73	28.47	4.00	1.08

thermal stratification, planktonic material (including the zooplankton) of high nutritive value can reach the sediment surface. This is in contrast to the situation in deeper, stratified lakes, where easily hydrolysable substances are recycled within the epilimnion and only detritus of poor food quality reaches the bottom (Hargrave, 1973, Jónasson, 1972).

Sedimentation of organic material is determined by the hydromechanics and the circulation pattern of the open lake (see Chapter 8). Soft mud forms in areas where sedimentation exceeds erosion, and organic material accumulates. Water- and organic content of the sediments in the open lake are well correlated (see Chapter 13). As outlined above, such zones yield the highest densities of benthos and benthos-feeding fish.

Macrophytic material is also important in the detritus of these soft mud

Table 23.14. Mean annual biomass (or biomass at period of maximal development of submerged macrophytes) and annual production of different trophic levels within the open lake. In values of g dry weight/m^2. Conversions: algae dry weight = 25 per cent of fresh weight, or 2 × the weight of carbon.

Values of phytoplankton and zooplankton, in the original papers given per m^3, have been converted to m^2 surface area by a factor of 1.3 (mean depth of the open lake zone in meters). Values for macrophytes, epipelic algae and zoobenthos are weighted means for the northern part of the lake (Fig. 23.5) according to a distinction of benthic zones as discussed above (p. 345).

Source of data:

phytoplankton, biomass: Dokulil 1979 – Chapter 17; production: Dokulil, 1974 (mean value for 1968–70, Dokulil, pers. commun.).

epipelic algae: Prosser, unpublished data – see Table 23.13 (values from May 71 – May 72).

submerged macrophytes biomass: Schiemer & Prosser 1976, see Chapter 19; production estimates based on a P/B_{max} ratio of 5.

zooplankton: Herzig 1979 – Chapter 22 (only Crustacea).

zoobenthos: this chapter, production estimates based on a P/B ratio of 15.

	biomass			production		
	1970	1971	1972	1970	1971	1972
Primary producers						
phytoplankton	0.24	0.32	0.72	142[2]	—	—
epipelic algae	—	0.45		—	—	—
submerged macrophytes	—	3.05	0.43	—	15.3	2.2
Secondary producers						
zooplankton	0.29	0.94	—	2.9	12.4	—
zoobenthos	—	0.11		—	1.6	

areas, especially if we consider that the pathways of macrophytic and microphytic detritus presumably differ within the food chain.

Algae of various origins are to some extent immediately used by benthic herbivores. Decaying macrophytes, however, together with the organic outdrift of the reed belt and allochthonous material contribute to a detritus pool which has to be processed by bacteria and fungi before becoming available to the zoobenthic community as food (see Hargrave, 1972, Bretschko, 1975). Decaying macrophytes which forms the basis for a prolonged bacterial activity have a stabilising effect on the benthic ecosystem (Fenchel, 1972).

So far only preliminary studies have been made on the numbers and biomass of bacteria in the Neusiedlersee sediments: During the period of 4.4.72–23.8.72 Dokulil (1975), using a membrane filter technique, found total numbers between 17.4 and 24.5 10^9 bacteria g^{-1} fresh weight of sediment in the macrophyte zone (450 m station, Fig. 23.5). These are population densities characteristic for the sediments of eutrophic lakes.

Between winter 1970 and spring 1972 analyses of dehydrogenase activity in sediments from different localities were carried out[1] in order to outline the zonation of benthic bacteria in the lake (Table 23.15). Although very preliminary, the data clearly demonstrate the strong decline of bacterial activity from the reed fringe towards the open lake. The data are too scanty, however, to allow any comment on seasonal changes.

1. These values have been kindly supplied by Dr. G. Malicky, Lunz/See.
2. Mean value for the years 1968–70.

Table 23.15. Dehydrogenase activity of Neusiedlersee sediments in different localities (see Fig. 23.5, macrobenthos sampling sites) and at different times throughout the year. The location 'reed belt' represents organically enriched, aerated bottoms in small open water areas not covered with *Phragmites*, approximately 50 m within the outer reed fringe.
The values represent relative absorption units (for details see Malicky – Schlatte, 1973).

	7.12.70	20.1.71	29.4.71	8.11.71	3.3.72
reed belt	261	644	872	—	800
10 m	—	296	160	220	—
450 m	146	41	64	35	40
2,000 m	—	19	—	37	—

Food relationships, especially their qualitative aspects, have so far received inadequate attention. The food links of the predominant species with respect to the spatial structure of the ecosystem of the open lake are outlined in Fig. 23.13.

Algivores: Several meiobenthic species like *Monhystera macramphis* (nematodes) and *Chaetogaster langi* (oligochaetes) feed to some extent on epipelic algae. *Hypsibius augusti*, a tardigrade species confined to the reed fringe zone, feeds predominantly on epipelic diatoms in Neusiedlersee (Dollfuss pers. commun.).

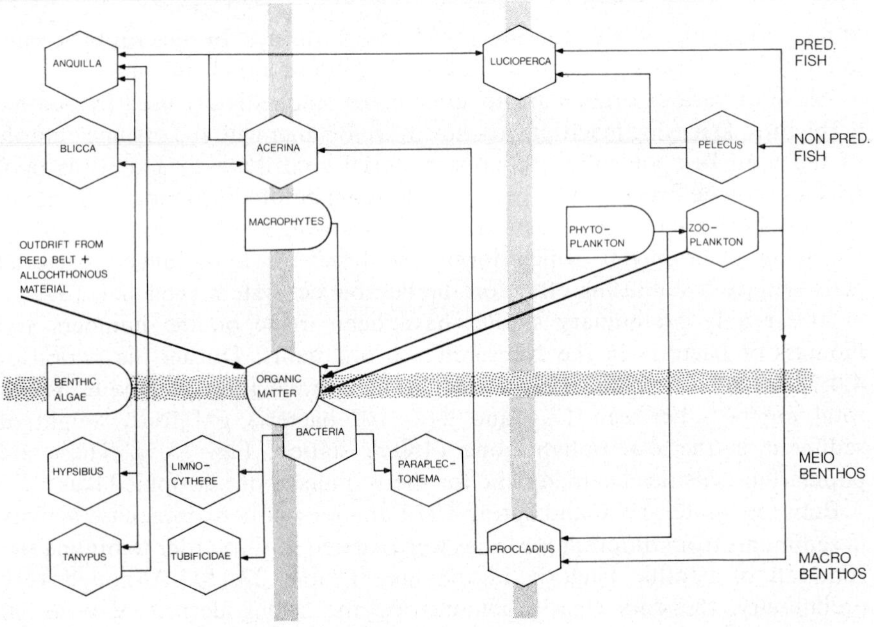

Fig. 23.13. Major food links in the benthos of Neusiedlersee. Three ecological zones are distinguished by two vertical shaded bars. Compartments positioned on bars are quantitatively important for adjacent zones. The horizontal bar represent the mud–water interphase.

Of the macrobenthic species, *Tanypus punctipennis*, the dominant form of the reed fringe zone, is a particle feeder whose diet contains a high proportion of algae. Olah (1976, here further references) found diatoms as the main food source of the species in Balaton. Preliminary analysis of gut contents of L4 larvae during summer 1971 demonstrated that epipelic algae are of paramount importance.

In view of the technical difficulties involved in the determination of the very small algal cells in the gut, statements concerning selectivity with regard to species should only be made with caution. It appears that larger algae like *Gyrosigma acuminatum* + *G.* sp. (cell volumes $15 \cdot 10^3 \mu^3$), *Nitzschia tryblionella* ($19 \cdot 10^3 \mu^3$), *Surirella peisonis* ($33 \cdot 10^3 \mu^3$), *Nitzschia sigmoidea* ($70 \cdot 10^3 \mu^3$) are preferred to the most numerous but small diatoms which dominate in the inshore zone during the summer months (*Navicula* spp. $0.7 \cdot 10^3 \mu^3$; *Nitzschia* spp. $0.25 \cdot 10^3 \mu^3$)[1].

Blicca björkna L., one of the most numerous fishes during the period of IBP, feeds predominantly by sucking off the surface layer of the soft mud near the reed fringe and in the macrophyte zone. Hacker (1974) (See Chapter 28) found on an average 40 per cent of the gut contents to consist of epipelic algae (mean value for the whole season and for all age classes, except the first 2 months of life). However, it may well be that the bacteria and microfauna of the sediment surface layer, which are difficult to quantify within the gut, are of major importance. Chironomids (7 per cent) and other macrobenthos (15 per cent) are of lesser significance. In addition, zooplankton can play a role during certain periods of the year, which indicates a flexibility in the feeding strategy of the species as compared, for example, with *Acerina cernua* (Winkler, 1875; see Chapter 28). The higher biomass of epipelic algae during winter may thus be due not only to better light conditions but also to less grazing.

Bacterivores/Detrivores: Most benthic species in Neusiedlersee can be characterized as particle feeders. Suitably sized particles of different origins, such as algae, bacteria and detritus are ingested. The ostracods *Limnocythere inopinata* and *Iliocypris bradyi*, as well as the benthic cladocerans *Macrothrix laticornis* and *Iliocryptus sordidus*, employ a feeding strategy of this type. Tubificids feed in deeper sediment layers and bacteria constitute the most nutritive part of their diet.

The number of specific bacteriovorous species are probably few. The nematode *Plectus palustris*, a species of insignificance in the benthos of Neusiedlersee, was studied experimentally with respect to bioenergetic parameters dependent upon densities of bacterial food (Schiemer et al. in press, Duncan et al. 1974). A second Plectidae, *Paraplectonema pedunculatum*, quantitatively the most important nematode species in Neusiedlersee, very probably is dependent on bacteria as food.

Predators: whereas meiofauna plays only a subsidiary role in the diet of *Tanypus punctipennis*, at least the last 2 larval stages of *Procladius* sp. are

1. Determination of diatoms in the guts of chironomids and the calculations of cell volumes were accomplished by M. Prosser.

predominantly carnivorous. Larvae move actively on the sediment surface in search of prey. Besides detritus particles of unrecognizable origin and algae, the remains of ostracods, cladocerans, chironomids and oligochaetes have been found in the gut of larger larvae.

The population maximum of *Procladius* sp. lies some distance from the reed fringe (see above), where benthos other than nematodes is scarce. Of the benthic animals, nematodes, rotifers, oligochaetes and other chironomids constitute possible items of food[1]. However, considering the distribution pattern of *Procladius* sp. and possible prey organisms it is impossible for the energy demand of the *Procladius* population to be met by benthic food. Thus, zooplankton passing the mud–water interphase must be considered to be the main food source.[2] This would explain the peculiar fact of a predominantly carnivorous benthic population in the open water of Neusiedlersee.

Among fish most of the cyprinids (except *Alburnus alburnus* (L.), and *Pelecus cultratus* (L.)) the eel (*Anquilla anquilla* (L.)), the perch (*Perca fluviatilis*), and the pope (*Acerina cernua* L.) feed to some extent on benthic food.

Because of its high abundance *Acerina cernua* is of particular importance. This species feeds predominantly on chironomids as is well documented from other lakes.

The proportion of chironomids in the diet of *A. cernua* in Neusiedlersee increases considerably during the first year of life but remains constant later. Similar results have been obtained from Balaton (Tölg, 1960).

The average percentage of chironomid larvae in the gut content of fish older than 1 year (>5 cm total length) in Neusiedlersee was found to be approximately 80 per cent (Meisriemler 1974). The chironomids in the fish guts, sampled simultaneously (same time and stations) with the macrobenthos in 1971–72 were analysed with respect to species composition and larval stages (Schiemer, unpubl.). The results of this detailed analysis allow us to draw some conclusions about the feeding strategy of the species. The predominant food is provided by the two surface-dwelling chironomids *Tanypus punctipennis* and *Procladius* sp. Orthocladiinae, dominant on the submerged macrophytes, have not been found in the fish guts at all.

The species association in the gut contents of fish caught in different zones varies in accordance with the benthic fauna typical of the respective zone (Table 23.16). *Tanypus punctipennis* dominates in fish from the reed fringe zone. *Procladius* sp. prevail in the gut content of fish from the 450 m and 2,000 m station, corresponding to the dominance pattern of the benthic fauna at these sites.[3]

Small Chironomini species (*Cryptotendipes usmaensis*, *Cryptocladopelma virescens*, *Harnischia pseudosimplex* and *Leptochironomus tener*) are pres-

1. In sorted samples kept in petri dishes larger larvae proved to be cannibalistic.

2. In cultures at 20°C the species completed the L_4 stage within 20 days on a diet of C_3 of *Arctodiaptomus spinosus*.

3. This finding also suggests that no extensive migrations of the fish take place over short periods of time.

Table 23.16. Percentage ratios of chironomid species in the gut content of *Acerina cernua* caught at 3 localities (see Fig. 23.5, macrobenthos stations), on different occasions. The values are compared with the corresponding benthic chironomid association. The percentage ratios for the gut contents are based on the number of chironomids (n) obtained from 3–6 fishes in each case, ranging from 5–10 cm total length. T.p. = *Tanypus punctipennis*, P.sp. = *Procladius* sp., Ch. spp. = small species of Chironomini, see above, Ch.pl. = *Chironomus plumosus*.

		T.p.	P.sp.	Ch.spp.	Ch.pl.
10 m station					
9.7.1971	fish gut (n = 124)	80	10	10	0
	benthos	71	12	2	15
11.9.1971	fish gut (n = 162)	95	2	3	0
	benthos	86	8	2	2
450 m station					
30.7.1971	fish gut (n = 19)	6	89	5	0
	benthos	2	95	3	0
11.9.1971	fish gut (n = 144)	0	85	15	0
	benthos	5	77	18	0
2,000 m station					
7.6.1971	fish gut (n = 69)	0	84	16	0
	benthos	0	80	20	0
11.9.1971	fish gut (n = 109)	1	84	15	0
	benthos	0	89	11	0

ent in the gut contents in ratios corresponding to those in the benthos. *Chironomus plumosus*, however, living in deeper strata of the soft mud areas, may be under-represented in the *A. cernua* prey in Neusiedlersee, whereas it is one of the main food organisms in Balaton (Tölg, 1960; Entz & Sebestyen, 1946).

Table 23.17 shows the food selectivity of *A. cernua* in respect to instar and size classes of larvae ingested. The data give clear evidence that almost exclusively L3 and L4 instars of *T. punctipennis* and *Procladius* sp. are selected. In terms of body length this means that instars from 2–3 mm upwards are preferred.

The larger L4 instar are clearly preferred to the L3 of both species. Considering only L3 and L4 instars and pooling the data for both species, a mean selectivity factor of 1.7 can be calculated for the 6 samples given in Table 23.17. This means that the ratio of L4:L3 in the gut contents is 1.7 times that in which they occur in the benthos.

The quantitative significance of the feeding of *A. cernua* for the population dynamics of *Procladius* sp. is as calculated for the macrophyte zone, using data from summer 1971. The consecutive generations of *Procladius* sp. can be distinguished although they overlap with respect to instar and length class frequencies (see p. 360). The developmental times of instars can be estimated

Table 23.17. Population structure (larval stages in per cent) of *Tanypus punctipennis* and *Procladius* sp. at the 10m, 450 m and 2,000 m stations (see Fig. 23.5). The values for different periods are compared with the instar composition in *Acerina* guts on the corresponding dates. The numbers of chironomids (n) in each case obtained from 3–6 fishes ranging from 5–10 cm in total length.

Tanypus punctipennis					
10 m station					
9.7.1971	fish gut (n = 100)	0	0	23	77
	benthos	8	23	36	33
11.9.1971	fish gut (n = 154)	0	0	60	40
	benthos	27	36	14	23
Procladius sp.					
450 m station					
30.7.1971	fish gut (n = 17)	0	0	6	94
	benthos	1	21	44	34
11.9.1971	fish gut (n = 120)	0	6	67	27
	benthos	0	45	50	5
Procladius sp.					
2,000 m station					
7.6.1971	fish gut (n = 58)	0	2	29	69
	benthos	21	38	21	20

from their temporal sequence during the summer months, i.e. 10 days for the L2, 15 days for L3, and 20 days for L4.

Quantitative changes in the individual larval stages to be expected between two sampling dates can be calculated using the following formula:

$$N_{b2} = {}_{\Delta}t(N_{a1}/D_a - N_{b1}/D_b) + N_{b1}$$

N_{b2} = the anticipated value for stage b at the 2nd date

N_{a1}, N_{b1} = population densities of consecutive stages a and b at the 1st date

D_a, D_b = the developmental time for a specific instar at temperature and food conditions prevailing between dates 1 and 2

${}_{\Delta}t$ = period between date 1 and 2 in days
(for a detailed discussion see Winberg, 1971, p. 100)

The difference between the expected frequencies of a specific instar and the actual one found can be taken as elimination within the interval ${}_{\Delta}t$. Table 23.18 shows a calculation of this kind for the first summer generation in 1971 at the 450 m station. The calculation is restricted to L3, L4, pupae + imagines, since the number of L1 larvae is likely to be underestimated and no values for number of eggs deposited are available. Since, however, *A. cernua*

Table 23.18. Numbers of consecutive larval stages of the summer generation of *Procladius* sp. at the 450 m station in 1971 (see p. 360) compared with expected values, calculated as described in the text (P + 1 = pupae + imagines). The difference between calculated and actual values of the L3 and L4 instars gives the elimination over the time period Δt (in days). The actual numbers are sums of 10 samples with a 16.6 cm^2 surface area.

	actual values				expected values			difference	
date	L1	L2	L3	L4	L3	L4	P + I	L3	L4
15.6.71	2	10	8	0					
25.6.	18	27	7	0	12.7	5.3	0	5.7	5.3
7.7.	24	47	17	0	33.8	5.6	0	16.8	5.6
21.7.	3	46	34	20	76.9	15.9	0	32.9	(+4.1)
1.8.	1	17	36	28	59.7	33.9	11.0	23.7	5.9
14.8.	0	0	19	17	26.9	41.0	18.2	7.9	24.0
26.8.	0	0	0	11	3.8	22.2	10.2	3.8	11.2
8.9.	0	0	0	6		4.4	6.6		(+1.6)
20.9.	0	0	0	0		2.4	3.6		2.4

prefers L3 and L4 larvae as food, this restriction is of minor importance in the quantification of the food chain.

The number of larvae eliminated during the period 15.6.1971–20.9.1971 was 5,470 L3 and 3,277 L4, while 2,988 pupated and emerged per m^2. Highest losses occurred between 7.7.–21.7. (124 ind m^{-2} day^{-1}), 21.7.–1.8. (162) and 1.8.–14.8. (147); during this period the densities of L3 and L4 larvae were at their highest. In percentage of the total larval population this represents losses of 2.1 per cent, 2.6 per cent and 3.0 per cent per day in terms of numbers. These are much higher losses than those found during the winter months (see page 359).

The elimination values can be compared with the food demands of the *A. cernua* population: Ingestion rates of the species have been determined by Gwahaba (1976) under laboratory as well as under field conditions (Neusiedlersee, July 1976).

The average ingestion was 174.1 mg dry weight per fish and per day in the field and 130 mg dry weight per fish per day in laboratory experiments (fishes ranging from 5–10 cm in total length).

Multiplying these ingestion rates by the population density estimated by Meisriemler (1974) for the beginning of August 1971 at the 450 m sampling station, a food demand of 10.1 mg dry weight per day m^{-2} is obtained. The estimated elimination of the chironomid population in terms of biomass is given in Table 23.19 for the period between 15.6.–8.9.1971.

Table 23.19. Biomass (a) and elimination of biomass (b) (mg d wt/m^2) of *Procladius* sp. for different periods of time based on values given in Table 23.18 (450 m station, summer generation 1971). Population density (c) (ind./ha) of *Acerina* at the 450 m station acc. to Meisriemler 1974.

15.6.		25.6		7.7.		21.7.		1.8.		14.8.		26.8.		8.9.	
117		82		57		160		142		133		86		82	a
	3.7		4.2		2.0		4.8		11.6		5.8		0		b
980		—		0		—		—		580		—		0	c

The similarity between calculated data of population losses and food demands indicates that feeding of *A. cernua* is the main parameter in population control of *Procladius* in the lake. A further hypothesis can be derived from a comparison of fish densities in offshore strata and the fluctuations of chironomid biomass, namely that higher fish densities are linked to biomass peaks (see Meisriemler, 1974). This means that inshore–offshore migrations of *A. cernua* may be correlated with the developmental pattern of *Procladius* sp. generations.

References

Andronikova, I. N., Drabkova, V. G., Kuzmenko, K. N., Michailova, N. F., Stravinskaya, E. A. 1972. Biological productivity of the main communities of the Red Lake.– In: Productivity problems of freshwaters. Warszawa-Krakow, 57–71.

Berczik, A. 1967. Zur Populationsdynamik des Makrobenthos imVelencer See.– Opusc. Zool. Budapest 6/22: 247–265.

Biro, K. 1973. Nematodes of lake Balaton IV. Seasonal qualitative and quantitative changes.– Annal. Biol. Tihany 40: 135–158.

Borutzskii, E. V., Sokolova, N. Y., Yablonskaja, E. A. 1971. A review of Soviet studies on production of estimates of chironomids.–Limnologica 8: 183–191.

Bretschko, G. 1975. Annual benthic biomass distribution in a high-mountain lake (Vorderer Finstertaler See, Tyrol, Austria).– Verh. Internat. Verein. Limnol. 19: 1279–1285.

Brinkhurst, R. O. 1969. Observations on the food of the aquatic Oligochaeta.– Verh. Int. Verein. Limnol. 17: 829–830.

Charles, W. N., East, K., Brown, D., Gray, M. C., Murray, T. D. 1974. The production of larval chironomidae in the mud at Loch Leven, Kinross.–Proc. R. Soc. Edinb., B, 74: 241–258.

Daday, I, 1891. Beiträge zur Süßwasserfauna Ungarns.– Természetrajzi Fürzetek, Budapest 14: 16–31.

Dietz, G. 1966. Jahreszyklische faunistische und ökologische Untersuchung der Ciliatenfauna der Natrongewässer am Ostufer des Neusiedlersee. Diss. Univ. Wien, pp. 121.

Dokulil, M. 1975. Bacteria in the water and mud of Neusiedlersee (Austria).– Symp. Biol. Hung. 15: 135–140.

Dokulil, M. 1979. Seasonal pattern of phytoplankton.–In: Neusiedlersee, ed. H. Löffler, Junk, The Hague.

Dollfus, H. 1971. Rämuliche und zeitliche Verteilung von *Hypsibius augsti* (Tardigrada) im Neusiedlersee.–unpubl. manuscript pp. 78.

Donner, J. 1968. Zwei neue Schlamm-Rotatorien aus dem Neusiedlersee, *Paradicranophorus sudzukii* und *Paradicranophorus sordidus*.–Anz. math.–nat. Kl., Österr. Akad. Wiss. 10: 1–8.

Donner, J. 1972. Rädertiere der Grenzschicht Wasser – Sediment aus dem Neusiedler See.– Sitz. Ber. Österr. Akad. Wiss., Math.– nat. Kl. I, 180: 49–63.

Duncan, A., Schiemer, F. & Klekowski, R. Z. 1974. A preliminary study of feeding rates on bacterial food by adult females of a benthic nematode, *Plectus palustris* De mann 1880.–Pol. Arch. Hydrobiol. 21, 2: 249–258.

Emszt, K. 1904. 2. Mitteilung aus dem chemischen Laboratorium der agro-geologischen Aufnahmeabteilung der kgl. ungar. Geologischen Anstalt.– Jahresber. d. kgl. ungar. Geol. Anst. für 1902: 212–224.

Entz, G. 1966. Benthic investigations in Lake Balaton.– Verh. Int. Verein. Limnol. 16: 228–232.

Entz, G., Sebestyen, O. 1946. Das Leben des Balaton-Sees.– Magy. Biol. Kut. Munk. 16: 179–411.

Fenchel, T. 1972. Aspects of decomposer of food chains in marine benthos.–Verh. Dtsch. Zool. Ges. 65: 14–22.

Findenegg, I. 1959. Die Gewässer Österreichs. Ein limnologischer Überblick.– 14. Internat. Limnologenkongreß, Biol. Stat. Lunz pp. 68.

Geyer, F., Mann, H. 1939. Limnologische und fischereibiologische Untersuchungen am ungarischen Teil des Fertö.– Annal. Biol. Tihany 11: 61–182.

Graefe, G., Hohorst, B., Hohorst, W., Zilch, A. 1972. Zur Molluskenfauna des Neusiedler Sees (Burgenland, Österreich).– Mitt. dtsch. malak. Ges. 2/23: 352–354.

Gwahaba, J. J. 1976. Feeding and digestion in *Acerina cernua* in Neusiedlersee.–UNESCO Training Course Rep., Ms.

Hacker, R. 1974. Produktionsbiologische und nahrungsökologische Untersuchungen an der Güster (*Blicca björkna* (L.)) des Neusiedlersees.–Diss. Univ. Wien pp. 93.

Hacker, R. 1979. Fishes and fishery in Neusiedlersee.–In: Neusiedlersee, ed. H. Löffler, Junk, The Hague.

Hacker, R., Herzig, A. 1970. Erstes Auftreten der Wandermuschel *Dreissena polymorpha* Pallas im Neusiedler See.– Anz. math.– nat. Kl., Österr. Akad. Wiss. 15: 265–267.

Hargrave, B. T. 1972. Prediction of egestion by the deposit feeding amphipod *Hyalella azteca*.– Oikos 23, 1: 116–124.

Hargrave, B. T. 1973. Coupling carbon flow through some pelagic and benthic communities.– J. Fish. Res. Board. Can. 30: 1317–1326.

Herzig, A. 1979. The zooplankton of the open lake.–In: Neusiedlersee, ed.: H. Löffler, Junk, The Hague.

Hustedt, F. 1959. Die Diatomeenflora des Neusiedler Sees im österreichischen Burgenland.– Österr. Bot. Z. 106: 390–430.

Imhof, G., Burian, K. 1972. Energy-flow studies in a wetland ecosystem (reed belt of lake Neusiedler See). Special publication of the Austrian Academy of Sciences for the International Biological Programme pp. 15.

Johnson, M. G., Brinkhurst, R. O. 1971. Production of benthic macroinvertebrates of Bay of Quinte and Lake Ontario.– J. Fish. Res. Bd. Canada 28: 1699–1714.

Jónasson, P. M. 1972. Ecology and production of the profundal benthos in relation to phytoplankton in Lake Esrom. Oikos suppl. 14: 1–148.

Jungwirth, M. 1979a. The superficial sediments of Neusiedlersee.–In: Neusiedlersee, ed. H. Löffler, Junk, The Hague.

Jungwirth, M. 1979b. *Limnocythere inopinata* (Baird) (Cytheridae, Ostracoda): Its distribution pattern and relation to the superficial sediments of Neusiedlersee.–In: Neusiedlersee, ed. H. Löffler, Junk, The Hague.

Kajak, Z., Bretschko, G., Schiemer, F., Leveque, C. (in press). Zoobenthos.–In: 'Contributions towards a theory of the aquatic ecosystem' – Cambridge University Press.

Kajak, Z., Rybak, J. I. 1966. Production and some general production regularities in several Mazurian lakes.– Verh. int. Ver. Limnol. 16: 441–451.

Kimerle, R. A., Anderson, N. H. 1971. Production and bioenergetic role of the midge *Glyptotendipes barbipes* (Steger) in a waste stabilization lagoon.– Limnol. Oceanogr. 16: 646–659.

Köfler, D. 1975. Zur Faunistik und Ökologie der Wassermilben des Neusiedlersee-Gebietes.–Diss. Univ. Graz pp. 67.

Kraus, H. 1965. Zur Turbellarienfauna des Seewinkels im Neusiedlerseegebiet.– Wiss. Arb. Bgld. 32: 60–115.

Lenz, F. 1926. Chironomiden aus dem Balatonsee.– Arch. Balatonicum 1: 129–144.

Löffler, H. 1961. Vorschlag zu einem automatischen Schlämmverfahren.– Int. Rev. ges. Hydrobiol. Hydrogr. 288–291.

Maitland, P. S. and Hudspith, P. M. G. 1974. The zoobenthos of Loch Leven, Kinross, and estimates of its production in the sandy littoral area during 1970 and 1971.– Proc. R. Soc. Edinb., B, 74: 219–239.

Malicky-Schlatte, G. 1973. Über die Dehydrogenaseaktivität im Sediment des Lunzer Untersees.– Arch. Hydrobiol. 72, 4: 525–532.

Meisriemler, P. 1974. Produktionsbiologische und nahrungsökologische Untersuchungen am Kaulbarsch (*Acerina cernua* (L)) im Neusiedlersee.–Diss. Univ. Wien pp. 92.

Milbrink, O. 1973. On the use of indicator communities of Tubificidae and some Lumbriculidae in the assessment of water pollution in Swedish lakes.– Zoon 1: 125–139.

Mortimer, C. H. 1942. The exchange of dissolved substances between mud and water in lakes.– J.

Ecol. 30: 147–201.
Neuhuber, F., Hammer, L. 1979. The oxygen conditions in Neusiedlersee.–In: Neusiedlersee, ed. H. Löffler, Junk, The Hague.
Olah, J. 1975. Metalimnion function in shallow lakes.– Symp. Biol. Hung. 15: 149–155.
Olah, J. 1976. Energy transformation by *Tanypus punctipennis* (Meig.) (Chironomidae) in Lake Balaton.– Annal. Biol. Tihany, 43: 83–92.
Ott, J. and Schiemer, F. 1973. Respiration and anaerobiosis of free living nemotodes from marine and limnic sediments.–Neth. J. Sea Res. 7: 233–243.
Pehofer, H. E. 1977. Bestand und Produktion benthischer Nematoden im Piburger See (Ötztal, Tirol).– Diss. Univ. Innsbruck pp. 87.
Pesta, O. 1954. Studien über die Entomostrakenfauna des Neusiedler Sees.– Wiss. Arb. Bgld. 2, pp. 84.
Ponyi, J. 1969. Quantitative investigations on mud-living crustaceans in the open water of Lake Balaton.–Annal. Biol., Tihany 35: 213–222.
Preisinger, A. 1979. Sediments in Neusiedlersee.–In: Neusiedlersee, ed. H. Löffler, Junk, The Hague.
Preys, K. 1977. The littoral and profundal benthic nematodes of lakes with different trophy.–Ekol. pol. 25: 21–30.
Prus, T. 1969. Distribution and age structure in populations of *Pelopia kraatzi* Kieff. (Tendipedidae, Pelopiinae) in Lake Wilkus.– Pol. Arch. Hydrobiol. 16, 1: 67–78.
Ruttner-Kolisko, A., Ruttner, F. 1959. Der Neusiedlersee.– Wiss. Arb. Bgld. 23: 195–201.
Sebestyen, O. 1937. Colonization of two new fauna elements of Pontus origin, *Dreissena polymorpha* Pall. and *Corophium curvispinum* G. O. Sars f. *devium* Wundsch in lake Balaton.–Verh. Int. Ver. Limnol., 5: 169–181.
Schiemer, F. 1965. Über einige Funde der Gattung *Monhystrella* (Nematoda, Monhysterinae) in binnenländischen athalassohalinen Salzgewässern.– Wiss. Arb. Bgld. 34: 59–66.
Schiemer, F., 1970. Das Benthos des Neusiedlersees unter besonderer Berücksichtigung der Nematoden. (Hung., German summary).–Hidrologiai Tájekoztato, Budapest 1970: 159–161.
Schiemer, F. 1978. Verteilung und Systematik der freilebenden Nematoden des Neusiedlersees.– Hydrobiologia 58: 167–194.
Schiemer, F. in press. A Contribution to the phenology of chironomids in Neusiedlersee.–Chir. Symp. Prag.
Schiemer, F. 1979. Submerged macrophytes in the open lake.–In: Neusiedlersee, ed. H. Löffler, Junk, The Hague.
Schiemer, F., Duncan, A. 1974. Oxygen consumption of a freshwater benthic nematode, *Tobrilus gracilis* Bastian.–Oecologia 15: 121–126.
Schiemer, F., Löffler, H., Dollfuss, H. 1969. The benthic communities of Neusiedlersee (Austria).– Verh. Int. Ver. Limnol. 17: 201–208.
Schiemer, F., Prosser, M. 1976. Distribution and biomass of submerged macrophytes in Neusiedlersee.– Aquatic Botany 2: 289–307.
Stanczykowska, A., Schiemer, F. in prep. *Dreissena polymorpha* (Pall.) im Neusiedlersee–Ausbreitung, Biotopwahl und Wachstum.
Tauber, A. F. 1959. Trübung und Sedimentverfrachtung im Neusiedlersee.– Wiss. Arb. Bgld. 23.
Tauber, A. F. & Wieden, P. 1959. Zur Sedimentschichtfolge im Neusiedler See.– Wiss. Arb. Bgld. 23: 68–73.
Tölg, I. 1960. Untersuchung der Nahrung von Kaulbarsch-Jungfischen (*Acerina cernua* L.) im Balaton.– Annal. Biol. Tihany 27: 147–164.
Winberg, G. G. 1971. Methods of estimating the production of aquatic animals.– 175 pp London-New York: Academic Press.
Winkler, H. 1975. Experimente zur Etho-Ökologie der Beziehungen zweier Fischarten (*Gymnocephalus cernua*, Percidae; *Blicca björkna*, Cyprinidae).–Z. Tierpsychol. 39: 33–38.
Zieba, J. 1971. Production of macrobenthos in fingerling ponds.– Pol. Arch. Hydrobiol. 18: 235–246.

4 *Limnocythere inopinata* (Baird) (Cytheridae, Ostracoda): its distribution pattern and relation to the superficial sediments of Neusiedlersee

M. Jungwirth

Limnocythere inopinata is the most abundant ostracod species of the open lake (Schiemer et al, 1969), whereas the ostracod fauna of the reed belt is mainly represented by Candoninae and Cyprininae (see Chapter 26). Like the other benthic fauna (Schiemer et al. 1969) *L. inopinata* shows a distinct distribution pattern.

Limnocythere inopinata is highly euryplastic in respect to various ecological parameters (Löffler, 1961, Nüchterlein, 1969). Benthos samples taken at two-week intervals from March 1974 to November 1975 showed that adults as well as juvenile instars occur throughout the year, even under ice cover. These observations are in agreement with the statement of Vesper (1975), who described *L. inopinata* as an 'eurychrone species'.

The relation between ostracod distribution and substrate conditions has been dealt with by various authors. Generally the nature of the bottom seems to be the major environmental factor influencing the distribution pattern of ostracods. Beside the grain size, which according to Kilenyi (1969) and Powell (1977) plays a decisive role, consistency and content of organic material are major factors (Kornicker, 1974). *L. inopinata* is found over a wide range of grain sizes, like clay, silt or sands (Vesper, 1975). In Neusiedlersee it occurs in the compact eastern shore zone as well as in the northern or western soft mud areas.

By means of a Gilson sampler (16.62 cm^2 aperture) ten samples were taken from 42 (Fig. 24.1) of the 94 positions within the open lake shown in Fig. 24.2 (comp. Fig. 1 in Chapter 13). After fixation with formaldehyde and washing through a 50 μm net individuals were counted under the microscope.

Fig. 24.1 shows the number of individuals per m^2 (arithmetic mean of 10 samples). Along the eastern shore less than 1,000 individuals per m^2 are found. Frequency increases slightly in the central lake basin to about 2,000 individuals per m^2. Towards the macrophyte zone in front of the northern and western reed belt the density increases to 15,000 individuals per m^2; the highest values can be observed in bays.

A comparison of the values in Fig. 24.1 and 24.2 reveals the following: Positions within the field indicated by X in Fig. 24.2 have a high water content (>74 per cent) and ignition loss (>5.3 per cent) as well as a correspondingly high number of individuals/m^2 (Fig. 24.1). The field indicated by O comprises sites of low water content (<41 per cent) and ignition loss (<2 per cent) and very low values for individuals per m^2. Table 24.1 shows the values for water content, ignition loss and average individuals/m^2 for the fields X and O. The

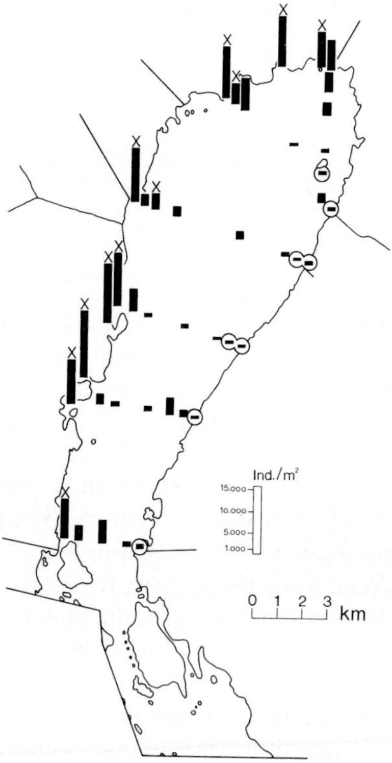

Fig. 24.1. Limnocythere inopinata: Average individuals/m^2 (arithmetic mean of ten samples per position). Positions marked by O correspond to the field O in Fig. 24.2. Positions marked by X correspond to the field X in Fig. 24.2.

mean values in field O are 33.8 per cent water content, 0.96 per cent ignition loss and 638 ind/m^2. The mean values in field X are 81.1 per cent water content, 6.18 per cent ignition loss and 10,106 individuals/m^2.

High individual densities of *L. inopinata* are found exclusively in connection with soft mud of high water content (> 80 per cent) and ignition loss (>6 per cent). This is the case in calm water areas with enhanced sedimentation and high detritus formation due to submerged vegetation. In eastern shore zones, where the composition of the superficial sediment is of mud, compact mud and/or sand and where the water turbulence is high, the number of ostracods is very low.

It is not yet clear to what extent high individual densities of *L. inopinata* in the soft bottom areas are due to active concentration e.g. on account of favourable food conditions (high detritus supply) or to accumulation by passive drifting. Probably the two factors are jointly responsible for the distinct and complex distribution pattern. In any case it seems reasonable to believe that *L. inopinata* is to a great extent influenced by currents and water movement in general. At the sediment surface it might be stirred up, carried off and accumulated at locations of low water movement. Nevertheless for an

Table 24.1. Sample and mean values from the fields O and X in Fig. 24.1 and Fig. 24.2

Field O: <41 per cent water content, <2.0 per cent ignition loss												
water content (%)			26.0	27.1	31.0	32.8	34.3	35.2	36.6	40.5	41.0	33.8
ignition loss (%)			0.30	0.37	0.92	1.25	0.60	1.14	0.88	0.99	1.99	0.96
individuals/m^2			460	120	900	720	180	1,260	840	960	300	638
Field X: >74 per cent water content, >5.3 per cent ignition loss												
water content (%)	74.2	75.5	77.9	80.4	81.0	81.8	83.2	83.2	84.2	85.3	87.1	81.1
ignition loss (%)	5.96	5.85	5.32	5.70	9.94	6.35	5.32	6.10	5.38	6.32	5.90	6.18
individuals/m^2	9,000	3,480	12,900	12,360	12,360	8,280	9,360	12,660	15,480	11.750	3,540	10,106

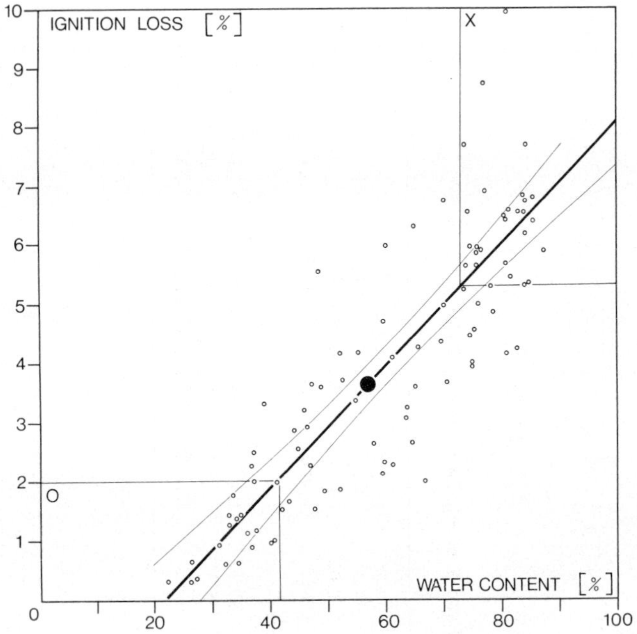

Fig. 24.2. Water content/ignition loss regression of the top centimeter at 94 positions (see Fig. 13.1) and its 99 per cent confidence limits. Field O: <41 per cent water content, <2.0 per cent ignition loss. Field X: >74 per cent water content, >5.3 per cent ignition loss. Compare Table 24.1.

exact explanation further and more detailed investigations will have to be carried out.

References

Imhof, G., 1967. Ökologische Gliederung des Schilfgürtels am Neusiedlersee und Übersicht über die Bodenfauna unter produktionsbiologischem Aspekt.– Sitz. Ber. Österr. Akad. Wiss., Math.– nat. Kl., Abt. I, 175: 219–235.

Kilenyi, T. I., 1969. The problems of ostracod ecology in the Thames estuary. 251–267, in: The taxonomy, morphology and ecology of recent Ostracoda. J. W. Neale, Edinburgh, 553 pp.

Kornicker, L. S., 1974. Ostracoda (Myocopodina) of Cape Cool Bay, Massachusetts.– Smithonian Contrib. to Zool. 173, 20 pp.

Löffler, H., 1961. Beiträge zur Kenntnis der Iranischen Binnengewässer.– Int. Rev. ges. Hydrobiol. 46: 309–406.

Nüchterlein, H., 1969. Süßwasserostrakoden aus Franken. Ein Beitrag zur Systematik und Ökologie der Ostrakoden.– Int. Rev. ges. Hydrobiol. 54, 1: 223–287.

Powell, S., 1977. Einige Aspekte der Beziehung zwischen Sedimenteigenschaften und der Fortbewegung benthischer Süßwasserostrakoden mit spezieller Berücksichtigung der *Cytherissa lacustris* (Sars).– Diss. Univ. Wien, 122 pp.

Schiemer, F. et al., 1969. The benthic communities of Neusiedlersee (Austria).– Verh. Internat. Verein. Limnol. 17: 201–208.

Vesper, B., 1975. Ein Beitrag zur Ostrakodenfauna Schleswig-Holsteins.– Mitt. Hamburg. Zool. Mus. Inst., 72: 97–108.

5 Arthropod communities connected with *Phragmites*

G. Imhof

The invertebrate fauna of the extensive reed stands of Neusiedlersee has been much neglected by zoologists in the past, particularly when compared with the attention paid to the unusual bird fauna of the region. This even applies to faunistic stock lists, as is clearly reflected in the latest review by Sauerzopf (1959). The reason why so few studies have been carried out is probably to be seen in the inaccessibility of the tall, dense and more or less inundated reed stands. Upon the suggestion of W. Kühnelt (University of Vienna) who had already drawn attention to ecological patterns of invertebrate communities in the *Phragmites* belt (Kühnelt, 1948), the reed belt of Neusiedlersee was in 1966 selected as offering a particularly good example of freshwater wetland for an ecological research project within the Terrestrial Productivity Section of the IBP programme. During eight years of research this biotope with its interwoven terrestrial and aquatic environments was shown to contain invertebrate communities of unusual abundance and diversity. So far, more is known about the spatial and temporal patterns exhibited by these communities than about their functional interrelations, which have been studied qualitatively rather than quantitatively. The main emphasis has been placed upon the study of life cycles, reproductive and nutritive biology and individual energy budgets of single species – mainly insects, almost exclusively within the 'macrofauna' – which are dominant or are considered to hold a key position in food chains.

As far as methods are concerned the invertebrate fauna in a wetland biotope can easly be divided into terrestrial and aquatic communities and research was mainly carried out along these lines. From an ecological point of view, however, a change of medium, such as is involved in emergence of aquatic insects, plays an important role in the dynamics of the ecosystem.

Most terrestrial work in the reed belt was confined to an insect community which is trophically related to *Phragmites communis.* In the area of Rust, a small town on the western shore of the lake where most investigations were carried out, 24 species of Homoptera, Lepidoptera and Diptera have been recorded, each of them specialized to a distinct part of the plant (Waitzbauer, 1969, 1972; Waitzbauer et al., 1973; Pruscha, 1973). As with tall semisubmerged grass-like Monocytyledon stands elsewhere, almost no grazing occurs on external tissues of *Phragmites* – not even by locusts which are reported to feed on inflorescences of *Spartina* in North American coastal marshes (Smalley, 1959; Davis & Grey, 1966) or *Papyrus* (Weber, 1942). The only phytophagous insect on *Phragmites* not concealed in the interior is an aphid. A coccid lives beneath the leaf seaths, but all others bore in the stems or form galls. An account of this community is given in the Table 25.1.

As expected, the life cycle of the stalk-inhabiting insects is closely related to

Table 25.1. 'Reed insects', trophically connected with Phragmites

Species	Biological remarks	
Acari Tarsonemidae		
Stenotarsonemus phragmitidis Schdl.	forms large leaf-galls at the top of the plant	1)
Homoptera		
Pseudococcidae		
Chaetococcus phragmitidis March.	beneath the leaf sheaths	2)
Aphididae		
Hyalopterus pruni Geoffr.	changes from plum-trees in winter to reeds in summer; sucking on the leaves	2)
Lepidoptera		
Cossidae		
Phragmatoecia castaneae HB	in submerged parts of the stalk; changes the host stalk during a 2 years' life cycle	4)
Noctuidae		
Archanara geminipunctata HW	in stalks up to the growth tip, which it destroys; life cycle as in the latter	2)
A. neurica HB.		4)
Leucania impura HB.		4)
L. obsoleta HB.		4)
Senta maritima TR.	partly feeding on parenchyma, partly predatory on other stalk inhabitants	4)
Pyralidae		
Schoenobius gigantellus Schiff.	in submerged parts of stalks and rhizomes, destroying several young shoots	4)
Chilo phragmitellus HB.	in all parts of older stalks	4)
Diptera		
Itonididae		
Perrisia inclusa Frfld.	isolated galls in the basal parts of the stem; prefers large shoot diameter	3)
P. incurvans Nijfeldt	several larvae in one gall at the top; stops growth	3)
Thomasiella flexuosa Wtz.	aggregations of free-living larvae (up to 250) in the top region of the stalk; prevents flowering	3)
T. arundinis Schiner	10–20 larvae in deformed lateral sprouts	3)
T. massa Erdös	numerous free-living larvae in the stalk: also feeds on fungi of the genus *Cladosporium* and *Tolula*	3)
Asynapta spec.	together with *Thomasiella flexuosa* or Chloropidae	3)
Dolichopodidae		
Thrypticus smaragdinus Gerst.	bores within the stem wall	3)
T. bellus Loew.	like the former	3)

Table 25.1 'Reed insects', trophically connected with Phragmites–*Contd.*

Chloropidae		
Lipara lucens MG.	1 larva in 1 stem only, forming a large top gall and preventing further growth	1) 3)
Platycephala planifrons F.	1 larva in the growth tip; partially saprophagous	3)
Calamoncosis minima Strobl	together with *L. lucens* above the gall between rolled-up leaves	3)
Haplegis diadema MG.	together with *Lipara lucens* and *P. planifrons* around the galls, feeding on dead leaves	3)
H. flavitarsis MG.	like the former	3)

1) Waitzbauer, 1969 2) Waitzbauer, 1972 3) Waitzbauer et al., 1973 4) Pruscha, 1973

that of their host plant. Most species are univoltine, only a few big Lepidoptera having a 2 years' life cycle, during which the larvae change the host stalk. Diptera undergo larval development during summer and emerge between April and June of the subsequent year. Although feeding ceases at the end of summer, due to the atrophy of nutritive tissues in the stalk and in galls, hibernation takes place at the larval stage in most cases; only a few species pupate before the onset of winter. Intensive studies of the life cycle and biology of *Lipara lucens* have been made by Waitzbauer (1969) and of the cossid *Phragmatoecia castaneae* by Pruscha (1973). The galls of *Lipara lucens* provide a further habitat for a number of inquilines, mainly saprophagous or omnivorous. Besides the Itonididae *Thomasiella flexuosa, Perrisie incurvans* and *P. inclusa* and the Chloropidae *Calamoncosis minima*, *Haplegis diadema* and *H. flavitarsis* already mentioned, Waitzbauer (1969) also listed in the anthomyzid fly *Anthomyza gracilis* Fall. and the tenthredinid *Selandria flavens* Kl., all of them developing completely in the first year in *Lipara* galls. Other arthropods such as the mite *Siteroptes graminum* Reuter feed on the rotting leaves around the galls, or, like the grasshoppers *Gonocephalus dorsalis* Latr. and *C. fuscus* Fabr., use the spaces between the stunted leaf-sheaths as a shelter for their offspring. In the subsequent year, too, the abandoned galls may serve as a breeding chamber for a number of Hymenoptera: *Prosopis kriechbaumeri* Först. (Apidae); *Osmia leucomelaena* K. and *Cenomus lethifer* Shuk (both Sphegidae). Finally, the *Lipara* larva itself serves as a host for several parasitic Hymenoptera, the following having been recorded from Neusiedlersee: *Polemon liparae* Gir. (Braconidae), *Scambus phragmitidis* Perkins and *Exeristes arundinis* Kriechb. (both Ichneumonidae). If the parasites of the various inquilines are also included a large community can be encountered in connection with *Lipara lucens*.

Waitzbauer (1973) also found a correlation between the site of development of certain endophagous Diptera and a peculiar morphological feature of the pupae, the so-called 'Bohrhörnchen' (boring horns) on the forehead. They take the form of a pair of conspicuous and robust horns in those species which live in the basal parts of the stems, free or in galls. The animals use their horns to pierce the walls when emerging. Species living in the middle parts of the plants have reduced boring horns since emerging pupae break through the wall at spots which have already been gnawed by the larvae. The pupae of species living near the growth tip do not bear horns at all and emerge through a larval tunnel.

A question of major interest during the IBP studies was the nature and extent of infestation of *Phragmites* by the different 'pest' insects and their impact on growth and production of the reed. In 1968 a quantitative sampling programme was carried out along a transect reaching 280 m from the inshore fringe to the centre of the reed belt. Altogether 38 sampling squares each of 4 m^2 were marked out over three zones within this transect. They are characterized as follows:

RI: landward reed fringe, only occasionally flooded during winter or spring; mixed stands of tall sedges, grasses and reed;
RII: regularly flooded during winter and spring, drying up during summer; reed stands dominated by *Phragmites*;
RIII: permanently flooded reed beds; uniform *Phragmites* areas, interspersed with open water patches and stands of *Typha.* The reed was harvested completely in all sampling squares in December and analysed as to stalk length and insect infestation (Waitzbauer et al. 1973). The results show characteristic differences in distribution of the various endophagous insects (Table 25.2): the density of gall midges decreases gradually from the margin towards the centre of the reed belt; chloropids are particularly concentrated in the extreme marginal zone. On the other hand some species only occur in

Table 25.2. Infestation rates and abundance of endophagous reed insects in different zones (after Waitzbauer et al., 1973 and Pruscha, 1973).

Species	% of infested stalks			individuals per square meter		
	RI	RII	RIII	RI	RII	RIII
Lepidoptera:						
Phragmatoecia castaneae	—	8.2	12.2	—	5	7
Pyralidae	8	7.8	1.0	5	5	0.7
Diptera – Itonididae:						
Perrisia inclusa	3.1	1.5	0.8	17.8	8.7	2.4
Thomasiella arundinis	1.8	1.2	—	30.0	15.4	—
Th. flexuosa	3.1	3.7	2.2	37.8	45.0	29.7
Th. massa	3.7	0.7	0.3	75.0	6.3	3.4
Asynapta spec.	1.9	0.7	—	13.0	5.0	—
Diptera–Dolichopodidae:						
Thrypticus bellus	1.8	1.5	—	9.6	7.2	—
Th. smaragdinus	—	0.4	1.3	—	2.7	7.0
Diptera – Chloropidae:						
Calamoncosis minima	3.0	0.7	—	2.0	0.5	—
Haplegis diadema *H. flavitarsis*	13.5	1.8	—	78.3	8.4	—
Lipara lucens	4.5	1.5	—	3.0	1.0	—
Platycephala planifrons	13.6	0.6	—	9.1	0.4	—

submersed stands, e.g. the dolichopodid *Thrypticus smaragdinus* and the Lepidoptera *Phragmatoecia castaneae* and *Schoenobius gigantellus*. The reason for the differences in spatial distribution is known in only a few cases, one of which is *Lipara lucens* from which Mook (1967, 1971) showed that the state of growth of the reed is the decisive environmental factor. Whereas the females prefer stalks belonging to the mean diameter class for oviposition, the survival of young larvae on their way to the growth tip is greatest in stalks of the thinnest diameter class. Thus the highest densities occur in stands with retarded growth under suboptimal conditions, as in marginal zones not regularly inundated. In contrast, the adult caterpillars of *Phragmatoecia* prefer the thickest stalks when changing their host plant in June.

Detailed studies have been carried out to assess the production and energy turnover of the two endophagous insects *Lipara lucens* (Waitzbauer, 1969) and *Phragmatoecia castaneae* (Pruscha, 1973). The results show that the withdrawal of organic matter from the Phragmites plant is almost negligible. For the larval growth period, net production (P) of *Lipara* was given as 0.1 Kcal/m^2/y in the zone of maximum density and food intake (C) as 0.17 Kcal/m^2/y; the corresponding figures for *Phragmatoecia* are $P = 2.9$ and $C = 13.9$ (for comparison, the net production of the reed in permanently flooded stands amounts to approx. 15,000 Kcal/m^2/y). Rough overall estimations suggest that the total organic matter consumed by insects trophically dependent on *Phragmites* is less than 1 per cent of the net production of the reed. Thus, the impact of this phytophagous community on the reed lies rather in specific effects like growth inhibition (all form top galls), prevention of flowering or mechanical damage. In the research area of Rust most damage was caused by the pyralid *Schoenobius gigantellus*, which destroyed about 5 per cent of all young shoots (Pruscha, 1973), whereas most abundant species like *Phragmatoecia* or the reed scale *Chaetococcus phragmitidis* do not cause any visible damage.

Recently, investigations were also started on the insect community ecologically associated with the cat's tail, *Typha augustifolia*, which forms localized stands in the reed belt, usually fringing open water patches. Apart from the numerous species of boring larvae of the families Noctuidae (Lepidoptera), Heleidae, Lycoriidae, Scatopsidae (Diptera) which live in the stalks, the male inflorescence appears to be the habitat of a considerable insect community. Waitzbauer (1976) recorded 14 species of Diptera and 3 spp. of Coleoptera trophically connected with the male flowers in May and June. The cryptophagid beetle *Telmatophilus schönherri* Gyll. is most closely bound to *Typha*. It undergoes its entire development from egg through three larval instars to the imago in the inflorescence, within the brief flowering period of only one month. All larval instars feed on pollen. How the adult beetle spends the winter after leaving the *Typha* inflorescence is not known. Since the inflorescences tend to disintegrate after maturation, high losses of larvae occur when they fall. Whereas an average number of 145 young larvae hatched from the eggs on one inflorescence only 2–10 adult beetles emerged.

Several species of hovering flies (Syrphidae) also appear to feed on the *Typha* pollen. Since they are mobile they also play an important role in pollination. The most frequent species, *Liops vittatus* Meig. also mates and deposits eggs on the male *Typha* inflorescence. After hatching, the young larvae fall into the water and pass their instar on the water lens *Lemna trisulca*, before entering deeper water layers. Another but less frequent species, *Eurinomyia transfuga* L., has a similar life cycle. Emergence catches of both species indicated the abundance of fully grown larvae to be three times higher in *Typha* stands than in *Phragmites* stands. The larvae of the following three hovering fly species occurring regularly on male *Typha* flowers are terrestrial, feeding on the reed aphid *Hyalopterus pruni*: *Platychirus clypeatus* Meig., *P. fluviventris* Macq., *P. perpallidus* Verr. Decaying flowers are also visited by the shore flies (Ephydridae) *Notiphila guttiventris* Stenh. and *N. riparia* Meig. which suck sap from rotting plant tissues. The other recorded species are only occasional visitors of *Typha* flowers as imagos; their larval development takes place in different habitats of the reed belt, where most of them represent characteristic constituents. These are Diptera: *Stratiomyia furcata* Fabr., *S. longicornis* Scop. (Stratiomyidae); *Eristalinus sepulcralis* L., *Eurinomyia lunulata* Meig. (Syrphidae); *Sepedon sphegeus* Fabr., *S. spinipes* Scop. (Sciomycidae); *Elachiptera austriaca* Duda (Chloropidae), and Coleoptera: *Helodes minuta* L., *Cyphon variabilis* Thunbg. (Helodiae).

Besides these phytophagous insects with trophic relations to particular emergent macrophytes, a number of mobile beetles can regularly be found associated with the latter: *Odacantha melanura* L., *Demetrias imperialis* Germ. (Carabidae); *Philonthus suffragani* Joy., *Paederus litoralis* Grav. (Staphylinidae); *Coccidula scutellata* Hrbst. (Coccinellidae); *Malachius spinosus* ER. (Cantharidae); *Dolopius marginatus* L. (Elateridae). Most of them are assumed to be predators, although details of their life histories are as yet unknown. They appear to be permanent inhabitants of the reed belt, hibernating preferably in broken reed stalks or in abandoned *Lipara* galls (Waitzbauer, 1969).

Considerable advances have been made in our knowledge of the composition and ecology of the spider fauna in the reed belt. Although attention was paid to it before most insect groups were studied (Nemenz, 1956, 1959, 1967), the present knowledge originates from the comprehensive work of Pühringer (1972, 1975), who listed nearly 50 species belonging to 16 families inhabiting the reed belt. Most of them occur, however, only in the transitional zone with mixed vegetation; they are supposed to have a wider distribution outside reed stands, or to be bound to *Carex* species, e.g. *Singa heri* (Hahn) (Araneidae), *Hypomma bituberculata* (Wider) (Micryphantidae), *Theridion pictum* (Walck.) (Theridionidae), *Eucta kaestneri* Crome (Tetragnatidae). On the other hand, some species are particularly adapted to living conditions in inundated tall reed stands and represent characteristic constituents of the animal community of this biotope. Examples are species hunting on the water suface, such as the frequently occurring lycosid *Pirata piraticus* (Cl.) and the peculiar big pisaurid *Dolomedes fimbriatus* (Cl.), or the very characteristic water spider *Argyroneta aquatica* (Cl.) (Agelenidae). Other species breed and pass part of all their life cycle in the panicles of *Phragmites*.

The latter, the so-called reed spiders, were the object of qualitative and quantitative studies on distribution, ecological requirements, life cycle, population dynamics and secondary production made by Pühringer (1972, 1975). In the reed belt near Rust he found altogether 8 species of this type, characterised by its occurrence in panicles at least for breeding (Table 25.3).

Table 25.3. Reed spiders

Species	Habitat requirement and zonal distribution	mean number of adult ind.m^{-2} in spring
Araneidae:		
Araneus folium Schrank (formerly syn. *A. cornutus*)	in panicles throughout the vegetation period, prefers dispersed stands, density decreasing from RI to RIV	RI: 0.66 RIV: 0.18
Singa phragmiteti Nemenz	preponderantly in panicles, prefers dense stands, only in RIV	0.35
Tetragnathidae:		
Arundognatha striata (L. Koch)	prefers dispersed stands, only in RIV	0.15
Clubionidae:		
Clubiona juvenis Simon	indifferent to structure of reed stands, density increasing from RI to RIV	RIV: 0.54
C. phragmitidis C. L. Koch	prefers dispersed stands, no zonal gradient	0.18
Dictynidae:		
Dictyna arundinacea (L.)	preponderantly in panicles, no zonal gradient	
Linyphiidae:		
Donacochora speciosa (Thorell)	mostly near the water, irregular in all zones	
Salticidae:		
Mithion canestrini (Ninni)	vagile, irregular in all zones	

For definition of zones RI to RIV see p. 25.5. RIV is the extension of RIII towards the lake, differing from the latter in having deeper water; the sampling area RIV is about 1 km from the landward fringe.

The spatial distribution of a species is not only governed by the general zonal features of the reed belt, but to a large degree by structure and growth patterns of the reed stands, including number of stalks per square unit, variation of thickness of stalks, regularity of spatial patterns, progress of break-down and so on. A common requirement of all species is the presence of standing reed stems which still bear panicles of the preceding vegetation period. This is a major limiting factor in areas where the reed is cut for commerical use during winter. On the basis of this requirement an estimation of the size of the population could be made by analysing the still upright panicles (about 5 per m^2 in undisturbed stands). This is considered to be the only possibility of a quantitative assessment. Most relevant figures are to be gained from spring samples, when nearly all reed spiders climb up to the panicles from their winter quarters either to pursue different activities or simply to warm up. Another concentration of individuals occurs in the panicles in autumn when juveniles are dispersed by wind action with the help of their spun threads. Before the onset of winter all spiders leave the panicles (which usually do not outlast a second winter) and pass the winter in sheltered places closer to the water surface, preferably in the interior of broken or cut stumps of *Phragmites.* Only *Araneus folium* could not be found there or in other suitable places during winter. The site of hibernation of this species is still unknown.

Detailed analysis of structure and dynamics of the populations could only be achieved for the two species of Araneidae and the two Clubionidae. They were based on the class distribution of the width of the cephalothorax, the interpretation of which is, however, difficult, due to an inconstant number of moulting stages and a high variability in their duration. The only species showing distinct generations is *Singa phragmiteti* which breeds in spring. *Araneus folium* has its main breeding period in May, and a second, but only partial one, in September. In contrast, adult clubionids appear throughout the entire vegetation period.

Figures for the energy budget have not yet been worked out in detail, but preliminary estimations can be given (for the whole reed spider community – referring to zone R IV): mean biomass: 100 cal/m^2; net production: 450 cal/m^2/year; consumption: 900 cal/m^2/year. Thus, the reed spiders are supposed to remove about 25 per cent of the small-sized flying insects emerging from the water (mainly Trichoptera, Ephemerotera and Diptera). On the other hand, the spider community forms an essential food basis for small birds in the reed belt, particularly reed warblers, as has been proved by gut analysis (Leisler, 1970) (see Chapter 29).

References

Davis, L. V. & Grey, I. E. 1966. Zonal and seasonal distribution of insects in North Carolina salt marshes.–Ecol. Monogr. 36: 275–295.

Kühnelt, W. 1948. Der Neusiedlersee als Lebensraum für die Tierwelt.– In: Der Neusiedlersee, ein Kleinod Österreichs. Österr. Naturschutzbund, Wien.

Leisler, B. 1971. Vergleichende Untersuchung zur ökologischen und systematischen Stellung des Mariskensängers am Neusiedlersee.– Diss. Univ. Wien.

Mook, J. H. 1967. Habitat selection by *Lipara lucens* Mg. (Dipt. Chlorop.) and its survival value.– Arch. Neerl. Zool. 17, 4: 469–549.

Mook, J. H. 1971. Influence of environment on some insects attacking common reed (*Phrag. comm.*)–Hidrobiologia 12: 305–312.

Nemenz, H. 1959. Arachnida.– In: Sauerzopf (Ed.) Landschaft Neusiedlersee. Wiss. Arb. Bgld. 23: 208 pp.

Nemenz, H. 1967. Einige interessante Spinnenfunde aus dem Neusiedlersee-gebiet.– Anz. math.– nat. Kl. Österr. Akad. Wiss. 6: 132–139.
Pruscha, H. 1973. Biologie und Produktionsbiologie des Rohrbohrers *Phragmatoecia castaneae* Hb. (Lepidoptera Cossidae).– Sitz. Ber. Österr. Akad. Wiss., Math.– nat. Kl. I, 181: 1–49.
Pühringer, G. 1975. Zur Faunistik und Populationsnynamik der Schilfspinnen des Neusiedlersees.– Sitz. Ber. Österr. Akad. Wiss., Math.– nat. Kl. I 184, 8–10: 379–419.
Sauerzopf, F. & Tauber, A. F. (editors) 1959. Landschaft Neusiedlersee.– Wiss. Arb. Bgld. 23: 208 pp.
Smalley, A. E. 1959. The growth cycle of *Spartina* and its relation to the insect population in the marsh.– Proc. Salt Marsh Conf. of Marine Inst. Univ. Georgia.
Waitzbauer, W. 1969. Lebensweise und Produktionsbiologie der Schilfgallenfliege *Lipara lucens* Mg (Diptera Chloropidae).– Stiz. Ber. Österr. Akad. Wiss., Math.– nat. K. I, 178: 175–242.
Waitzbauer, W. 1972. Aspects on productivity in reed-feeding insects.– Verh. dt. Zool. Ges. 65: 116–119.
Waitzbauer, W. 1976. Die Insektenfauna männlicher Blutenstände von *Typha augustifolia*.– Zool. Anz. 196, 1/2: 9–15.
Waitzbauer, W., Pruscha, H., Picher, O. 1973. Faunistisch ökologische Untersuchungen an schilfbewohnenden Dipteren im Schilfgürtel des Neusiedlersees.– Stiz. Ber. Österr. Akad. Wiss., Math.– nat. Kl. I, 181: 11–136.
Weber, N. A. 1942. A biocoenose of *Papyrus* heads (*Cyperus papyrus*).–Ecology 123: 115–119.

6 The crustacean fauna of the *Phragmites* belt (Neusiedlersee)

H. Löffler

26.1 Introduction

The crustacean fauna of reed belts has been given little attention.[1] Some information exists about the Danube delta region, and from various parts of Europe and Asia, although it is lacking from one of the most prominent reed areas of Lake Hamun at the border of Iran and Afghanistan. From the data available there is little doubt that investigations of any large *Phragmites* area, permanently or periodically flooded, will reveal not only similarity but also variety in composition of the crustacean fauna. This is probably due to the rich spatial structure combined with rapidly changing environmental parameters.

26.2 Methods

The samples discussed in this chapter were collected near Rust (Fig. 26.1) throughout 1971 and 1972, since they were from the same sites as those mentioned in Chapter 25 and Imhof (1966). Mixed ten litres samples taken at different depths and not from the same but nearby spots they can scarcely be considered to fulfil quantitative requirements although they allow an estimate

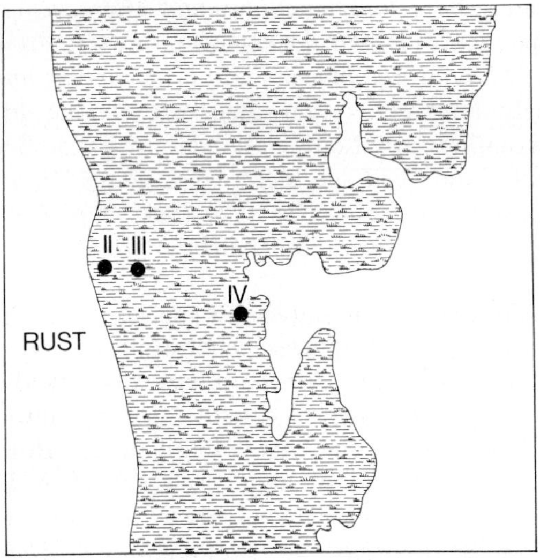

Fig. 26.1. Sampling positions near Rust.

[1]Very recently appeared: Ponyi & Dévai (1979): The Crustacea of the Hungarian area of Lake Fertö.

of the per cent composition. It is, however, very doubtful whether the area of Rust may be considered representative for the whole reed belt. The paper by Pesta (1954) indicates that long-term and spatial differences in composition may be expected. The material revealed the relatively large number of 58 taxa. In addition four species (*Daphnia magna*, *Oxyurella tenuicaudis*, *Tretocephala ambigua*, *Diaptomus kupelwieseri*) were described from the reed belt in an earlier paper (Pesta 1954) and one observed on a previous occasion (*Nitocra hibernica*). Thus a total of 63 crustaceans can be listed.[1] This figure corresponds roughly to the number of species found in the macrophytic zones of the Seewinkel bordering Neusiedlersee in the east (Löffler 1959) and is roughly the same as some of the macrophytic zones of temperate climates like Mikolajskie. On the other hand the striking contrast to the open portion of Neusiedlersee (Chapter 22) becomes obvious; except in the fringe zone where water movement exerts a mutual influence almost no species are common to both. This phenomenon also applies to other animal groups of the two portions of the lake (Chapter 22, 23, 27) and shows that the two systems which are connected only by exchange of soluble and particulate matters and by hydrology (water budget etc) are relatively independent.

26.3 Species composition

Of the 19 species of Cladocera found in the *Phragmites* belt only six (*Ceriodaphnia laticaudata*, *Graptoleberis testudinaria*, *Acroperus elongatus* *Tretocephala ambigua* and *Polyphemus pediculus*) have not been described from the Seewinkel, whereas eleven of the total of 26 copepods have not yet been reported from this region. Besides the genera *Macrocyclops* and *Thermocyclops* and *Diacyclops languidus* and *Microcyclops varicans* these are mainly the harpacticoids (*Phyllognathopus viguieri*, *Attheyella trispinosa*, *Attheyella crassa*, *Elaphoidella bidens* and *Bryocamptus minutus*) which, due to their relatively long generation time, are unable to cope with the considerable fluctuations in salinity and alkalinity of the Seewinkel lakelets. Of the ostracods (a total of 13 species) it is the genus *Candona* which differs in species composition. Four species (*Candona marchica*, *Candona sucki*, *Candona acuminata* and *Candona fabaeformis*) are absent from the Seewinkel.

As mentioned above there are only a few species common to both portions of the lake. These are either species penetrating the open lake (*Daphnia pulex*, *Acanthocyclops robustus*, *Thermocyclops crassus* and *Diaptomus bacillifer*) or open lake species occasionally drifting into the reed belt (*Diaphanosoma brachyurum*, *Daphnia longispina* s.l., *Bosmina longirostris* and *Diaptomus spinosus*) but found only in samples taken closest to the open lake (R IV). The only two species which seem to be common to both parts are the copepods *Cyclops strenuus* and *Mesocyclops leuckarti*, which occur throughout the *Phragmites* belt. However, the latter species was only observed throughout the open lake during 1970 whereas before (Pesta 1954)

[1]Including the open lake and Ponyi & Dévai's results: 78 species.

and afterwards it has invariably been restricted to the neighbourhood of the *Phragmites* belt. As pointed out by Herzig (Chapter 22) a general decline in *Cyclops* has also been observed since 1970, this may be connected with the decrease of the submersed macrophytic vegetation although there is no evidence for such a correlation.

With respect to the species composition of the reed belt the following genera common in the open lake are lacking: *Diaphanosoma*, *Macrothrix*, *Ilyocryptus*, *Ilyocypris* and *Limnocythere*. This is also true for the currently absent genus *Moina* (see Chapter 22). Remains of all of these organisms in sediments of the *Phragmites* belt can be safely used as an indication of water movement or of conditions before the establishment of the reed belt. No investigation of this sort has so far been undertaken. Possible reasons for the distinctness of littoral crustaceans (zooplankton) have been discussed recently by Gliwicz and Rybak (1976). Food and physico-chemical conditions (fluctuations in O_2 concentration) have been considered important. In the case of the reed belt the reduced water movement favours neuston feeders. It is yet unknown to what extent avoidance of shore may be involved in determining the species present. Finally, practically nothing is known about the possible influence of fish in the reed belt on the composition of invertebrates. Only a few data so far exist on the predation pressure of planktivorous fish in the open lake.

26.4 Seasonal aspects of composition (Fig. 26.2)

The seasonal change in the crustacean fauna of the *Phragmites* belt is much more pronounced than in the open lake, where only *Diaphanosoma brachyurum* (and in earlier times also *Moina rectirostris* and *Mesocyclops leuckarti*) disappear during winter whereas merely a decrease of biomass is registered in the other species in the cold season. If the reed belt is frozen over for several weeks, all species with the exception of *Asellus aquaticus* may be killed due to a total lack of oxygen and accumulation of H_2S, CO_2, and most likely also CH_4 in some places. Only resting stages and probably some of the subitaneous eggs may survive. There are, however, places where, due to a rich algal flora and adequate optical properties, some species tolerating occasional anerobic conditions (cyclopoids) may survive even under ice cover. Otherwise the cold season is characterized by *Arctodiaptomus bacillifer*, *Cyclops strenuus* and probably also by *Canthocamptus staphylinus*.

There is little doubt that some common species like *Daphnia pulex*, *Chydorus sphaericus*, *Eucyclops serrulatus*, *Megacyclops viridis* and *Asellus aquaticus* occur throughout the year and in most parts of the reed belt.

The majority of species appear in the warm season or reach a maximum biomass at this time. A few, however, occur predominantly during springtime (rarely, but not in Neusiedlersee, in autumn as well) like *Cypris pubera* and *Cyprois marginata*. In contrast to these, quite a few Cladocera like *Ceriodaphnia laticaudata* and *reticulata* tend to attain their biomass maxima in early autumn. Again it must be admitted that neither the sampling frequency nor the number of sampling sites can be considered representative enough to

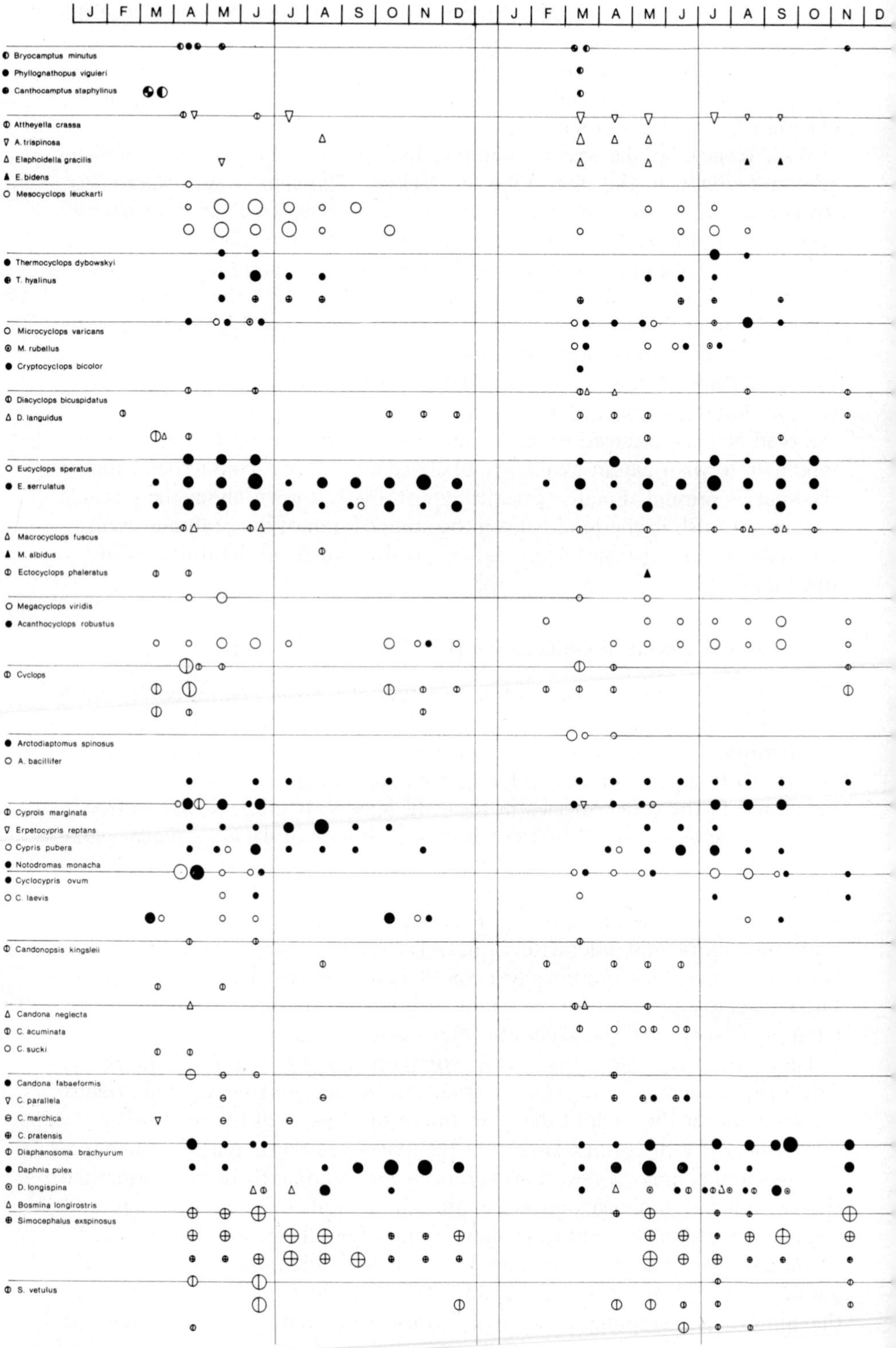
J F M A M J J A S O N D J F M A M J J A S O N D
Bryocamptus minutus
Phyllognathopus viguieri
Canthocamptus staphylinus
Attheyella crassa
A. trispinosa
Elaphoidella gracilis
E. bidens
Mesocyclops leuckarti
Thermocyclops dybowskyi
T. hyalinus
Microcyclops varicans
M. rubellus
Cryptocyclops bicolor
Diacyclops bicuspidatus
D. languidus
Eucyclops speratus
E. serrulatus
Macrocyclops fuscus
M. albidus
Ectocyclops phaleratus
Megacyclops viridis
Acanthocyclops robustus
Cyclops
Arctodiaptomus spinosus
A. bacillifer
Cyprois marginata
Erpetocypris reptans
Cypris pubera
Notodromas monacha
Cyclocypris ovum
C. laevis
Candonopsis kingsleii
Candona neglecta
C. acuminata
C. sucki
Candona fabaeformis
C. parallela
C. marchica
C. pratensis
Diaphanosoma brachyurum
Daphnia pulex
D. longispina
Bosmina longirostris
Simocephalus exspinosus
S. vetulus

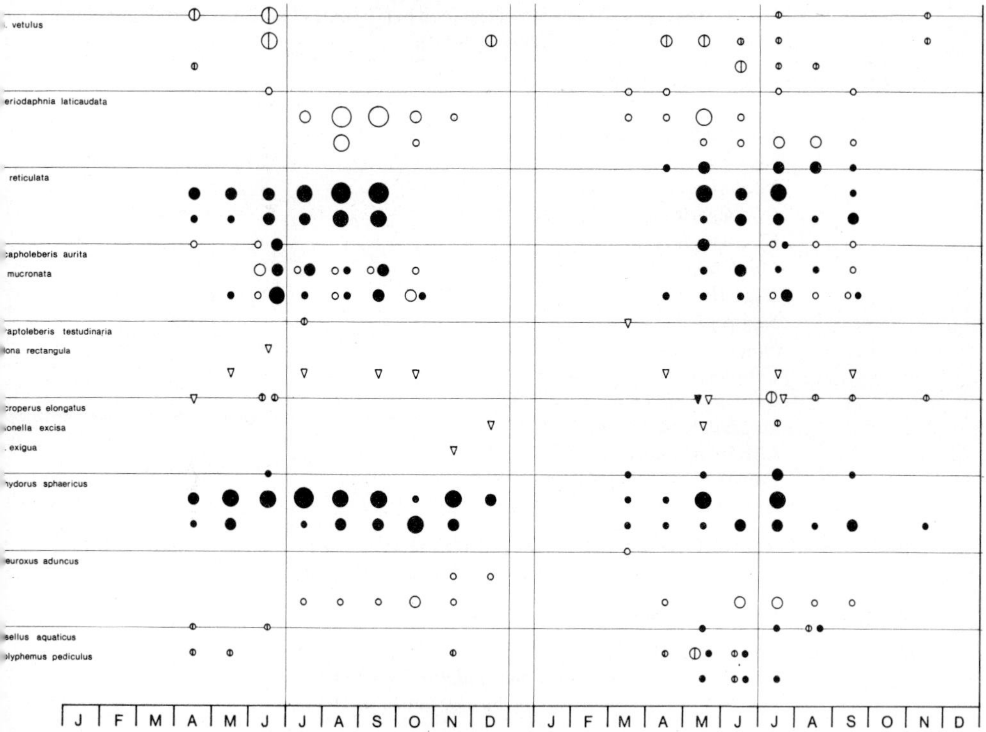

Fig. 26.2. Spatial and seasonal distribution of crustaceans in the *Phragmites* belt. The line indicates zone II followed by zones III and IV. The size classes of the diameters indicate the number of individuals per 10 l (1–10, 10–100, 100–1,000 and 1,000–10,000). *Thermocyclops hyalinus* = acc. to Kiefer (1978) *Th. crassus*

permit a detailed description of the seasonal development or even of population dynamics. It would also be of interest to learn to what extent this seasonal development is synchronised in different parts of the reed belt.

Spatial distribution: Since sampling along the profile near Rust was carried out for two years it was of interest to learn to what extent a random distribution or a zonation in different species exists. As mentioned above there is obviously a group of species behaving indifferently with respect to space (Table 26.1). But beside this relatively small group the other species, as far as can be judged from their frequency, seem to have distinct preferences, as expressed in Table 26.1. Thus species of the open lake can frequently be found in the outermost zone due to passive drifting. With respect to *Thermocyclops crassus*, also restricted to IV but not typical of the open lake. it seems, however, that otherwise planktonic species (and this may be true for *Mesocyclops leuckarti* too) may be limited to the fringe of the reed belt due to unfavourable conditions in the open lake (turbidity and excessive turbulence): They only occasionally invade the lake beyond the reed belt, which was the case with *Mesocyclops leuckarti* in 1970. No explanation can be given with respect to the genus *Elaphoidella* and to the Cladocera which are more or less

Table 26.1. Spatial distribution of Crustacea in the reed belt (species found less than 5 times have been omitted).

no preference:	*Bryocamptus minutus*
	Eucyclops serrulatus
	Megacyclops viridis
	Cyclops strenuus
	Notodromas monacha
	Cyclocypris ovum
	Cyclocypris laevis
	Candonopsis kingsleyi
	Daphnia pulex
	Simocephalus vetulus
	Simocephalus exspinosus
	Ceriodaphnia reticulata 1972
	Ceriodaphnia laticaudata 1972
	Scapholeberis aurita
	Chydorus sphaericus 1972
	Polyphemus pediculus (only 1972)
	Asellus aquaticus
species predominant in II:	*Attheyella trispinosa* (once in IV)
	Ectocyclops phaleratus (once in III, twice in IV)
species predominant in IV:	*Alona rectangula* (once in II, once in III)
species predominant in II, III:	*Thermocyclops dybowskii* (once in IV)
	Diacyclops bicuspidatus (twice in IV)
	Cryptocyclops bicolor (once in IV)
species predominant in III, IV:	*Mesocyclops leuckarti* (once in II)
	Ceriodaphnia laticaudata 1971 (once in II)
	Ceriodaphnia reticulata 1971
	Scapholeberis mucronata (three times in II)
	Pleuroxus aduncus (once in II)
	Chydorus sphaericus 1971 (once in II)
species found only in II:	*Macrocyclops fuscus*
	Arctodiaptomus bacillifer
species found only in IV:	*Thermocyclops crassus*
	Arctodiaptomus spinosus
	Diaphanosoma brachyurum
	Daphnia longispina
	Bosmina longirostris
species found only in II, III:	*Microcyclops varicans*
	Microcyclops rubellus
	Acroperus elongatus
species found only in III, IV:	genus *Elaphoidella*

strictly confined to III and IV. Pressure by predators cannot be excluded as a possible reason for the restriction of large copepods (*Arctodiaptomus bacillifer* and *Macrocyclops fuscus*) to zone II. However, both species are relatively rare and much more information would be needed to confirm this idea. On the other hand, five species of copepods restricted to or predominant in zones II and III are small, but with the exception of *Diacyclops bicuspidatus* prefer warm water. It therefore cannot be excluded that the thermal regime of the reed belt and the submerged macrophytes may also play an important role. No species restricted to III has so far been detected. Of the neuston feeders, the occurrence of *Scapholeberis aurita* and *Scapholeberis mucronata* has been discussed by Müller (1976). He has shown that the non-pigmented *S. aurita* prefers dense stands of *Phragmites* whereas only instars of the ventrally pigmented *S. mucronata*, which is more tolerant of short wave radiation, have the same preference, and the adults live predominantly in open water. As shown in Table 26.1 no distinct zonation could be observed in either species.

Table 1 also indicates the abundance of the species. The sampling method probably accounts partly for the low abundance of non-swimming species like *Candona* and some of the harpacticoids. Of the other copepods only four species tend to have a high abundance (*Eucyclops serrulatus*, *Cyclops strenuus*, *Megacyclops viridis* and *Mesocyclops leuckarti*). Only three ostracod species attain high numbers of individuals (*Notodromas monacha*: the abundance is mainly due to instars, *Cyclocypris ovum* and *Cyclocypris laevis*). By far the most species with high abundance occuring among the Cladocera (*Daphnia pulex, Ceriodaphnia reticulata, Ceriodaphnia laticaudata, Simocephalus vetulus*, *Simocephalus exspinosus* and *Chydorus sphaericus*). Table 26.2 presents the abundance of *Asellus aquaticus* (instars of more than 2 mm) obtained with a corer of 0.05 m^2. These figures are of course more reliable than those obtained by scooped samples. It is regrettable that data from the other zones are lacking.

Table 26.2. Abundance of *Asellus aquaticus* in III

1971			1972		
26.3.	RIII	202 ind./m^2	4. 3.	RIII	204 ind./m^2
19. 4.	,,	140 ,,	12. 4.	,,	246 ,,
11. 5.	,,	52 ,,	2. 6.	,,	340 ,,
15. 6.	,,	64 ,,	21. 7.	,,	28 ,,
15. 7.	,,	136 ,,	17. 8.	,,	48 ,,
6. 8.	,,	76 ,,	27. 9.	,,	120 ,,
20.10	,,	812 ,,	25.11.	,,	28 ,,
4.12.	,,	204 ,,			

The nature of the sampling procedure renders it impossible to describe the life cycle and population dynamics of any species of the reed belt. Only *Asellus aquaticus* has so far attracted attention (Imhof, unpubl.). In this species the reproduction period lasts from the end of March until the end of September. From May onwards all developmental stages are present, which indicates a succession of several overlapping generations. Instars from autumn which do not reach maturity in the same year grow until springtime and attain a considerable body size (males up to 14 mm, females up to 10 mm). Females which become adult during summer attain a length of only 4 mm. There are obviously considerable fluctuations in body size as well as phenology and abundance from year to year, most likely dependent upon the hydrological and climatological conditions in the reed belt.

Knowledge of the reed belt of Neusiedlersee is still very scarce. This certainly is due in part to the enormous difficulties presented by the mosaic of different habitats making up the large ecosystem. A more dynamic and quantitative approach presents a challenging task.

References

Banu, A. C. 1967. Limnologia Sectorului Românescu al Dunarii. Ed. Acad. Rep. Soc. Rom., 651 pp.

Imhof, G. 1979. Arthropod communities connected with *Phragmites*. In: Neusiedlersee, ed. H. Löffler, Junk, The Hague.

Löffler, H. 1959. Zur Limnologie, Entomostraken- und Rotatorienfauna des Seewinkelgebietes (Burgenland, Österreich). Sitz. Ber. Österr. Akad. Wiss. math.– nat. Kl., I, 168, 4–5: 315–362.

Müller, G. 1976. Probleme der Pigmentierung der Gattung *Scapholeberis* (Crustacea, Cladocera). Diss. Univ. Wien, 90 pp.

Negrea, St. 1964. Citeva aspecte ale studiului cladocerilor (Crustacea, Cladocera) din complexul de balti crapina-ji jila. Hidrobiol. 5: 137–156.

Pesta, O. 1954. Studien über die Entomostrakenfauna des Neusiedler Sees. Wiss. Arb. Bgld. 2: 1–83.

Pieczynska, E. 1976. Selected problems of lake littoral ecology. Drukarnia ZW CZSR Inowroclaw. Zam. 238 pp.

Ponyi, J. E. and I. Dévai. 1979. The Crustacea of the Hungarian area of Lake Fertö. Opusc. Zool. Budapest, XVI, 1–2, 107–127.

26.5 Synopsis of planktonic and epiphytic microfauna in wetlands[1]

G. Imhof

Zooplankton is as a rule only considered to be an important constituent of the animal community of open water bodies and little attention has been paid to it in flooded areas with emergent macrophytes. More recently some quantitative studies on zooplankton have been carried out in littorals of backwaters and fish ponds (Straskraba 1963, 1965, 1967; Postolkova 1967; Losos & Hetesa 1971; Amoros 1973). Although mainly focused on the economic aspects of fish culture (e.g. influence of fish population on composition and dynamics of zooplankton communities; changes of plankton composition caused by pond management), such studies give some insight into the structure and dynamics of zooplankton in emergent stands. Crustacea (Cladocera and Copepoda) form the most important component of the zooplankton of the macrophyte stands, numerically exceeding the Rotatoria during the vegetational period when zooplankton abundance is high. Straskraba (1965) reports a Crustacea: Rotatoria ratio in moderately dense emergent vegetation of 2–5 : 1 and 5–15 : 1 in dense submerged vegetation – a result which was earlier confirmed by Reinisch (1925). Only during winter do Rotatoria prevail.

An inherent difficulty in the definition of zooplankton in littoral zones is presented by the fact that the representatives of the above-mentioned groups are connected in differing degrees with the aquatic plants. Straskraba (1967) thus distinguishes 4 ecological groups:

1. This chapter has already been submitted for publication in the IBP synthesis volume: Ecology and productivity of Wetlands (editor: Květ et al.)

1. Littoral species. Although capable of swimming, they mostly crawl on vegetation, scraping out periphyton or preying on organisms sitting or moving on plants; larger forms prevail.

2. Tycholimnetic species. They usually swim freely amongst vegetation but also occur in open water regions; mostly filter feeders dependent on phytoplankton; generally smaller species.

3. Pelagic species. They mainly inhabit the pelagic region of water bodies, but may be common in the littoral region, particularly in sparse stands of emergents, when phytoplankton is abundant.

4. Benthic species.

Sharp boundaries, however, cannot be drawn between these groups, as is shown by the fact that some species belong to different groups in different water bodies.

In keeping with these differences in biology and trophic relations, a rather great diversity of species may occur in littoral zones. Straskraba (1967) reports altogether 77 species of 'zooplankton' in a small backwater of the river Elbe in Bohemia, of which 34 are Cladocera. Numbers of Cladocera species in comparable littoral habitats range from 13 in an English moorland pond (Smyly 1952) to 59 in a eutrophic pond in Silesia (Langhans 1911). In narrow littoral belts with sparse vegetation the species composition shows a considerable degree of resemblance to that of the open water region. But in denser and extensive stands where little exchange with the open water occurs the littoral species become dominant: they seem to replace the tycholimnetic species. Străskraba (1965) found that this is also expressed by different seasonal cycles: whereas tycholimnetic species have their maximum in summer, littoral species show bimodal cycles with one well defined maximum in spring and one less well defined in autumn. In the extensive and uniform reed beds of Neusiedlersee a distinct zonation of zooplankton composition has been ascertained from the inshore margin to the lake-side stands over a distance of more than 1 km. So far it has not been possible to attribute this zonation to environmental conditions. In the weed beds of Lake Chad, Dejoux & Saint-Jean (1972) report the crustacean fauna inhabiting the dense stands of *Ceratophyllum* and *Potamogeton* as being original in regard to the benthic and pelagic fauna of open water places, even inside the weed areas.

In addition to the presence of higher vegetation other abiotic and biotic factors can strongly influence the composition and seasonal dynamics of zooplankton species. In unstable waters like drained fish ponds a selective factor is represented by the ability to develop ecophases resistant to desiccation (ephippia of Cladocera, resting eggs of Copepoda or even certain copepodid stages. This ability also ensures a rapid repopulation when the pond is refilled. Amoros (1973) who studied fish ponds with different drainage cycles in France stressed the importance of this phenomenon in fish culture: those species which resist drainage, treatment and cultivation of the pond soil, belong to the larger size classes preferred by carp as food, whereas in ponds left undrained for several years the small species become dominant. Fish themselves influence species composition by selective feeding as has been shown by fencing experiments (Straskraba 1965b, Postolkova 1967) or by total removal of the fish fauna (Vladimirova 1963; Stråskraba 1967).

Junk (1973) carried out an intensive investigation on the fauna of floating meadows in the Amazonian flood plains. He found that the populations of Cladocera and Copepoda inhabiting the interstitial spaces within the tight root entanglement of the floating mats respond quickly to changes in environmental conditions. After the dry period, rising water containing organic turbidity as well as phytoplankton, leads to a rapid increase of zooplankton abundance. This sharply decreases when the current speeds up and white water loaded with suspended inorganic material penetrates the foodplains; after white water flushing has stopped the zooplankton abundance increases again until the mats become dry. This somewhat simplified scheme is modified at different types of sites according to the local water regime. Under such unstable conditions zooplankton plays an important role; after a rapid increase of abundance, the zooplankton biomass can attain as much as half that of macrofauna and crustacean microfauna together.

Several authors have estimated zooplankton abundance and biomass in the littoral of lakes, ponds and reservoirs. Besides the problem of phytophilous species, spatial variations such as swarming complicate the results. Particularly in narrow littoral zones of small water bodies dense swarms of pelagic zooplankton may appear near the shoreline. Sládeček (1973) reports a density of ca. 4,000 ind./l Crustacea in a swarm in the Nesyt fish pond littoral, whereas at the same time only 73 ind./l were found at a comparable site. In the Poltruba backwater a density of even 27,000 ind./l was recorded as compared with a normal figure of 2,000 during maximum abundance (Stråskraba 1967). Abundance and biomass figures most refer to volume units. Using such a scale, zooplankton abundance and standing crop in littorals are commonly higher (up to tenfold) than those of the respective pelagial (Lipin & Lipina 1950; Morduchai-Boltovskoj et al. 1958; Rudescu et al. 1965; Stråskraba 1965, 1967). Based on surface units, however, standing crop in the littoral is equal to or less than that in the pelagial because zooplankton tends to accumulate near the water surface between vegetation, or because of the shallowness of littoral water itself.

Biomass figures per square meter surface, which are more suitable for comparisons between different wetland types, are rather rare. As an example of the littoral of a large basin in the temperate region, Zimbalevskaja (1972) reports from macrophyte stands in the Kiev reservoir a mean zooplankton biomass of about 8 $Kcal/m^2$ during the vegetational period. Similar values were found by Junk (1973) in the root mats of floating meadows in the Amazonian region, with peaks up to about 20 $Kcal/m^2$.

The surface of submerged parts of macrophytes is colonized by a larger variety of microfaunal elements than represented in the zooplankton. In temperate freshwater the fauna of the periphyton community ('Aufwuchs') comprises Ciliata, Porifera, Hydroidea, Rotatoria, Nematoda, Oligochaeta, Bryozoa, Ostracoda, Copepoda, Cladocera and larvae of different insect groups, mainly chironomids. Species composition and ecology of this community has long been studied (e.g. Duplakov 1933; Meuche 1939). Qualitative composition and colonization density is dependent on several ecological conditions of which the duration of flooding in unstable water bodies and the life span of macrophytes serving as substrate are of basic importance. The colonization density of different submerged macrophytes (*Potamogeton*, *Myriophyllum, Elodea*) which was analysed by Bownik (1970) showed a maximum in autumn on plant species decaying during the winter (*Potamogeton*), and in autumn and spring on surviving macrophytes (*Myriophyllum*, *Elodea*). The main components colonizing submerged macrophytes are Nematoda, Oligochaeta and Chironomidae. Sessile forms such as Spongia and Bryozoa are restricted to stalks of emergent plants that have been standing for several years. Many papers dealing with periphyton colonization discuss the problem of whether the chemical and mechanical properties of the substratum influence the composition of the periphyton community. Experiments performed by Pieczynska & Spodniewska (1963) using different natural and artificial substrates revealed no significant differences in the composition and dominance structure of at least periphyton animals (Nematoda, Oligochaeta, Chironomidae). These findings support the validity of using glass plates for investigations of periphyton production. Considerable differences have been found, however, on substrates of the same kind in different lakes and even in different sites of the same lake. This was also confirmed by investigations of periphyton on emergents in several Mazurian lakes (Pieczynska 1970). Trophic conditions seem to be the main factor controlling spatial distribution and seasonal variation, which may be considerable. Pieczynska (1970) demonstrated that 3 trophic types of periphyton can replace one another in different sites and at different times: (1) an autotrophic type with less than 10 per cent animal biomass, (2) an autotrophic-heterotrophic type with approximately the same animal and algal biomass and at least partly dependent on detritus, (3) a heterotrophic type with more than 90 per cent of animals, often consisting of monospecific masses of Spongia or Bryozoa. The heterotrophic type, however, prevails during autumn and winter. Also the abundance peaks found during this season by Bownik (1970) on submerged plants are partly attributed to the supply of decaying plant material colonized by bacteria – detrital complexes, which seem to represent the main nutrition of the periphytic microfauna.

Figures of zooperiphyton abundance and biomass present the same problems regarding comparability as the phytophilous macrofauna. The surface area of solid substrata is the relevant dimension governing spatial limitation of periphyton development. Values for surface area units are more easily obtainable from emergents and are more suitable for comparisons than those involving weight units as used for submerged macrophytes of which the surface is often difficult to determine.

References

Amoros, C. 1973. Evolution des populations de cladocères et copépodes dans trois étangs piscicoles de la Dombes.–Ann. limnol. 9, 2: 135–155.

Bownik, L. J. 1970. The periphyton of the submerged macrophytes of Mikotajskie lake.– Ekol. Pol. 18, 24: 503–520.

Dejoux, C. & Saint-Jean, L. 1972. Etudes des communautés invertebrés d'herbiers du lac Tchad: recherches préliminaries.– Cah. Orstrom sér Hydrobiol. 6, 1: 67–83.
Duplakov, S. N. 1933. Materialy k izuceni ju perifitona (Materialien zur Erforschung des Periphytons).– Trud. Limnol. Sta Kosino 16: 5–160.
Gliwicz, Z. M. & Rybak, J. 1976. Selected problems of lake littoral ecology.
Junk, W. 1973. Investigations on the ecology and production-biology of the floating meadows in the Middle Amazon: II. The Aquatic fauna in the root zone of floating vegetation.– Amazoniana IV, 1: 9–102.
Langhans, V. 1911. Untersuchungen über die Fauna des Hirschberger Großteichs.–Monogr. Int. Rev. Hydrobiol. 3: 101 pp.
Lipin, A. N. & Lipina, N. N. 1950. (Macroflora of standing waters and its relations to aquatic animals).– Trudy vseroch. nautshno-issledov. Inst. prud. rybn. chozja. 5: 1–270.
Losos, B. & Hetesa, J. 1971. Hydrobiol. Studies on the Lednicke Rubniky ponds.– Acta Sci. Nat. Brno 5, 10: 1–54.
Meuche, A. 1939. Die Fauna im Algenbewuchs. Nach Untersuchungen im Litoral ostholsteinischer Seen.– Arch. Hydrobiol. 34: 349–520.
Mortuchai-Boltovskoi, F. D., et al. 1958. (Fauna of the littoral region of the Rybinskoe reservoir).– Trudy biol. Stancii Borok 3: 142–194.
Pieczyńska, E. 1970. Periphyton in the trophic structure of freshwater ecosystems.– Pol. Arch. Hydrobiol. 17, 1/2: 141–147.
Pieczynska, E. & Spodniewska, I. 1963. Occurrence and colonization of periphyton organisms in accordance with the type of substrate.– Ekologia Polska ser A, 11, 22: 533–545.
Postolkova, M. 1967. Comparison of the zooplankton amount and primary production of the fenced and unfenced littoral regions of Smyslov pond.– In: Contributions to the productivity of the littoral region of ponds and pools.– Rozpr. CSAV, R. Mat. Prir. 77, 11: 63–79.
Reinisch, F. K. 1925. Die Entomostrakenfauna in ihrer Beziehung zur Makroflora der Teiche.– Arch. Hydrobiol. 15: 253–279.
Rudescu, L., Niculescu, C., Chivu, I. P. 1965. Monographia stufului din Delta Dunarii.– Editura Academiei Republicii Socialiste Romania, 542 pp.
Sládeček, V. 1973. A swarm of zooplankton in the Nesyt fishpond. In: Littoral of the Nesyt Fishpond. Academia Praha, 71–81.
Smyly, W. J. P. 1952. The entomostraca of the weeds of a moorland pond.– J. Anim. Ecol. 21: 1–11.
Straskraba, M. 1963. Share of the littoral region in the productivity of two fishponds in Southern Bohemia.– Rozpravy CSAV, rada Mov 73, 13: 3–58.
Straskraba, M. 1965. The effect of fish on the number of invertebrates in ponds and streams.– Mitt. Int. Verein. Limnol. 13: 106–127.
Straskraba, M. 1967. Quantitative study on the littoral zooplankton of the Poltruba backwater with an attempt to disclose the effect of fish. In: Contributions to the productivity of the littoral region of ponds and pools.– Rozpravy CSAV, R. Mat. Prir. 77, 11: 7–35.
Vladimirova, T. M. 1963. (Alterations of zooplankton and zoobenthos of Zemtchusnoe Lake treated with polychloryphen).– Izv. gosud. nautchno-issledov. inst. ozern. i rechnogo rybnogo chozj. 55: 70–83.
Zimbalevskaja, I. N. 1972. Zooplankton in water vegetation thickets and its role in productivity of the reservoir shallow waters.– In: Tseeb, Ya. Ya. & Maistrenko, Yu. G. (Ed.) The Kiev Reservoir – Hydrochemistry, Biology, Productivity. Naukova Dumka Publ., Kiev. 1972.

27 The Rotifers of Neusiedlersee

J. Donner

Most of the data so far available concerning the occurrence of rotifers in Neusiedlersee result from observations made by Varga (1926, 1934) (see Chapter 22). The present author's own investigations to be described in this chapter, have yielded considerable additional information on the faunistics and distribution of rotifers (Donner 1953, 1959, 1968, 1972, 1975). The collections so far made are listed in Tables 1 and 2 and comprise 163 taxa, including 23 bdelloids.

27.1 Discussion

The species *Paradicranophorus sordidus* and *P. sudzukii* are of limited distribution in Neusiedlersee. They inhabit the benthos and are incapable of swimming upwards. In fact ciliary movements occasionally cease altogether in *P. sudzukii.* The halophilic species *Rhinoglena fertöensis* (Varga), on the other hand, has been found in vast quantities by Althaus in the freshwater lake near Eisleben (1956) (See Chapter 22), and earlier reported in Germany by Remane in 1934/36 (Althaus 1956). Concerning her find, the author comments: 'The question as to whether this is in fact Varga's *Rh. fertöensis* cannot be answered with certainty.' The smooth-toed *Cephalodella stenroosi*, described from Neusiedlersee (Donner 1972), has also been encounted in an old arm of the Danube near Wallsee, Austria, and given the name *var. austriaca* (Donner, 1978). *Philodina amethystina* was found between ground mosses in Bohemia by Bartoš (1951), and has so far only been reported from Neusiedlersee. *Collotheca bilfingeri* Berzins is known from Württemberg (Bilfinger 1894, named *Floscularia calva*), from Neusiedlersee and from northern Germany (1971). The only two sites from which *Lepadella nympha* has been reported are the mud of a pond in southern Moravia (Donner 1943) and Neusiedlersee. *Colurella oblonga glabra* was described by Klement (1959) from a small ephemeral body of water in Württemberg, and has also been found in a gorge near the Salzach (Donner 1970, 1975) whereas the type form was discovered in very shallow ponds in South Moravia (Donner 1943).

Neusiedlersee is apparently the only place in Europe where *Squatinella rostrum var. myersi* Voigt is to be encountered. It has been described and drawn from American material by Voigt (1957).

The following can be listed as rarely-occurring species: *Aspelta psitta*, *Cephalodella tinca*, *Collotheca cyclops*, *Encentrum nesites*, *Eosphora anthadis*, *Floscularia pedunculata*, *Mniobia armata*, *Rotaria laticeps*, *Habrotrocha praelonga*, *Keratella tropica*, *Pleurotrocha* (*Notommata*) *trypeta.* Others, although widespread, invariably occur as isolated specimens: *Adineta oculata*,

Table 27.1. Utricularia (in the Phragmites belt)

Symbols used in Table 1 and 2:
1 – single record
v – rare
+ – abundant
ö – frequently
h – more frequent
sh – very abundant
m – mass occurrence

		acc. to VARGA 1926, 1934	offshore 10.5.1970	offshore 24.8.1970	offshore 7.12.1970	offshore 18.5.1971	offshore 28.7.1971	offshore 26.9.1967	offshore VII. 1971	inshore 18.5.1971	inshore 28.9.1971
Brachionus	*quadrident. v. splendid.* Don.			h			v		v		
	quadrident. v. brevisp. (Ehr.)	+		v							
Cephalodella	● *auriculata* (O. F. Müller)		sh			1	v				
	● *catellina* (O. F. Müller)		1				h				sh
	delicata Wulfert				v						
	● *exigua* (Goose)		ö								
	● *forficata* (Ehrenberg)		1			v					
	● *forficula* (Ehrenberg)		1			v	h		v		
	● *gibba* (Ehrenberg)		ö	v	v	ö	v	+			v
	globata (Gosse)		v		ö						
	gobio Wulfert			v							
	● *gracilis* (Ehrenberg)		v		v						
	hoodi (Gosse)	+	v	1		v					
	● *megalocephala* (Glascott)		ö	v			+				+
	● *misgurnus* Wulfert			v		1	+				
	rigida Donner			v							
	sterea (Gosse)		m	h	v	ö	sh	+	v		h
	● *tinca* (Wulfert)						1				
	ventripes (Dixon-Nuttall)		m	m	v	ö	v	+	1		

Collotheca	●	*algicola algicola* (Hudson)		v	v				+		
		bilfingeri Berzins				+		1			
		campanulata (Dobie)						ö			
		cyclops (Cubitt)				v					
		ornata ornata (Ehrenb.)	+	v		v	ö	+	+		
		ornata v. cornuta (Dobie)	+	1	h	v	ö	v			
		trilobata (Collins)			v			1			
Colurella	●	*adriatica* Ehrenberg		h			h	v		v	
	●	*colurus* (Ehrenberg)					1				
	●	*obtusa* (Gosse)		h	ö	v	h	ö	+	v	
	●	*uncinata uncinata* (O. F. Müller)				ö	sh	sh		sh	
	●	*uncinata f. bicuspid.* (Ehrenb.)		ö	sh		1				
	●	*uncinata f. deflexa* (Ehrenb.)			1				+		
		sp.							+		
Dicranophorus		*caudatus* (Ehrenberg)						+			
	●	*forcipatus* (O. F. Müller)		v			1		+		
Encentrum	●	*diglandula* (Zawadowski)						+			
		mustela (Milne)	+			+					
		nesites Harring et Myers				v	ö	h	+		
Eosphora	●	*anthadis* Harring et Myers			ö						
		najas Ehrenberg			ö	+	h	v			v
Euchlanis		*calpidia* (Myers)					ö				
	●	*dilatata* Ehrenb. with Keel						v		v	
	●	*dilatata* Ehrenb. without Keel					sh	h		ö	
		incisa Carlin			h		h				
	●	*oropha* Gosse		h	m				+		
Floscularia		*melicerta* (Ehrenb.)		h			sh				
		pedunculata (Joliet)					v				
	●	*ringens* (Linné)		h	1		m	+			
Itura	●	*aurita* (Ehrenberg)		v	1			+			
		myersi Wulfert						+			
		viridis (Stenroos)					1				
Keratella	●	*quadrata quadrata* (O.F.M.)		1							
	●	*quadrata f. valgoid.* EDM et HU.			1						

		acc. to VARGA 1926, 1934	offshore							inshore	
			10.5.1970	24.8.1970	7.12.1970	18.5.1971	28.7.1971	26.9.1967	VII. 1971	18.5.1971	28.9.1971
Lecane (L.)	*inermis* (Bryce)			v							
	● *luna* (O. F. Müller)		1	sh		1	sh	+	h		
	ungulata (Gosse)	+						+			
(M.)	● *bulla* (Gosse)		v	sh		1	v	+	v		v
	● *closterocerca* (Schmarda)		h	m	v	h	h	+	h		sh
	● *goniata* (Harring et Myers)			+							ö
	● *perplexa* (Ahlstrom)							+			
Lepadella	● *patella patella* (O. F. Müll.)		v	sh	v	ö	sh	+	ö		h
	● *(patella) persimilis* (Lucks)		sh	h	ö	ö	ö		ö		sh
	● *triptera* Ehrenberg		v	sh	+	ö	sh	+	ö		sh
Limnias	*ceratophylli* Schrank						sh				
Lindia	*torulosa* Dujardin			v							
Lophocharis	● *oxysternon* (Gosse)		v	sh	m		+		sh	v	sh
	● *salpina* (Ehrenberg)				ö						
Microcodides	*robustus* (Glascott)					+		+			
Monommata	● *dentata* Wulfert		ö	v	v	1		+			
Mytilina	● *mucronata* (O. F. Müller)		v	h	1	1	v		v		
	ventralis v. brevispina Ehr.	+	1	v			v				+
Notholca	● *acuminata* (Ehrenberg)		ö		ö	ö					
	● *squamula* (O. F. Müller)				+						
Notommata	*contorta* (Stokes)			1				+			
	copeus Ehrenberg	+		h			v				
	● *cyrtopus* Gosse			ö							
	glyphura Wulfert		1	v	+						
Pleurotrocha	● *petromyzon* Ehrenb.		1	ö	+	v					ö
Polyarthra	● *dolichoptera* Idelson								ö		
Proales	*alba* Wulfert						+				
	● *fallaciosa* Wulfert		v	ö	+		+				

Ptygura		*beauchampi* Edmondson	sh	sh	+	ö	h				
		brachiata (Hudson)					v				
		longicornis (Davis)		v							
		melicerta (Ehrenberg)	ö	h		ö					
Resticula		*melandocus* (Gosse)	v			+			1		
Scaridium	●	*longicaudum* (O. F. Müller)		v							
Squatinella		*rostrum rostrum* (Schm.)	v	1				+			h
Synchaeta	●	*oblonga* Ehrenberg	v								
	●	*tremula* (O. F. Müller)	ö		v						
Testudinella		*emarginula* (Stenroos)		+	v	ö	v				
	●	*patina* (Hermann)	v	ö	v	ö	sh		v		sh
Trichocera (T.)	●	*rattus* (O. F. Müller)	v	v	v	ö			1		+
Trichotria	●	*pocillum* (O. F. Müller)	h	sh	ö	h	v		v		
Bdelloidea											
Dissotrocha		*aculeata* (Ehrenberg)		v		v					
		aculeata medio-acul. (Jans.)			+						
Habrotrocha	●	*bidens* (Gosse)	1								
		praelonga de Koning			+						
Macrotrachela		*insolita* de Koning			+						
Mniobia		*armata* (Murray)					v				
Otostephanos		*donneri* Bartos		v	+	v		+			v
		monteti Milne			+						
Philodina		*acuticornis odiosa* Milne	ö	v		1		+			
	●	*megalotrocha* Ehrenberg	v	sh	v	ö	sh				sh
Pleuretra		*brycei* (Weber)							1		
Rotaria		*citrina* (Ehrenberg)	ö	sh		ö	ö				
		laticeps Wulfert			+						
		macrocera (Gosse)	ö	1		1	v				
		rotatoria (Pallas)	1	v	ö	v		+		m	
		socialis Kellicot			+						
		tardigrada (Ehrenberg)				v					
		tridens (Montet)			+	v					
Bdelloid		*g. sp.*							v		

Table 27.2. Phragmites belt

		V. 1970–IX. 1971, Neus. Pharagmites	III. 1971–V. 1972, Rust, Pharagmites	DON. 1972, near Pharagmites, W u. N	DON. 1972, Seemitte (midlake station)	DON. 1972, Ostufer (eastern shore)	DON. 1953. 1972, 1975	acc. to VARGA 1926, 1934
Anuraeopsis	*fissa* (Gosse)		+					+
Aspelta	*psitta* Harring et Myers			+				
Asplanchna	*girodi* de Guerne			+				
Brachionus	*angularis* Gosse		+					+
	calyciflorus v. dorcas (Gosse)			+				+
	calyciflorus v. pala (Ehrenb.)		+					+
	leydigii v. trident. tripart. (Leissl.)			+	+	+		
Cephalodella	*auriculata* (O. F. Müller)	+						+
	catellina (O. F. Müller)	+	+					
	catellina f. major (Zawadowski)			+				
	exigua (Gosse)	+						+
	forficata (Ehrenberg)	+	+					+
	forficula (Ehrenberg)			+	+			+
	gibba (Ehrenberg)	+	+		+			+
	gracilis (Ehrenberg)	+			+			+
	megalocephala Glascott	+						
	misgurnus Wulfert	+						
	stenroosi Wulfert			+				
	stenroosi smooth toed			+	+			
	sterea (Gosse)	+	+					
	tenuiseta (Burn)						+	

Collotheca	*algicola algicola* (Hudson)						+	
Colurella	*adriatica* Ehrenberg	+	+	+	+	+		+
	colurus (Ehrenberg)	+		+	+			+
	oblonga v. glabra Klement		+					
	obtusa (Glosse)		+					
	uncinata uncin. (O. F. Müller)	+	+					+
	uncinata f. bicusp. (Ehrenb.)	+	+					+
	uncinata f. deflexa (Ehr.)		+					+
Dicranophorus	*forcipatus* (O. F. Müller)		+	+				
	uncinatus (Milne)			+	+			
Encentrum	*diglandula* (Zawadowski)			+				
	incisum Wulfert	+						
	putorius putorius Wulfert		+	+				
	sp. B			+				
Eosphora	*anthadis* Harring et Myers		+					
	najas Ehrenberg	+	+					+
Euchlanis	*deflexa* Gosse		+					
	dilatata Ehrenberg		+					+
	oropha Gosse		+					+
	triquetra Ehrenberg		+					+
Filinia	*longiseta* (Ehrenberg)		+					+
Floscularia	*ringens* (Linne)		+					
Itura	*aurita* (Ehrenberg)			+				+
Keratella	*cochlearis* (Gosse)	+	+	+	+	+		+
	quadrata quadrata (O. F. Müller)	+	+	+	+	+		+
	quadrata f. valgoides Edm. et Hutch.			+				
	tropica f. asymmetrica Barr. et Dad.		+					
	tropica f. monstrosa Barr. et Dad.		+					
	stenroosi (Meissner)			+				
Lepadella	*acuminata* (Ehrenberg)		+					
	nympha Donner		+					
	ovalis (O. F. Müller)		+					
	patella patella (O. F. Müller)		+					
	(patella) persimilis (Lucks)	+	+					
	quadricarinata (Stenroos)		+					
	triptera Ehrenberg		+					+

		V. 1970–IX. 1971, Neus. Pharagmites	III. 1971–V. 1972, Rust, Pharagmites	DON. 1972, near Pharagmites, W u. N	DON. 1972, Seemitte (midlake station)	DON. 1972, Ostufer (eastern shore)	DON. 1953, 1972, 1975	acc. to VARGA 1926, 1934
Lophocharis	*oxysternon* (Gosse)	+	+					+
	salpina (Ehrenberg)		+					
Monommata	*dentata* Wulfert		+					
	sp.		+					
Mytilina	*mucronata* (O. F. Müller)		+					+
	trigona (Goose)		+					+
Notholca	*acuminata* (Ehrenberg)	+	+					+
	squamula (O. F. Müller)		+	+	+	+		+
Notommata	*cyrtopus* Gosse			+				
Paradicrano-phorus	*sordidus* Donner			+				
	sudzukii Donner			+	+			
Parencentrum	*saundersiae* (Hudson)			+				
Platyias	*quadricornis* (Ehrenberg)		+					
Pleurotrocha	*petromyzon* Ehrenberg		+		+			+
	(Notommata) trypeta Harr. et M.		+					
Polyarthra	*dolichoptera dol.* Idelson		+					
	dolichoptera dissim. Nipkow		+					
	vulgaris Carlin		+					
Proales	*fallaciosa* Wulfert		+					
Ptygura	*sp.*	+						
Resticula	*nyssa* Harring et Myers			+				

Squatinella	*rostrum rostr.* (Schmarda)		+					+
	rostrum v. myersi Voigt		+					
	tridentata v. mutica (Ehrenberg)		+					
Lecane (L)	*agilis* or *inermis*			+				
	flexilis (Gosse)		+					
	luna (O. F. Müller)		+	+	+			+
	nana (Murray)		+					
(M)	*bulla* (Gosse)	+	+					+
	closterocera (Schmards)	+	+					+
	goniata (Harring et Myers)		+					
	hamata (Stokes)		+					
	harringi (Ahlstrom)		+					
	perplexa (Ahlstrom)					+		
Synchaeta	*oblonga* Ehrenberg	+	+	+		+		+
	pectinata Ehrenberg		+					+
	tremula (O. F. Müller)		?	+				+
Testudinella	*patina* (Hermann)	+	+					+
Trichocerca (T.)	*rattus* (O. F. Müller)		+					+
(D.)	*ruttneri* Donner						+	
(D.)	*weberi (Jennings)*		+					+
Trichotria	*pocillum* (O. F. Müller)		+	+				+
Bdelloidea								
Adineta	*oculata* (Milne)	+						
Dissotrocha	*aculeata* (Ehrenberg)		+					+
Habrotrocha	*bidens* (Gosse)	+						
Philodina	*amethystina* Bartos		+					
	megalotrocha Ehrenberg	+		+				
Philodinavus	*paradoxus* (Murray)	+						
Rotaria	*citrina* (Ehrenberg)	+						+
	cf. curtipes Murray			+				
	laticeps Wulfert			+				
	neptunia (Ehrenberg)			+				+
	rotatoria (Pallas)	+		+				+
	rotatoria with diverging spurs			+				
	tardigrada (Ehrenberg)	+		+				
	tridens Montet			+				
Bdelloid	*g. sp.*	+	+	+				

Proales alba, *Collotheca trilobata*, *Itura myersi*. They are also found in Neusiedlersee. *Rotaria laticeps* and *Lecane* (L.) *nana* are saltwater forms, the latter also listed by Althaus. *Mytilina trigona* lives on mud.

Certain peculiarities in dominance can be observed, such as, for example, the predominance of *Trichocerca rattus* and *Lophocharis oxysternon*, both occasionally occurring in large numbers near Rust. *Philodinavus paradoxus* seems to be an outsider in this vicinity. Just as with crustaceans, there appear to be very few sodabiontic species amongst the rotifers. Of the latter, *Hexarthra polydonta* is probably the most frequently encountered and is found in the open water portion of Neusiedlersee. It does not occur in the reed belt.

By far the majority of rotifers in Neusiedlersee are eurytopic. Only a small group, not to be found in similar or even in neighbouring bodies of water, lends a certain degree of uniqueness to the lake. This is confirmed by the results of other authors quoted above.

References

Althaus, B., 1956/57. Faunistisch-ökologische Studien an Rotatorien salzhaltiger Gewässer Mitteldeutschlands. Wiss. Z. Martin-Luth.– Univ. Halle-Wittenb., VI, 1: 117–157.

Bartoš, E., 1951. Korenonozci, Vířníci a Zelvusky mechu Sumavskych Predhorí. Cas. Nar. Mus., odd. Prírod., 50–69.

Bilfinger, L., 1894. Zur Rotatorienfauna Württembergs. 2. Beitrag. Jahr. Ver. vaterl. Naturk. Württ., 35–65.

Carlin, B., 1939. Über die Rotatorien einiger Seen bei Aneboda. Medd. Lunds Univ. Limnol. Inst., 2: 1–68.

Daday, E. von. 1894. Beiträge zur Kenntnis der Mikrofauna der Natrongewässer des Alföldes. Math. Naturw. Ber. Ungarn, 11: 286–321.

Donner, J., 1943. Zur Rotatorienfauna Südmährens (III). Zool. Anz. 143, 7/8: 172–179.

Donner, J., 1953. *Trichocerca* (*Diurella*) *ruttneri* nov. spec., ein Rädertier aus Insulinde, Indien und dem Neusiedlersee. Österr. Zool. Z. IV, 1/2: 19–22.

Donner, J., 1959. Bemerkungen zur Rädertierart *Synchaeta oblonga* Ehrenb. Verh. Zool. Bot. Ges. Wien, 98 und 99: 26–30.

Donner, J., 1968. Zwei neue Schlamm-Rotatorien aus dem Neusidler See, *Paradicranophorus sudzukii* und *Paradicranophorus sordidus*. Anz. math. nat. Kl. Österr. Akad. Wiss., 10: 224–232.

Donner, J., 1970. Die Rädertierbestände submerser Moose der Salzach und anderer Wasser-Biotope des Flußgebietes. Arch. Hydrob./Suppl. XXXVI (Donauforschung IV), 2/3: 109–254.

Donner, J., 1972. Rädertiere der Grenzschicht Wasser – Sediment aus dem Neusiedler See. Sitz. Ber. Österr. Akad. Wiss. math. nat. Kl., Abt. I, 180, 1–4: 49–63.

Donner, J., 1975. Seltene und auffallende seessile und notommatide Rotatorien aud dem Schilfgürtel des Neusiedler Sees. ibid. 183, 4–7: 131–148.

Donner, J., 1978. Material zur saprobiologischen Beurteilung mehrerer gewässer des Donau-Systemes bei Wallsee und in der Lobau, Österreich, mit besonderer Berücksichtigung der litoralen Rotatorien Arch. Hydrobiol. Suppl. 52 (Donauforschung 6) 2/3, 117–228.

Kertész, G., 1960. Die Rotatorien des Péteri-Sees. Annal. Univ. Sci. Budap. de Rol. Eötv. nom., Sectio biol., 3: 243–251.

Klement, V., 1959. Zur Rotatorienfauna des Monrepos-Teiches bei Ludwigsburg. 2. Beitrag. Jh. Ver. vaterl. Naturk. Württemberg, 114: 193–221.

Koste, W., 1971. Das Rädertier-Porträt. Die Rädertiergattung *Collotheca* – Mitteleuropäische Arten mit besonders auffallenden Koronalfortsätzen. Mikrokosm. Heft 6 Juni, 161–167.

Löffler, H., 1957. Vergleichende limnologische Untersuchungen an den Gewässern des Seewin-

kels (Burgenland). Verh. Zoo. Bot. Ges. Wien, 97: 27–52.
Löffler, H., 1959. Entomostraken- und Rotatorienfauna der Seewinkel-Gewässer. Landsch. Neusiedlersee, Wiss. Arb. Bgld. 23: 138–139.
Nógrádi, T., 1957. Beiträge zur Limnologie und Rädertierfauna Ungarischer Natrongewässer. Hydrobiologia IX. 4: 348–360.
Pejler, B., 1962. On the Taxonomy and Ecology of Benthic and Periphytic Rotatoria. Zool. Bidr. Uppsala 33: 327–422.
Varga, L., 1926. Die Rotatorien des Fertö (Neusiedler See). Arch. Balaton I, 181–225.
Varga, L., 1934. Neue Beiträge zur Kenntnis der Rotatorien-Fauna des Neusiedler Sees. Allat. Közlem. Budapest 31: 139–150.
Voigt, M., 1956/57. Rotatoria. Die Rädertiere Mitteleuropas. Bd. I und II.

28 Fishes and fishery in Neusiedlersee

R. Hacker

28.1 Introduction

The earliest known records concerning fish and fishery in Neusiedlersee date back to the 16th century. Apart from a few publications concerning identification (e.g. Heckel & Kner 1858) most reports are concerned purely with faunistics, occasionally supplemented by details on the importance, maximum recorded lengths and weights of species of commercial interest e.g. Peteny (MS.), Roditzky, Pernt (1894), Haempel (1926, 1927), Mika & Breuer (1928), Blöch (1941), Benda (1950), Herman (1887) etc. (further bibliography see Aumüller 1965, and Láslóffy 1972).

Most nonfaunistic literature from the 20th century is devoted to growth investigations. Quantitative information, as for example with regard to annual fishery yields, is altogether rare, e.g. Varga (1932), Varga & Mika (1937), Stundl (1947), Bruschek (1971), Liepolt (1972), Sauerzopf & Hofbauer (1959), and in many cases incomplete and unrealistic (see also Löffler, 1974).

The first attempt to elucidate biological relationships was made by Geyer & Mann (1939), who published data on the food and growth of fish in the Hungarian part of the lake. These authors even conjectured as to the existence of regional peculiarities in the feeding habits of three species. This idea has been confirmed for pope (*Acerina cernua*) and white bream (*Blicca björkna*) in the northern part of the lake. Nevertheless, the significance of Geyer & Mann's work should not be overestimated since their information was obtained from a single investigation in October 1938 involving a very limited number of fish.[1]

During the 1950's two more detailed investigations were made on the biology (e.g. spawning habits) and growth of the two species at that time commercially most important: Nawratil (1953) compared growth rates of northern pike (*Esox lucius*) from Neusiedlersee and Attersee in Upper Austria. Unterüberbacher (1958) did the same for domestic common carp (*Cyprinus carpio*) and its 'wild form'[2] within Neusiedlersee.

Towards the end of the International Biological Programme (1968–1972) investigations largely dealing with ecological aspects were carried out (Meisriemler 1974, Hacker, 1974). These activities are being continued on a limited scale in the UNESCO-MaB-programme (Man and Biosphere).

1. Gut analysis from 156 specimens involving 9 species; growth investigations on 112 specimens involving 10 species.

2. After Berg (1941): *Cyprinus carpio morpha hungaricus* = syn. *C. hungaricus* Heckel (1836), Heckel & Kner (1858) = syn. *C. c. typus hungaricus* Unger = *C. carpio* L.

28.2 Faunistics and stocking

An outline of faunistic studies on Neusiedlersee is given in Table 28.1. The figures at the end of each column indicate the total number of species mentioned by the different authors. The relatively wide range of species should not be misinterpreted as being indicative of actual changes in species compositions. Conversely, similar numbers should not be mistakenly held to indicate a 'static' situation. For a variety of reasons the numbers involved are only comparable to a very limited extent: thus many authors have not clearly defined the area to which their lists apply, and several investigators e.g. Sauerzopf & Hofbauer (1959), Mika (1962), included the ichthyofauna of Neusiedlersee affluents and of the Seewinkel pans. Uncritical acceptance of such lists has led to misinterpretations by later workers. Secondly, contrary to claims made in later publications, some authors have in fact never given a separate faunal list for Neusiedlersee. Thus Heckel & Kner's book was originally designed as an identification key and a catalogue of the fish of the former Austrian Empire, so that remarks on distribution of individual species apply to a much wider territory than Neusiedlersee alone. The lake is only mentioned in connection with the origin of material used for morphological and taxonomical comparisons. Similarly, Neusiedlersee is only mentioned specifically in a few cases in Kähsbauer's (1961) article, which is in fact a catalogue of the ichthyofauna of the whole of Austria.

One of the few cases in which a comparison between the results of two investigations would appear to be acceptable involves the studies of Mika & Breuer (1928), and Varga & Mika (1937). The latter paper reports the effects of an extensive catastrophe involving the lake in winter 1928/29. At this time the water depth was below 0.6 m and most of the lake froze down to the bottom (Varga 1932). The most remarkable result of Varga & Mika's comparison is that even such drastic events as those described above seemed to affect species composition quantitatively rather than qualitatively. Thus Mika & Breuer (1928) found 23 species permanently inhabiting the lake and Varga & Mika (1937) reported 24. The annual fishery yield in the Hungarian part of the lake, however, dropped from 80–100 metric tons (1924–1929) to 4–50 t (1929–1935).

Nothing more than speculation can be offered regarding the recovery of the lake by means of natural restocking; some suggest that at least part of the recovery process had its source in the tributaries and artificial outflow: thus Varga & Mika (1937) report on chub (*Squalius cephalus*) – according to Huet (1962) a running water species – as 'a new and quite common faunal component'. They also record that, in 1931, about 4–5 t of common carp, i.e. approx. 50 per cent of the carp harvest, or approx. 17 percent of the total fishery yield in the Hungarian part of the lake, entered through the Hanság canal. As a result of exceptional floods, Neusiedlersee was supplied at that time with water from the Rivers Danube, Raba, Repce, and Ikva via the Hanság canal, which is normally the only (artificial) outflow from the lake. It is very likely that at irregular intervals several species of fish entered the lake by this route (also later) e.g. tubenose goby (*Proterorhinus marmoratus;* see Mika & Varga 1940, Bauer & Schubert 1957), or razor fish (*Pelecus cultratus*).

Human activities, so far unchecked, including voluntary and accidental stocking with new species, both in historical and modern times, are in fact responsible for a substantial part of the qualitative and quantitative change in the Neusiedlersee ichthyofauna. Serious attempts of control and management, although still not organized on a scientific basis, were initiated in 1975. Hungarian fishery officials devised a plan for stocking the lake with species of commercial interest. In 1975 an international treaty, binding for both Hungarian and Austrian fishery, was drawn up for the implementation of the above plan, as well as for coordinating e.g. size limits, quantity and type of fishing gear, observance of closed seasons (for species of interest in commercial fishery).

Apart from its positive aspects, the plan might, however, involve some severe disadvantages:

1. The likelihood of overstocking with species of current economic value.
2. Inconsiderate stocking with exotic species currently in fashion.
3. The introduction of pests and of new fish diseases can not be excluded.

Ad 1: The plan for stocking was basically designed for the Hungarian part of the lake and was extrapolated for equivalent units of area of the Austrian part (= approx. 4/5 of total area), without taking into account regional productivity peculiarities of the latter (see Chapter 23).

Ad. 2: During the last years two phytophagous exotic species – grass carp (*Ctenopharyngodon idella*) and silver carp (*Hypophthalmichthys molitrix*) – have been introduced deliberately into Neusiedlersee. The irresponsible introduction of foreign species affecting the primary production complex of an ecosystem is known from many examples to be extremely dangerous. So far no evaluation of the effects resulting from these activities, in particular the introduction of grass carp, can be given. The observed steady decline of macrophyte biomass (Schiemer & Prosser 1976) during recent years is not necessarily caused by grass carp only, since, according to Schiemer (1979, see Chapter 19) long-term fluctuations of macrophyte abundance are quite common in this lake. On the other hand, signs of feeding observed on *Phragmites communis* (Maier pers. commun.) suggest a shortage of soft plants such as *Myriophyllum spicatum* and *Potamogeton pectinatus* since these are known to be preferred by grass carp. The production of 30 t grass carp in 1976 would require a consumption of approx. 300 t of macrophytes (according to Fischer 1972, 1973). In view of the preference of soft plants a considerable reduction in the macrophyte belt could be expected. Gut analysis from 9 specimens caught in central lake regions have in fact shown: 7 fish with *P. pectinatus* exclusively, 2 fish with mixed diet mainly consisting of *Ph. communis* and a little *P. pectinatus* and *M. spicatum*. Soft plants in the reed belt (particularly *Utricularia vulgaris*) have also undergone a considerable reduction (see Chapter 21). Even a partial destruction of the macrophyte belt would bring in its wake a series of consequences for the

Table 28.1. Comparison of fishes inhabiting Neusiedlersee found by different authors.

	1	2	3	4	5	6	7	8	9	10	11	12	13	14
Salmonidae														
Salmo gairdneri Rich.									ø					
Esocidae														
Esox Lucius L.	+	o	o	o	o	o	o	o	o	A	o	o	o	o
Umbridae														
Umbra krameri Walb.	o_1	o	o	x	o	o	o	+		+			?	
Anguillidae														
Anguilla anguilla (L.)						x		o	o	o	o	o	o	o
Cyprinidae														
Abramis ballerus (L.)	o		o	o	o	o	o	o	o	o	o			
Abramis brama (L.)	o	o	o	o	o	o	o	o	o	A	o	o	o	o
Alburnus alburnus (L.)	o	o	o	o	o	o	o	o	o	A	o	o	o	o
Aspius aspius (L.)	o		o			o		o	o			o	o	o
Bliccca björkna (L.)	o	o	o			o		o	o		o		o	o
Barbus barbus (L.)						x		o	o	A			?	
Carassius carassius (L.)	o	o	o	o	o	o	o	o	o	A	o		o	o
C. (auratus) gibelio (Bloch)	o					o_2		o						
Ctenopharyngodon idella (Val.)														
Cyprinus carpio L.	o	o	o	o	o	o	o	o	o	A	o	o	o	o
Gobio gobio (L.)	o		o	o	o	o	o		o				?	
Hypophthalmichthys molitrix (Val.)														o
Idus idus (L.)						o		o	o			x	?	
Leucaspius delineatus (Heckel)						o		$+_{10}$	o			o	o	o
Pelecus cultratus (L.)								+	o	o		x	?	o
Rhodeus amarus sericeus Bloch				o	o	o	o	o	o				o	o
Rutilus rutilus (L.)	o	o	o	o	o	o	o	o	o	A			o	o
Scardinius erythrophthalmus (L.)	o	o	o		o	o	o	o	o	A		o	o	o
Squalius cephalus (L.)				o	o		o	o	o	A		x_3	?	o
Tinca tinca (L.)		o	o	o	o	o	o	o	o	A		o	o	o

Cobitis taenia L.		o	o			o		o	o	o		o	?	
Misgurungs fossilis (L.)	o	o	o		o	o	o	o	o	o		o	o	o
Nemachilus barbatulus (L.)								o	o				?	
Siluridae														
Silurus glanis L.		o	o			x+?		$+x_4$	o	A	$+_5$	x	o	o
Gobiidae														
Proterorhinus marmoratus (Pallas)						o		o	o	o	o	o	?	o
Percidae														
Acerina cernua (L.)		o	o	o		o		o	o		o_5	o	o	o
Lucioperca lucioperca (L.)		o		o	x	o	o	o	o	o	o	o	o	
Perca fluviatilis L.		o	o	o		o	o	o	o	A		o	o	o
Centrarchidae														
Lepomis gibbosus (L.)												o		o
Cottidae														
Cottus gobio (L.)									o					
Gadidae														
Lota lota (L.)		o	o			o		o	o	A		x	?	
Number of species	12	16	20	17	14	23_6	15	24	29	6_7	15	24	18_8	23_9

1 = Heckel + Kner 1958
2 = Herman 1887
3 = Vutskits 1896
4 = Seligo 1926
5 = Haempel 1926
6 = Mika & Breuer 1928
7 = Haempel 1929
8 = Varga & Mika 1937
9 = Sauerzopf & Hofbauer 1959
10 = Kähsbauer 1961
11 = Koenig 1964
12 = Kritscher 1973
13 = Hacker & Meisriemler, in Löffler 1974
14 = present status

+ = extinct
o = occurring
x = rare – not regular inhabitant
A = after Kähsbauer (1961), general for Austria
? = uncertain – no reference specimen
ø = stocking experiments in Seewinkel-pans
1 = only for surroundings of Neusiedlersee
2 = as a variety of Carassius carassius
3 = not to be found in the lake itself
4 = common before 1900
5 = before 1917
6 = 23 permanent inhabitant + 4 occasional imigrants
7 = 6 specially recorded for Neusiedlersee
8 = 18 certain + 10 uncertain
9 = 23 certain
o = before 1929

benthos. The resultant loss of spawning substrate would play a considerable role in the natural occurrence of several species of fish, e.g. pike perch (*Lucioperca lucioperca*) (Dmitrjev 1977).

A third species introduced regularly is the eel (*Anguilla anguilla*), which is not a natural inhabitant of the Danube river system.

28.3 Growth of Neusiedlersee fish

In view of the general lack of information concerning overall fish production, presentation and discussion of growth data may provide us with a few useful means of predicting biological changes (due to increased stocking or the introduction of exotic species). Growth patterns, both annual and in two cases seasonal, are therefore presented in the Table 28.2, 28.3, 28.4, 28.5.

The data of Geyer & Mann are derived from a single investigation made in October 1928. Geyer & Mann made the assumption that in October practically all fish had already reached a length corresponding to that at winter ring formation (= end of growing season), admitting the possibility of

Table 28.2. Annual growth of Neusiedlersee fish, Geyer & Mann (1939), (number in brackets indicates number of individuals examined)

species	total length (mm) at age classes				
	0(≃1)	I(≃2)	II(≃3)	III(≃4)	IV(≃5)
E. lucius		339 (23)	495 (2)		
A. brama		183 (4)	190 (3)		
A. alburnus	44 (33)	83 (70)	97 (21)		
B. björkna			153 (17)		
C. carassius		108 (15)	125 (4)		
C. carpio		173 (32)	259 (15)	331 (7)	369 (3)
R. rutilus	41 (3)	88 (3)	147 (1)	179 (29)	
S. erythrophthalmus		76 (33)	129 (39)	164 (13)	174 (5)
A. cernua		134 (4)			
P. fluviatilis		174 (24)	213 (2)		

Table 28.3. Annual growth of northern pike (*Esox lucius*) in Neusiedlersee (Navratil, O. 1952)

age	calculated mean of L_t (mm)	range of obtained L_t (mm)	calculated mean of W_t (g)
1	360	250 to 470	314
2	490	340 to 720	808
3	620	450 to 870	1,750

Table 28.4. Comparison of annual growth of two varieties of common carp (*Cyprinus carpio*) in Neusiedlersee: domesticated (scaled & mirror) carp – 'wild carp' (*C. c. t. hungaricus*), Unterüberbacher (1958)

total length (mm) at age	1	2	3	4	5	6
domesticated carp (both scaled and mirror)						
back calculated	132	239	351	421	480	516
measured	—	252	367	424	471	520
$W = 2.9 \cdot 10^{-5} \cdot L^{2.92}$						
'wild carp' (*C. c. t. hungaricus*)						
back calculated	125	313	448	523	578	—
measured	—	330	440	525	585	650
$W = 3.8 \cdot 10^{-5} \cdot L^{2.82}$						

slight but negligible length increments up to actual ring formation (a definable stage). This assumption is not generally acceptable, since at least the weight data cannot be regarded as representative of the actual stage, i.e. as being the weight at time of winter ring formation. Thus Figs. 28.1a,b and 28.2a,b show for example for two unexploited species, that considerable changes in length and especially in weight can occur between October and the end of February (in Neusiedlersee the time at which in most species more than 90 percent of populations have completed annulus formation). It is obvious that these assumptions, due to the rather small number of individuals examined, do not permit comparison of Geyer & Mann's work with other data.

The first reliable comparative information on annual growth rates was provided by Nawratil (1953) on northern pike (*Esox lucius*), and Unterüberbacher (1958) on common carp (*Cyprinus carpio*). Nawratil compared growth of pike from Neusiedlersee and from Attersee (Table 28.3). He showed that Neusiedlersee pike grew more slowly for the first three years, but tended to accelerate, and even 'overtake' Attersee pike with increasing age. Unterüberbacher gives a comparison of growth intensity for two varieties of common carp, 'domestic' carp (scaled and mirror) and 'wild' carp (*C. c. typus hungaricus*) (Table 28.4). In 1948 the lake was stocked with domestic carp for the first time; prior to this the only carp inhabiting Neusiedlersee was the

Table 28.5. Annual growth of some Neusiedlersee fish, present status, L_1–L_8 = total length (in mm) of fish at formation of winterring 1–8

species		L_1	L_2	L_3	L_4	L_5	L_6	L_7	L_8
Alburnus	mean	60	81	107					
alburnus[1]	s	10.6	12.8	12.07					
	n	290	199	33					
$W = 2.93 \cdot 10^{-6} \cdot L^{3.19}$		r = 0.91							
Blicca	mean	52	77	97	116				
björkna[2]	s	9.1	10.2	11.1	15.1				
	n	964	546	124	24				
$W = 4.2 \cdot 10^{-6} \cdot L^{3.2}$		r = 0.99							
Acerina	mean	48	71	86					
cernua[2]	s	10.2	11.3	12.1					
	n	570	137	57					
$W = 8.8 \cdot 10^{-6} \cdot L^{3.1}$		r = 0.995							
Anguilla	mean	124	207	305	403	493	561	631	713
anguilla[3]	s	20.1	29.9	46.4	53	51	49.4	56.4	96.7
	n	93	93	93	91	77	58	32	14
$W = 1.1 \cdot 10^{-5} \cdot L^{2.69}$		r = 0.99							
Pelecus	mean	147	233	281					
cultratus[2]	s	13.8	28.7	11.1					
	n	59	142	15					
$W = 2.38 \cdot 10^{-6} \cdot L^{2.33}$		r = 0.98							
Lucioperca	mean	83	191	275	358	420	477		
lucioperca	s	2.8		36.6	50.8	64.6	61		
	n								
$W = 5.8 \cdot 10^{-5} \cdot L^{3.05}$		r = 0.99							

s = standard deviation
n = number of fish examined
W = total weight of fish (in g) r = correlation coefficient
1. mean value of lengths actually measured from fish at time of ring formation
2. data from combining measured and backcalculated values using mainly scale readings plus calibration curves
3. data by backcalculation from otoliths exclusively

'wild' carp. Unterüberbacher's aim was to evaluate the growth intensity of the different forms with a view to further stocking. In fact he found that all domestic forms were superior, especially in weight increase per unit time, to the poorly growing 'wild' form. Consequently continuous stocking with domestic carp (almost all breeds available) is now practised.

Most recent data on annual growth rates are shown in Table 28.5 for the two species currently of most commercial importance, eel (*Anguilla anguilla*) and pike perch (*Lucioperca lucioperca*), as well as for food of their potential prey fish: bleak (*Alburnus alburnus*), white bream (*Blicca björkna*), ruff (*Acerina cernua*), and razor fish (*Pelecus cultratus*). Annual growth rates were

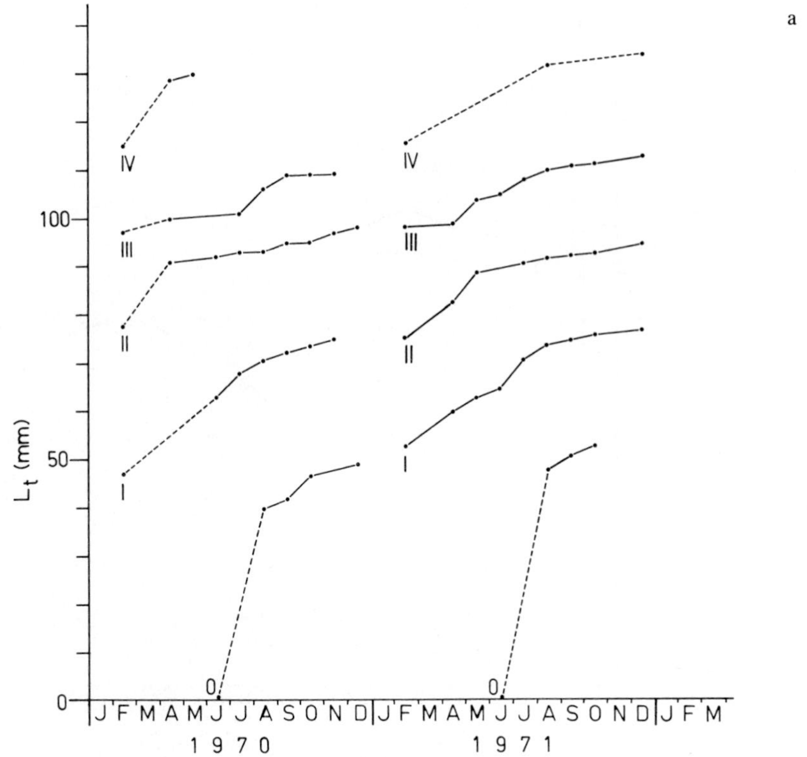

Fig. 28.1. Seasonal changes in length of two unexploited species in Neusiedlersee (a) for *Blicca björkna* (b) for *Acerina cernua* (acc. to Hacker 1974, Meisriemler 1974; I to IV = age classes)

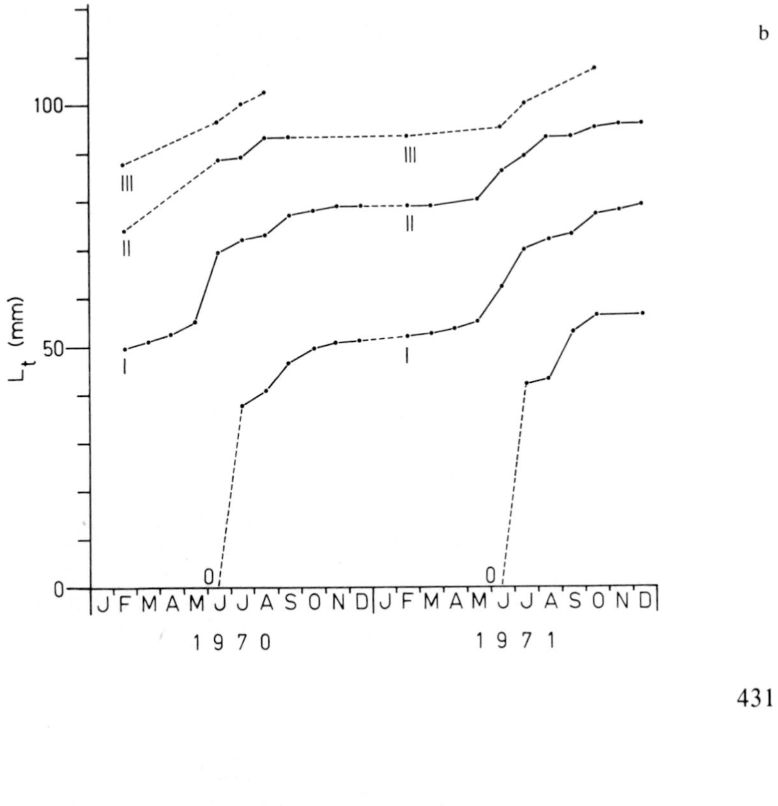

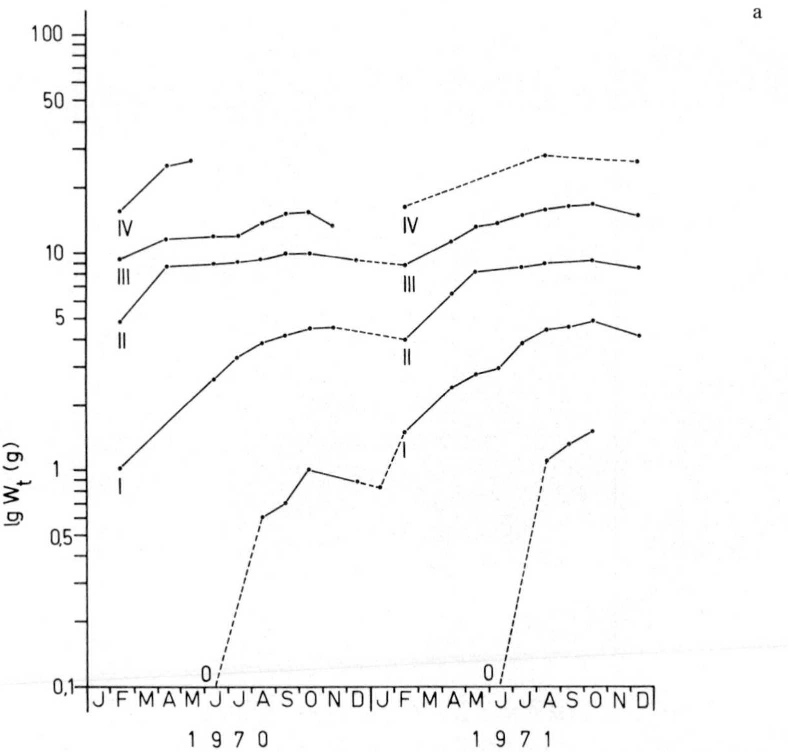

Fig. 28.2. Seasonal changes in weights of two unexploited species in Neusiedlersee (a) for *Blicca björkna* (b) for *Acerina cernua* (acc. to Hacker 1974, Meisriemler 1974; I to IV = age classes)

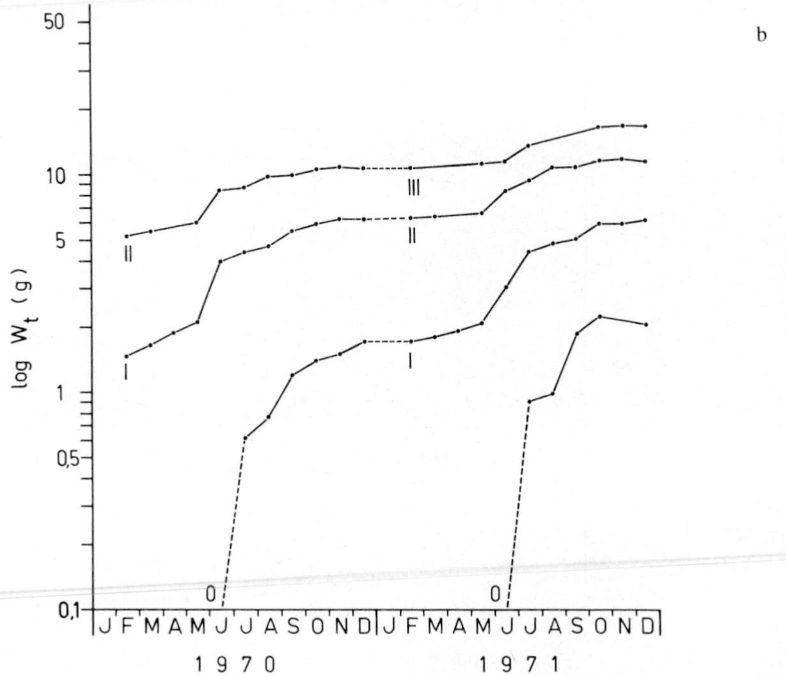

determined by calculation from otolith- and scale-readings and were also directly measured in specimens at time of annulus formation. In the case of scale reading, 'calibration curves' were constructed from actual scale sizes and body lengths (total length), and converted into units of weight by employing L/W-regressions parameters see Table 28.5).

Although earlier growth data as e.g. Geyer & Mann's are comparable to only a very limited extent, a long-term comparison within the lake appears to be justifiable for bleak, white bream, and ruff. Thus annual growth for bleak in 1938, assuming a seasonal growth pattern similar to that found in 1970 and 1971 for white bream (Fig. 28.1a and 28.2a), or ruff (Fig. 28.1b and 28.2b), does not appear to be very different from the present situation, whereas white bream and ruff now both exhibit much poorer annual growth: e.g. maximal lengths found in 1970–72 for ruff did not exceed the mean lengths of two-year-old animals in 1938 (Meisriemler 1974). The same applies to white bream, whose length for 1938, age class II was markedly above that for four-year-old 1971 specimens (Hacker 1974).

A comparison of growth data (Length) from various European waters is given for pike perch (Table 28.6) and European eel (Table 28.7), at present the two species of most commercial importance. It can be seen that pike perch in Neusiedlersee grows relatively poor, whereas eel is in the top European field (Hacker & Meisriemler 1978).

Since growth and feeding ecology of both species were studied extensively from 1970 to 1972 these growth data will be discussed in some detail in the following:

Sampling was done by purse seining and by trapping (mesh stretched 5 mm in both gears) at both regular (every fortnight or monthly) and irregular intervals at least at four sites, each representing one of the lake zones: site B – sheltered bay within the reed belt, site I – 'reed fringe', site II – 'macrophyte zone', and site III – 'open lake').

Sexual dimorphism in growth was found to be statistically not significant (except in white bream older than four years) nor could significant regional differences be detected. Monthly weighted mean values for lengths and weights presented in Fig. 28.1a,b and 28.2a,b were determined in two ways:

1. Lengths and weights of individuals of the 0-age class were measured.

2. For all older stages mean monthly length and weight data obtained from catches were compared, a. with mean monthly length increment calculated from scale readings, b. with weight increase derived from calculated length increment converted into weight (employing separate L/W-regressions for the monthly catch involved). Since no statistically significant differences could be found between calculated and measured values all 'starting'-values (indicating the onset of a growth season) may be taken as significant (except 0-age class).

For *Blicca björkna*, in age classes 0+ and 1+, neither of which was spawning, length increases relatively smoothly up to the end of summer; the growth curve then becomes increasingly flat towards autumn and reaches a plateau in winter.

Table 28.6. Annual growth of eel (*Anguilla anguilla*) in various European water bodies (L_1–L_{10} = total length in mm at time of formation of annulus[1] 1–10)

name of water body	L_1	L_2	L_3	L_4	L_5	L_6	L_7	L_8	L_9	L_{10}	author
Efrau stream (Wales)	160	195	225	262	295	330					
Rhyd-hir stream (Wales)	138	162	205	240	300	325					Sinha & Jones 1966, 1967
Severn river (England) ♂		135	155	175	250	270	280	320	340	390	
♀		135	155	175		270	290	320	370		
Gulf of Finland (eastern portion)					311	402	470	567	625	683	Berg 1949
Unterelbe, Alster (Germany)	95	120	150	195	260	330	380	440	520	580	Hohendorf 1966
Nordsee, Elbe (Germany)	94	155	232	264	300	337	386	439	528	548	Penáz & Tesch 1970
Niederelbe (Germany) ♂	90	120	150	190	250	310	350	390			Marcus & Smolian 1920
♀	90	120	150	190	260	340	390	450	610		
closed Northern Germany lakes		100 to 150	220	270	350 to 400	330 to 660	390 to 700	420 to 720	520 to 720	710 to 780	Wundsch 1949–51
Serventsee (Masuren)					420	470	490	490	510		Marcus & Smolian 1920
Neusiedlersee (Austria)	124	207	305	403	493	561	631	713			Hacker & Meisriemler 1978

1. only freshwater years

Table 28.7. Annual growth of pike perch (*Lucioperca lucioperca*) in various European water bodies (L_1–L_{12} = total length in mm at time of formation of annulus 1–12)

name of water body	L_1	L_2	L_3	L_4	L_5	L_6	L_7	L_8	L_9	L_{10}	L_{11}	L_{12}	author
mean value from Swedish, Finnish and Russian lakes							460	486	507	549	592	618	Nagieć 1961
Onega Lake (USSR)	128	194	257	323	382	430	478	514	551	586	625	641	Berg 1949
Ilmen Lake (USSR)	166	286	392	489	572	652	727	782	827				Berg 1949
Kurisches Haff (USSR)	129	258	392		533	609		650		730			Marre 1933
Frisches Haff (Poland) Mouth of Vistula	122	237	367	468	575		591	613	677	749	781		Filuk 1955
Stettiner Haff (Poland)	149	308	472	558	642	682	700	722	880	790	840		Neuhaus 1934
	169	300	390	510	610	645	665						Wiktor 1954
		444	562	647	665	687	726	776		815			Neuhaus 1934
mean value from 24 lakes of Northern Germany	130	240	340	430	490	550	560	570		700			Bauch 1955
Lake Jissel (The Netherlands)	120 to 145	280 to 310	400 to 450	500	550								Willemsen 1969
Velumeer (The Netherlands)	130	320	450										Willemsen 1969
Tjeukemeer (The Netherlands)	120	290	390										Willemsen 1969
Oder Estuary (DDR)	160	310	470	560	650								Unger 1934
mean value from 8 lakes of Central Germany	132	267	401	470	543	567	590						Helfer 1944
Balaton (Hungary)	200	286	394	471	570	550	620		730	680	730		Unger 1931
	204	289	358	423	482	535	576	609					Biró 1970
Raselmlakes (Rumania)		256	411	517	556	632	648						Grimalschi 1938
Asowsches Meer (USSR) 'Don-Zander'	198	368	426	486	546	620	694	776					Berg 1949
'Kuban-Zander'	197	428	499	561	638	699	793	828					Berg 1949
Aral Lake (USSR)	178	306	423	513	594	641							Berg 1949
Wolga (USSR)	241	322	391	474	534	604	674	743	775				Berg 1949
Neusiedlersee (Austria)	83	191	275	358	420	477							this paper

Age class 2+ displays a relatively steep increase in growth from February to April; from April (main spawning season to end of May) the curve remains flat (almost horizontal) for the rest of the year. The marked break in the age class 2 growth curves (length and weight) in April (1970 and 1971) may be due to the first full spawning activity (see Lagler et al. 1962). 95 per cent of 2+ individuals are spawning for the first time.

For age class 3+ the growth curve is relatively flat until June (end of May = end of spawning), becomes steeper during summer and (as in the other age classes) flattens towards fall and winter. Hence it appears that older *Blicca björkna* may recover more readily from the strain of spawning than individuals that have spawned for the first time. Unfortunately the material for 4+ age class is insufficient to provide support for this hypothesis. Diet as a possible source of differences in the growth pattern between the various age classes can be excluded: no significant differences in diet could be found with the exception of the first two month of 0-age class.

Growth rates of *Acerina cernua* less than one year old (= 0+ age class) were found to be highest during summer (especially August). For *Acerina cernua* older than one year, as opposed to *Blicca björkna*, already spawning (50 per cent of individuals examined), growth curves (length and weight) remain relatively flat until the end of spawning in April. Highest growth rates are observed immediately after spawning in summer. Again, as in *Blicca björkna*, length growth curves flatten towards fall and winter.

Zero and even negative weight-growth rates can be found from October to December (no January values available).

References

Aumüller, S., 1965. Algemeine Bibliographie des Burgenlandes, II. Teil.– Naturwissenschaften, Eisenstadt, 93 pp.

Bauch, G., 1966. Die einheimischen Süßwasserfische.– Neumann-Neudamm, Melsungen, 200 pp.

Bauer, K. & Schubert, P., 1957. *Proterorhinus marmoratus* Pall. (Gobiidae)–ein für die österreichische Fauna neuer Fisch., Bgld. Heimatbl. (Eisenstadt) 19, 6–9.

Benda, H., 1950. Fischereibiologisches über den Neusiedlersee.– Österr. Fischerei 3, 8/9: 189–194.

Berg, L. S., 1949. Freshwater Fishes of the USSR and Adjacent Countries.– Transl. from Russian by: Israel Progr. f. Scient. Transl., Jerusalem 1962, 1381 pp.

Biró, P., 1970. Investigation of growth of pike-perch (*Lucioperca lucioperca* L.) in Lake Balaton.– Ann. Biol. Tihany 37: 145–164.

Blöch, F., 1941. Der Neusiedlersee und seine Fischerei.– Allg. FZ. München 66.

Bruschek, E., 1971. Die Situation der österreichischen Fischerei in der Gegenwart.– Proc. Symp. 'New Ways of Freshwater-Fishery Intensification', Vodňany, Czechoslovakia 1971, 22–39.

Deelder, C. L., 1957. On the growth of eels in Ijsselmeer.–J. Cons. Int. Explor. Mer 23-83–88.

Dmitrijev, B., 1977. (Pflanzenfresser im Stau).– Rybovodosstvo i Rybolovsstvo 20/1977, 3, 15–16 (Russ.), (referred to in: Österr. Fischerei 31/1, 21, 1978.

Einsele, W., 1961. Über das Wachstum des Aales in österreichischen Gewässern.– Österr. Fischerei 14, 7–9: 136–138.

Filuk, J., 1962. Studie nad biologia i polowami sandacza Zalewu Wiślanego.– Prace Morskiego Institutu Ryb. Gdynia 11/A., 225–274.

Fischer, Z., 1972. The elements of energy balance in grass carp (*Ctenopharyngodon idella* Val.)

Part III. Assimilability of proteins, carbohydrates, and lipids by fish fed with plants and animal food.– Pol. Arch. Hydriobiol. 19, 1: 83–95.
Fischer, Z., 1973. The elements of energy balance in grass carp (*Ctenopharyngodon idella* Val.) Part IV. Consumption rate of grass carp feed on different type of food.– Pol. Arch. Hydrobiol. 20, 2: 309–318.
Frost, W. E., 1945. The age and growth of eels (*Anguilla anguilla*) from the Windermere catchment area.– J. Anim. Ecol. 14: 26–36, 106–124.
Frost, W. E., 1950. The eel fisheries of the River Bann, Northern Ireland, and observations on the silver eels.– J. Conseil perm. int. Explor. Mer 16: 358–383.
Frost, W. E., 1951. Einige Beobachtungen über Aale aus dem Windermere Fanggebiet, England.– Z. Fisch. 10 N.F., 599–607.
Gandolfi-Hornyold, A. 1921. Determinacion de la edad en algunas anguilas platedas (maresas) de la Albufera de Valencia.– Ann. Inst. General. Tecn. Valencia Trab. Lab. Hidrobiol. espan. 11: 27 pp.
Gandolfi-Hornyold, A., 1930. Recherches sur l'age, la croissance et la sexe de la petite Auguille argentée du Lac de Tunis.– Bull. Stat. oceanogr. Salambo 17: 1–50.
Gandolfi-Hornyold, A., 1934. Observation sur la taille et la sexe de deux cents petites anguilles argentées des Valli de Comachio.– Bull. Soc. centr. Agricult. Peche 4–5, 8 pp.
Gemzöe, K. J., 1906. Age and rate of growth of the eel.– Rpt. Dan. Biol. Stat. 14: 10–39.
Geyer, F. & Mann, H., 1939. Limnologische und fischereibiologische Untersuchungen am ungarischen Teil des Fertö (Neusiedlersee).– Arb. ungar. biol. Forschungsinst. Tihany 11: 61–182.
Grimalschi, W., 1938. Über das Wachstum und die Nahrung des Zanders aus den Raselmseen.– Bull. Sect. Scient. L'Acad. Roumaine 22: 180–187.
Hacker, R., 1974. Produktionsbiologische und nahrungsökologische Untersuchungen an der Güster (*Blicca björkna* (L.)) des Neusiedlersees.– Diss. Univ. Wien 93 pp.
Hacker, R. & Meisriemler, P. 1978. Vorläufiger Bericht über Wachstumsuntersuchungen am Aal (*Anguilla anguilla*) des Neusiedler Sees.– Öst. Fischerei 31, 2/3: 29–36.
Haempel, O., 1926. Der Neusiedlersee und seine Fischereiverhältnisse.–Österr. Fischereiz. 20: 23.
Haempel, O., 1929. Fische und Fischerei des Neusiedlersees.– Int. Rev. ges. Hydrobiol. 22: 445–452.
Haemptel, O. & Neresheimer, E., 1914. Über Altersbestimmung und Wachstum des Aales.– Z. Fisch. 14: 265–285.
Heckel, J. & Kner, R., 1858. Die Süßwasserfische der Österreichischen Monarchie.– W. Engelmann Leipzig, 388 pp.
Helfer, H., 1944. Beiträge zur Kenntnis des Zanders (*Lucioperca sandra* C. V.) in deutschen Binnengewässern.– Z. Fisch. 42: 67–119.
Herman, O., 1887. A Magyar Halászat Könyve.– Budapest 860 pp.
Hohendorf, K., 1966. Eine Diskussion der Bertalanffy-Funktionen und ihre Anwendung zur Charakterisierung des Wachstums von Fischen.– Kieler Meeresf. 22: 70–97.
Huet, M., 1962. Influence du courant sur la distribution des poissons dans les eaux courantes.– Schweiz. Z. Hydrol. 24: 413–432.
Kähsbauer, P., 1961. Catalogus Faunae Austriae, Teil XXIaa: Cyclostomata Teleostomi (Pisces).– Wien 56 pp.
Koenig, O., 1964. Führer rund um den Neusiedler See.– Jugend & Volk, Wien-München, 133 pp.
Kritscher, E., 1973. Die Fische des Neusiedler Sees und ihre Parasiten.– Ann. Naturhist. Mus. Wien 77: 289–297.
Lagler, K. F., Bardach, J. E. & Miller, R. R., 1962. Ichthyology: The study of Fishes.– J. Wiley, New York – London, 545 pp.
Lászlóffy, W., 1976. Bibliographie des Neusiedlersee Gebietes, Györ, 294 pp.
Lea, E., 1910. On the methods used in herring investigations.– Publs. Circonst. Consp. int. Explor. Mer 53: 7–25.
Liepolt, R., 1972. Die Fischerei in Österreich. Österr. Wasserwirtsch. 24, 9/10, 180–183.
Löffler, H. 1974. Der Neusiedlersee. Naturgeschichte eines Steppensees.– 175 pp. Molden Wien.
Marcus, K. & Smolian, K., 1920. In Smolian, K.: Handbuch der Binnenfischerei Mitteleuropas, Radbeul-Berlin, 2 Bde.

Marre, G., 1933. Untersuchungen über die Zanderfischerei im Kurischen Haff.– Z. Fisch. 31: 309–343.
Meisriemler, P., 1974. Produktionsbiologische und nahrungsökologische Untersuchungen am Kaulbarsch (*Acerina cernua* (L.)) des Neusiedlersees.– Diss. Univ. Wien, 98 pp.
Mika, F., 1962. Sopron város vizeinek halfaunája és a fertö halászat gazdasági jelentösége. Különlenyomat hydrologiai tájékozató. (Fishfauna of the waters around Sopron and the commercial importance of Neusiedlersee fishery).– Repr. from a Hydrological Guide).
Mika, F. & Breuer, G., 1928. Der Neusiedlersee vom Standpunkte der Fischereiwirtschaft.– Arch. Balatonic. II, 121–131.
Mika, F. & Varga, L., 1941. Gobiius marmoratus Pall. in Ungarn und Nachbargebieten. Int. Rev. ges. Hydrob.
Nagieć, M., 1961. Wzrost sandacza (*Lucioperca lucioperca* L.) w jeziorach połncnej Polski.– Roczn. Nauk. roln. 77–B, 549–580.
Nawratil, O., 1953. Zur Biologie des Hechtes im Neusiedlersee und Attersee.– Österr. Zool. Z. 4: 4–5.
Neuhaus, E., 1934. Studien über das Stettiner Haff und seine Nebengewässer. III. Untersuchungen über den Zander.– Z. Fisch. 32: 599–634.
Peňáz, M. & Tesch, F. W., 1970. Geschlechterverhältnis und Wachstum beim Aal (*Anguilla anguilla*) an verschiedenen Lokalitäten von Nordsee und Elbe.– Ber. Dtsch. Wiss. Komm. Meeresf. 21: 190–310.
Pernt, J., 1894. Der Neusiedlersee und seine Fischerei.– Mitt. Österr. Fischereiverb. 46–52: 74–82.
Roditzky, J. B. Mosonvármegye Leirasa. az: Ostrak. Magyar Laásban és Képben.
Sauerzopf, F., 1965. Beitrag zur Fischfauna des Burgenlandes.– Wiss. Arb. Bgld. 32: 142–146.
Sauerzopf, F. & Hofbauer, E., 1959. Fische und Fischerei im Neusiedlersee.– Wiss. Arb. Bgld. 23: 195–201.
Schiemer, F. & Prosser, M. V., 1976. Distribution and biomass of submerged macrophytes in Neusiedlersee.– Aquat. Botany 2: 289–307.
Seligo, A., 1926. Die Fischerei in den Flüssen, Seen und Strandgewässern Mitteleuropas.– Handb. Binnenfischerei Mitteleuropas 5: 319–321.
Sinha, V. R. P. & Jones, J. W., 1966. On the sex and distribution of the freshwater eel (*Anguilla anguilla*).– J. Zool. Lond. 150: 371–385.
Sinha, V. R. P. & Jones, J. W., 1967. On the age growth of the freshwater eel (*Anguilla anguilla*).– J. Zool. Lond. 153: 99–117.
Stundl, K., 1947. Die Fischerei des Neusiedlersees und die Möglichkeiten ihrer Ertragssteigerung.– Bgld. Heimatbl. (Eisenstadt) 1: 8–27.
Tesch, F. W., 1973. Der Aal, Biologie und Fischerei.– Paul Parey, Hamburg-Berlin, 306 pp.
Unger, E., 1931. Alter und Wachstum der zwei Zanderarten des Balaton-Sees.– Verh. int. Ver. Limnol. 5: 415–430.
Unterüberbacher, H., 1958. Über Wachstum und Lebensweise des Karpfen im Neusiedlersee.– Diss. Univ. Wien.
Varga, L., 1932. Katastrophen der Biozenosen des Fertö. Int. Rev. ges. Hydrob. 19, 189–294.
Varga, L. & Mika, F., 1937. Die jüngsten Katastrophen des Neusiedlersees und ihre Auswirkungen auf den Fischbestand des Sees. Arch. Hydrob. 31, 527–546.
Vutskits, G., 1896. Fauna Regni Hungariae, Pisces, Budapest, königl. ung. Naturw. Ges.
Wiktor, J. 1954. Analiza stada sandaeza w Zalewie Szczecińskim. Prace MIR 7.
Willemsen, J., 1969. Food and growth of pike-perch in Holland. Proc. 4th Coarse Fish Conf. 1969, 72–78.
Wundsch, H. H., 1949–1951. Abhandl. aus Fischerei und deren Hilfswissenschaften.

29 Birds of Neusiedlersee

Friedrich Böck

29.1 Introduction

Despite the fact that Neusiedlersee is a bird sanctuary of international importance few of the available scientific publications contain data on more than one species or a group of related species. Zimmermann (1943) and Bauer, Freundl & Lugitsch (1955) covered the whole of northern Burgenland in their summarizing articles and were primarily concerned with faunistic aspects, whilst Koenig (1952) was specifically interested in the birds of the reed belt. However, in the last decade a series of more extensive investigations has been devoted to the biology of individual species (Bauer 1965, Leisler 1971, Spitzer 1972 and van den Elzen 1971).

Since a comprehensive picture of the, undoubtedly substantial, role of birds in the ecosystem as a whole is still lacking, the present chapter is necessarily a compilation.

Since 1974 the long-established tradition of bird-ringing on Neusiedlersee has been continued in Illmitz (Berthold & Schlenker 1975).[1]

29.1.1 *Spatial and Temporal Aspects*

Two main regions can be distinguished in connection with the birdlife of Neusiedlersee. These are the open lake and the reed belt, and this chapter will be limited to species that play a larger role in one or other of these two environments. The grassland on the landward side of the reed belt will be ignored, this excluding the limicoles, which were undoubtedly also encountered in large numbers on the lake in former times, when the reed belt did not cover nearly the whole shore as it does now.

The significance of the lake itself is to be seen in its function as a resting and sleeping place for geese, gulls and ducks, and as a source of food for the latter and for the divers. The reed belt offers not only suitable nesting places for some of the above-mentioned forms (grey-lag goose, ducks, divers and gulls) but, in addition, it constitutes the main environment and source of food of a whole series of species such as the herons, water rails, reed warblers and the bearded tit which are more or less well adapted to its structural peculiarities. The plentiful supply of insects in the reed belt also attracts such flying-insect catchers as swallows, swifts and the red-footed falcon from quite different regions. In addition, the dense reed is used as a sleeping place by a number of

1. The 'Österreichische Gesellschaft für Vogelkunde' has led bird-ringing activities in Neusiedl/See since the Fifties. I am obliged to the 'Österreichische Gesellschaft für Vogelkunde' for putting unpublished data at my disposal.

song birds that are in no way specially adapted to it (starlings, fringillids such as linnets and greenfinches, wagtails, yellow wagtails and swallows). The reason for this preference may be that the vertical structural elements, growing as they do in water, are not easily climbed by enemies.

The above-mentioned properties, however, are strongly influenced by seasonal changes. As a general rule the lake is frozen for 2–3 months each year, and thus is of no interest whatever in winter to species entirely dependent upon water, i.e. all swimming birds (ducks, geese, divers, coots). Living conditions in the reed belt also change for the species that seek their food on the water's surface and are not migrants. The ice appears to represent a limiting factor to the moustached warbler (*Acrocephalus melanopogon*/Temm.), which remains longest of all the reed warblers and migrates least far to the south (Mediterranean). The bearded tit (*Panurus biarmicus* L.), a non-migrant, survives the winter months in a different way (see p. 451).

Seasonal changes in the appearance of the reed belt (defoliation, seed ripening, resting phase and fresh growth in spring) along with the drop in insect production, above all of flying forms in the winter months, imply alterations in living conditions for the inhabitants of the dense reed. In some cases this leads to their moving out, or at least to a change of diet (bearded tit). A partial compensation is provided by the dead reed stalks which are ideal quarters for overwintering arthropods (Waitzbauer, Pruscha & Picher 1973). The latter then serve as a food source even for species that otherwise live at a considerable distance from the reed belt.

29.1.2 *Meteorological Factors*

Of all the factors which exert an influence upon birdlife the wind (mainly from the northwest and southeast) is by far the most important. The long, uninterrupted approach across the water surface is highly favourable to wave formation and this, combined with the extreme shallowness of the lake, leads to erosion of the bottom (see Chapter 15) and thus to permanent turbidity. This is probably the reason why the common merganser and cormorants are encountered only seldom and in small numbers on the lake as compared with the nearby Danube (about 20–30 km). Further, the strength and direction of the wind considerably influence the ducks' choice of resting places. In heavy storms they seek the slight shelter provided in the lee of the reed belt or even retire into the dense reed itself.

Indirectly, via the water level, the wind exerts a particularly strong influence on birds breeding in the reed belt. The large, open expanses of water favour the formation of considerable seiches which can bring about a rise or fall in water level of 30–40 cm within a short time. A large part of a batch of eggs or nestlings of the song birds nesting in the reed can in this way be washed away and lost. Thus the weather has a far greater influence here on the success of breeding than is usually exerted by normal factors such as rainy periods.

Wind and snowfall in winter are also responsible for the formation of the 'bent layer', a structure consisting of bent and matted plant stalks, mainly reed stems and small-leaved spikes. Many species (moustached warbler, bearded tit, little crake) prefer this layer to the purely vertical structure of the undamaged reed for nesting and seeking food.

On the other hand, the reed itself often acts as a wind break, so that a relatively favourable local climate prevails in its interior. This enables certain prey species to become active earlier than outside the reed, thus providing some of the bird species with more favourable feeding prospects.

29.2 The open lake

The open lake fulfils an important function as a resting place for migratory ducks, geese, grebes and gulls, particularly in autumn and spring. In winter, as already mentioned, it is of little interest from this point of view due to ice, although certain places, such as the entry of the Wulka, or the vicinity of the 'Podersdorfer Schoppen', a reed island, 4.5 km north of Podersdorf, usually remain free of ice and in this case 1,000–1,500 ducks, mostly mallards, collect there.

This is a very small number when compared with the Danube only 30 km away, where on a 40 km stretch the average over a considerable number of years amounts to 4,000–8,000 mallards. It can be assumed that some of them are from the Neusiedlersee region (Böck & Scherzinger 1975). In some years the white-tailed eagle (*Haliaetus albicilla* L.) is to be seen trailing the duck community on the lake for a time, taking advantage of birds that have been shot and wounded, or muskrats that venture too far out of the reed (Spitzer 1966).

Due to the extreme disturbance caused by pleasure boats and commercial fishery very few ducks are seen on the lake during the summer months. They also prefer the protection of the reed belt during the moulting season. It is almost impossible to obtain exact figures for the numbers of ducks except in the winter, since the larger portion are no doubt on the patches of open water within the reed belt. Nevertheless, occasional counts made from boats give a reasonable good overall estimate, but an inventory over several years, similar to those already made for aquatic birds of other regions, would be of great value.

The species spectrum is liable to change rapidly in connection with the various stages of migration, but the mallard (*Anas platyrhynchos* L.) obviously plays the predominant role (about 8,000 have been reported for the relatively undisturbed southern part of the lake in mid-October). Large groups of gadwall (*Anas strepera* L.), garganeys (*A. querquedula* L.), teals (*A. crecca* L.) and pochards (*Aythya ferina* L.) have also been noted in recent years (at some counts in groups of 600–800). Smaller numbers of shovelers (*Anas clypeata* L.), wigeon (*Anas penelope* L.), tufted duck (*Aythya fuligula* L.), goldeneye (*Bucephala clangula* L.) and the three mergus species are observed sporadically. Sea ducks that have strayed inland are occasionally seen. Data are insufficient to permit statements as to whether the various species prefer to avoid any particular part of the lake. In general it can merely be said that although depending upon strength and direction of the wind (see above) the favourite resting places are the open water especially in the quiet southern portion of the lake between Sandeck and Neudegg, the entry of the Wulka, the bays between 'Hölle' and Illmitz, the entry of the 'Golser Kanal' and the Podersdorfer Schoppen (see Fig. 29.1). In addition to the wind, bathing, sailing and commercial fishing activities constitute a considerable disturbance to the resting places of the ducks.

For migratory geese, chiefly the bean goose (*Anser fabalis* Latham) and the white-fronted goose (*Anser albifrons* Scop.), the lake and the traditional resting place in the 'Lange Lacke' of the Seewinkel are most frequented. On

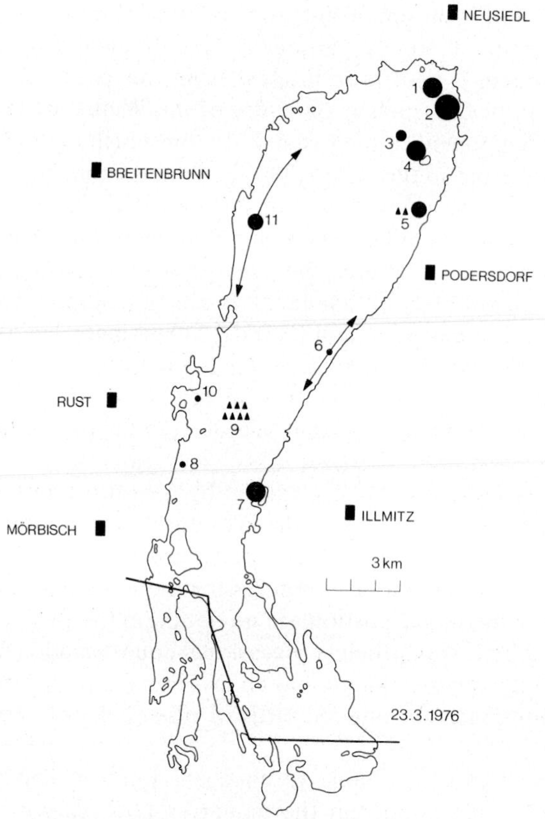

(a)

Fig. 29.1. Distribution of resting geese, ducks and coot (black spots) and great crested grebes (black triangles) on three different days on Neusiedlersee (the area south of the line Illmitz – Mörbisch has not been covered). The data from the 10 November 1976 collected by G. Aubrecht

(a) 1: 50 mallards (*Anas platyrhynchos*), 10 garganeys (*Anas querquedula*), 30 pochards (*Aythya ferina*), 15 tufted ducks (*Aythya fuligula*), 80 coot (*Fulica atra*)
2: 220 mallards, 5 pochards, 20 coot
3: 30 geese (indet.)
4: 54 mallards, 40 teal (*Anas crecca*), 44 coot
5: 56 mallards, 6 teal, 4 tufted ducks, 23 coot
6: 7 mallards, 10 goldeneyes (*Bucephala clangula*)
3 tufted ducks

Fig. 29.1 continued

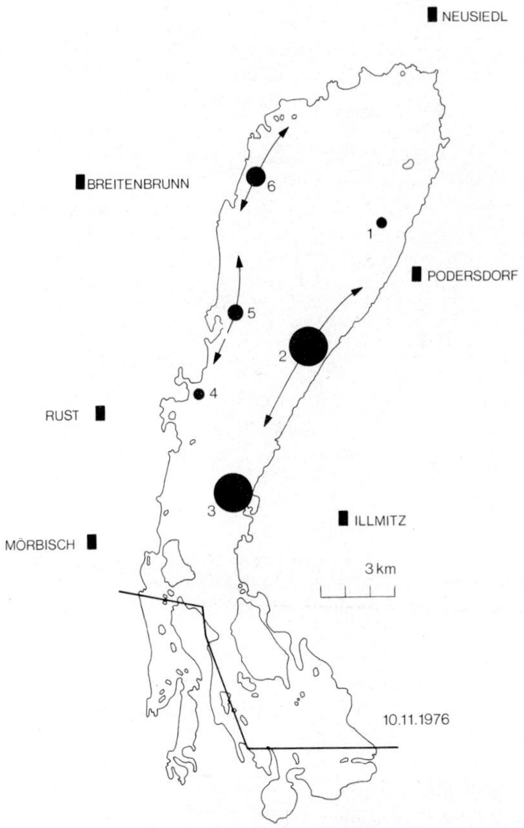

(b)

7: 20 mallards, 45 pochard, 5 goldeneyes, 45 coot
8: 6 mallards
9: 7 great crested grebes (*Podiceps cristatus*)
10: 5 goldeneyes
11: 36 pochards, 2 smew (*Mergus albellus*), 32 coot

(b) 1: 31 geese (indet)
2: 20 mallards, 550 small *Anas* sp. (probably teal)
3: 300 mallards, 300 small *Anas* sp. (as above)
4: 50 mallards
5: 22 mallards, 30 small *Anas* sp. (as above)
6: 200 small *Anas sp.* (as above)

(c) 1: 15 mallards
2: 4 mallards, 5 garganeys, 1 tufted duck, 2 coot
3: 30 mallards, 150 teals, 5 garganeys
4: 4 mallards, 2 shovelers (*Anas clypeata*), 2 gadwalls (*Anas strepera*), 6 garganeys, 25 tufted ducks, 2 coot
5: 14 mallards, 4 shovelers, 77 teal, 1 garganey, 44 pochards, 19 tufted ducks, 10 goldeneyes, 8 scaups (*Aythya marila*), 51 grey-lag geese, 8 coot
6: 6 mallards, 10 teals, 2 shoveler, 84 pochards, 3 tufted ducks, 9 goldeneyes, 5 grey-lay geese, 2 coot
7: 3 mallards, 52 pochards, 4 tufted ducks

Fig. 29.1 continued

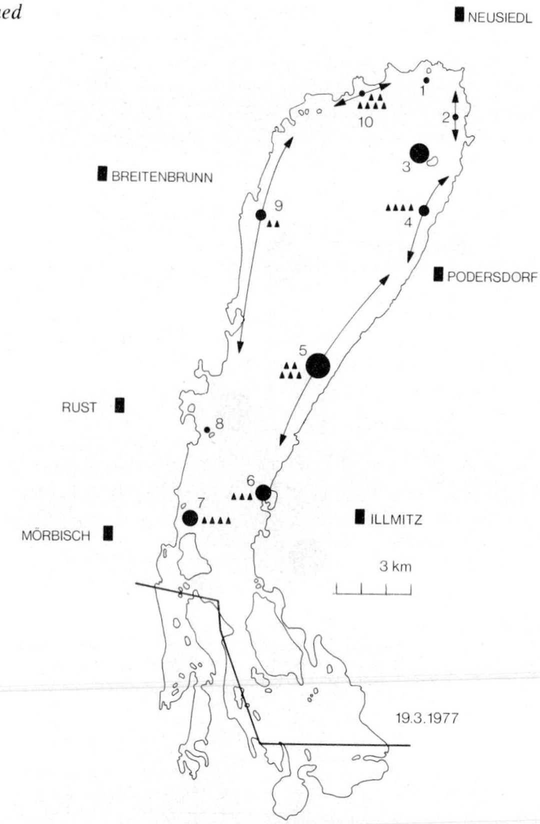

(c)

8: 1 mallard, 7 pochards, 1 coot
9: 27 mallards, 2 teals, 2 garganeys, 8 coot
10: 6 mallards

the whole they choose the same part of the lake as the ducks, i.e. an area west of the Podersdorfer Schoppen, where they gather at night after spending the day feeding on the Parndorfer plateau (about 10,000 in mid-November) or in the southern part of the lake (similar numbers). Two smaller resting places situated west of the 'Hölle' and off Podersdorf (Leisler 1969) are still sometimes used when the lake is frozen over. More detailed information concerning the grey-lag goose (*Anser anser* L.) is given in the section on the birds of the reed belt.

In recent years another representative of the Anatidae, the mute swan (*Cygnus olor* Gm.), has been introduced to the lake. As long ago as 1955 Bauer et al. reported attempts to introduce the species near Neusiedl and Rust. Although one or two broods are successfully reared each year on the lake the numbers have not so far increased noticeably (March 1977: one pair each near Neusiedl, Weiden and Illmitz; two near Rust).

Gulls also use the lake, principally as a resting place. In the evenings almost all the year round (as long as there are open stretches of water) large flocks of black-headed (*Larus ridibundus* L.) gulls can be seen flying in to the northern shore near Neusiedl and Weiden after spending the day in search of food on the Danube and on the intervening fields of the Parndorfer plateau. In the depths of winter their numbers are noticeably reduced. The largest breeding colonies of the species in the region of Neusiedlersee are found on the 'Lange Lacke' and on the 'Wörthenlacke' in the centre of the Seewinkel, although there are smaller colonies in other places as well. Sometimes at least, colonies occur in the reed belt of the western shore, in pools that have been cleared of reed by fire or by harvesting. In spring and autumn a number of common gulls (*Larus canus* L.) regularly land with the flocks (often several hundred) of black-headed gulls from the north, and can sometimes be seen on the open lake during the day.

The number of mediterranean herring gulls (*Larus argentatus michahellis*) flying into the entire lake region has considerably increased since the fifties. In recent years, according to Spitzer (verbal commun.), 1,300 specimens spend the summer in eastern Austria. They, too, principally seek their food by day on the Danube, but favour the margins of the pools of the Seewinkel or even the lake itself for sleeping. A few can be observed on the lake during the daytime. The food situation on the lake is particularly bad for gulls nowadays due to the lack of a shore zone with washed-up drift.

In the course of their flights to Central Europe, skuas (*Stercorarius* sp.) occasionally land on Neusiedlersee in winter. This was the case in the winter of 1976–77, but since the animals still had their juvenile plumage and attempts to secure specimens were unsuccessful precise information as to the species involved cannot be given.

A characteristic bird of the open lake during the summer months is the common tern (*Sterna hirundo* L.), which seeks its food both near the reed belt and on patches of open water within the reed. Like the black-headed gull it breeds on the pools of the Seewinkel (200 pairs according to Bauer 1965). It is a diving species and obviously prefers to seek its food in those parts of the lake with less turbidity. By investigating pellets Bauer (1965) found that fish accounted for only 30 per cent of the food of the common tern and this was mainly bleak (*Alburnus alburnus*), a species that swims near the surface. The remaining 70 per cent consist largely of larvae of the water beetles *Dytiscus* sp. and *Cybister* sp.

The black tern (*Chlidonias nigra* L.) which is regularly encountered in quiet bays at the margin of the reed during its migratory flights and throughout the summer, used previously to breed in the reed belt south of Weiden (Bauer 1965). The favourite hunting ground of this mainly insectivorous species was at that time the reed margin west of Neusiedl am See.

Just as occasional sea ducks and *Mergus* sp. are sometimes observed, a number of divers (*Gavia arctica* L. and *G. stellata* Pontopp.) are reported each year, but these are as a rule involuntary visitors, since the lake is just as unsuitable an environment for them as for *Mergus* sp. and cormorants (*Phalacrocorax carbo*) (see above).

The most frequently encountered species of grebe (Podicipidae) and the form most likely to be seen on the open lake is the great crested grebe (*Podiceps cristatus* L.). In spring its very striking display can regularly be observed on the water at the edge of the reed. It is almost impossible to estimate the numbers of breeding pairs, however, since it, too, lives on the patches of open water within the reed belt. Observations made by Spitzer (1968, verbal commun.: about 15 adults in June between Breitenbrunn and Rust, 9 juveniles and 2 immature animals, by Prokop (March 1974 about 15 specimens off the bathing beach at Neusiedl, verbal commun.), Staudinger (in the same year, at least 4 breeding pairs north of Rust, verbal commun.) and by the present author, at the southern portion of the lake (1975 at least 20 specimens on one day in April), suggest that the species is considerably more common than assumed by Bauer et al. (1955).

The larger part of the food of the great crested grebe consists of fish, in contrast to other grebes. From studies made on Swiss specimens it appears that cyprinids (chiefly *Alburnus alburnus* and *Rutilus rutilus*) make up 69 per cent of the fish consumed (Bauer & Glutz 1966). Comparable investigations have not been carried out for Neusiedlersee, but it seems justifiable to assume that the situation here is similar.

29.3 The reed belt and its significance for birds

The reed belt is by no means the uniform environment that it appears to be from the outside (see Chapter 20). The water level, fire and reed cutting, and the resulting plant societies produce a very distinct mosaic of widely differing types of smaller environment, each providing ideal possibilities for different species to occupy the various niches. Only a few plant species are involved in these environments. The dominant species, although present in a large variety of growth forms, is *Phragmites communis.* Other species of importance to the bird fauna include the lesser bulrush (*Typha angustifolia*), the great fen-sedge (*Cladium mariscus),* various species of sedge (*Carex acutiformis, C. riparia*) as well as *Bolboschoenus maritimus*, *Solanum dulcamara* and *Salix cinerea*, and a submerged macrophyte, *Utricularia vulgaris*, the bladderwort (Weisser 1970) (see Chapter 20).

Horizontal zones within the reed belt are not everywhere developed to the same degree, but are often masked by a variety of factors such as lake causeways, canals and surface inflow like the Wulka (Koenig 1952). Nevertheless the zonation suggested by Imhof (1966, see also Chapter 25) for the vicinity of Rust can be taken to be more or less characteristic for the entire reed belt. The lakeward reed stands are flooded throughout the year, and submersed vegetation (usually *Utricularia*) is commonly encountered between the usually very thick reed stalks. Along canals or on larger areas of open water stands of narrow-leaved *Tyhpa angustifolia* are found. Towards the landward margins the reed becomes less dense and its stalks thinner. In such places the reed dries out in some seasons, and there is a dense undergrowth of rushes and giant sedges. In a few places on the lake this zone contains large stands of *Cladium mariscus.* The degree of flooding undergoes very considerable fluctuations due to the action of the prevailing winds. In places, stands of the grey willow, *Salix cinerea*, border the extreme landward edges of the reed, and some even grow in the part of the reed interspersed with sedges. In all of these zones, from the water's edge to the landward margins, fire, harvesting and other artificial measures as well as the activities of the muskrat (*Ondathra zibethica*) and the grey-lag goose, both of which eat the reed plants, can lead to thinning out of the reed or even to the development of patches of open water. Following a strict succession in which *Typha angustifolia* plays a leading role, the damaged areas of the *Phragmites* stands are gradually restored to their original state (Leisler 1971).

A preference for or the avoidance of a particular environment by birds is by no means primarily dependent upon the plant sociological conditions, but rather upon the structural elements resulting from them. Thus seen, the reed belt is a macroenvironment with a very distinctive vertical structure (according to Waitzbauer, Pruscha and Picher 1973 about 65–75 stalks per m^2), although not rising to a height of more than 4.3 m above the water surface (Weisser 1970). Deviations from the basic plan can arise from varying conditions of reed growth (stalk length, density and diameter) or the presence of other plant species, and the so-called 'bent layer' (a densely interwoven and matted layer formed of the bent upper parts of the reed stalks) which interrupts the purely vertical structure. The distribution of some species of birds in the reed belt shown in Fig. 29.2 should only be regarded as a rough plan, since under certain conditions the structure may result from other factors than those shown in the zonation outline, and provide a suitable environment for a different species of bird. In 1952 Koenig reported that the great reed warbler prefers reed stalks of a particular diameter and is thus most

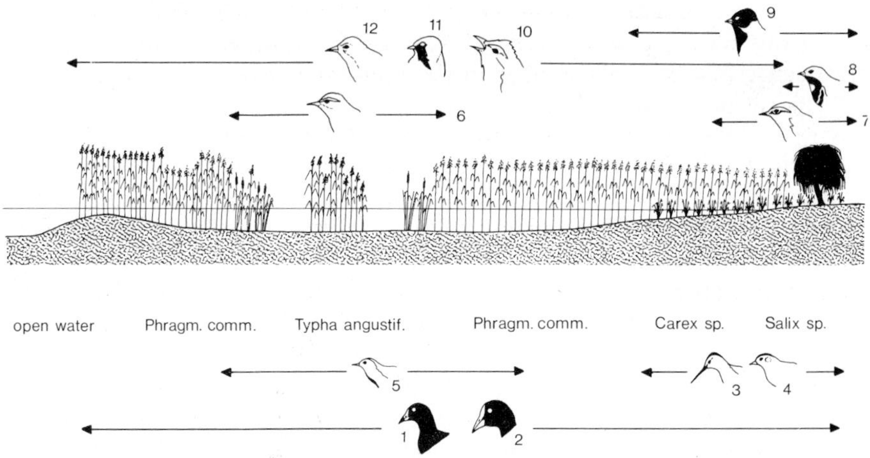

Fig. 29.2. Schematic distribution of some characteristic species along a transect through the reed belt from the open water to the dry land during the breeding season. 1: moorhen (*Gallinula chloropus*); 2: coot (*Fulica atra*); 3: water-rail (*Rallus aquaticus*); 4: spotted crake (*Porzana porzana*); 5: little crake (*Porzana parva*); 6: moustached warbler (*Acrocephalus melanopogon*); 7: sedge warbler (*Acrocephalus schoenobaenus*); 8: bluethroat (*Luscinia svecica cyanecula*); 9: reed bunting (*Emberiza schoeniclus*); 10: great reed warbler (*Acrocephalus arundinaceus*); 11: bearded tit (*Panurus biarmicus*); 12: reed warbler (*Acrocephalus scirpaceus*).

frequently encountered in the lakeward portions of the reed belt where the stalks are especially thick. It is, however, found just as often further landward in spots where deeper water resulting from bomb craters also favours the development of thicker stalks.

A series of studies carried out in the region was concerned with adaptations in a variety of often closely related species to the special living conditions encountered in the reed belt. Investigations have covered both morphological (particularly Leisler 1975 and 1977) and behavioural adaptations (Koenig 1951, 1952, Van den Elzen 1971). Jilka and Leisler (1974) studied the adaptations of three species of warbler to different environments on the basis of the frequency spectrum of their songs.

According to earlier authors the red-necked grebe (*Podiceps griseigena* Bodd.) (Bauer, Freundl & Lugitsch 1955, Zimmermann 1943) bred on Neusiedlersee in the last century. Observations made towards the end of the sixties in the vicinity of Illmitz suggested the presence of a brood. According to observations made by Leisler (verbal) and Winkler (verb. commun.) and by the author in the neighbourhood of Neusiedl it seems that the black-necked grebe (*Podiceps nigricollis*) and the little grebe (*P. ruficollis* Pal.) in fact breed more frequently in the reed belt than was assumed by Bauer, Freundl and Lugitsch (1955). Any degree of accuracy in estimating the numbers of these two species is almost impossible in the reed belt. Unlike the crested grebe and the red-necked grebe they feed almost exclusively on insects. A few specimens are also occasionally seen on the open lake during migration and in winter.

The reed belt offers an ideal breeding ground for various species of duck, by far the most numerous being the mallard. Many observations suggest, that of the species breeding in the reed belt, the next most abundant is the ferruginous duck (*Aythya nyroca* Güldenst.). According to Bauer, Freundl and Lugitsch (1955) the number in the region has risen very considerably since the second World War.

In view of the enormous quantities of plant matter produced annually in the reed belt it is surprising that so few vertebrates take advantage of it as a source of food. Of the mammals, only the muskrat is of any importance in this respect (Knoflacher 1974) and among the birds the coot (*Fulica atra* L.) and the grey-lag-goose (*Anser anser* L.) are the chief plant consumers, although in winter the bearded tit (*Panurus biarmicus*) eats the *Phragmites* seeds. After maximum dry weight has been attained in July, however, the food value of the plant drops rapidly as the stalks quickly harden. According to Geisselhofer and Burian (1970) (see Chapter 20) the maximum fresh weight production is probably attained even one month sooner, so that in fact the aerial parts of the reed plants represent an abundant and readily available source of food for vertebrates only from April until June. Species relying upon it have then to take recourse to different food or to move elsewhere. Koenig (1952) pointed out that the sprouting of the reed shoots coincides with the hatching of the grey-lag goose, for which the young reed shoots represent the main source of food in the first weeks of April. Leisler (1969) reported about 230 pairs of grey-lag goose breeding in the entire reed belt, in the years 1966 to 1968, of which 60 lived in the much thinner reed on the eastern shore of the lake. From mid-April onwards, assuming an average of three offspring per pair, there must be about 1,150 grey-lag geese within the reed plus another 140–200 non-breeding individuals. Since the grey-lag goose usually prefers the less dense areas within the reed, or its margins, it can, like the muskrat, exert an influence on the surroundings by preventing the small pools and canals from becoming overgrown. In addition, the geese peck off the old reed within a radius of 10 m of their nests. The reed is thus thinned out and offers some smaller species of birds a more suitable environment than the homogeneous stands. During the moulting season which follows the period of hatching and rearing the young, the birds are even incapable of flight for a short time and withdraw to the inaccessible parts of the reed. In mid-July the grey-lag geese begin to leave the lake and to collect in the centre of the Seewinkel.

Little is known so far about the coot on Neusiedlersee. This duck-like rail is found everywhere in the regularly flooded regions of the reed belt. It, too, relies mainly upon plant matter for food, including young, freshly sprouting shoots from the lakeward margin of the reed belt besides the inevitable reed itself. The floating nests are built mainly in the old reed near open water. No figures are available, but numbers probably depend upon the water level, since rising water means an increase in area.

Very little is known about the other species of rail in the region. The moorhen (*Gallinula chloropus* L.), at present nests in large numbers in the reed belt and occasionally spends the winter there. Its occurrence seems to be independent of water level, but its numbers have increased: whereas Zimmermann (1944) reported it as nesting in small numbers and only in the region of the western shore of the lake, later authors (Koenig 1952, Bauer, Freundl and Lugitsch 1955) list it as being quite common.

In areas with deep water one of the most commonly encountered species is the little crake (*Porzana parva* Scop.). A condition for its occurrence is dense vegetation of the type provided by *Typha.* Its food consists mainly of insects and invertebrates.

The spotted crake (*Porzana porzana* L.) and the water rail (*Rallus aquaticus* L.) prefer the landward zones of the reed belt, although the rail is apparently the less rigid of the two with respect to biotope requirements, being also found in more deeply flooded areas. It is a predator, feeding preferably on amphibia, fish, insects and their larvae and other invertebrates, and even on young birds and small mammals. All previous authors agree that the water rail regularly spends the winter in the lake region. With the rising water level the numbers of spotted crake appear to have decreased somewhat. Baillon's crake (*Porzana pusilla* Herm.) rarely nests in the area and is only occasionally encountered.

Herons, egrets and spoonbills (*Platalea leucocordia* L.) always attracted more attention than any of the other birds of the Neusiedlersee region. The purple heron (*Ardea purpurea* L.) is seen in considerable numbers whereas *Ardea cinerea* is less common, but the numbers of the great white egret (*Casmerodius albus* L.) nesting here are the largest in the whole of Central

Europe. Since the older studies of Seitz (1937a, b) and Koenig (1952) the situation has clearly altered since there are now only four nesting colonies on the western shore instead of the 8 previously reported. The greater part of the nesting animals is concentrated in one large colony on the south-eastern shore (Festetics & Leisler pers. commun.). This is probably a reaction to the increasing changes in their biotope occasioned by the reed harvesting (colonies are preferably set up in old, undisturbed reed) and to disturbances due to the ever-increasing bathing and boating in the vicinity (e.g. Rust, Breitenbrunn, Neusiedl). Fortunately, the overall numbers do not appear to have decreased, except in the case of *Ardea cinerea*, (see Table 29.2).

The latter heron, *Ardea cinerea*, is in fact a species of the river woodlands whereas in the Neusiedlersee region it builds reed nests just like other species. In any case, it seems that the numbers reported earlier, particularly by Bauer, Freundl and Lugitsch, were too high. The purple heron is the best adapted of all to life in the reed. Its colour provides an excellent camouflage; its long toes enable it to grasp several stalks at a time and thus to progress from one reed clump to the next. Its behaviour is also well adapted in that when faced with danger it adopts an attitude similar to the well-known 'pillar position' of the bittern, and if disturbed in any way the young take refuge beneath the nest (Koenig 1952). It is one of the last to arrive in the breeding grounds, only slightly before the little bittern in early April, although the herons and great white egrets have usually arrived in March, or even in February in mild winters. It normally seeks its food within the reed belt whereas the other species are accustomed to cover long distances between nesting places and feeding grounds. The purple heron consumes a relatively higher proportion of insects and other invertebrates than the other large heron species (Bauer & Glutz 1966). Herons and great white egrets hunt their food in pools within the reed, on the edges of ponds, and in ditches and fields. The little egret (*Egretta garzetta* L.) is a regular summer visitor to Neusiedlersee, but so far has not nested. The squacco heron (*Ardeola ralloides* Scop.) is rarer, although it is suspected that it does sometimes nest. The night heron (*Nycticorax nycticorax* L.) is a regular visitor but on account of its secretive habits is only seldom seen on the lake. It definitely nested in the area up to the beginning of the thirties, and has probably done so occasionally since (Bauer & Glutz 1966).

Other Ardeidae characteristic of the reed belt are the bittern (*Botaurus*

Table 29.1. Number of breeding pairs of herons and spoonbills on Neusiedlersee. Based on data from Bauer, Freundl & Lugitsch (1955), Koenig (1961) and Festetics & Leisler (pers. commun.).

	1952	1960	1970–74
grey heron (*Ardea cinerea* L.)	180	93	20
purple heron (*A. purpurea* L.)	250	273	300
great white heron (*Casmerodius albus* L.)	120–140	329	325
spoonbill (*Platalea leucorodia* L.)	200–250	179	140
total number	750–820	874	785

stellaris L.) and the little bittern (*Ixobrychus minutus* L.). The former feeds on small vertebrates whereas the little bittern mainly prefer aquatic insects and their larvae. Both species are genuine reed-dwellers, both in colouring and mode of walking. The bittern only occurs in scattered pairs in the reed, each usually claiming a very large territory, whereas the little bittern may even form loose colonies. The latter is a migratory species and is not encountered before mid-April, whereas the bittern, according to Bauer, Freundl and Lugitsch (1955), sometimes remains in the lake region during the winter.

The spoonbill (*Platalea leucorodia*) also nests close to the heron colonies, fishing in the shallow waters for the small crustaceans that make up the larger part of its diet, so that even more than other species it is nowadays obliged to resort to the pools of the Seewinkel as an alternative food source. Although in the 1920's large numbers of the glossy ibis (*Plegadis falcinellus* L.) nested in the area it disappeared in the thirties and has only been reported on isolated occasions since: apparently some pairs have attempted to build nests (Bauer & Glutz 1966).

Raptors play a very small role in the lake and its immediate surroundings. Only one species, the marsh harrier (*Circus aeruginosus* L.), is a genuine reed-dweller. The reed provides it with the small mammals, birds, amphibia and reptiles that constitute its food and also serves it as a nesting place. Next to the kestrel, which is hardly ever seen above the reed, it is the most frequently encountered raptor of the region. It, too, ranges far from the lake, over meadows and fields, in its search for food. The ground squirrel (*Citellus citellus*) seems to be a favourite prey there. The breeding population is estimated by Glutz, Bauer & Bezzel (1971) to be about 27–30 pairs. In dry years, the Montagu's harrier (*Circus pygarus* L.) sometimes builds its nest in the landward parts of the reed. During summer only one other raptor, the red-footed falcon (*Falco vespertinus* L.), is of any ecological importance, occasionally nesting in groups of trees throughout the Neusiedlersee region. It is a specialized insect hunter and consumes in flight the dragonflies that it catches above the reed. The osprey (*Pandion haliaetus* L.) is regularly observed during migration, and in winter the bald eagle (*Haliaetus albicilla* L.) can sometimes be observed (see above).

The song birds of the region obviously include a whole series of forms distinctly specialized to life in the reed (warblers and bearded tit). In addition to studies on the biology of certain species the mist-net trappings carried out by Samwald at the ornithological station in Neusiedl am See 1957–1966 have provided comparable data on the relative frequency of the various species at different seasons. It should be remembered that the data shown in Fig. 29.4,5 and Table 29.1 refer only to the immediate vicinity of the station, i.e. the lakeward reed margin on the northern shore of the lake.

The bearded tit (*Panurus biarmicus*), like the moustached warbler (*Acrocephalus melanopogon* Temm.) which will be dealt with further on, is an ornithological pecularity of the Neusiedlersee region. Thanks to the studies of Koenig (1951), Spitzer (1972–1974) and Van den Elzen (1971) we are well acquainted with its habits. Like the moustached warbler it prefers a biotope

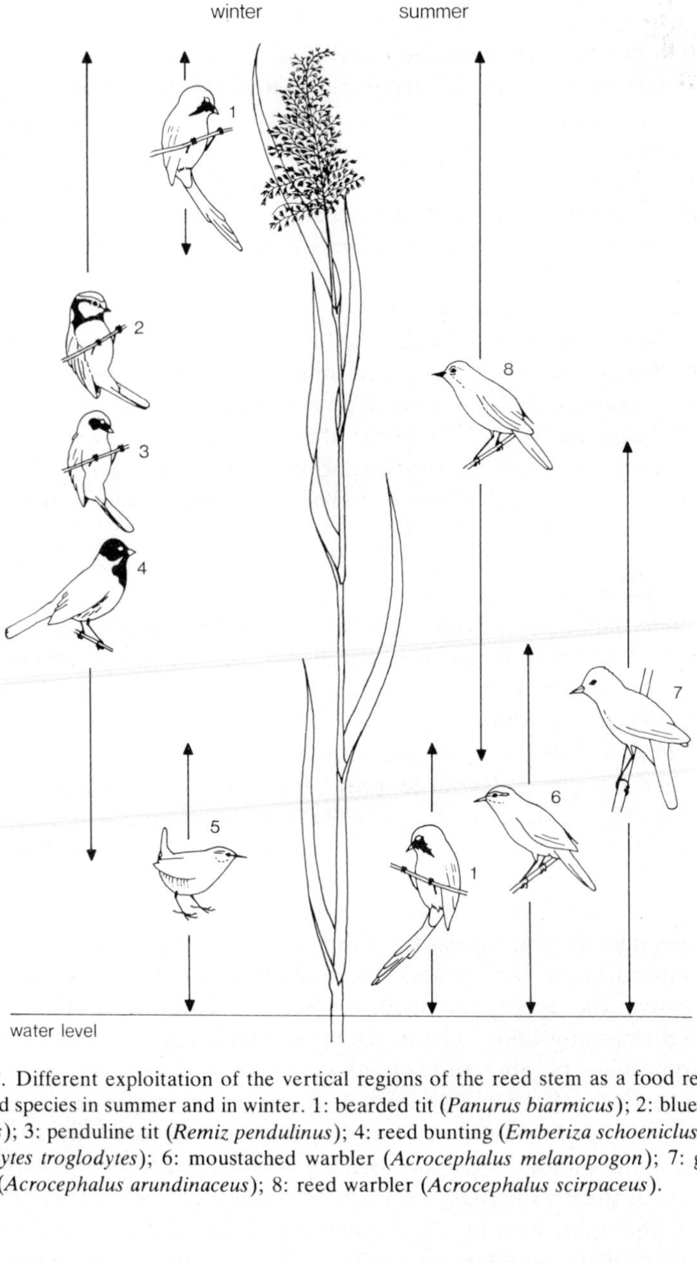

Fig. 29.3. Different exploitation of the vertical regions of the reed stem as a food resource by some bird species in summer and in winter. 1: bearded tit (*Panurus biarmicus*); 2: blue tit (*Parus caeruleus*); 3: penduline tit (*Remiz pendulinus*); 4: reed bunting (*Emberiza schoeniclus*); 5: wren (*Troglodytes troglodytes*); 6: moustached warbler (*Acrocephalus melanopogon*); 7: great reed warbler (*Acrocephalus arundinaceus*); 8: reed warbler (*Acrocephalus scirpaceus*).

with some variety in vegetation but it is also found in the large expanses of old reed, especially if there is a 'bent layer'. Koenig (1952) and other observers confirmed that it uses heron's nests. The peculiarity of the species lies in the fact that its summer and winter nutrition differ completely (Table 29.3 and Table 29.4). It was shown by Van den Elzen that in summer bearded tits mainly eat beetles, preferably water beetles. Their breeding and feeding territories are often in different places because although the former may be

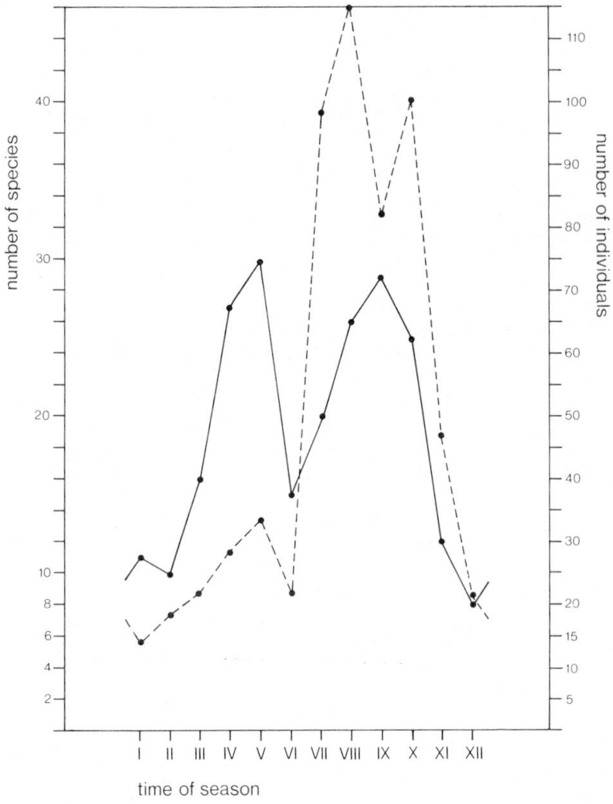

Fig. 29.4. Seasonal distribution of the number of caught species and the mean value of caught individuals per day for the site of the old bird-banding station at Neusiedl/See (near the open water). The data are based on the catches by Th. Samwald from 1962 until 1966. Caught species ——, mean value of individuals – – – –.

on dry ground the latter is invariably over water. In winter the situation is completely different, and the bearded tit consumes little else but seeds, chiefly those of the reed plants, and covers large distances in search of the best places. Spitzer (1972) demonstrated that the changeover of diet is paralleled by a marked alteration in the gizzard. The gizzard is heavier in winter than in summer due to an increase in musculature. Like other grain-eating birds the bearded tit also takes up grit in winter which assists in the grinding of the hard grains in the gizzard. What sets off these changes in the gizzard is not yet known, nor is the exact course of events understood. A possible explanation of the collapses in population characteristic of the species might be that following warmer periods in early spring the gizzard changes over to its summer condition and is then unable to cope with the reed grains which the birds are once more obliged to eat if another cold period sets in. The gizzard muscles are already too weak and the birds die of starvation.

Ecologically the reed bunting (*Emberiza schoeniclus*) most resembles the bearded tit. It nests in large numbers in the sedgy parts of the reed, but is

Table 29.2. Seasonal distributions of the relative abundance values of all passerine birds caught at the bird-banding station in Neusiedl/See from 1962 until 1966 by Th. Samwald

	Jan.	Feb.	March	April	May	June	July	Aug.	Sept.	Oct.	Nov.	Dec.
sand martin (*Riparia riparia*)					0.64			0.16				
swallow (*Hirundo rustica*)					0.88	1.44	0.02	1.18	0.56			
house martin (*Delichon urbica*)					0.29			0.02				
yellow wagtail (*Motacilla flava*)							0.24	0.83	0.41			
white wagtail (*Motacilla alba*)				0.08	1.70	6.71	1.69	1.32	0.31	0.39		
tree pipit (*Anthus trivialis*)								0.02				
meadow pipit (*Anthus pratensis*)									0.02			
red-backed shrike (*Lanius collurio*)									0.05			
wren (*Troglodytes troglodytes*)	1.65	0.26	0.49	0.08						0.29	0.54	0.75
dunnock (*Prunella modularis*)			1.47	0.70					0.02	0.12		
Savi's warbler (*Locustella lusinioides*)				3.48	5.03	2.79	1.69	1.12	0.85	0.08		
grasshopper warbler (*Locustella naevia*)					0.06							
moustached warbler (*Acrocephalus melanopogon*)			0.98	0.85	1.93	6.19	3.85	4.34	2.50	1.94	0.27	
sedge warbler (*Acrocephalus schoenobaenus*)				4.80	7.19	0.62	4.88	12.58	10.25	0.90		
aquatic warbler (*Acrocephalus paludicola*)					0.06							

marsh warbler (*Acrocephalus palustris*)				0.02	0.05	0.02			
reed warbler (*Acrocephalus scirpaceus*)		13.23	37.25	28.28	46.04	49.78	35.13	3.17	
great reed warbler (*Acrocephalus arundinaceus*)		6.34	18.07	9.80	14.11	8.34	1.94	0.16	
icterine warbler (*Hippolais icterina*)			0.12		0.02	0.02			
garden warbler (*Sylvia borin*)			0.35	0.10		0.02	0.02		
blackcap (*Sylvia atricapilla*)	0.16				0.02	0.02	0.02	0.02	
lesser whitethroat (*Sylvia curruca*)		0.31	0.06			0.02			
whitethroat (*Sylvia communis*)		0.08	0.64			0.09	0.05		
chiffchaff (*Phylloscopus collybita*)	3.27	1.39	0.18	0.10	0.04	0.05	0.44	1.14	0.07
willow warbler (*Phylloscopus trochilus*)	0.33	0.62	0.12		0.02				
wood warbler (*Phylloscopus sibilatrix*)					0.04	0.06			
goldcrest (*Regulus regulus*)							0.02	0.08	
spotted flycatcher (*Muscicapa striata*)			0.06						
pied flycatcher (*Ficedula hypoleuca*)		0.31	0.18	0.03			0.02		
stonechat (*Saxicola torquata*)							0.07		
whinchat (*Saxicola rubetra*)		0.08	0.06				0.02		
redstart (*Phoenicurus phoenicurus*)	0.70	0.29	0.02			0.12	0.06		
black redstart (*Phoenicurus ochruros*)							0.02		

	Jan.	Febr.	March	April	May	June	July	Aug.	Sept.	Oct.	Nov.	Dec.
nightingale (*Luscinia megarhynchos*)				0.31								
bluethroat (*Luscinia svecica*)			0.16	1.55	0.99	1.24	0.65	0.59	0.19	0.02		
robin (*Erithacus rubecula*)	0.33	0.26	8.02	7.42	0.06				0.56	0.67	0.67	0.50
song thrush (*Turdus philomelos*)			0.33							0.04		
bearded tit (*Panurus biarmicus*)	24.42	35.40	43.70	43.93	21.40	28.69	16.68	10.21	27.98	63.19	56.28	36.00
peduline tit (*Remiz pendulinus*)	36.96	26.61	23.08	9.67	0.18	12.18	8.66	7.80	15.02	15.78	14.57	10.75
blue tit (*Parus caeruleus*)	25.74	29.72	11.95	0.08		0.21	0.06	0.08	1.84	6.62	13.70	25.50
great tit (*Parus major*)	1.32	0.26	0.16						0.07	0.78	0.67	
short toed treecreeper (*Certhia brachydactyla*)										0.02		
reed bunting (*Emberiza schoeniclus*)	5.94	5.94	5.40	3.40	1.93	1.55	1.25	1.20	1.43	3.27	10.54	15.50
greenfinch (*Carduelis chloris*)	0.33	0.26	0.33	0.15								
goldfinch (*Carduelis carduelis*)	0.99			0.15	0.06		0.02		0.05	0.25	0.13	
siskin (*Carduelis spinus*)	1.98	1.03	0.16	0.06						0.82	2.28	2.25
linnet (*Carduelis cannabina*)				0.06								
tree sparrow (*Passer montanus*)	0.33	0.26		0.15	0.06	0.21				0.18	0.34	8.75
starling (*Sturnus vulgaris*)				0.18								

found throughout the belt outside the breeding season. Like the bearded tit and the penduline tit it is found in the Neusiedlersee region all the year round. Bauer, Freundl and Lugitsch (1955) pointed out that individuals spending the winter in the region, or those merely in transit, are members of other subspecies than the breeding population which leaves in winter. In summer the reed bunting lives on insects, but the winter population eats reed grains and, like the other birds, takes advantages of the plentiful supply of Diptera larvae and pupae in the reed stalks.

Another characteristic small bird of the region and one that is present all the year round is the penduline tit (*Remiz pendulinus* L.). Its conspicuous nests can be found from mid-May onwards in poplars and willows of the drier landward zone. Franke (1937) was able to demonstrate that it also builds reed nests. The nesting season commences uncommonly late, at the end of May to June. Small insects and occasionally plant matter serve as food, and despite the statements of Bauer, Freundl and Lugitsch (1955) it frequently spends the winter in the region, as could be demonstrated by trappings in recent years. It also eats the larvae living inside the reed stalks, but unlike the reed bunting that splits open the stalks, or the blue tits that peck them open, the penduline tit prizes the stalk open by inserting the closed tip of its bill and then opening it.

Acrocephalus warblers make up the larger part of the insect-eating songbirds in the reed belt in summer. The first to appear in spring is the moustached warbler (*Acrocephalus melanopogon* Temm.), which is at its northernmost European nesting limit in the Neusiedlersee region. It begins to breed at the beginning of April, a month earlier than the other species. Although it prefers to seek its food in the lower third of the vegetation, fishing directly out of the water or on its surface, it also combs the 'bent layer'. As

Table 29.3. Food of the most important breeding passerine birds of the reed belt of Neusiedlersee in summer. The numbers show the percentage of stomachs in which the different kinds of food had been found. Based on data from Leisler (1971) and van den Elzen (1971). A.m.: *Acrocephalus melanopogon* (moustached warbler), A. sch.: *Acrocephalus schoenobaenus* (sedge warbler), A.s.: *Acrocephalus scirpaceus* (reed warbler). A.a.: *Acrocephalus arundinaceus* (great reed warbler), P.b.: *Panurus biarmicus* (bearded tit)

	A.m. n = 25	A.sch. n = 34	A.s. n = 51	A.a. n = 18	P.b. n = 16
Coleoptera	92	77	69	78	100
Hymenoptera	55	44	80	67	10
Diptera	44	60	79	28	24
Rhynchota	60	20	20	34	25
Lepidoptera	8				
Odonata	4	23	3	17	6
Araneae	72	38	45	45	62
Larvae	16		10	34	43
Mollusca	4			5	31
indet. anim.	28	6	1		25

Table 29.4. Food of the most important passerine birds of the reed belt of Neusiedlersee in winter. The numbers show the percentage of stomachs in which the different kinds of food has been found. Based on data from Spitzer (1972), Leisler and the author (unpublished). P.c.: *Parus caeruleus* (blue tit), R.p.: *Remiz pendulinus* (penduline tit), E.sch.: *Emberiza schoeniclus* (reed bunting), P.b.: *Panurus biarmicus* (bearded tit)

	P.c. n = 42	R.p. n = 8	E.sch. n = 8	P.b. n = 15
Coleoptera	47.66		37.5	
Hymenoptera		12.00		
Diptera	4.7			7
Rhynchota	9.5			
Araneae	59.5	75	25	
Larvae	57	75	62.5	
Mollusca				7
indet. anim.	7.1			
Phragmites seeds			62.5	100
Cyperaceae				17
Salix				1
Typha				1

Table 29.5. Food of the nestlings of some important breeding passerine birds of the reed belt of Neusiedlersee (collected by putting a ring around the nestling's neck, to stop it gulping its food). The numbers show the percentage of samples in which the different kinds of food was found. Based on data from Leisler (1971) and van den Elzen (1971). A.m.: *Acrocephalus melanopogon* (moustached warbler), A.s.: *Acrocephalus scirpaceus* (reed warbler), A.a.: *Acrocephalus arundinaceus* (great reed warbler), P.b.: *Panurus biarmicus* (bearded tit)

	A.p. n = 796	A.s. n = 399	A.a. n = 142	P.b. n = 15
Coleoptera	4	4	4	
Hymenoptera		2	2	1
Diptera	54	88	40	46
Rhynchota	7	4	4	
Lepidoptera	9	8	7	11
Odonata	2	6	22	15
Araneae	60	28	22	23
Larvae	44	32	65	50
Mollusca	13	4	18	
Trichoptera	0.5	0.5	14	
Aphidoidea	5	25		
indet.anim.	2	0.5	2	

would be excepted, the species composition of its food, which contains a remarkably high proportion of spiders, is accordingly very different from that of other warblers. Its successful adaptation to fishing was demonstrated by Van den Elzen (1971); if trapped specimens were offered food under water they could retrieve it from a depth of 9 cm, whereas other warblers could only reach down to a depth of 1.8 cm (similar values were found for bearded tit and reed-bunting). For its nest the moustached warbler usually selects a bush that stands out conspicuously from its surroundings, as a rule in mixed vegetation.

Reed warbler (*Acrocephalus scirpaceus* Herm.) and great reed warbler (*Acrocephalus arundinaceus* L.) arrive from April onwards, and whereas the latter is already rare by September the former is still encountered in October. The reed warbler mainly seeks insects on the reed stalks but also catches a large part of its food in flight. The great reed warbler, on the other hand, rarely catches in flight but often fishes its food out of the water (even small fish, but mostly the larvae of water-beetles or dragonflies, see Table 29.3). Both species are distributed throughout the reed belt but are the only species regularly occurring in large numbers in the pure reed stands. Stalk thickness (not less than 6.5 mm) is probably a limiting factor for the great reed warbler on account of its size. The reed warbler may even form sparsely populated colonies in particularly favourable spots.

The marsh warbler (*Acrocephalus palustris* Bechst.) does not nest in the reed (or at least only on the extreme landward margins) although regularly encountered at other seasons in small numbers. The sedge warbler (*Acrocephalus schoenobaenus* L.) is a very common nesting bird on the landward margins of the reed with a dominant undergrowth of sedge and a sprinkling of willow bushes. At all other times it is common in the entire reed belt. The rare aquatic warbler (*Acrocephalus paludicola* Viell.), on the other hand, seldom breeds, and then only on the landwards edges of the reed.

A close relative of the warblers, and also a characteristic inhabitant of Neusiedlersee, is Savi's warbler (*Locustella luscinoides* Savi). *Locustella spp.* have adapted to moving their legs alternately and walking with a smooth gait in contrast to the hopping and grasping of the warblers. Savi's warbler deviates slightly from the other *Locustella spp.* in this respect in that its movements resemble more closely those of the warblers (Leisler 1977). It is most often encountered in the less flooded regions with an undergrowth of sedge, as well as in places where the pressure of snow has caused accumulations of slanting to horizontal structures.

On the landward margins of the reed where the plant societies are especially varied due to the presence of bushes and a number of plants not truly belonging to the reed belt, the blue throat (*Luscinia svecica cyanecula* Wolf) is a commonly encountered nesting species. It, too, is also found in other parts of the reed at other seasons.

Due to lack of an adequate supply of food apart from the insects in the reed stalks this wealth of insect-eating birds is replaced in winter mainly by forms that are able to exploit the larvae and pupae within the stalks (see above and Fig. 29.5). The Most common insect-eater of the reed belt in winter is the blue tit (*Parus caeruleus* L.), which is in fact not a reed form in the true sense of the term. This was already pointed out by Koenig (1952) and Bauer, Freundl and Lugitsch (1955). According to daily trapping figures from 1971 and 1972 the blue tits, reed buntings and penduline tits equal in numbers the insect-eating birds present in the reed during the nesting season, so that apparently the insect reserves of the reed in winter are sufficient to support such forms as are capable exploiting them. Although large numbers of blue tits have been ringed over many years there have been no reports of finds which might throw light on the origin of the birds that spend the winter in the Neusiedlersee region. The large numbers suggest that birds from more distant regions join the breeding population of the neighbouring Leitha Mountains that moves into the reed in winter. The great tit (*Parus major* L.) is far less frequent, and the solitary specimen of the azure tit (*Parus cyanus*) observed on one occasion was no doubt an accidental visitor. A regular winter visitor is the wren (*Troglodytes troglodytes* L.). Although its numbers are insignificant it can in some ways be regarded as replacing the moustached warbler, in that it seeks insects that spend the winter in the leaf axis, particularly in the lowest layer of the vegetation (Leisler 1971). A few specimens of the robin (*Erithacus rubecula* L.) are seen in winter, but during the migratory seasons, especially in spring, it may turn up in the reed belt in considerable numbers (see Table 29.1). During the spring and autumn migrations a number of species use the reed as a temporary resting place, but presumably play no significant role in the economy of the region (see Fig. 29.2 and Table 29.1).

A different situation is presented by species such as swallows and wagtails which nest outside the reed belt (the former, however, also breeds in the reed in connection with the bathing huts) but exploit it as a source of food, and by species such as the starling (*Sturnus vulgaris* L.) which use the reed in vast numbers as sleeping quarters.

29.4 Migrations and changes of locality

Due to its geographical situation the Neusiedlersee region constitutes an important resting place for passing and overwintering populations of a variety of aquatic and reed species of bird. For some species (great white egret and moustached warbler) it is their northernmost breeding ground. An analysis of ringing results may throw light upon the broader geographical connections that are established with the region by migrating birds including ecological (coincidence of arrival with optimum food conditions) as well as epidemiological aspects (e.g. spread of viruses by migrating birds, see Aspöck et al. 1973).

Four groups of birds (populations) can be distinguished on the basis of type of translocatory movements:

1. Populations that breed in the region but spend the winter in the south (all warblers, *Locustella*, herons etc)
2. Species that breed in more northerly regions and spend the winter in the Neusiedlersee area (populations of penduline tits, reed bunting)

3. Populations that breed in northerly regions and regularly use the Neusiedlersee region as a resting place during migratory flights in spring and autumn (geese, ducks)
4. Populations that rest anywhere in the Pannonian Basin including occasionally the Neusiedlersee region, as well as birds that breed in the latter region, but otherwise inhabit the surrounding country and do not range too far (e.g. bearded tit)

Ringed individuals of only three species of heron have been recovered in any larger numbers. It seems that the purple heron and grey heron (*A. cinerea*) travel furthest and even reach West Africa (see Fig. 29.6), whereas

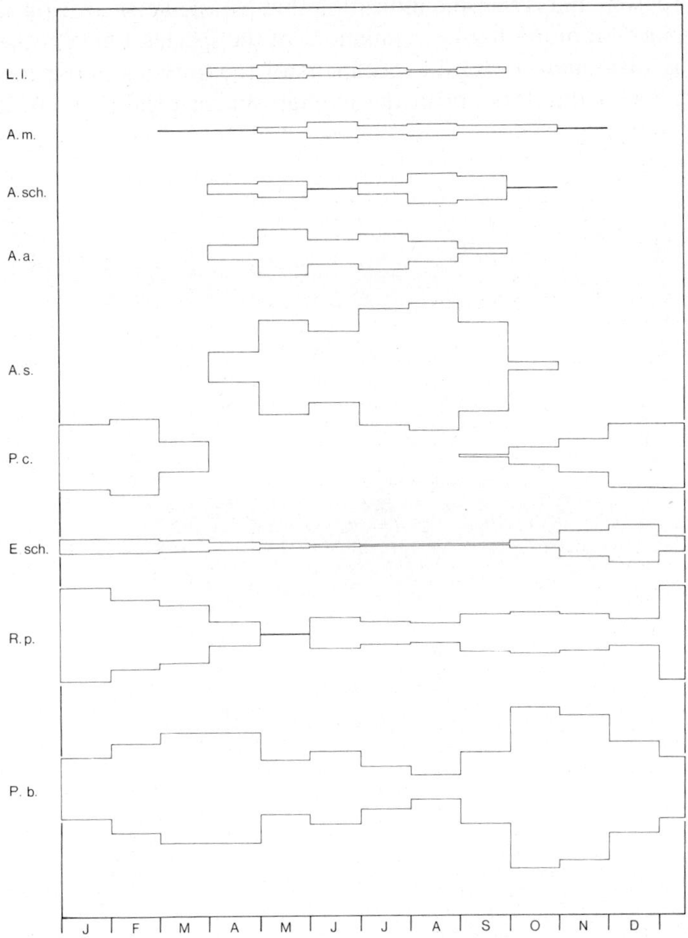

Fig. 29.5. Seasonal distribution of the relative abundance values of the most important bird species caught at the bird-ringing station in Neusiedl/See from 1962 until 1966 by Th. Samwald. L.l.: *Locustella luscinoides* (Savi's warbler); A.m.: *Acrocephalus melanopogon* (moustached warbler); A.sch.: *Acrocephalus schoenobaenus* (sedge warbler); A.a.: *Acrocephalus arundinaceus* (great reed warbler); A.s.: *Acrocephalus scirpaceus* (reed warbler); P.c.: *Parus caeruleus* (blue tit); E.sch.: *Emberiza schoeniclus* (reed bunting); R.p.: *Remiz pendulinus* (penduline tit); P.b. *Panurus biarmicus* (bearded tit).

the great white egret covers only small distances by comparison (see Zink 1976). A characteristic of all species of herons is the undirected type of migration shortly after the end of the breeding season. Birds ringed on Neusiedlersee have been found to have taken the opposite direction from the main migratory routes. The grey heron regularly spends the winter in central Europe whereas the great white egret is seldom encountered at this season. The spoonbill appears to pass the winters mainly in North Africa (Fig. 29.6) although some individuals, probably exceptions, have been observed in the Neusiedlersee region in winter (Leisler 1966). The only anatids for which extensive data are available are the grey-lag goose and the mallard (Fig. 29.7). One ringed specimen of mallard has been reported from Russia in the breeding season, and is the sole indication that Neusiedlersee region serves as a resting place for more distant populations of the species. On the other hand, the finding of numerous ringed specimens of the grey-lag goose has shown that their winter quarters are in the mediterranean region, as far as North Africa.

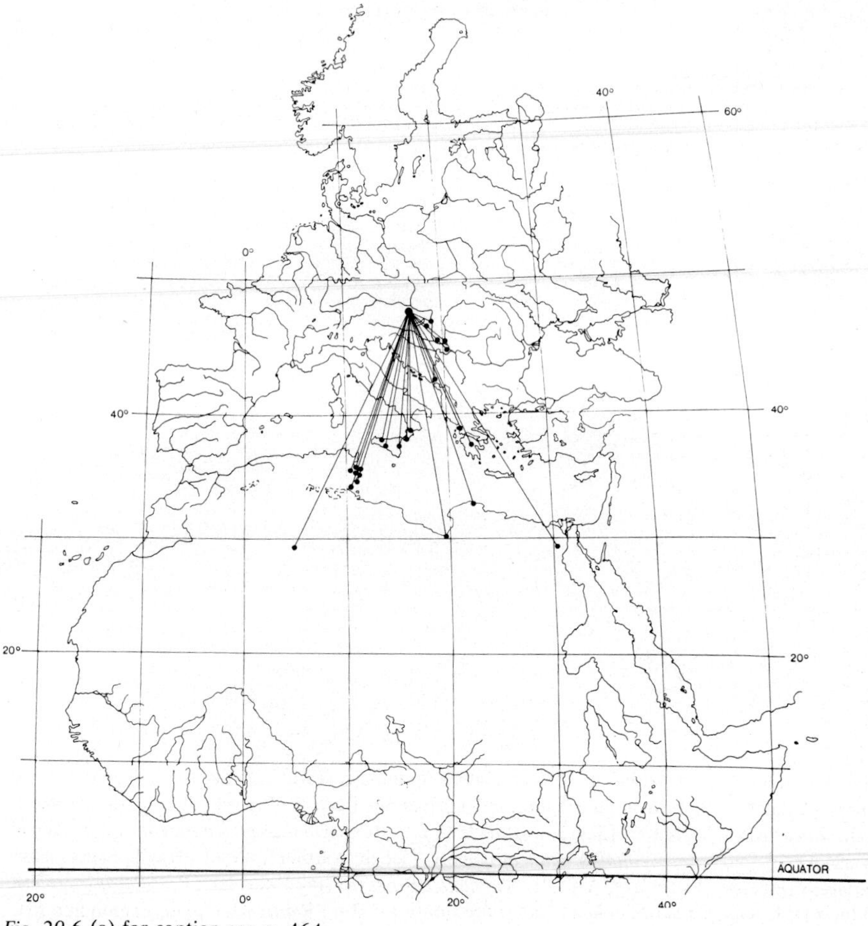

Fig. 29.6 (a) for caption see p. 464.

Fig. 29.6 (b) for caption see p. 464.

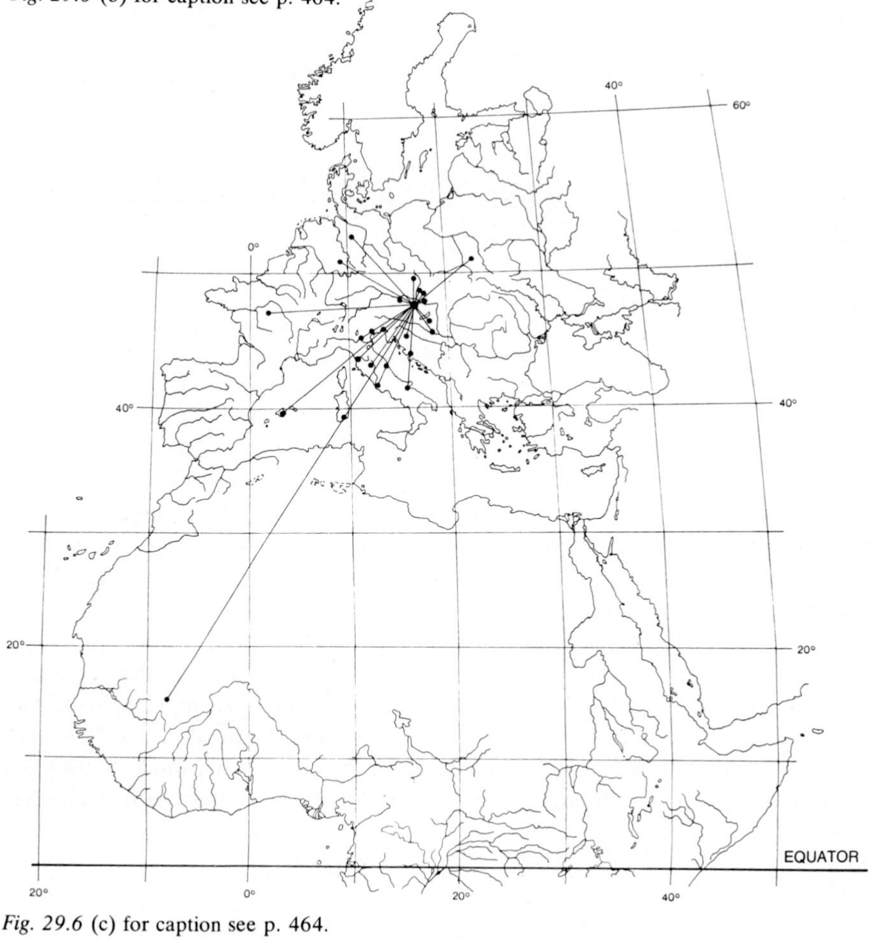

Fig. 29.6 (c) for caption see p. 464.

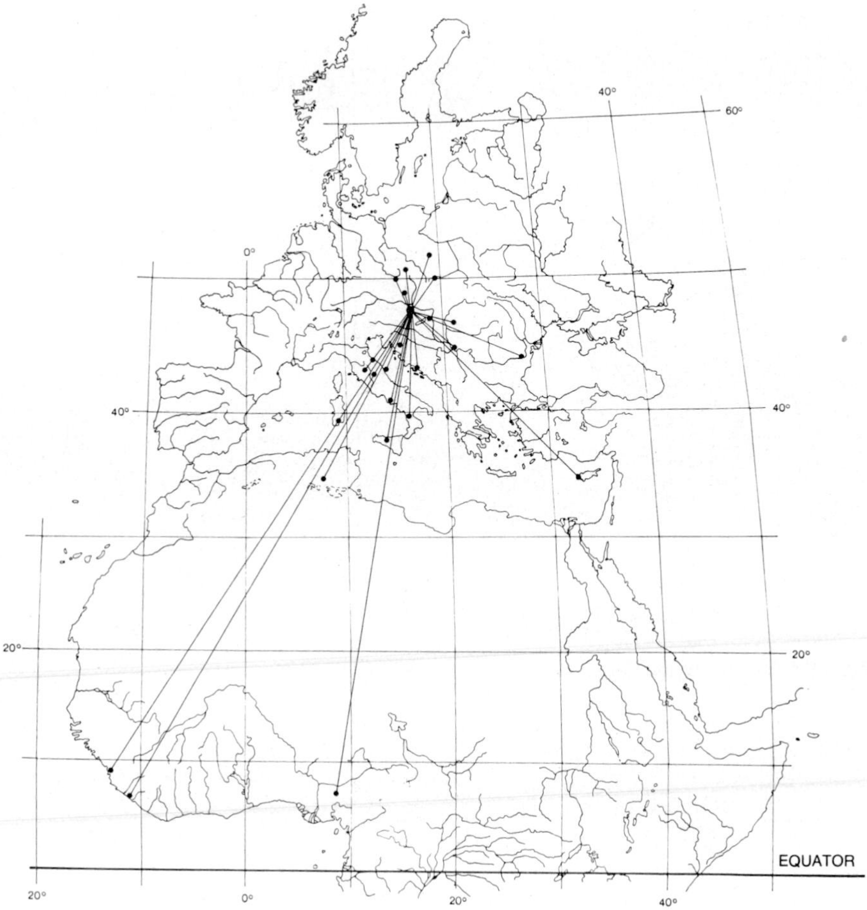

(d)

Fig. 29.6. Recoveries of birds, ringed in Neusiedlersee area (nestlings). a: *Platalea leucorodia* (spoonbill); b: *Casmerodius albus* (great white egret); c: *Ardea cinerea* (grey heron); d: *Ardea purpurea* (purple heron). Most of the ringing of herons and spoonbills was carried out by Koenig and his collaborators.

Water rails, which may also spend the winter on Neusiedlersee, tend to migrate towards the southwest and none has been found further north. Two ringed specimens of the spotted crake, on the other hand, have been recorded in Lithuania and Lettland, which means that Neusiedlersee is used as a resting place for north-eastern populations of the species during migration. However, the number of finds is very low for both species (Fig. 29.8).

The bearded tit is not a great traveller, although it seems to be connected with other suitable biotopes in the Pannonian Basin and in the pond region of southern Bohemia (Fig. 29.9 see also Spitzer 1974). The migrations of the penduline tit take it somewhat further since ringed species of Polish and Czechoslovakian breeding populations have mainly been found in Italy in winter (Fig. 29.9). As already mentioned, contrary to earlier opinion (Bauer

Fig. 29.7 (a) for caption see p. 466.

Fig. 29.7 (b) for caption see p. 466.

Fig. 29.7. Recoveries of birds, ringed in the Neusiedlersee area (a, c) or ringed in other places and recovered at Neusiedlersee (b). a: *Anser anser* (grey-lag goose) ringed in the Neusiedlersee area; b: *Anser Anser* (grey-lag goose) shot in the Neusiedlersee area; c: *Anas platyrhynchos* (mallard) ringed in the Neusiedlersee area. The ringing of grey-lay geese and mallard was carried out by R. Triebl.

et al. 1955), the penduline tit is also found in considerable numbers in the reed in winter, although no information is available.

Nearly all the finds of reed bunting ringed in the Neusiedlersee region have been made in Italy. Bauer et al. (1955) pointed out that different populations (subspecies) are present on Neusiedlersee at different times of year.

A reasonable amount of data is available for only three species of warblers. The recovery of ringed reed warblers is particularly interesting since it shows that whereas populations from the Neusiedlersee region migrate exclusively towards the southeast, populations of the same species from other parts of central Europe travel southwestwards (Zink 1973). The great reed warbler of the region migrates in a southwesterly direction, differing in this only slightly from other European populations (Zink 1973). Of all the warblers the moustached warbler covers the least distance on its migratory flights, spending the winter in the northern Mediterranean region (see Leisler 1973). Almost no ringed specimens of reed warbler or great reed warbler have been reported from their winter quarters (with the exception of one great reed warbler in West Africa). (see Fig. 29.10).

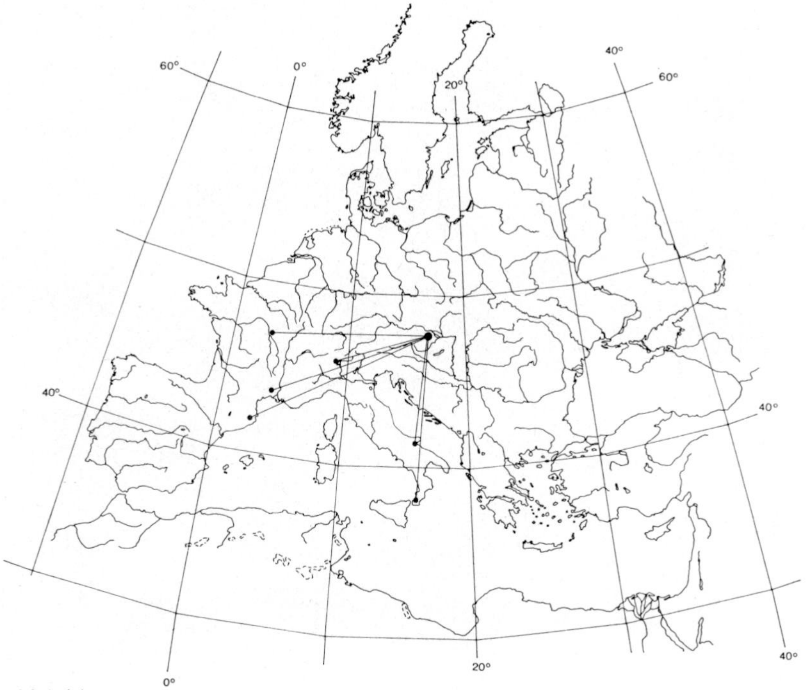

Fig. 29.8 (a)

Fig. 29.8 (b)

Fig. 29.8. Reports of birds, ringed in the Neusiedlersee area a: *Rallus aquaticus* (water rail); b: *Porzana porzana* (spotted crake).

Fig. 29.9 (a) for caption see p. 469.

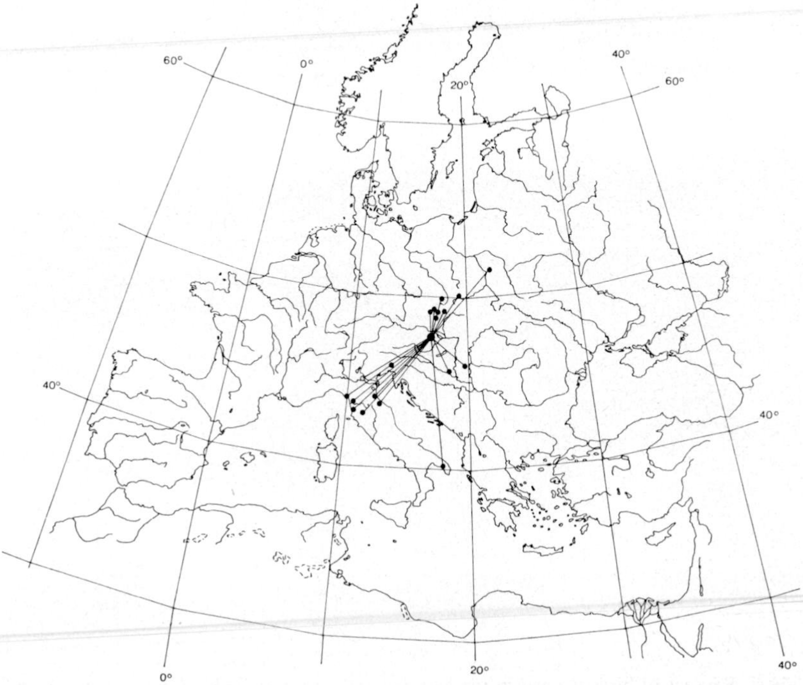

Fig. 29.9 (b) for caption see p. 469

Fig. 29.9 (c)

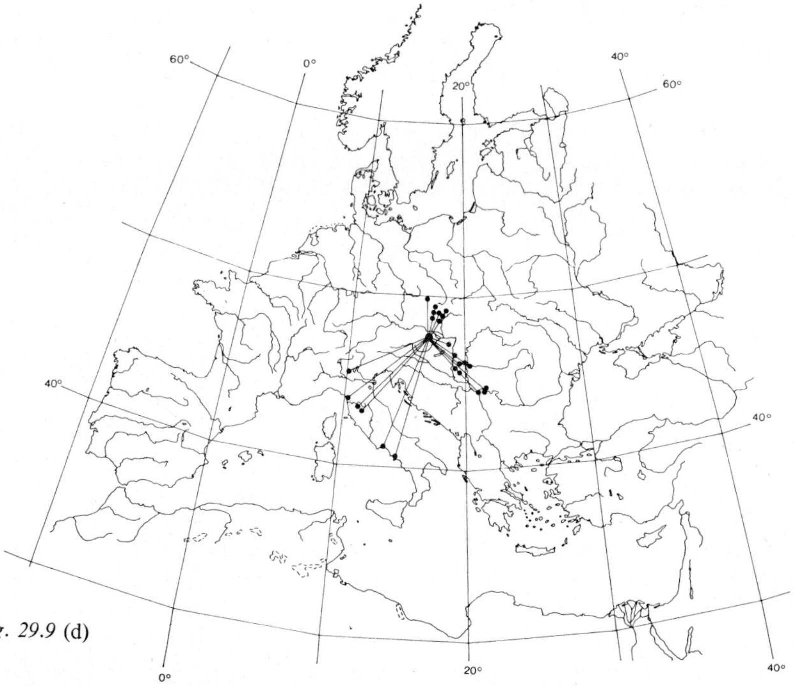

Fig. 29.9 (d)

Fig. 29.9. Reports of birds, ringed in the Neusiedlersee area (b, c, d) or ringed in other places and recovered at Neusiedlersee (a). a: *Remiz pendulinus* (penduline tit), recovered at Neusiedlersee; b: *Remiz pendulinus* (penduline tit) ringed at Neusiedlersee; c: *Emberiza schoeniclus* (reed bunting) ringed at Neusiedlersee; d: *Panurus biarmicus* (bearded tit) ringed at Neusiedlersee.

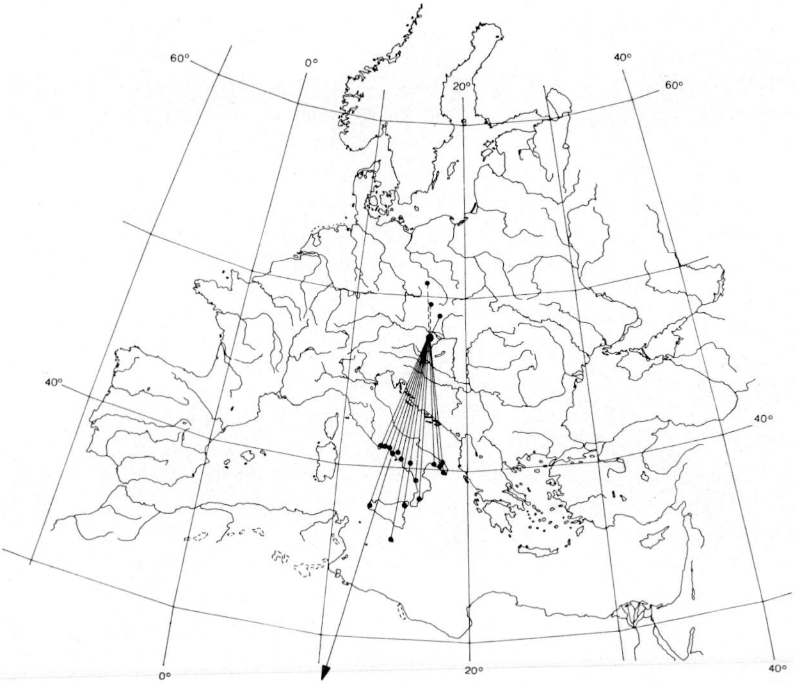

Fig. 29.10 (a) for caption see p. 471.

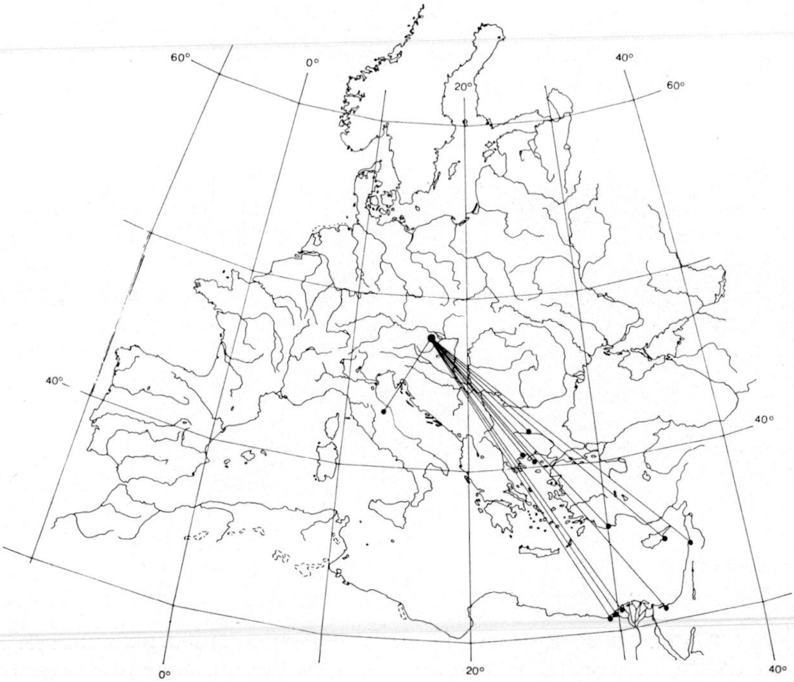

Fig. 29.10 (b) for caption see p. 471.

(c)

Fig. 29.10. Reports of birds, ringed in Neusiedlersee area (a, b, c, d) or ringed in other places and recovered at the Neusiedlersee (a). a: *Acrocephalus arundinaceus* (great reed warbler) ringed at Neusiedlersee, also including a few recaptures of birds ringed in other places (broken line); b: *Acrocephalus scirpaecus* Term. (reed warbler); c: *Acrocephalus melanopogon* (moustached warbler).

29.5 Discussion of production biology

Quantitative data regarding the role of birds as consumers in the ecosystem are so far of a speculative nature. Nevertheless a brief consideration of the problem may lead to a rough estimate of the contribution of birds to the overall figures for the ecosystem worked out by Imhof & Burian 1972. One of the first attempts to estimate the amount of food consumed by herons was made by Koenig (1961). Taking as a basis the breeding population for 1960 (see Table 29.1) and assuming a daily food consumption of 333 g per heron (great white egret, purple heron and grey heron) and 500 g per spoonbill he arrived at a total of 642.3 kg per day for the 1,748 birds, and a consumption of 115.620 kg for the duration of their stay (end of March until end of September, i.e. 6 months), for the breeding population. Assuming an average of two fledglings per pair this amounts to a total food consumption of 173.430 kg fresh weight per season for the entire population, if the young are assumed to eat half as much as the adults. According to Koenig, the food consists to one-third of small mammals and reptiles, one-third water insects and their larvae and one-third fish, frogs and newts. Most of the fish species consumed are present in large numbers and are of no economic value e.g. *Alburnus alburnus.* Herons, and particularly the spoonbills, also rove far afield in search of food and are very frequent visitors to the pools of Seewinkel and the swampy meadows of the Hanság from July onwards. Thus not all of their food comes from the immediate neighbourhood of Neusiedlersee.

Furthermore, when the young are fully fledged a so-called intermediary migration takes place, at least of the great white egret, and begins with a disorderly departure of part of the population. The losses seem to be made good, however, by herons arriving from other breeding grounds in the course of similar migrations. In all probability part of the population of the heron that breeds in the river woodlands of the Danube and March spends the summer in the Neusiedlersee region. It can also be assumed that the proportion of fish consumed in the reed belt is higher than the figures given by Koenig. Despite the data available there is apparently no way of arriving at a satisfactory estimate of the role played by herons in the energetics of the reed-belt ecosystem, mainly because the radius of action of these species extends far beyond the lake region proper.

The situation is slightly more hopeful in the case of the grey-lag goose, using the data given by Leisler (1969) as a basis for further calculations: Assuming 3 fledglings for each of the 213 breeding pairs, plus approximately 150 non-breeding individuals (for the reed belt) this means a population of about 1,300 geese at the end of the breeding season. Assuming a daily requirement of 1 kg green stuff per day and animal (Leisler 1969) for a period of about four months, and allowing half as much for the young for half of this time, the total required by the entire goose population would be 114.000 kg fresh weight per season. A large part of this is no doubt accounted for by reed. Assuming a production of 100 kg excrement (dry weight) per day for the same length of time, 11,460 kg dry weight of excrement are produced during this period in the whole reed belt. Although the geese regularly graze on meadows adjoining the reed belt the larger part of food uptake and the deposition of excrements can be assumed to take place in the reed itself.

It would be of great interest to gather similar data for the far larger number of song-birds, but unfortunately even less material is available for them than for the other species. The following is an attempt to calculate the numbers of the two most common species of warbler: using the data concerning size of territory of the reed warbler complied by Leisler (1971) from the literature and assuming that the mean value calculated therefore is somewhere near correct (on the lakeward areas of the reed conditions are undoubtedly so favourable that population densities are at their highest and the individual territories smallest, whereas in other parts of the reed the reed warbler exhibits a much lower population density) the number of breeding pairs in the Austrian part of the reed belt can be put at about 210,000. The same method cannot be applied for estimating numbers of the great reed warbler since they are unable to occupy large portions of the reed where the stalks are too thin (see above). Using the ratio of great reed warblers to reed warblers trapped during the breeding season, which is 1:3, a figure of 70,000 pairs results for the same area.

There are no data available about the breeding success in this area, which depends on various factors as for instance the influence of wind and water level. The author's data on quantitative food consumption are insufficient for further calculations until now.

The aim of this short synopsis of the few production data is to show that birds may be of greater importance for the ecosystem than was estimated by Imhof & Burian (1972). One of the most important facts is that it is possible for more vagile migratory species like birds to change their habitat depending on the supply of food. The time of reproduction and the maximal quantity of food are often seen to coincide as mentioned above for the grey-lag geese and the reed. Another example may be seen in the increasing number of insect-eating birds in July, the month where production of chironomids, one of the main food resources of the smaller species, is at a maximum (Imhof 1972). Much work on birds and their position in the ecosystem still remains to be done in this area.

References

Aspöck, H., Picher, O., Kunz, Chr. & Böck, F. 1973. Virologische und serologische Untersuchungen über die Rolle von Vögeln als Wirte von Arboviren in Österreich.– Zbl. Hyg., I. Abt. Orig. A224, 156–167.

Bauer, K. 1965. Zur Nahrungsökologie einer binnenländischen Population der Flußseeschwalbe (*Sterna hirundo*).–Egretta 8, 2: 35–51.

Bauer, K., Freundl, H. & Lugitsch, R. 1955. Weitere Beiträge zur Kenntnis der Vogelwelt des Neusiedlersee-Gebietes.– Wiss. Arb. Bgld. 7: 123 pp.

Bauer, K. & Glutz, U. 1966. Handbuch der Vögel Mitteleuropas.– Bd. 1, Akademische Verlagsges., Frankfurt am Main. 483 pp.

Berthold, P. & Schlenker, R. 1975. Das Metnau-Reit-Illmitz-Programm' – ein langsfristiges Vogelfangprogramm der Vogelwarte Radolfzell mit vielfältiger Fragestellung.– Vogelwarte 28, 2: 97–122.

Böck, F. & Scherzinger, W. 1975. Ergebnisse der Wasservogelzählungen in Niederösterreich und Wien aus den Jahren 1964–65 bis 1971–72.– Egretta 18, 2: 34–53.

Elzen, R. van den 1971. Nahrung und Nahrungswahl der Bartmeise (*Panurus biarmicus*).– Diss. Univ. Wien.

Franke, H. 1937. Aus dem Leben der Beutelmeise.– Beitr. Fortpfl. Biol. Vögel 13: 1–4, 1–18.

Geisslhofer, M. & Burian, K. 1970. Biometrische Untersuchungen im geschlossenen Schilfbestand des Neusiedler Sees.– Oikos 21: 248–254.

Glutz, U., Bauer, K. & Bezzel, E. 1971. Handbuch der Vögel Mitteleuropas.–Bud. 4, Akademische Verlagsges. Frankfurt am. Main. 943 pp.

Imhof, G. 1966. Ökologische Gliederung des Schilfgürtels am Neusiedler See und Übersicht über die Bodenfauna unter produktionsbiologischem Aspekt.– Sitz. Ber. Österr. Akad. Wiss., Math.– nat. Kl. I, 175: 219–235.

Imhof, G. 1972. Quantitative Aufsammlung schlüpfender Fluginsekten in einem semiterrestrischen Lebensraum mittels flächenbezogener Eklektoren.– Verh. Dtsch. Zool. Ges. 65: 120–123.

Imhof, G. & Burian, K. 1972. Energy-Flow studies in a Wetland Ecosystem.– Spec. Publ. Austr. Acad. Sc. IBP, 15 pp.

Jilka, A. & Leisler, B. 1974. Die Einpassung dreier Rohrsängerarten (*Acrocephalus schoenobaenus*, *A. scirpaceus*, *A. arundinaceus*) in ihre Lebensräume in Bezug auf das Frequenzspektrum ihrer Reviergesänge.J. Orn. 115, 2: 192–212.

Koenig, O. 1951. Das Aktionssystem der Bartmeise (*Panurus biarmicus*). Teil 1 und 2.–Österr. Zool. Zschr. III, 1–82, 247–325.

Koenig, O. 1952. Ökologie und Verhalten der Vögel des Neusiedlersee-Schilfgürtels.– J. Orn. 93: 207–289.

Koenig, O. 1961. Das Buch vom Neusiedlersee.– 288 pp. Wollzeilenverlag, Wien.

Knoflacher, H. 1974. Zur Ethologie und Produktionsbiologie der Bisamratte (*Ondathra zibethica*).–Diss. Univ. Wien.

Leisler, B. 1966. Ein Überwinterungsversuch des Löfflers (*Platalea leucorodia*) am Neusiedlersee.– Egretta 9, 2: 53–54.

Leisler, B. 1969. Beiträge zur Kenntnis der Ökologie der Anatiden des Seewinkels (Burgenland). Teil I, Gänse.– Egretta 12, 1. u. 2: 1–52.

Leisler, B. 1971: Vergleichende Untersuchungen zur ökologischen und systematischen Stellung des Mariskensängers (*Acrocephalus* (*Lusciniola*) *melanopogon*), ausgeführt am Neusiedler See.–Diss. Univ. Wien.

Leisler, B. 1973. Die Jahresverbreitung des Mariskensängers (*Acrocephalus melanopogon*) nach Beobachtungen und Ringfunden.– Vogelwarte 27, 1: 24–38.

Leisler, B. 1975. Die Bedeutung der Fußmorphologie für die ökologische Sonderung mitteleuropäischer Rohrsänger (*Acrocephalus*) und Schwirle (*Locustella*).–J. Orn. 116: 117–153.

Leisler, B. 1977. Die ökologische Bedeutung der Lokomotion mittel-europäischer Schwirle (*Locustella*).–Egretta 20: 1–25.

Seitz, A. 1937. Von den Reiherkolonien am Neusiedler See.– Beitr. Fortpfl. Biol. Vögel. 10: 228–229.

Seitz, A. 1937. Beobachtungen in den Reiherkolonien des Neusiedler Sees.– Beitr. Fortpfl. Biol. Vögel 13: 13–22.

Spitzer, G. 1966. Das Vorkommen des Seeadlers (*Haliaetus albicilla*) an der niederösterreichischen Donau.– Egretta 9, 2: 43–52.

Spitzer, G. 1972. Jahreszeitliche Aspekte der Biologie der Bartmeise (*Panurus biarmicus*).– J. Orn. 113, 3: 241–275.

Spitzer, G. 1974. Zum Emigrationsverhalten der osteuropäischen Bartmeise (*Panurus biarmicus russicus*) – Eine Diskussion der Fernfunde Neusiedler Bartmeisen.– Vogelwarte 27, 3: 186–193. 186–193.

Waitzbauer, W., Pruscha, H. & Picher, O. 1973. Faunistisch-ökologische Untersuchungen an schilfbewohnenden Dipteren im Schilfgürtel des Neusiedler Sees.– Sitz. Ber. Österr. Akad. Wiss., math.– nat. Kl. I, 181: 111–136.

Weisser, P. 1970. Die Vegetationsverhältnisse des Neusiedler Sees. Pflanzensoziologische und ökologische Studien.– Wiss. Arb. Bgld. 45: 1–83.

Zink, G. 1973. Der Zug der europäischen Singvögel. Ein Atlas der Wiederfunde beringter Vögel.– Vogelwarte Radolfzell.

Zink, G. 1976. Ringfundergebnisse bei den Silberreihern (*Casmerodius albus*) des mittleren Donauraumes.– In: Critti in memoria di Augusto Toschi. Supplemento alle Ricerche di Biologie della Selvaggina, Vol. VII, Bologna.

Zimmermann, R. 1943. Beiträge zur Kenntnis der Vogelwelt des Neusiedlersee-Gebietes.– Ann. Nat. hist. Mus. Wien 54.

0 Projects involving Neusiedlersee

F. Sauerzopf

Events such as floods, droughts, freezing solid, typical of a shallow lake, have been responsible for a multitude of projects (see Cholnoky, 1922; Goll, 1907; Kovacs, 1901; Roth-Fuchs, 1929; Sauerzopf, 1959; Varga, 1937; Wendelberger, 1951; Winkler, 1923). Conflicting opinions became evident at each planning stage, some advocating a complete drainage of the lake, whilst others demanded its total or partial preservation. The measures so far put into force, and with only a single exception all of the earlier projects, aimed at drainage of the lake. Not until the peace treaty of St. Germain allocated the major part of the lake to Austria on account of the German-speaking population of the district, was complete or partial preservation held to be desirable. Prior to this it had been hoped to gain an equivalent of approximately 50 per cent of the area already available to the lake-side communities by totally draining the lake. However, the doubtful suitability of the soil beneath large areas of the lake was completely overlooked, despite the fact that relevant studies on this subject were already in existence from the years 1865–68, when the lake was dried-out (Moser, 1866).

A historical survey of the engineering developments involving the Waasen, the Ikva, the Rabnitz and the Kleine Raab forms an indispensable background to a consideration of the various lake projects. As early as 1699 the Austrian ruler Leopold I ordered a study of the inundations in the Raab area. He charged Count Franz Nàdasdy with the task. In 1762 a Viennese engineer, Maximilian Freman, drew up a plan for the regulation of the Raab and the Hanság. He proposed a drainage canal between the Hanság and the Danube, and planned to drain the wetlands by regulating the Rabnitz and the Kleine Raab. In 1780 the Council of Governors at Sopron charged Samuel Krieger, the county engineer, with the compilation of new recommendations. Their fate was that of all previous plans in that they were never put into effect. At about the same time, Hegedüs completed his plan for the drainage of the Hanság and Neusiedlersee commissioned by Prince Esterházy and the counties concerned. The project comprised a main canal draining the Hanság and a subsidiary one to receive the water of the Ikva, the Répce and the Kleine Raab. Work on the main canal was begun in 1775, the cost of its 8.9 km, with a width of 7.5 m, amounting to 6,300 florins. An extension and the regulation of the affluents involved an additional 13,843 florins, the larger part of which was paid by Prince Esterházy, to whom the drained lands belonged. Maintenance, however, was neglected and thus the canal was completely silted-up and inoperable by 1820. Subsequently, A. Wittmann, steward of the Esterházy estates, presented a new regulation plan in which he proposed the construction of two canals to contain the Waasen, and to be connected by a drainage network. His plans were approved by the governmental commission, but there the matter remained. In 1826 J. Beczedes was invited to submit a new proposal. He suggested a main canal and two marginal canals passing through the lake to contain the Wulka water. Assuming a construction time of 5 or 6 years, and employing 4,000 labourers the project would have cost 2.4 million florins and was expected to yield 3,000 Joch of land. Another plan, that of Kecskes, was to take over Hegedüs' Esterhazy canal and to construct a second one passing through the middle of the Hanság. Since in most cases nothing more than the plan was completed the floods in the Hanság had to be combatted with temporary measures. For example, at the beginning of the eighteen-thirties K. Bocies constructed a canal along the deepest parts of the Waasen up to 2 km west of the Pamhagen dam which had been completed in 1780. This so-called Fertö canal did not reach Neusiedlersee. In 1873 a decisive step was taken in the constitution of the 'Raabregulierungsgesellschaft' (Society for the regulation of the Raab) with the object of regulating the Raab and draining the lake. The society's first plan provided for a canal across the lake and along the eastern shore leading to the Hanság. Because of the expense, however, only the Vorflut-region, i.e. the Raab and its tributaries, was regulated. A modification of the project in the year 1880 provided a canal from Neusiedl am See right through the middle of the lake. Yet another plan involved, besides a canal through the middle of the lake, another from the Wulka to the Ikva. This plan was approved by the Hungarian Department of Commerce in 1889, but the representative of the Hungarian provinces did not approve the drainage project and refused to meet the expense of 5 million florins. Following the drawing-up of entirely new plans an act was passed on 14 June 1895 sanctioning the construction of the so-called Einserkanal. Prior to this, however, the existing canal leading through the Hanság had been reconstructed as far as the Pamhagen bridge at a cost of 2 million florins, but the canal was not linked up with the lake due to lack of funds. Not until the turn of the century did the numerous planners and interested parties begin to question the expediency of draining the lake. In 1902 the Hungarian Department of Agriculture delegated a commission to study the yield to be expected from a drainage of Neusiedlersee. It transpired that the eastern parts of the lake were completely unsuitable for agricultural purposes and would better be kept under water, whereas the western margin offered the possibility of cultivation. A similar opinion had already been expressed by Moser in 1866. These findings were taken into account in the plans drawn up by the Fertö Commission. Between 1908 and 1910, in accordance with a proposal made by Jenö Kvassay in 1905, the Einserkanal was connected with the lake and a lock was constructed in the vicinity of Pamhagen. Although the effect of these measures at first appeared to be negligible, and the lake remained undrained, it became apparent some decades later that the lake did indeed lose 2/3 of its water in this way, the fact being masked, however, by the fluctuating water level. A comparison of the water balance made by Kopf (1963, 1964, 1974) with that of Swarowski (1920) using the measurements of 1901 shows a considerable difference (Kopf, 1967). The fact is that the maximum water level where the Raab empties into the Danube is 20 cm higher than the bottom of the lake in the area of the outflow. This explains why the consequences were not immediately recognizable. A plan that was not made public until 1933 originated from Sándor Károly and dated back to

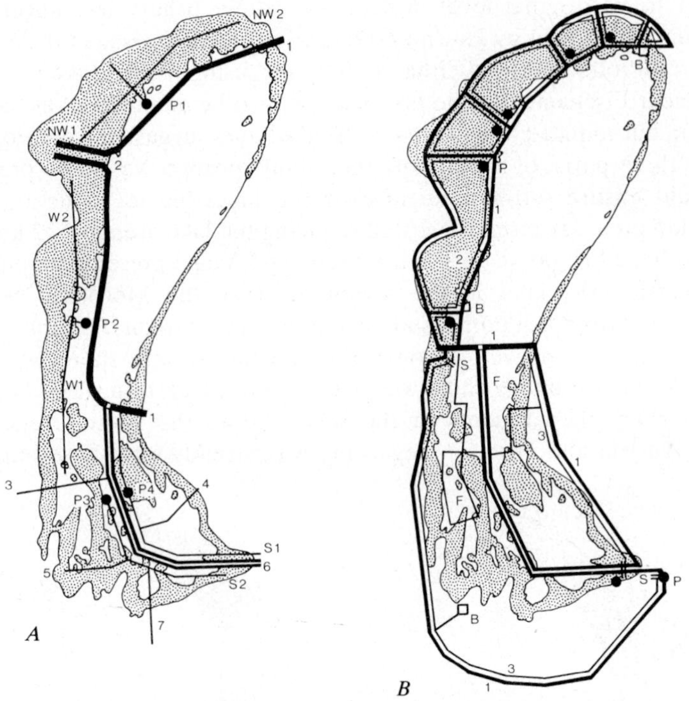

Fig. 30.1A. Project Károlyi, 1920. 1 – dams, NW1/NW2 – north-western drainage channel 1 and 2; W1/W2 – western drainage channel 1 and 2; S1/S2 – southern drainage channel 1 and 2; 2 – Wulka-Kanal; 3 – Kroisbach-Kanal; 4 – eastern drainage channel; 5 – Wolfs-Kanal; 6 – Einser Kanal; 7 – Klein-Andräer-Kanal; P_1–P_4 – pumping stations; remaining lake area: 116.3 km^2

Fig. 30.1B. Project Vogel & Sárkány, 1931. 1 – dams; 2, 3 – channels; B – public baths; F – fish pond; P – pumping station; S – lock; remaining lake area: 119.2 km^2

the time prior to the division of the lake between Austria and Hungary. The plan envisaged a dam from Weiden to Illmitz to cut off the western part of the lake, which would then be drained. Dams would serve to guide the Wulka into the remaining part of the lake and a pumping station would effect the drainage. Proposals made after the major part of the lake had been allocated to Austria revealed divergent tendencies, ranging from maintenance of the existing lake area to reduction of the lake area and addition of water, or reduction without water inflow. Merlicek suggested restoring the lake to its original size, drawing the water needed for this purpose from the Leitha near Katzelsdorf. About 20 m^3/sec was to be led through tunnels and a canal crossing the watershed, and collected in three reservoirs. At a gradient of 133 m, 17 million Kwh would accrue, and the level of the lake would be raised by a further meter. Kadnar hoped to gain 8,000–10,000 Joch of land in the western lake area but to stabilize the water level in the remaining portion to 1.7 m by supplying water from the Leitha at Bruck. A similar plan was proposed by Reinhardt in 1939 and Neresheimer planned merely to restore the water to its original level in order to render fishery less uncertain. A highly complex project by Hoffman-Desperis in 1929 envisaged the construction of reservoirs in the Leitha mountains, using water taken from the Danube near Fischamend. The lake itself was to be dammed so as to rise by about 2 m, but reduced to half its area. Schlarbaum suggested a division of the lake into three parts, of which only the middle portion was to be preserved. This would ensure self-maintenance of the lake due to reduction of the evaporation area. Hainisch suggested reducing the lake area to 127 km², thus creating a huge fish pond. In 1931 Sárkány and Vogel presented a plan for a dam traversing the lake in the region of Rust and Mörbisch. With the exception of several fish ponds and an area for reed exploitation the southern portion of the lake, as well as some areas on the western shore, were to be drained. A main drainage canal was planned on the eastern shore, as well as canals to carry the affluents in the west and in the south. Negotiations between Austria and Hungary regarding water rights were based upon this

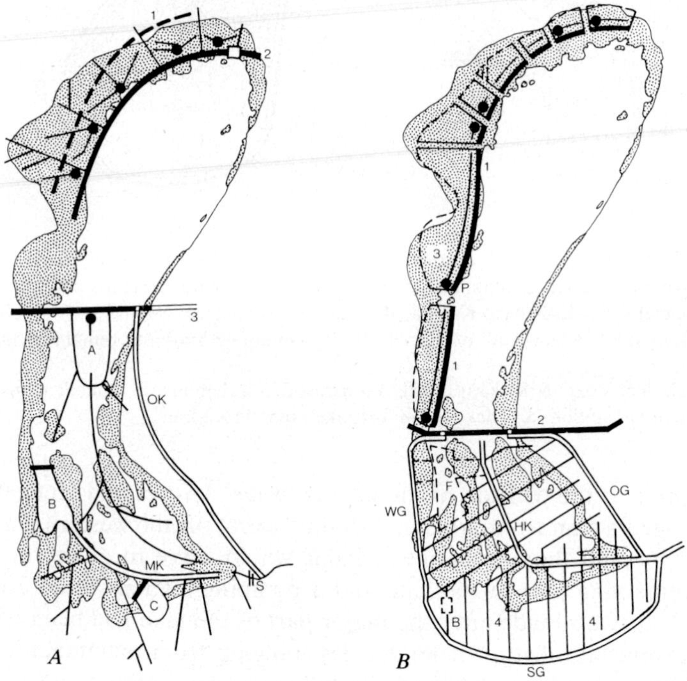

Fig. 30.2A. Project Goldemund, 1933. 1 – planned dam; 2, 3 – dams; A, B, C – ponds; ● – pumping station; MK – channel; S – lock; remaining lake area: 116.6 km²

Fig. 30.2B. Project Riedinger, 1940. 1, 2 – dam; 3, 4 – drained area; B – public baths; P – pumping station; HK – main drainage channel; OG – eastern drainage channel; SG – southern drainage channel; WG – western drainage channel; F – fish pond; remaining lake area: 162 km² plan and were even approved by both states, but financial support was not forthcoming. In 1933 Goldemund published a similar project differing from that of Sárkány and Vogel only in the provision of a pumping station for draining the western part. The project of Grünhut-Bartoletti was a purely Austrian proposal, envisaging two dams in a north-south direction, bounding the 112.5 km^2 of the lake that were to be preserved. During World War II, which put an end to all of these projects, the only studies made were Schuster's little-known doctoral thesis and a so-called general water economy plan drawn up by Riedinger (1940), summarizing various former plans. The latter visualized a dam crossing from Mörbisch to Apetlon. To the north, the reed belt (54 km^2) was to be cut off from the open lake and drained, whilst the latter was to be reduced to an area of 162 km^2. The greater part of the southern portion (121 km^2) was also to be drained, with only a small area being preserved for reed cultivation and fishery. The first post-war project in the early fifties had nothing to do with the regulation of Neusiedlersee but involved a bridge crossing the lake on concrete piers (Schlarbaum) and located on a line north of Oggau towards Podersdorf. For financial reasons the project was quietly dropped. In about 1953 the plan of a traversing dam near Mörbisch was discussed, with the aim of stabilizing the water level and thus providing better conditions for fishery. It was intended to finance this project with Marshall Aid, but the situation in eastern Austria, at that time still under Russian occupation, prevented its fulfilment. Later on, the question of a dam across the lake, requested and promised ever since Burgenland became Austrian in 1921, was repeatedly brought up. Considerations and proposals on the subject were contained in an unpublished project 'Studie Seewinkel' 1962 (Kopf, 1963). Following a detailed consideration, including studies of the water balance, the project of a lake dam was dropped on ecological grounds: the effects on the ecological equilibrium of the lake could not be foreseen and the idea of a bridge was again taken up. Since the project was given much attention by press, radio and television, and aroused considerable controversy of an almost purely emotional nature, it warrants mention here. A road bridge south of the existing bathing beach, and leading from Mörbisch to Illmitz, would provide a link between the Seewinkel east of Neusiedlersee and the Eisenstadt region, with its central schools, hospital and administration. It would reduce the distance by 50 per cent and provide a rapid communication route to the industrial area of Upper Styria via the Wiener Neustadt region (Schreiber 1972). The plan for the bridge, which was to be a pillar construction approximately 5 m above the mean water level is depicted in Ofner's publication of 1972. To combat the danger of eutrophication fully biological water treatment plants have been constructed in the lake vicinity and in the more remote catchment area (Wulka Basin) (see Chapters 1, 10). The plants in Rust, Oggau, Breitenbrunn, Neusiedl am See, Weiden and Podersdorf, as well as some minor waste water treatment plants are already in operation. The remainder, together with the central waste plant for the Wulka Basin, are under construction. Some hope is also given by the fact that Neusiedlersee was declared a Biosphere Reserve by UNESCO in 1977.

References

Cholnoky, J. 1922. A folyók és tavak vizállásáról (Vom Wasserstand der Flüsse und Seen).– Hidrogr. Közl.

Goll, K., 1907. Die Schwankungen des Neusiedlersees.– Jahresber. deut. Realsch. Triest.

Grünhut-Bartoletti, 1935. Die Regulierung und Nutzbarmachung des Neusiedler Sees auf österreichischem Gebiet.– WW. u. Techn.

Gutachten des Hydrographischen Zentralbüros über das genenerelle Projekt der Ingenieure Sárkány und Vogel in Budapest vom November 1931 zur Regulierung des Neusiedlersees.

Hainisch, S., 1925. Der Neusiedlersee.– Die Wasserwirtschaft.

Hoffman, F. u. Deperis, G., 1929. Wasserkraftwerk an der Donau und am Neusiedlersee.– Die Wasserwirtschaft.

Kadnar, A., 1930. Über die Regulierung des Neusiedlersees.– Die Wasserwirtschaft.

Károly, S., 1930. A Fertö szabállyozása (Die Regulierung des Neusiedlersees)– Sopron Hirl. 17: 253.

Kopf, F., 1963. Studie Seewinkel Wien 1963.– unpubl.

Kopf, F., 1963. Wasserwirtschaftliche Probleme des Neusiedler Sees und des Seewinkels.– Österr. Wasserwirtschaft 15, 9/10: 190–203.

Kopf, F., Die wahren Ausmaße des Neusiedler Sees.– Österr. Wasserwirtschaft 16, 11/12: 255–262.

Kopf, F., 1967. Die Rettung des Neusiedler Sees.– Österr. Wasserwirtschaft 19, 7/8: 139–151.

Kopf, F., 1974. Der neue Wasserhaushalt des Neusiedler Sees.– Österr. Wasserwirtschaft 26, 7/8: 169–180.

Kovács, V., 1901. A Fertö tavának ingacozása (Die Schwankungen des Neusiedlersees).– Jb. d. Gymn. in Estergom.

Merlicek, E., 1922. Technische Aufgaben im Burgenland.– Die Wasserwirtschaft 4.

Moser, I., 1866. Der abgetrocknete Boden des Neusiedler Sees.– Jhb. d.k.u.k. geol. R. A. Wien.

Neresheimer, E., 1925. Zur Nutzbarmachung des Neusiedlersees.– Öst. Fztg. 19.

Ofner, K., 1972. Technische Aspekte einer Neusiedlerseeüberquerung.– Wiss. Arb. Bgld. 48: 5–17.

Roth-Fuchs, G., 1929. Beiträge zum Problem Neusiedlersee.– Mitt. Geogr. Ges. Wien, 72.

Roth-Fuchs, G., 1929. Beobachtungen über Wasserschwankungen am Neusiedlersee.– Mitt. Geogr. Ges. Wien, 76.

Sauerzopf, F., 1956. Probleme und Projekte am Neusiedlersee.– Bgld. Heimatbl. 18/4.

Sauerzopf, F., 1972. Die biologischen Aspekte einer Seequerverbindung Illmitz-Mörbisch.– Wiss. Arb. Bgld. 48: 27–40.

Schreiber, G., 1972. Die Brücke über den Neusiedlersee aus regionaler Sicht.– Wiss. Arb. Bgld. 48: 19–26.

Schuster, F., 1943. Das Regulierungsproblem des Neusiedlersees.– Diss. Techn. Hochsch. München.

Swarowski, A., 1920. Die hydrographischen Verhältnisse des Burgenlandes.– Bgld. Festschrift, Wien.

Szekendi, F., 1938. Hanság és a Fertö lecsapolási kisérleteinek története (Geschichte der Versuche zur Entwässerung des Neusiedlersees und des Waasens).– Magyarórár.

Szekenyi, B., 1903. Fertöszabállyozási tervezet (Regulierungsprojekt Neusiedlersee)– Nagycek.

Varga, L., 1932. Katastrophen in der Biocoenose des Fertö (Neusiedlersee).– Int. Rev. Ges. Hydrobiol. 27, 130–150.

Varga, L., Mika, F., 1937. Die jüngsten Katastrophen des Neusiedlersees und ihre Einwirkungen auf den Fischbestand des Sees.– Archiv Hydrobiol. 31: 527–546.

Wendelberger, G., 1951. Die Wasserstandsschwankungen des Neusiedlersees.– Natur und Land 37: 6.

Winkler, A., 1923. Die Zisterzienser am Neusiedlersee und die Geschichte dieses Sees.– Mödling.

General Bibliography

Bechtle, W. 1976: Der Neusiedler See in Farbe, Stuttgart, Kosmos, Franckh'sche Verlagsbuchhandlung, 71 pp.

Koenig, O. 1961: Das Buch vom Neusiedlersee, Wien, Wollzeilen Verlag

Koenig, O. 1964: Führer rund um den Neusiedlersee, Wien, Verlag Jugend und Volk.

Leisler, B. 1979: Neusiedler See, Kilder-Verlag, Greven, 62 pp.

Löffler, H. 1974: Der Neusiedlersee, Wien, Verlag Fritz Molden, 175 pp.

Laszloffy, W. 1972: A Fertö-Táj Bibliográfiája. (Bibliographie des Neusiedlersee-Gebietes), Kiadja: Györ-Sopron Megye Tanácsa, 294 pp.

Mazek-Fiala, K. 1947: Die österreichische Seesteppe. Leben und Landschaft am Neusiedlersee (Az osztrák sztyepp. Élet es táj a Fertö tónál), Wien, Verlag K. Kühne, 104 pp.

Pichler, J. 1979: A Fertö-Táj. Tudományos Kutatási Terve, Kézirat Közreadja Györ-Sopron megye Tanácsa, 419 pp.

Sauerzopf, F., Tauber, A. F. (eds.) (1959): Landschaft Neusiedlersee, Burgenl. Landesmuseum Eisenstadt, 208 pp.

List of species of organisms in Neusiedlersee

(* : extinct in the lake)

With the exception of birds (listed from the entire catchment area) only animals known from the lake and of aquatic character have been considered. However, the list includes also terrestrial organisms which have been mentioned in some of the chapters.

For animals the family is indicated. Authors of groups of organisms have been mentioned only if the information has not yet been published. Several groups are still poorly known, such as Protozoa, Plathelminthes, Gastrotricha, Hirudinea, Acari, Hemiptera, Bryozoa.

Phylum Bacteriophyta

Acinetobacter sp. (= Achromobacter)
Beggiatoa sp., 175
Rhodopseudomonas gelatinosa?

Phylum Mycophyta

Cladosporium sp.
Torula sp.

Phylum Cyanophyta

Anabaena aequalis Borge
A. flos-aquae (Lyngbye) Breb., 177, 206
A. cf. torulosa (Carm.) Lagerh.
A. viguieri Denis et Frémy, 177, 206
Aphanizomenon gracile Lemm., 177, 206
Aphanothece sp.
Calothrix braunii Bornet et Flahault
C. parietina (Naeg.) Thuret, 175, 177
Chamaesiphon subaequalis Geitler, 177
Chroococcus limneticus Lemm.
Ch. minutus (Kütz.) Naeg., 177, 204, 205, 206, 209, 210
Ch. turgidus (Kütz.) Naeg., 177
Coelosphaerium kuetzingianum Naeg., 177
C. naegelianum Unger
Eucapsis alpina Clements et Shantz
Gomphosphaeria aponina Kütz.
G. lacustris Chodat
Lampropedia hyalina Schroeter, 173, 176, 177
Lyngbya aestuarii (Mert.) Liebm.
L. hieronymusii Lemm., 177, 178
L. kuetzingii Schmidle
L. limnetica Lemm., 177, 206
L. martensiana Menegh., 177
Merismopedia glauca (Ehr.) Naeg., 177
M. punctata Meyen, 177, 204, 205, 206, 209, 210
M. tenuissima Lemm.
Microcystis aeruginosa Kütz.
M. pulverea (Wood) Migula, 177, 205, 206, 209, 210
Nodularia spumigena Mertens
Nostoc planctonicum Poretzky et Tschern., 177
Oscillatoria amphibia Ag., 177
O. borneti Zukal
O. brevis Kütz., 177
O. chalybaea Mertens, 177, 178
O. chlorina Kütz., 177, 178
O. cortiana (Menegh.) Gomont
O. lauterbornii Schmidle, 177
O. limosa Kütz., 177, 178
O. minima Gicklhorn
O. okeni (Ag.) Gomont, 177
O. princeps Vaucher, 177
O. pseudoacutissima Geitler, 177
O. splendida Greville
O. subtilissima Kütz., 177
O. tenuis Ag. var. tergestina (Kütz.) Rabenh., 177, 178
O. trichoides Szafer
Pseudanabaena catenata Lauterb., 177, 178
Spirulina jenneri (Stizenb.) Geitler, 177, 206
Sp. labyrinthiformis Menegh.
Sp. major Kütz. 177, 178
Sp. raphidioides Geitler, 177
Sp. subsalsa Oerst., 177, 178
Sp. subtilissima Kütz., 177
Tolypothrix lanata Wartm.

Phylum Phycophyta

CLASS CHRYSOPHYCEAE

Order Chrysomonadales Chrysocapsales, Chrysotrichales

Anthophysa vegetans (O.F.M.) Stein
Apistonema expansum Geitler, 189
Chromulina sp.
Ch. biococca Schiller, 189
Chrysocapsa granifera (Mack) Bourr. = Sarcinochrysis granifera, 189, 206
Dinobryon sertularia Ehr., 189
Mallomonas sp., 189
Ochromonas sp.
Phaeothamnium confervicola Lagerh., 178, 189
Synura sp., 206
S. uvella E., 191
Sarcinochrysis granifera (Mack) Tscherm.-Woess, 178–189

Order Diatomales (Bacillariophyceae, Bacilliorophyta)

Achnanthes brevipes Ag., 193
A. biasolettiana Ag.
A. exigua Grun.
Achnanthes hungarica Grun.
A. lanceolata Bréb.
A. minutissima Kütz., 193
Amphipleura pellucida Kütz., 175, 190, 194
A. rutilans (Trentepohl) Cleve, 175, 194
Amphiprora paludosa W. Smith, 195, 206
A. costata Hust., 173, 192, 193, 195, 204, 206
Amphora coffeaeformis Ag.
A. commutata Grun., 195
A. ovalis Kütz., 195, 204, 206
A. ovalis var. pediculus Kütz.
A. perpusilla Grun.
A. veneta Kütz., 195
Anomoeoneis costata (Kütz.) Hust., 194
A. serians (Bréb.) Cleve
A. sphaerophora (Kütz.) Pfitz., 174, 190, 194
Bacillaria paradoxa Gmelin, 171, 173, 174, 175, 176, 192, 196, 206
Caloneis amphisbaena (Bory) Cleve, 174, 190, 195
C. bacillum (Grun) Mereschk., 195
C. permagna (Bail.) Cleve, 190, 193, 195
C. schumanniana f. biconstricta Grun., 195
C. silicula (Ehr.) Cleve, 195
C. silicula var. peisonis Hust., 195
C. silicula fa. truncatula Grun.
Campylodiscus clypeus Ehr., 37, 38, 172, 173, 174, 193, 197, 204, 205, 206
C. clypeus var. bicostata (W. Smith) Hust., 197, 206
C. sp. 38
Chaetoceros muelleri Lemm. 193, 205, 206
Cocconeis pediculus Ehr.
C. placentula Ehr., 193
Cyclotella comta (Ehr.) Kütz., 173, 193
C. kuetzingiana Thwaites
C. meneghiniana Kütz., 190, 193, 205, 206, 215
C. stelligera, 205
C. tenuistriata Hust.
Cyclotella sp., 204
Cylindrotheca gracilis (Bréb.) Grun., 174, 190, 193, 196
Cymatopleura solea (Bréb.) W. Smith, 173, 174, 197, 206
Cymbella affinis Kütz., 196
C. aspera (Ehr.) Cleve, 196
C. cistula (Hempr.) Grun., 196
C. hungarica (Grun.) Pantocs.
Cymbella lacustris (Ag.) Cleve, 174, 175, 192, 196
C. lanceolata (Ehr.) V. Heurck, 196
C. microcephala Grun.
C. pusilla Grun., 193, 195
C. prostrata (Berk.) Cleve, 175, 195
Diatoma elongatus Ag., 193, 206
D. vulgare Bory, 193
Diploneis ovalis (Hilse) Cleve
Epithemia argus Kütz., 196
E. sorex Kütz., 196
E. turgida (Ehr.) Kütz., 196
E. zebra (Ehr.) Kütz., 196
E. zebra var. porcellus (Kütz.) Grun.
Eunotia lunaris (Ehr.) Grun., 193
Fragilaria brevistriata Grun., 206
Fr. capucina Desm., 193
Fr. capucina fo. mesolepta (Rabenh.) Grun.
Fr. construens (Ehr.) Grun.
Fr. crotonensis Kitton, 193
Fr. pinnata Ehr.
Fr. sp. 204, 205
Gomphonema acuminatum Ehr.
G. acuminatum fo. brebissonii (Kütz.) Cleve
G. angustatum (Kütz.) Rabenh.
G. constrictum Ehr., 196
G. constrictum fo. capitatum (Ehr.) Cleve
G. intricatum Kütz. var. pumila Grun.
G. intricatum var. vibrio (Ehr.) Cleve
G. longiceps Ehr. fo. subclavatum Grun.
G. olivaceum (Lyngbye) Kütz., 174, 196
G. olivaceum var. calcarea Cleve, 175, 193, 196
G. parvulum (Kütz.) Grun., 196

Gyrosigma acuminatum (Kütz.) Rabenh., 195, 206, 377
G. attenuatum (Kütz.) Rabenh., 195, 206
G. macrum (W. Smith) Cleve, 171, 173, 174, 190, 191, 195, 205, 206
G. peisonis (Grun.) Hust., 190, 193, 195, 205, 206
G. spencerii (W. Smith) Cleve, 173, 174, 195
Hantzschia amphioxys (Ehr.) Grun., 196
H. spectabilis
H. vivax (W. Smith) Hust., 196
Hantzschia spectabilis (Ehr.) Hust., 196
Mastogloia dansei (Thw.) W. Smith
M. grevillei W. Smith var. subconstricta Pantocs.
M. smithii Thwait., 174, 194
Melosira italica (Ehr.) Kütz.
M. varians Ag., 193
Navicula accommoda Hust.
N. capitata Hust.
N. cincta (Ehr.) Ralfs, 194
N. cryptocephala Kütz., 192, 194
N. cuspidata Kütz., 174, 194
N. cuspidata var. ambigua (Ehr.) Cleve
N. cuspidata var. heribaudii Perag.
N. gracilis Ehr., 192, 195
N. halophila (Grun.) Cleve, 193, 194
N. halophilioides Hust. 194
N. hungarica Grun., 195
N. hungarica fo. capitata (Ehr.) Cleve
N. mutica Kütz.
N. oblonga Kütz., 174, 176, 192, 195, 204
N. protracta Grun., 195, 206
N. pupula Kütz.
N. pupula fo. pygmaea
N. pygmaea Kütz., 174, 190, 195
N. radiosa Kütz., 174, 176, 192, 195, 206
N. reinhardtii Grun., 195
N. rhynchocephala Kütz.
N. salinarum Grun., 195
N. stundlii Hust., 195
N. subrhynchocephala Hust.
N. subsulcatoides Hust.
N. tenuipunctata Hust., 195
N. viridula Kütz.
N. sp. 173, 377, 204
Neidium iridis (Ehr.) Cleve
Nitzschia acicularis W. Smith, 197
N. acicularioides Hust., 172, 196
N. admissa Hust., 196
N. amphibia Grun., 196
Nitzschia apiculata (Greg.) Grun., 196
N. communis Rabenh., 196
N. dissipata (Kütz.) Grun., 196
N. filiformis (W. Smith) Hust., 175, 192, 197
N. frustulum (Kütz.) Grun., 196
N. frustulum var. perpusilla (Kütz.) Grun.
N. geitleri Hust., 197
N. hungarica Grun., 196
N. hungarica var. linearis Grun.
N. jugiformis Hust., 196
N. kuetzingioides Hust., 196
N. legleri Hust., 196
N. linearis W. Smith
N. lorenziana Grun. (var. subtilis Grun.), 174, 190, 197
N. navicularis (Bréb.) Grun., 194, 196
N. obtusa W. Smith, 197
N. palea (Kütz.) W. Smith, 197, 206
N. peisonis Pantocs., 196
N. salinarum Grun., 196
N. sigma (Kütz.) W. Smith
N. sigmoidea (Ehr.) W. Smith 173, 197, 206, 377
N. spectabilis (Ehr.) Ralfs, 206
N. subamphioxoides Hust., 196
N. subcapitellata Hust., 196
N. subtiloides Hust., 197
N. tryblionella Hantzsch, 174, 190, 196, 377
N. vitrea Norm., 196
N. sp., 173, 176, 204, 205, 377
Pinnularia gentilis (Donkin) Cleve
P. kneuckeri Hust., 195
P. major Kütz, 195
P. microstauron (Ehr.) Cleve var. brebissonii (Kütz). Hust., 195
P. nobilis Ehr.
P. viridis (Nitzsch) Ehr.
Pleurosigma elongatum W. Smith, 195, 206
Rhoicosphenia curvata (Kütz.) Grun., 174, 175, 192, 193
Rhopalodia gibba (Ehr.) O. Müller, 174, 196, 204, 206
Rh. gibba fo. ventricosa (Ehr.) Grun.
Rh. gibba var. peisonis (Pant.) Hust.
Rh. gibberula (Ehr.) O. Müller, 196
Scoliopleura peisonis, 205
Scoliotropis peisonis (Grun.) Hust., 192, 193, 195, 205, 206
Stauroneis spicula Dickie = Navicula spicula (Dickie) Cleve, 194
St. wislouchii Poretzky et Anisimowa, 194
Stephanodiscus astraea (Ehr.) Grun.
St. hantzschii Grun., 193, 205, 206
Surirella angusta Kütz.
S. ovata Kütz., 197, 206
S. peisonis Pantocs., 171, 172, 173, 174, 192, 193, 197, 206, 206, 204, 206
S. robusta Ehr. var. splendida (Ehr.) v. Heurck, 197
S. splendida (Ehr.) Kütz.
S. tenera Greg., 197

S. tenera fo. cristata
Synedra acus Kütz., 193, 204, 206
S. acus var. radians (Kütz.) Hust.
S. acus var. angustissima Grun.
S. affinis Kütz = S. tabulata, 175
S. amphicephala Kütz.
S. capitata Ehr., 193, 204, 206
S. nana Meister, 206
S. parasitica W. Smith
S. pulchella Kütz., 193
S. tabulata (Ag.) Kütz. = S. affinis Kütz., 193, 206
S. ulna (Nitzsch) Ehr. 193, 204, 206
S. ulna fo. biceps (Kütz.) v. Schönf.
S. ulna fo. danica (Kütz.) Grun.
S. ulna fo. aequalis (Kütz.) Hust.
S. ulna fo. spathulifera Grun.
S. vaucheriae Kütz.
S. sp. 205

CLASS XANTHOPHYCEAE

Characiopsis sp.
Chlorobotrys sp.
Goniochloris laevis Pascher
Ophiocytium cochleare A. Braun
O. majus Naeg., 188, 189, 199
O. parvulum A. Braun
Mischococcus confervicola Naeg., 200
Tribonema bombycinum Derbès et Sol.
Tribonema elegans Pascher
Tr. minus Hazen
Tr. regulare
Tr. vulgare Pascher
Vaucheria sp., 200

CLASS PYRRHOPHYCEAE

Order Dinophycales

Amphidinium bidentatum Schiller, 181
A. caerulescens Schiller, 183
A. eucephalum Schiller
A. glaucovirescens Schiller
A. inconstans Schiller
A. lacustre Stein.
A. multiplex Schiller
A. obliquum Schiller
A. sauerzopfii Schiller, 183
A. vorax Schiller, 180, 183
Exuviaella peisonis Schiller
Glenodinium amphiconicum Schiller, 180, 183
Gl. armatum Levander
Gl. bieblii Schiller, 180, 183
Gl. cinctum Ehr.
Gl. coronatum Wolosz., 183
Gl. denticulatum Schiller
Gl. fungiforme Schiller
Gl. gessneri Schiller
Gl. kampneri Schiller
Gl. peisonis Schiller
Gl. sciculiferum Schiller
Gymnodinium absumens Schiller
G. achroum Schiller
G. amphiconicoides Schiller
G. baumeisteri Schiller
G. caerulescens Schiller
G. coronatum Wolosz., 181
G. cyaneum Schiller, 180, 181
G. devorans Schiller
G. glaucum Schiller, 181
G. granii Schiller
G. knollii Schiller, 180, 181
Gymnodinium legicon veniens Schiller
G. mitratum Schiller, 181
G. paradoxiforme Schiller, 180, 181
G. pascheri (Suchlandt) Schiller, 176, 180, 181, 207
G. peisonis Schiller
G. schuettii Schiller
G. submontanum Schiller
G. tatricum Wolosz.
G. veris Linderm. 181 = G. pascheri status veris, 180
G. viridaliut Schiller
G. wawrikae Schiller
G. wigrense Schiller
Gyrodinium elongatum Schiller
G. pallidum Schiller
Massartia crassifilum Schiller
M. edax Schiller
M. fungiformis (Anissomowa) Schiller
M. hiemalis Schiller, 183
Peridinium borgei Lemm.
P. bipes Stein
P. cinctum (Müller) Ehr., 180, 183, 207, 208
P. cunningtonii Lemm., 183
P. hiemale Schiller, 180, 183
P. inconspicuum Lemm., 180, 183
P. palatinum Lauterb., 183
P. peisonis Schiller (?)
P. penardiforme Lindem.
Prorocentrum tubiferum Schiller

Order Cryptophycales

Chilomonas naviculaeformis Schiller
Ch. paramaecium Ehr.
Chroomonas acuta Utermöhl
Ch. austriaca Schiller
Ch. breviflexa Schiller
Ch. caerulea Skuja, 178, 179

Ch. cor Schiller, 179
Ch. curvata Schiller, 179
Ch. cyanea Schiller, 179
Ch. elegans Schiller
Ch. minor Nyg.
Ch. nana Schiller, 179
Ch. reflexa Kissel.
Ch. unamacula Schiller, 178, 179
Cryptochrysis minor Nyg. 179
Cryptomonas appendiculata Schiller, 179
Cr. brevis Schiller
Cr. conicoides Schiller
Cr. constricta Schiller
Cr. erosa Ehr., 181, 208
Cr. inaequalis Schiller
Cr. komma Schiller, 181
Cr. lens Schiller
Cr. marssonii Skuja, 181
Cr. navicula Schiller
Cr. obovata Skuja, 181
Cr. ornatopaux Schiller
Cr. ovata Ehr. et var. sursumexstans Schiller, 178, 181
Cr. peisonis Schiller, 181
Cr. perimpleta Schiller var. cordiformis Schiller
Cr. piriformis Schiller
Cr. platyuris Skuja, 179, 181
Cr. pochmanni Schiller
Cr. postunguis Schiller
Cr. procera Schiller
Cr. pyrenoidifera Geitler
Cr. reflexa (Marsson) Skuja, 178, 181
Cr. rostrata Skuja, 181
Cr. rusti Schiller
Cr. sinuosa Schiller
Cr. sphaerafaux Schiller
Cr. spiroides Schiller
Cr. violacea Schiller, 181
Cyanomonas caudiculata Schiller
C. curvata Schiller, 179
C. decernens Schiller
C. effecta Schiller
C. procedens Schiller
Rhodomonas lacustris Pascher et Ruttner, 204, 208, 210, 213
R. tenuis Skuja, 178, 179

CLASS EUGLENOPHYCEAE

Anisonema acinus DUJ., 178, 185
A. marinum Skuja
A. platysomum Skuja
A. prosgeobium Skuja
Colacium cyclopicola (Gicklhorn) Bourr., 180, 185
Colacium sideropus Skuja
C. vesiculosum Ehr.
Entosiphon sulcatum (Duj.) Stein, 178, 185
Euglena aculeata Schiller
Eu. acus Ehr. 172, 182, 183, 204, 208
Eu. adunca Schiller = Eu. rostrata Schiller
Eu. adhaerens Matvienko
Eu. aequabilis Schiller
Eu. aestivalis Schiller, 183
Eu. agilis, 185
Eu. anquis Schiller?
Eu. aspera Schiller?
Eu. aumülleri Schiller?
Eu. bichloris Schiller
Eu. caudata Hübner
Eu. chlamydophora Mainx
Eu. choretes Schiller
Eu. chromanularis Schiller
Eu. cicutaria Schiller
Eu. clavata Skuja
Eu. conglacians Schiller
Eu. copula Schiller
Eu. diskusii Schiller
Eu. ehrenbergii Klebs, 174, 183, 185
Eu. elastica Prescott
Eu. estonica Mölder
Eu. fiebigeri Schiller
Eu. flava Dang.
Eu. gibbosa Schiller
Eu. glacialis Schiller
Eu. gracilis Klebs, 182, 183
Eu. granulata (Klebs) Schmitz
Eu. halophila Schiller
Eu. hemichromata Skuja
Eu. heteroformis Schiller
Eu. höfleri Schiller
Eu. ignobilis L. P. Johnson
Eu. impleta Schiller et var. sparsocolorata Schiller, 183
Eu. intermedia (Klebs) Schmitz et var. klebsii Lemm., 182, 183
Eu. kalleides Schiller
Eu. laciniata Pringsheim
Eu. limaciformis Schiller
Eu. limnophila Lemm., 183
Eu. longoflagellata Schiller
Eu. machurae Schiller
Eu. minutomucronata Schiller
Eu. mobilis Schiller, 182, 185
Eu. multiformis Schiller
Eu. mutabilis Schmitz
Eu. nastriformis Schiller?
Eu. oblonga Schmitz, 182
Eu. oxyuris Schmarda var. charkowiensis (Swirenko) Chu. = Eu. charkowiensis Swir., 174, 182, 185, 204, 208

CLASS CHLOROPHYCEAE (except charales)

Sc. ecornis (Ralfs) Chod. var. disciformis Chod. = Sc. disciformis (Chod.) Fott et Kom. 186, 197, 207
Sc. ovalternus Chod., 186, 198
Sc. obliquus (Turp.) Kütz. = S. acutus
Sc. opoliensis P. Richt., 186, 198
Sc. platydiscus (G. M. Smith) Chod. 186, 198
Sc. quadricauda (Turp.) Bréb., 186, 198, 207
Sc. spinosus Chod., 186, 198
Sc. sp. 173, 197, 204, 208, 210, 213, 215
Sorastrum spinulosum Naeg.
Sphaerocystis schroeteri Chodat, 186, 198, 204, 207, 210, 213, 215
Spirogyra fluviatilis Hilse
Sp. lagerheimii Wittr.
Sp. mirabilis (Hass.) Kütz.
Sp. varians (Hass.) Kütz.
Staurastrum alternans Bréb., 188, 199
St. alternans fo. biradiata, 188, 199
St. dilatatum Ehr., 188, 199
St. hexacerum (Ehr.) Wittr., 188, 199
St. cf. polymorphum Breb.
Stigeoclonium farctum Berth.
St. subsecundum Kütz.
Tetraedron caudatum (Corda) Hansg.
T. minimum (A. Braun) Hansg., 186, 197, 204, 207, 210, 213–215
T. muticum (A. Braun) Hansg.
T. regulare Kütz.
T. trilobatum (Reinsch) Hansg.
T. trigonum (Naeg.) Hansg., 207, 214
Ulothrix amphigranulata Skuja, 186, 198
U. subtilissima Rabenh.
Zygnema sp.

Order Charales

Chara ceratophylla Wallr., 200, 235, 236
Ch. crinita Wallr., 200
Ch. delicatula Ag., 200
Ch. foetida A. Br., 200
Ch. foetida fo. paragymnophylla Migula, 200
Ch. fragilis Desvaux, 200
Ch. intermedia A. Br., 200
Ch. tenuispina A. Br., 200
Ch. spp., 235

CLASS RHODOPHYCEAE

Porphyridium aerugineum Geitler, 200
P. griseum Geitler, 200

PHYLUM BRYOPHYTA

Drepanocladus aduncus Mönkem. var. kneiffii Warnst.
Dr. aduncus var. polycarpus B. S. G.
Eurhynchium speciosum Jur.

Phylum Spermatophyta

Agrostis stolonifera L.
Alnus glutinosa (L.) Gaertn.
Alisma plantago-aquatica L.
Bolboschoenus maritimus (L.) Palla = Scirpus maritimus L., 200, 251, 446
Calystegia sepium R. Br.
Carex acutiformis Ehrh., 252, 273, 446
C. distans L., 252
C. disticha Huds., 252
C. elata All., 252
C. flacca Schreb., 252
C. gracilis Curt., 252
C. nutans Host., 252
C. otrubae Podb., 252
C. panices L., 252
C. paniculata L., 252
C. pseudocyperus L., 252
C. riparia Curt., VII, 252, 269, 270, 273, 446
C. tomentosa L., 252
C. spp. 252, 253, 269, 278, 394
Ceratophyllum demersum L., 235, 236, 407
Cladium mariscus (L.) Pohl., 446
Eleocharis palustris (L.) R. et Sch. ssp. uniglumis (Link.) Hartm.
Eupatorium cannabinum L.
Humulus lupulus L.
Juncus articulatus L.
J. effusus L.
J. gerardii Loisel.
J. inflexus L.
Lemna minor L.
L. trisulca L., 394
L. spp. 254
Lycopus europaeus L.
Lysimachia nummularia L.
L. vulgaris L.
Lythrum salicaria L.
Mentha aquatica L.
Myriophyllum spicatum L., VII, 174, 189, 235–241, 244–246, 254, 307, 345, 355, 425
Najas marina L., 235, 236
Phragmites communis Trin., VII–IX, 36–38, 43–46, 162, 163, 174, 235, 251–255, 257, 259, 261, 264, 266, 268, 269, 273, 306, 307, 376, 389, 391–396, 399–401, 405, 412, 416, 425, 446, 447, 448, 458
Polygonum amphibium L.
Potamogeton crispus L.

P. pectinatus L. ssp. balatonicus (Gams) Sóo, VII, 174, 199, 235–241, 245, 246, 254, 307, 345, 425
Rumex hydrolapathum Huds.
Salix alba L., 458
S. cinerea L., 446
S. rubens Schrank
Sambucus nigra L.
Schoenoplectus lacustris (L.) Palla, 254, 264
Sch. litoralis Schrader
Sch. tabernae montani (C. C. Gmel.) Palla
Scirpus lacustris L (= Schoenoplectus lacustris (L.)) Palla, 254, 264
Solanum dulcamara L., 446
Sparganium erectum L.
Symphytum officinale L.
Triglochin maritimum L.
Typha angustifolia L., 252, 254, 264, 268, 393, 446, 447, 449, 458
T. latifolia L., 254
T. sp., sp. 251, 253, 254, 392, 393, 394
Utricularia vulgaris L., VII, 124, 130, 174, 176, 185, 189, 197–200, 240, 252–255, 267–269, 273–278, 425, 446
Valeriana dioica L.

Phylum Rhizopoda

CLASS AMOEBINA

Cyphoderia sp. (Cyphoderiidae), 337
Difflugia sp. (Difflugiidae), 337

CLASS HELIOZOA

Actinophrys sp. (Actinophryidae), 337

Phylum Ciliata

CLASS EUCILIATA (G. Dietz, Vienna)

Aristerostoma minuta Kahl
Aspidisca costata (Dujardin) Kent (Aspidiscidae, Spirotricha)
A. lynceus Ehrb.
Blepharisma steini Kahl (Spirostomidae, Spirotricha)
Caenomorpha medusula Perthy (Metopidae, Spirotricha)
Chilodonella cucullula O.F.M. (Dysteriidae, Holotricha)
Ch. uncinata Ehrbg.
Chilodontopsis vorax Stokes (Nassulidae, Holotricha)
Cinetochilum margaritaceum (Ehrb.) Perty (Frontoniidae, Holotricha)
Cohnilembus pusillus Quennerstedt (Cohnilembidae, Holotricha)
Coleps hirtus Nitzsch (Colepidae, Holotricha), 338
Colpoda steini Maupas (Colpodidae, Holotricha)
Craspedothorax gracilis Kahl (Trichopelmidae, Holotricha)
Cristigera phoenix Penard (Pleuronematidae, Holotricha)
C. setosa Kahl
Cyclidium citrullus Cohn (Pleuronematidae, Holotricha)
C. glaucoma O.F.M.
C. heptatrichium Schew.
C. pellucidum Kahl
Cyphoderia sp., 337
Didinium balbiani Bütschli (Didinidae, Holotricha)
Dysteria minuta Roux (Dysteridae, Holotricha)
Epalxis bidens Kahl (Epalcidae, Spirotricha)
Euplotes affinis Dujardin (Euplotidae, Spirotricha)
E. patella (O.F.M.) Ehrbg.
Frontonia elliptica Beardsley (Frontoniidae, Holotricha)
F. leucas Ehrbg.
Halteria grandinella O.F.M. (Halteriidae, Spirotricha)
Lacrymaria olor O.F.M. (Holophryidae, Holotricha)
Lionotus lamella Ehrbg., Schewiakoff (Amphileptidae, Holotricha)
L. procerus Penard
Loxodes striatus Penard
Loxophyllum niemeccense Stein (Amphileptidae, Hologricha)
Metopus contortus Penard (Metopidae, Spirotricha)
M. es O.F.M.
M. galeatus Kahl
M. spiralis Smith
Microthorax simulans Kahl (Trichopelmidae, Holotricha)
Mylestoma anatium (Penard) Kahl (Mylestomidae, Spirotricha)
Nassula transpeisonica n.sp. Dietz (Nassulidae, Holotricha)
Ophryoglena inqieta Kahl (Ophryoglenidae, Holotricha)

Paramecium caudatum Erhbg. (Parameciidae, Holotricha)
P. trichium Stokes
Paraspathidium fuscum Kahl (Spathidiidae, Holotricha)
Pleuronema coronatum Kent (Pleuronematidae, Holotricha)
Prorodon teres Ehrbg. (Holophryidae, Holotricha)
Spirostomum minus Roux (Spirostomidae, Spirotricha)
S. teres Clap. & Lachm.
Stentor coeruleus Ehrbg. (Stentoridae, Spirotricha)
S. mülleri Ehrbg.
S. polymorphus (O.F.M.) Ehrbg.-Stein
S. sp. 338
Stylonychia sp. 338
Thylacidium truncatum Schew. (Bursariidae, Spirotricha)
Trochilioides recta Kahl (Dysteriidae, Holotricha)
Urocentrum turbo O.F.M. (Frontoniidae, Holotricha)
Uronema marinum Dujardin (Frontoniidae, Holotricha)
Uropedalium pyriformae Kahl
Vorticella convallaria Linnée-Noland (Vorticellidae, Peritricha)
V. microstoma Ehrg.
V. octava Stokes
V. similis Stokes-Noland
Zoothamnium parasiticum Stein (Vorticellidae, Peritricha)

Phylum Cnidaria

CLASS HYDROZOA

Chlorohydra sp.? (Hydridae)

Phylum Porifera

CLASS SILICEA

Spongilla sp. (Spongillidae)

Phylum Plathelminthes

CLASS TURBELLARIA (E. Reisinger +, Graz)

Castrada intermedia (Volz) (Typhloplanidae)
Castrella truncata (Abildgaard) (Dalyelliidae)
Dendrocoelum lacteum (Müll.) (Dendrocoelidae)
Dugesia lugubris (Schmidt) (Dugesiidae)
Gieysztoria cuspidata (Schmidt) (Dalyelliidae)
G. ornata maritima (Luth.)
Gyatrix hermaphroditus Ehrbg. (Polycystidae)
Macrostomum appendiculatum (O. Fab.) ? (Macrostomidae)
M. viride (E. Bened.)?
Mesostoma lingua (Abildgaard) (Typhloplanidae)
Microdalyellia armigera (Schmidt) (Dalyelliidae)
Microstomum lineare (Müll.) (Microstomidae)
Opisthocystis goettei (Bressl.) (Polycystidae)
Phaenocora unipunctata (Örst.) (Typhloplanidae)
Planaria torva (Müller) (Planariidae)
Polycelis nigra (Müller) (Planariidae)
Rhynchomesostoma rostratum (Müll.) (Typhloplanidae)
Stenostomum leucops (Ant. Dug.) (Stenostomidae)
Strongylostoma radiatum (Müll.) (Typhloplanidae)

CLASS MONOGENEA

Ancyrocephalus sp. (Dactylogyridae)
Dactylogyrus sp. (Dactlylogyridae)
Diplozoon paradoxum Nordm. (Dactylogyridae)

CLASS DIGENEA

Bilharziella polonica (Kowalewski) (Ornithobilharziidae)
Neodiplostomum cutricola C. (Diplostomidae)
Trichobilharzia szidati Neuhaus (Ornithobilharziidae)

CLASS CESTODA

Caryophyllaeus laticeps Pall. (Caryophyllidae)
Triaenophorus nodulosus Pall. (Trianophoridae)

CLASS NEMATODA (F. Schiemer, Vienna)

Amphidelus lemani longicaudatus Schiemer (Alaimidae)

CLASS ROTATORIA (J. Donner, Katzelsdorf)

CLASS GASTROTRICHA

CLASS ACANTHOCEPHALA

Phylum Mollusca

CLASS GASTROPODA

CLASS BIVALVIA

Phylum Annelida

CLASS CLITELLATA

Chateogaster langi Bret. (Naididae), 340, 348, 353, 376
Glossiphonia complanata (L.) (Glossiphonidae)
Haemopis sanguisuga (L.) (Herudidae)
Helobdella stagnalis (L.) (Glossiphonidae)
Hirudo medicinalis L. (Herudidae)
Homochaeta naidina Bret. (Naididae), 340
Limnodrilus profundicola (Verril) (Tubificidae), 340
L. udekemianus (Claparède), 340, 343
Piscicola geometra (L.) (Piscicolidae)
Potamothrix bavaricus (Oschmann) (Tubificidae), 340
Psammoryctides barbatus (Grube) (Tubificidae), 340, 343

Phylum Tardigrada (H. Dollfuss, St. Pölten)

Hypsibius augusti (J. Murr.) (Macrobiotidae), 340, 348, 350, 352, 353, 356, 376

Phylum Arthropoda

CLASS ARACHNIDA

Order Aranea

Araneus folium Schrank (formerly syn. A, cornutus) (Araneidea), 395, 396
A. cornutus (= A. folium Schrank), 395
Argyroneta aquatica (Cl.) (Agelenidae), 394
Arundognatha striata L. Koch) (Tetragnathidae), 395
Clubiona juvenis Simon (Clubionidae), 395
C. phragmitidis C. L. Koch, 395
Dictyna arundinacea (L.) (Diyctnidae), 395
Dolomedes fimbriatus (Cl.), 394
Doncacochora speciosa (Thorell) (Linphiidae), 395
Eucta kaestneri Crome (Tetragnatidae), 394
Hypomma bituberculata (Wider) (Micryphantidae), 394
Mithion canestrini (Ninni) (Silticidae), 395
Pirata piraticus (Cl.), 394
Singa heri (Hahn) (Araneidae), 394
S. phragmiteti Nemenz (Araneidae), 395, 396
Theridion Pictum (Walck.) (Theridionidae), 394

Order Acari (Hydracarina, D. Köfler, Graz)

Arrenurus bruzelii Koen. (Arrenuridae)
A. claviger Koen.
A. cuspidifer Piersig
A. fimbriatus Koen.
A. globator O.F.M.
A. pustulator Szalay
A. radiatus Piersig
Eylais extendens O.F.M. (Eylaidae)
E. hamata Koen.
E. infundibulifera infundibulifera Koen.
Georgella helvetica Haller (Hydryphantidae)
G. koenikei (Maglio)
Hydrachna conjecta O.F.M. (Hydrachnidae)
H. geographica O.F.M.
H. globosa Geer
Hydrodroma despiciens (Mull.) (Hydromidae)
Hydryphantes thoni Piersig (Hydryphantidae)
Hygrobates longipalpis Szalay (Arrenuridae)
Limnesia fulgida Szalay (Arrenuridae)
Piona coccinea Koch (Pionidae)
P. uncata controversiosa Piersig
P. uncata uncata Koen.
Pionopsis lutescens Szalay (Arrenuridae)
Siteroptes graminum Reuter, 391
Stenotarsonemus phragmitides Schldl., 390
Unionicola crassipes (Müll.) (Uniocolidae)

CLASS CRUSTACEA (A. Herzig, H. Löffler. Vienna)

Acanthocyclops robustus (Sars) (Cyclopidae), 291, 306, 400, 402
A. vernalis (= robustus) (Fischer), 341
Acroperus elongatus (Sars) (Chydoridae), 400, 403, 404
Alona rectangula Sars (Chydoridae), 37, 340, 403, 404
Alonella excisa (Fischer) (Chydoridae), 403
A. exigua (Lilljeborg), 403
Arctodiaptomus bacillifer (Koelbel) (Diaptomidae), 322, 324, 400, 401, 402, 403, 404
A. spinosus (Daday), VII, 291, 292, 293, 294, 296, 297, 298, 299, 300, 308, 309, 310, 311, 312, 316, 317, 318, 319, 320, 322, 323, 325, 327, 342, 400, 402, 404
Asellus aquaticus L. (Asellidae), 401, 403, 404, 405
Astacus leptodactylus (Esch.) (Astacidae)
Attheyella crassa (Sars) (Canthocamptidae), 340, 400, 402
A. trispinosa (Braday), 400, 402, 404

Thermocyclops dybowskyi (Lande) (Cyclopidae), 402, 404
T. crassus (Fischer), 291, 306, 320, 322, 400, 404
T. hyalinus (Rehb.) (= T. crassus (Fischer)), 402
Tretocephala ambigua (Lilljeborg) (Chydoridae), 400

CLASS HEXAPODA

Order Ephemeroptera

Cloeon dipterum L. (Baetidae), 342
Caenis horaria L. (Caenidae), 342

Order Odonata (M. Lödl, Langenzersdorf, W. Kühnelt, Vienna)

Aeshna affinis Van der Linden (Aeshnidae)
Ae. mixta (Latreille)
Anaciaeschna isosceles (Müller) (Aeshnidae)
Anax imperator Leach (Aeshnidae)
A. parthenope Selys
Brachytron pratense (Müller) (Aeshnidae)
Calopteryx splendens (Harris) (Calopterygidae)
C. virgo (L.)
Coenagrion lunulatum (Charpentier) (Coenagrionidae)
C. ornatum (Selys)
C. puella (L.)
C. pulchellum (Van der Linden)
Cordulia aenea (L.) (Corduliidae)
Crocothemis erythraea (Brullé) (Libelluidae)
Enallagma cyathigerum (Charpentier) (Coenagrionidae)
Erythromma najas (Hansemann) (Coenagrionidae), 342
E. viridulum (Charpentier)
Gomphus flavipes (Charp.) (Gomphidae)
G. vulgatissimus (L.)
Ischnura elegans elegans (Van Der Linden) (Coenagrionidae)
I. pumilio (Charpentier), 342
Lestes barbarus (F.) (Lestidae)
L. dryas Kirby
L. macrostigma (Eversmann)
L. sponsa (Hansemann)
L. virens vestalis Rambur
L. viridis (Van Der Linden)
Leucorrhinia pectoralis (Charpentier) (Libellulidae)
Libellula depressa L. (Libellulidae)
L. fulva Müller
L. quadrimaculata L.
Onychogomphus forcipatus (L.) (Gomphidae)
Ophiogomphus serpentinus (Charpentier) (Gomphidae)
Orthetrum albistylum (Seyls) (Libelluidae)
O. brunneum (Fonscolombe)
O. cancellatum (L.)
O. coerulescens (F.)
Platycnemis pennipes (Pallas) (Platycnemididae)
Somatochlora metallica (Van Der Linden) (Corduliidae)
Sympecma fusca (Van Der Linden) (Lestidae)
Sympetrum danae (Sulzer, 1776) (Libellulidae)
S. flaveolum (L., 1758)
S. fonscolombei (Selys)
S. meridionale (Selys)
S. pedemontanum (Allioni)
S. sanguineum (Müller)
S. striolatum (Charpentier)
S. vulgatum (L.)

Order Orthoptera

Gonocephalus dorsalis Latr. (Tettigoniidae), 391
G. fuscus Fabr., 391

Order Hemiptera (G. Imhof, Vienna, H. Metz, Illmitz)

Chaetococcus phragmitidis March. (Pseudococcidae), 390, 393
Corixa dentipes (Thoms.) (Corixidae)
C. punctata (Illig.)
Gerris sp. (Gerridae)
Hesperocorixa linnei (Fieb.) (Corixidae)
Hyalopterus pruni Geofr. (Aphididae), 390, 393
Ilyocoris cimicoides (L.) (Corixidae)
Micronecta pusilla Horv. (Micronectinae), 341
M. sp.
Nepa cinerea L. (Nepidae)
Notonecta glauca L. (Notonectidae)
Plea leachi Mc. Greg. & Kirk (Pleidae)
Ranatra linaris (L.) (Ranatrinae)
Sigara striata (Fieb.) (Corixidae)

Order Hymenoptera

Cenomus lethifer Suk (Sphegidae), 391
Exeristes arundinis Kriechb. (Ichneumonidae), 391
Osmia leucomelaena K. (Sphegidae), 391
Polemon liparae Gir. (Braconidae), 391
Prosopis kriechbaumeri Först. (Apidae), 391

Scambus phragmitidis Perkins (Ichneumonidae), 391
Selandria flavens KL. (Tenthredinidae), 391

Order Coleoptera

Acilius canaliculatus (Nicol.) (Dytiscidae)
A. sulcatus L.
Agabus (Gaurodytes) affinis (Payk.) (Colymbetidae)
A. (G.) biguttatus v. nitidus (F.)
A. (G.) bipustulatus L.
A. (Eriglenus) labiatus (Brahm.)
A. (Gaurodytes) subtilis (Er.)
A. (G.) uliginosus (L.)
Bagous argillaceus Gyll. (Curculionidae)
B. collignensis Hbst.
B. glabrirostris Hbst.
B. lutulentus Gyll.
Berosus luridus L. (Hydrophilidae)
B. bispina Reiche.
B. signaticollis Sharp.
B. spinulosus Stev.
Bidessus geminus F. (Hydroporidae)
B. nasutus Sharp.
B. unistriatus (Schr.)
Brychius elevatus Pan. (Haliplidae)
B. flavicollis Strm.
B. laminatus Schlall.
B. lineatocollis Marsh.
B. obliquus F.
B. ruficollis De G.
Cercyon convexiusculus Rey (Hydrophilidae)
C. impressum Strm
C. subsculatus Rey
C. tristis Illig
C. unipunctatus L.
Chaetarthria seminulum Hbst (Chaetarthriidae)
Coccidula scutellata Hrbst (Coccinellidae), 394
Coelambus confluens (F.) (Hydroporidae)
C. impressopunctatus (Schall.)
C. lernaeus (Schaum)
C. parallelogrammus (Ahr.)
Coleostoma orbiculare F. (Sphaeridiidae)
Colymbetes fuscus (L.)
Copelatus (Liopterus) haemorrhoidalis (F.) (Colymbetidae)
Cybister late lateral imarginalis De G. (Dytiscidae)
Cymbiodyta marginella F. (Hydrobiidae)
Cyphon phragmieticola Nyholm (Helodidae)
C. pubescens Sharp.
C. variabilis Thunbg, 394
Demetrias imperialis Germ. (Carabidae), 394
Deronectes (Scarodytes) halensis F. (Hydroporidae)
Dicranthus elegans Fabr. (Curculionidae)
Dolopius marginatus L. (Elateridae), 394
Donacia aquatica L. (Chrysomelidae)
D. clavipes Fb.
D. semicuprea Pz.
D. simplex Fb.
D. thalassina Germ.
Dryops auriculatus Geoffr. (Dryopidae)
D. ernesti Gozis
Dytiscus circumcinctus (Ahr.) (Dytiscidae)
Enochrus bicolor F. (Hydrobiidae)
E. caspius Kuw.
E. coarctatus Gredl.
E. melanocephalus Ol
E. quadripunctatus Hbst.
E. testaceus F.
Eubria palustris Germ. (Psephenidae)
Graphoderus austriacus (Strm.) (Dytiscidae)
Graptodytes bilineatus (Strm.) (Hydroporidae)
G. granularis (L.)
G. lineatus F.
G. pictus (F.)
G. varius (Aubé)
Gyrinus paykulli Ochs (Gyrinidae)
G. substriatus Steph.
G. suffriani Scriba
Helichus substriatus Müll. (Dryopidae)
Helochares lividus Forst. (Hydrobiidae)
Helodes minuta L. (Helodidae), 394
Helophorus aquaticus L. (Helophoridae)
H. granularis L.
H. guttulus f. typ. Motsch.
H. longitarsis Woll.
H. micans Fald.
H. nanus Strm.
H. nubilus F.
Hydaticus seminiger De G. (Dytiscidae)
H. transversalis Pontopp
Hydraena palustris Ex. (Hydraenidae)
Hydrobius fuscipes L. (Hydrobiidae)
Hydrochus angustatus Germ. (Hydrochidae)
Hydrocyphon deflexicollis Müll. (Helodidae)
Hydronomus alismatis Marsh (Curculionidae)
Hydrophilus flavipes Stev. (Hydrophilidae)
Hydroporus angustatus Sturm (Hydroporidae)
H. discretus Fairm.
H. dorsalis (F.)
H. palustris L.
H. planus (Fab.)
H. tristis (Payk)
Hydrous aterrimus Esch. (Hydrophilidae)
H. piceus L.
Hygrobia tarda Hbst. (Hygrobiidae)

Hygrotus decoratus (Gyll.) (Hydroporidae)
H. inaequalis (F.)
H. versicolor (Schall.)
Hyphydrus aubei Glgb. (Hydroporidae)
H. ovatus (L.)
Ilybius ater (Deg.) (Colymbetidae)
I. fenestratus (F.)
I. fuliginosus (F.)
I. obscurus Marsh
I. subaeneus Er.
Laccobius atratus Rottb. (Hydrobiidae)
L. bipunctatus (F.)
L. gracilis Motsch.
L. striatulus F.
Laccophilus hyalinus (De G.) (Laccophilidae)
L. minutus (L.)
L. variegatus (Germ.)
Limnebius atomus Duft. (Limnebiidae)
L. crinifer Rey
L. stagnalis Guilleb.
Malachius spinosus Er. (Cantharidae), 394
Noterus clavicornis De G. (Noteridae)
N. crassicornis Müll.
Ochthebius bicolor Germ. (Ochthebiidae)
O. foveolatus Germ.
O. marinus Payk
O. minimus F.
O. pallidipennis Cast.
O. peisonis Gglb.
Oedacantha melanura L. (Carabidae), 394
Paederus litoralis Grav. (Staphylinidae), 394
Peltodytes caesus (Duft) (Haliplidae)
Philonthus suffragani Joy. (Staphylinidae), 394
Plateumaris consimilis Schrk. (Chrysomelidae)
Rhantus bistriatus (Bergstr.) (Colymbetidae)
R. notaticollis Aubé
R. notatus F.
R. punctatus Fourer
R. suturellus (Harris)
Scirtes hemisphaericus L. (Helodidae)
Spercheus emarginatus Schall. (Spercheidae)
Telmatophilus schönherri Gyll. (Chrysomelidae)

Order Planipennia

Sisyra jutlandica Esb.-Pet. (Sisyridae)

Order Megaloptera (H. Asböck, Vienna)

Sialis fuliginosa Pictet (Sialidae)
S. lutaria L.?

Order Trichoptera (H. Malicky, Lunz)

Agrypina pagetana Curtis (Phryganeidae), 341
A. varia F.
Ceraclea senilis Burmeister (Leptoceridae)
Cyrnus crenaticornis Kolenati (Polycentropodidae), 341
Ecnomus tenellus Rambur (Ecnomidae), 342
Holocentropus picicornis Stephens (Polycentropodidae), 342
Limnephilus binotatus Curtis (= xanthodes McL.) (Limnephilidae)
L. decipiens Kolenati
L. flavicornis F.
Oecetis furva Rambur (Leptoceridae), 342
O. ochracea Curtis, 342
Orthotrichia costalis Curtis (Hydroptilidae), 342

Order Lepidoptera

Archanara geminipunctata Hw (Noctuidae), 390
A. neurica Hb., 390
Chilo phragmitellus Hb (Pyralidae), 390
Leucania impura Hb. (Noctuidae), 390
L. obsoleta Hb., 390
Phragmatoecia castaneae Hb (Cossidae), 390, 391, 392, 393
Schoenobius gigantellus Schiff. (Pyralidae), 390, 392, 393
Senta maritima Tr. (Noctuidae), 390

Order Diptera (G. Imhof, F. Schiemer, Vienna)

Ablabesmyia longistyla Fittk. (Chironomidae)
Acricotopus lucidus (Staeg.) (Chironomidae)
Aedes spp. (Culicidae)
Anopheles maculipennis (Culicidae)
Anthomyza gracilis Fall. (Anthomyzidae), 391
Asynapta sp. Itonididae), 390, 392
Atylotus gigas Hrbst. (Tabanidae)
Bischofia simplex Fall (Sciomyzidae)
Calamoncosis minima Strobl (Chloropidae), 390, 391, 392
Calobaea bifasciella Fall. (Sciomyzidae)
Camptochironomus pallidivittatus Mall. (Chironomidae)
(Ceratopogonidae) spp.
Chironomus plumosus L. (Chironomidae), 341, 343, 351, 352, 353, 360, 366, 367, 368, 371, 373, 379
Chrysops caecutiens L. (Tabanidae)
Chrysozona italica Meig. (Tabanidae)
Ch. pluvialis L.
Corynoneura scutellata (Winn.) (Chironomidae)
Cricotopus punctipennis (Meig.) (Chironomidae)

Tanypus punctipennis (Meig.) (Chironomidae), 341, 343, 351, 352, 353, 355–359, 365–370, 377–380
T. kraatzi (K.)
Tetanocera elata Fall. (Sciomyzidae)
T. ferruginea Fall.
Theobaldia annulata Schr. (Culicidae)
Therioplectes montanus Meig. (Tabanidae)
T. solstitialis Schin.
Thomasiella arundinis Schiner (Itonididae), 390, 391, 392
T. flexuosa Wtz., 390, 391, 392
T. incurtans, 391
T. massa Erdös, 390, 392
Thrypticus bellus Loew. (Dolichopodidae), 390, 392
T. smaragdinus Gerst. 390, 392
Tubifera pendula L. (Syrphidae)
T. trivittata Fabr.
Xenopelopia falcigera (Kieff.) (Chironomidae)
X. nigricans Fittkau

Phylum Chordata

CLASS TELEOSTOMI

Abramis ballerus (L.) Cyprinidae, 426
A. brama (L.), 426, 428
Acerina cernua (L.) (Percidae), 376–381, 423, 427, 428, 430–432, 436
Alburnus alburnus (L.) (Cyprinidae), 378, 426, 428, 430, 445, 446, 471
Anguilla anguilla (L.) (Anguillidae), 376, 426, 428, 430, 434
Aspius aspius (L.) (Cyprinidae), 426
Barbus barbus (L.) (Cyprinidae), 426
Blicca björkna (L.) (Cyprinidae), 367, 377, 423, 426, 428, 430–433, 436
Carassius carassius (L.) (Cyprinidae), 426, 428
C. (auratus) gibelio (Bloch), 426
Cobitis taenia L. (Cobitidae), 427
Cottus gobio (L.) (Cottidae), 427
Ctenopharyngodon idella (Val.) (Cyprinidae), 425, 426
Cyprinus carpio L. (Cyprinidae), 423, 426, 428, 429
C. carpio typ. hungaricus, 423, 429
Esox Lucius L. (Esocidae), 423, 426, 428, 429
Gobio gobio (L.) (Cyprinidae), 426
Hypophthalmichthys molitrix (Val.) (Cyprinidae). 425, 426
Idus idus (L.) (Cyprinidae), 426
Lepomis gibbosus (L.) (Centrarchidae), 427
Leucaspius delineatus (Heckel) (Cyprinidae), 426
Lota lota (L.) (Gadidae), 427
Lucioperca lucioperca (L.) (Percidae), 376, 427, 428, 430, 435
Misgurnus fossilis (L.) (Cobitidae), 427
Nemachilus barbatulus (L.) (Cobitidae), 427
Pelecus cultratus (L.) (Cyprinidae), 376, 378, 425, 426, 430
Perca fluviatilis L. (Percidae), 378, 427, 428
Proterorhinus marmoratus (Pallas) (Gobiidae), 425, 427
Rhodeus amarus sericeus Bloch (Cyprinidae), 426
Rutilus rutilus (L.) (Cyprinidae), 426, 428, 446
Salmo gairdneri Rich. (Salmonidae), 426
Scardinius erythrophthalmus (L.) (Cyprinidae), 426, 428
Silurus glanis L. (Siluridae), 427
Squalius cephalus (L.) (Cyprinidae), 424, 426
Tinca tinca (L.) (Cyprinidae), 426
Umbra krameri Walb. (Umbridae), 426

CLASS AMPHIBIA (H. Tunner, Vienna)

Bombina bombina (Linnaeus) (Discoglossidae)
Bufo bufo bufo (Linnaeus) (Bufonidae)
B. viridis viridis (Laurenti)
Hyla arborea arborea (Linneaus) (Hylidae)
Pelobates fuscus fuscus (Laurenti) (Pelobatidae)
Rana arvalis wolterstorffi (Fejervary) (Ranidae)
R. dalmatina (Bonaparte)
R. lessonae (Camerano)
R. esculenta (Linnaeus)
Triturus cristatus dobrogicus (Kiritzescu) (Salamandridae)
T. vulgaris vulgaris (Linnaeus)

CLASS REPTILIA

Natrix natrix natrix Linne (Colubridae)
N. tesselata Laurenti

CLASS AVES (F. Böck, P. Prokop, Chr. & M. Staudinger, Vienna)

Acanthis flammea (L.) (Fringillidae)
A. flavirostris (L.)
Accipiter gentilis (L.) (Accipitridae)

T. merula L.
T. philomelos C. L. Brehm, 456
T. pilaris L.
T. torquatus L.
T. viscivorus L.
Tyto alba (Scopoli) (Strigidae)
Vanellus vanellus (L.) (Charadriidae)
Upopa epops L. (Upopidae)
Uria aalge (Pontoppidan) (Alcidae)

CLASS MAMMALIA

Arvicola terrestris terrestris (Linné) (Cricetidae)
Micromys minutus pratensis (Osckay) (Muridae)
Microtus oeconomus mehelyi (Ehik) (Cricetidae)
Mustela erminea aestiva (Kerr) (Mustellidae)
Neomys anomalus milleri (Mottaz) (Soricidae)
N. fodiens fodiens (Pennant)
Ondatra zibethicus Linné (Cricetidae), 447
Rattus norvegicus norvegicus (Berkenhout) (Muridae)
Sorex araneus wettsteini (Bauer) (Soricidae)
S. minutus minutus (Linné)

Index of *genera* and *species* of organisms

Subject Index

Abbot's Pond, 226
abrasion terraces, 7
absorption, 57
  at 430 mm, 227
abundance, 363, 405
  Arctodiaptomus spinosus, 298
  bacteria, 328
  Bosmina, 303
  Cyclops spp., 306
  of larvae, 365
  rotifers, 282
  zooplankton, 325
Acari tarsonemini, 390
acid waters, 198
accumulation of nutrients, 119
adective energy supply, 73
Adriatic, 80
adsorption, 107
adults, 306, 312
  Arctodiaptomus spinosus, 300, 308
  copepods, 314
  Cyclops spp., 307
  emergence, 364
aeolian and limnic forces, 8
aeolian (aeolic) transport of minerals, 137, 148
Aeolosomatidea, 339, 440
aerenchymatic assimilating organs, 237
aerodynamic profile methods, 74
aerogenic decalcification, 95
aerogenic precipitation of calcium, 90
Afghanistan, 399
Agelenidae, 394
aggregation, 308
agrarian (population of Burgenland), 12
agriculture, promotion of, 17
air temperature, 48, 50
Al, 146
albedo, 56, 151, 154, 263
albite chlorite gneisses, 25
Alföld, 10
algae, 171, 327
  as components of the sediment, 133
  as food, 325
  (epipelic), 375
  food of *Tanypus punctipennis* (epipelic), 377
  of the reed belt, 252
  (Rhodophyta), 200
  utilization of Phar, 264
algal
  biomass, 233
  communities, 172
  concentration, horizontal differences, 225
  composition, 216
  flora of Neusiedlersee, 203
  food, 327
  growth and nutrient supply, 128
  in Balaton, 329
  species as fresh weight, 210
  species, dimensions of, 204
  vegetation of Neusiedler See, 171
algivores, 376
alkali
  carbonate, 99
  feldspar, 145
  salt soils, 9
  silicate solutions, 99
alkaline
  earth, 89
  earth, carbonates of, 105
  earth-carbonates, precipitation of, 107
  lakes east of Neusiedler See, 313
alkalinity (alkalinities), 89, 91, 93–95, 312
  indication of higher, 38
  tolerance limit for, 312
  tolerance of higher, 292
Alleröd, 34
allochthonous
  components of the sediment, 131, 133, 135, 371
  (in food limits of Neusiedler See), 376
  input of, 328, 375
  input (of L. Krugloe), 329
  material, 371
alpine lakes, 89
Alps, 3, 23, 24
  levelling processes, 6
Alster, 434
Althof, 65, 66
Amazonian
  flood plains, 408
  region, 408
American Weather Bureau, 66
aminoacids, 108
  in the sediment, 102, 108
ammonium, 116
  content of the lake water, 117
ammonium oxidation, 103
amphibia, 451
anaerobic
  condition of the lake under ice, 128

Cyclopoida (Cyclpidae, cyclpoids)
abundance, 296
biomass, 297
relative abundance, 293
resting stages, 325
vertical stratification, 311
Cyperacea, 458
Cyprinidae (cyprinids), 248, 378, 426, 446
Cyprininae, 385
Czechoslovakia, ČSSR, 222, 464

daily
course of phytosynthesis, 258
production, 216, 317
Danube, 4–8, 28, 33, 44, 411, 424, 428, 440, 441, 445, 472, 475, 476
basin, 47
delta region, 399
gravels, 6
Darscho Lacke, 199
death rate of *Arctodiaptomus spinosus*, 292
decline of *Utricularia vulgaris*, 278
decrease in abundance, 359
DDR, 435
deflation, 33
theory, 6
dehydrogenase activity, 375, 376
denudation, 7
deoxygenation, 352, 359
desiccation
ecophases resistant to, 407
habitats inshore, 278
Neusiedlersee, 37, 45, 79, 80, 94, 171
Desmidiaceae (desmid, Desmidiales), 175, 198, 208, 213
in the benthos, 199
destruction by man, 79
detritus, 326
as food (benthic species), 377
as food (chironomids), 378
detritus, 108
as food (*Diaptomus* and *Diaphanosoma*), 327
as food (zooplankton), 325
energy content of organic, 267
food chain, 326, 328, 329
from macrophytes, 345
from submerged vegetation, 139
macrophytic and microphytic, 375
Detrivores, 377
diagenetical alteration, 145
Diaptomidae
mean daily P/B, 311, 323
P/B-values, 321
diatom
debris, 146
flora, 191, 342
diatoms (Diatomeae), 171, 173–176, 191, 205, 208, 219, 222, 255
biomass, chlorophyll-a, chlorophyll content of biomass and carbon to chlorophyll ratio, 223
chlorophyll content of fresh weight biomass and carbon chlorophyll ratio (various lakes), 225
food of *Tanypus punctipennis*, 377
percentage contribution to total biomass, 215
seasonal section, 210
dicotyledon, 251
Dictynidae, 395
dimictic lakes, 34
Dinoflagellata (dinoflagellates), 171, 173, 179, 208
Dinophyceae, percentage of algal classes, 209
Diptera, 389–391, 393, 394, 396
Chlorophidae, 392
Dolichopodidae, 392
food of passerine birds, 457, 458, 472
Tonididae, 392
dispersion, 58
coefficient (light climate), 47
dissolved
Kj-N, 116
organic carbon, 247
phosphorus, 103, 110, 111
distribution (pattern), 348, 369, 385
macrophytes, 235, 236
Myriophyllum spicatum, *Potamogeton pectinatus*, 237
divers, 439, 440, 445
diversity, 226
of algae, 226, 228, 290
evenness of algae, 227
index of algae, 226
richness of algae, 227
dm = decimeter = 0.01 m^2, 259
dolichopodid, 392
'dolomictic', 90
dolomite, 133, 134, 139, 145, 148
'domestic' carp, 429
domestic waste, 105
Donnerskirchen, 10–12, 14, 16, 27
Don-Zander, 435
downwelling water, 308
downwind concentration, 308
dragonflies, 451, 459
drainage of Neusiedlersee, 80–82, 475
draw wells, 11
drift of turions of *Utricularia vulgaris*, 277
drifting sand, 9
drying out of the lake bed, 131
dry weight (d.w.), 238, 240, 241, 248, 258,

planktonic algae (see also phytoplankton), 329
  and abundance, fecundity of zooplankton, 325
plant
  consumers, 448
  sociological survey, 235
plateau, 6
Pleistocene, 5, 7, 33, 34
  gravels, 28
  processes, 6
  surface systems, 6
Pliocene, 5, 33, 45
Plymouth Sea Water, 160
P_O, 103–105, 109, 110, 114
PO_4 (see also phosphorus), 102, 107, 217
pochard, 441–444
Podersdorf, 11, 12, 14–16, 30, 36–38, 45, 49, 55, 61, 65–72, 88, 106, 110, 125, 172, 174, 175, 181, 198, 444, 479
'Podersdorfer Schoppen', 441, 442, 444
Podicipidae, 445
Poland, 241
  (Mazurian lakes), 322
Polish breeding population of the bearded tit, 463
Polish lakes, 367
pollen, 34, 345, 393
  analysis, 9, 35
Poltruba, 408
polymictic, 326, 329
  Lake Krugloe, 322
polythermic, 342
  species, 342
pond management, 406
pope, 423
poplars, 453, 457
population, 324
  Diaptomus spinosus, 300, 316
  density, 11, 381
  dynamics, 357, 395, 396, 401, 405
  estimate, 282
  increase, algae, 217
  human, 11, 12
  macrobenthos, 356
  movements, 12, 13
  Procladius, 358, 362
  structure, 364, 380
  Tanypus punctipennis, 365
Porifera, 408
positively buoyant, 308
postembryonic development
  Cladocera, 324
  times, 310
postembryonic stages, 316
P_p (see also particulate phosphorus concentration), 112, 113
  content, 112
  local distribution, 113
  temporal variation, 116
Prater terrace, 8
precipitation, 31, 58, 59, 66, 73, 80–84, 94, 96, 103, 107, 117, 245
  alkaline earth-carbonates, 107
  annual average, 28
  mean annual total, 83
predation by fish, 318
predators, 377, 403
preservation, 17, 475
  as natural reserves, 102
Pressburg, 11
pre Würm stage, 28
prey
  organisms, 378
  selection, 353
primary
  production, 233, 265, 275
  productivity, 233
producer, 325
production, 203, 316–320, 322, 325, 326, 371, 375, 472
  algae, 233
  Arctodiaptomus spinosus, 320
  Arctodiaptomus spinosus, Diaphanosoma brachyurum, 315
  biomass, 316
  capacity, algae, 221
  conditions, *Carex riparia*, 269
  Diaphanosoma brachyurum, 273, 319, 321, 317
  Diaptomus, 320, 322
  estimates planktonic crustaceans, 316
  macrophytes, 235
  of calcite, 248
  period, *Phragmites*, 263, 267
  Phragmites, 264
  Utricularia vulgaris, 274
  values, 258
  zone, 251
  zooplankton, 314, 325
productivity, 344
Project
  Goldemund, 478
  Karolyi, 477
  Riedinger, 478
  Vogel & Särkany, 477
promotion, *Phragmites*, 255
protein content, *Phragmites*, 262
Protozoa, 337
protodolomite, 133, 135, 137, 139, 145, 148, 344
  component of sediment, 133, 134
  growth etc., 136
protophyte vegetation, 176
P_p, 103, 104, 114

reproduction
  Arctodiaptomus spinosus, 300
  Diaphanosoma brachyurum, 327
  macrophytes, 236
reservoirs, 327, 328
residents, 12
resting eggs, 34, 300, 302
  Arctodiaptomus spinosus, 300
  Bosmina, 305
  Copepoda, 407
  Diaphanosoma brachyurum, 302, 325
  Hexarthra fennica, 287
  Leptodora kindti, 291
resting
  place, 441, 442, 444
  sleeping place, 439
  stages, 345, 401
  Arctodiaptomus spinosus, 300
  Diaphanosoma brachyürum, 325
Rhabdocoela, 338
rhizomes, 236
rhizomes, *Phragmites*, 251
Rhizopoda, 337, 338
Rhodophycea, 200
rhombohedral habit, 147
Rhyd-hir-stream, 434
Rhynchota, 457, 458, 472
rhythm of assimilation, 260
Richardson number, 50
richness, diversity, 227
ring, *Potamogeton pectinatus*, 238, 239, 242, 245
  centre, *Potamogeton pectinatus*, 238
  diameter, *Potamogeton pectinatus*, 238
  formations, *Potamogeton pectinatus*, 237, 245
  growth, *Potamogeton pectinatus*, 239, 245
Riss
  glaciation, 33
  gravels, 7, 8
  ice age, 28
  terrace, 8
  Würm, 9
  Würm interglacial salt horizon, 7
  Würm soils, 8
road bridge, 479
robin, 456, 460
Rohrbach
  Drassburg, 24
  forest, 22
Rohrlacke, 123, 124
Rosalia, 5, 14, 22, 23
  mountains, 3, 26
Rotatoria (see also rotifers), 338, 408
  distribution pattern, 348
rotifer, 281, 282, 289, 314, 316, 337, 342, 378, 411, 420
  abundance, 290
  fauna, 281, 342
rotifer
  numbers, 288, 289, 291
  numerical relation between the abundance of rotifers and crustaceans, 315
  plankton, 281, 287
  total number, 285
Rucás, 122, 123
ruff, 359, 430, 433
runoff coefficient, 81
Russia, 462
Russian lakes, 435
Russian occupation, 479
Rust, 3, 5, 7, 10–12, 14, 16, 23–27, 30, 33, 38, 49, 65–68, 91, 93, 97, 98, 106, 110, 123, 125, 131, 139, 173, 175, 176, 179, 181, 183, 185, 187, 189, 191, 197–200, 245, 255, 257, 277, 389, 393, 395, 399, 400, 403, 420, 444, 446–489
  gravels, 23, 25, 26
  hills, 24
  Untere Wiese, 145
Ruttner sampler, 103
Rybinsk Reservoir, 314, 327
rye, 14

salient algae, 199
saliferous
  horizon, 9, 10
  layer, 11
saline
  alkaline soils, 9
  bitter lakes, 191
  lakelets, 191
  soils, 5, 8–10
  upwelling groundwater, 39
  waters, 38
salinity, 171, 200
  1830–1973
  origin, 38
  tolerance limit of crustaceans, 312
Salmonidae, 426
salt
  alkali soils, 9
  concentration, 171, 174
  content, 312, 313
  lakelets, 10
  pans, 93, 338
  salt soils, 98
Salticidae, 395
saltwater forms, 420
Salzkammergut, ix, 58
Samuel Krieger, 475
sampling strategy, 281, 282
Sandeck, 442

Plates

1. Frequent winds from NW and SE are one of the most obvious features of the lake

2. Neusiedlersee seen from its eastern shore (Weiden).

3. Channel through the Phragmites belt near Illmitz (background: the Alps, Schneeberg).

4. Formerly cattle and animal farming in general around Neusiedlersee was almost the only activity and certainly has influenced the growth of the reed.

5. 'Seemitte' station used during the International Hydrological Decade (IHD).

6. Due to the wind long fissures of the ice cover, running parallel to the lake axis, are frequently observed.

7. Ice shoves have contributed to the forming of the ‘Seedamm’ (lake dam) along the eastern shore.

8. During the last years ice never accumulated to ice shoves.

9. Background harvested reed.

10. Traditional ice-sailing on the lake.

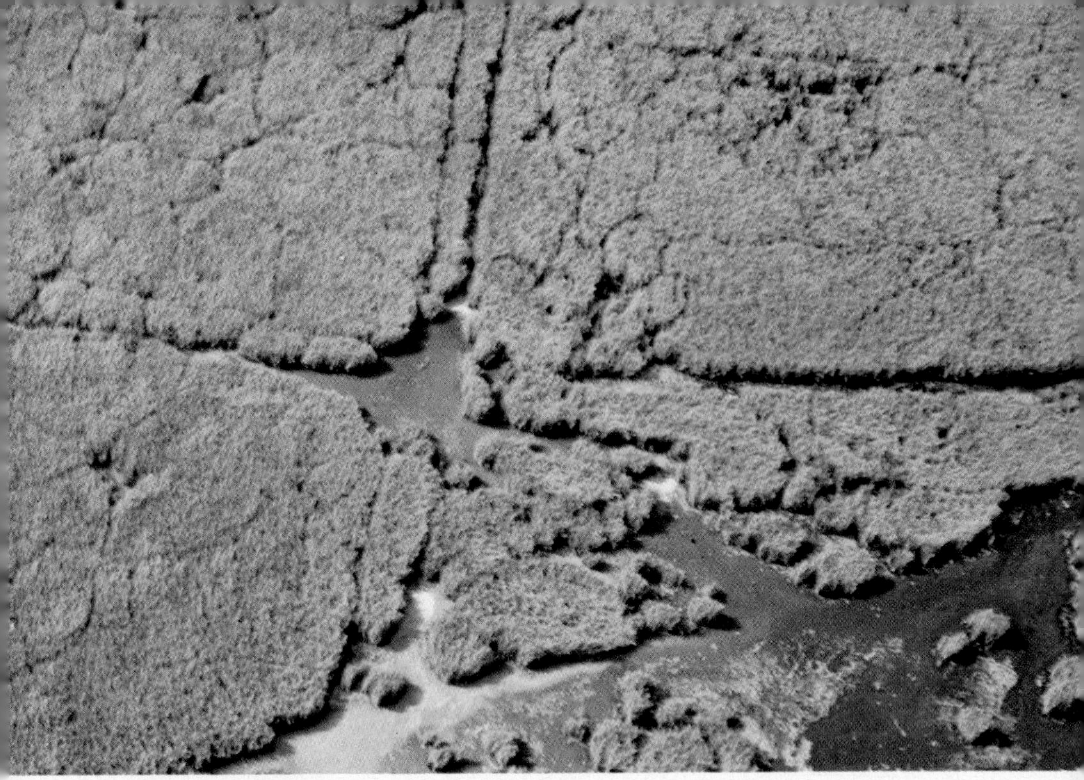

11. The Phragmites belt extends over 150 km^2; channels for fishery must be maintained.

12. Cutting of Phragmites is carried out when the lake is frozen over.

13. Platalea leucorodia nests close to the heron colonie
14. Nesting Casmerodius alba.
15. The conspicuous nesting Remiz pendulinus can b found in poplars and willows of the drier landwa zone.

16. Uncontrolled building activities are still going on along the reed belt fringe.

17. Building activities near Breitenbrunn.

18. Large sections of the reed belt are sacrificed to recreation: holiday settlement near Weiden.

19. Holiday houses of Weiden – a bad imitation of old styling.

20. Old wine cellars in Breitenbrunn.

21. Solontschak, a saline soil, is typical of the area.

22. The Einser Kanal, the artificial outlet of the lake runs along the Hungarian frontier (indicated by the watch-tower). Background: Alder forest of Kapuvar, an important habitat for carnivorous birds.

23. At an early stage during the late Pleistocene the lake covered the Hanság and then shifted westwards. The area at present is a sanctuary of bustards.